THE MATHEMATICAL STRUCTURE
OF STABLE PHYSICAL SYSTEMS

DR. MARTIN CONCOYLE & G.P. COATMUNDI M.S.

Order this book online at www.trafford.com
or email orders@trafford.com

Most Trafford titles are also available at major online book retailers.

© Copyright 2014 Dr. Martin Concoyle & G.P. Coatmundi M.S.

All rights reserved. No part of this publication may be reproduced, stored in a retrieval system, or transmitted, in any form or by any means, electronic, mechanical, photocopying, recording, or otherwise, without the written prior permission of the author.

Printed in the United States of America.

ISBN: 978-1-4907-2364-8 (sc)
ISBN: 978-1-4907-2365-5 (e)

Because of the dynamic nature of the Internet, any web addresses or links contained in this book may have changed since publication and may no longer be valid. The views expressed in this work are solely those of the author and do not necessarily reflect the views of the publisher, and the publisher hereby disclaims any responsibility for them.

Any people depicted in stock imagery provided by Thinkstock are models, and such images are being used for illustrative purposes only.
Certain stock imagery © Thinkstock.

Trafford rev. 01/11/2014

www.trafford.com
North America & international
toll-free: 1 888 232 4444 (USA & Canada)
fax: 812 355 4082

This book is an introduction to the simple math patterns used to describe fundamental, stable spectral-orbital physical systems (represented as discrete hyperbolic shapes), the containment set has many-dimensions, and these dimensions possess macroscopic geometric properties (which are also discrete hyperbolic shapes). Thus, it is a description which transcends the idea of materialism (ie it is higher-dimensional), and it can also be used to model a life-form as a unified, high-dimension, geometric construct, which generates its own energy, and which has a natural structure for memory, where this construct is made in relation to the main property of the description being, in fact, the spectral properties of both material systems and of the metric-spaces which contain the material systems, where material is simply a lower dimension metric-space, and where both material-components and metric-spaces are in resonance with the containing space. Partial differential equations are defined on the many metric-spaces of this description, but their main function is to act on either the, usually, unimportant free-material components (to most often cause non-linear dynamics) or to perturb the orbits of the, quite often condensed, material trapped by (or within) the stable orbits of a very stable hyperbolic metric-space shape. (see back-page)

This book is dedicated to my wife M. B. and to my mom and dad

Note this book is equivalent to other books which were originally put onto Scribd.com
Where one of the old books was:

Introduction to the stability of math constructs;
and a subsequent general and accurate as well as
practically useful description of stable material systems
By Martin Concoyle Ph. D.
And
With essays by G. P. Coatimundi
Copyright 2012, Martin Concoyle, and G P Coatimundi

And the other old book was:

A Book of Essays II:
Science history, and the shapes which are stable,
and the subspaces, and finite spectra,
of a high-dimension containment space
or
Science History; An Introduction to the stability of math constructs; the Shapes which Are Stable, and thus, Define a Finite Spectra in a High-dimension Containment Space, and a subsequent general, and accurate, and practically useful set of descriptions for the observed stable material systems (Various essays about stable math constructs, material interactions, and finite quantitative structures)

Copyright 2012, Martin Concoyle

CONTENTS

Preface xi

PART I

Chapter 1	A New Paradigm for Physics 3
Chapter 2	Review I 108
Chapter 3	Review II 131

PART II

Chapter 4	Topology and geometry 145
Chapter 5	Unitary transformations: 163
Chapter 6	New Math 210
Chapter 7	New ideas 230
Chapter 8	New context for life 256
Chapter 9	Model of intent 276

PART III

Chapter 10	Science history 309
Chapter 11	Science 325
Chapter 12	Science and religion 344
Chapter 13	Quantum physics 366
Chapter 14	Particle-physics 397
Chapter 15	Basis for physics 401
Chapter 16	Abstracts 420
Chapter 17	New Basis for physics 452
Chapter 18	New math categories 481
Chapter 19	New math ideas 501

PART IV

Chapter 20	Geometry	527
Chapter 21	Interactions	564
Chapter 22	Higher dimensional interaction space-form	576
Chapter 23	Material-interactions	587
Chapter 24	A further look at cubical simplexes	596
Chapter 25	Cubical simplex	620
Chapter 26	Cubical world	634
Appendix I		649
Appendix II		651
Chapter 27	Diagrams	652
References		675
Index		679

PREFACE

This book is pieced together from over 100 essays. There may be repetitions (sorry) but they are titled differently so the repetitions are hard to detect and eliminate.

In our society science has been categorized, and its authoritative truth "absolutely" determined . . . , at least this is what the propaganda-education system tries very hard to instill in the public . . . , but this categorization of science and math, and its supposed highly-valued-truth, is only defined in regard to a particular range of, technical activities, which are associated to the, supposed, authoritative truths which, in turn, are associated to the commercial categories into which the investment-class places science and math truth, so that the ruling investor-class only relates the, so called, scientific and mathematical truth of society to the investment interests of the banking-class, eg physics is about nuclear-weapons engineering, etc. The investment class has been successful at doing this, but the obvious failure of this supposed highly-valued-truth in regard to being able to describe the very large number of very fundamental physical systems which are very stable physical systems . . . . such as the relatively stable physical systems of: nuclei, atoms, molecules, the chemistry which occurs within living-cells, and the relation of this cell-chemistry to DNA, the properties of crystals, as well as describing the relatively stable solar-system all go without valid descriptions based on the so called laws of physics . . . . And this failure should lead any rational person to question these authoritative dogmas identified by the investor-class through the controlling context of peer-review . . . , where it should be noted that peer-review defines a contest for determining job-selection for the high-end engineering projects of the investor-class . . . , and subsequently to try to understand these stable physical systems in new mathematical ways.

Science was split off of religion by Copernicus and Galileo

But now (2013) the same way in which intellectual authority was then (1550) defined by the church, now intellectual authority has now come to be defined by the authority of science and math, but this type of authority is related to the narrowly defined practical use of this technical knowledge by the investor-class, and the investor-class defines high-value within society, by means of their control over the propaganda-education system, where now (2013) high-valued science and math is defined by peer-review, note that, Copernicus would never be allowed to publish in peer-review (and Galileo was thrown in jail for not being peer-reviewed), thus there is peer-reviewed science and math, where people compete for high-paying jobs which are related to technical projects which the bankers define (and into which the bankers have invested, ie it is conservative and of little risk), and there is "real science and math" which is defined by the same inquisitive spirit as expressed by Copernicus and Galileo, where the inquirer has the power to use new assumptions and new descriptive contexts etc etc.

Note

These essays span 2004 to 2012 and some old ideas (along these same lines) expressed around 2004 may not be ideas considered correct by myself today (2012), but I have not re-edited these old expressions (true for both concoyle and coatimundi).

Ideas are worth expressing, and the development of ideas can have interesting histories, and old ideas can be re-considered.

Today there exist experts of "dogmatic authority" who only try to extend dogmas, and who are represented in the propaganda system (as well as in educational institutions which also serve the interests of the owners of society, ie the modern day Roman Emperors) as being "always correct," yet they fail to be able to describe the stability of fundamental physical systems, and their descriptions have no relation to being related to practical creative development, since they are descriptive constructs based on probability and non-linearity, and deal with systems made-up of only a few components which possess unstable properties.

"Peer review" checks for the "dogmatic purity" of its contributors. However, such a situation in science and mathematics does not express the (true) spirit of knowledge. Knowledge is related to "practical" creativity, and knowledge is about equal free-inquiry with an "eye on" "what one wants to create." In this context knowledge should be as much about re-formulating, and re-organizing technical (precise) language as about learning from the current expressions of knowledge [with its narrow range of creativity associated to itself].

A community of "dogmatically pure" scientists and mathematicians is not about knowledge, but rather about the power structure of society (a society with a power structure which is essentially the same as that of the Holy-Roman-Empire, ie fundamentally based on extreme violence) and the scientists and mathematicians are serving the owners of society (the new Emperors, ie the investor-class) by competing in a narrow dogmatic structure of authority, so as to form a hierarchical array of talent to be selected from by the Emperor, and then used within the narrowly defined ranges of creativity, about which the owners of society want attended.

That is, scientists function in society as elite wage-slaves for the owners of society, and they are trained experts, similar to trained lap-dogs (and they are also arrogant in their fancy threads, as are lap-dogs).

This structure of knowledge is the opposite of "valid" knowledge, which should be related to a wide range of creative efforts, by many people, expressing many diverse interests.

It is good to express a wide range of ideas.

Whereas, "being correct" is associated to a "false," or at best, limited "knowledge," which serves "the owners of society" and is mostly used to express in the propaganda system, the (false) idea, that people are not equal. It is this type of idea which the Committee on Un-American Activities should investigate, since the US Declaration of Independence states that all people are equal, and this should be the basis for US law, and not: property rights, and minority rule, as well as the extreme violence needed to establish and maintain these ownership and social-control relationships within society, and which is the same basis as for Roman-Empire Law.

Copyrights

These new ideas put existence into a new context, a context for both manipulating and adjusting material properties in new ways, but also a context in which life and creativity (practical creativity, ie intentionally adjusting the properties of existence) are not confined to the traditional context of "material existence," and material manipulations, where materialism has traditionally defined the containment of material-existence in either 3-space or within space-time.

Thus, since copyrights are supposed to give the author of the ideas the rights over the relation of the new ideas to creativity [whereas copyrights have traditionally been about the relation that the owners of society have to the new ideas of others, and the culture itself, namely, the right of the owners to steal these ideas for themselves, often by payment to the "wage-slave authors," so as to gain selfish advantages from the new ideas, for they themselves, the owners, in a society where the economics (flow of money, and the definition of social value) serves the power which the owners of society, unjustly, possess within society].

Thus, the relation of these new ideas to creativity is (are) as follows:

These ideas cannot be used to make things (material or otherwise) which destroy or harm the earth or other lives.

These new ideas cannot be used to make things for a person's selfish advantage, ie only a 1% or 2% profit in relation to costs and sales (revenues).

These new ideas can only be used to create helpful, non-destructive things, for both the earth and society, eg resources cannot be exploited to make material things whose creation depends on the use of these new ideas, and the things which are made, based on these new ideas, must be done in a social context of selflessness, wherein people are equal creators, and the condition of either wage-slavery, or oppressive intellectual authority, does not exist, but their creations cannot be used in destructive, or selfish, ways.

PREFACE

The math descriptions, about which what this book is about, are about using math patterns within measurable descriptions of the properties of existence which are: stable, quantitatively consistent, geometrically based, and many-dimensional, which are used to model of existence, within which materialism, and its apparent indefinable randomness, is a proper subset.

In regard to the partial differential equations which are used to describe stable "material" systems they are: linear, metric-invariant [ie isometry (SO, as well as spin) and unitary (SU) fiber groups], separable, commutative (the coordinates remain globally, continuously independent), and solvable.

The metric-spaces, of various dimensions and various metric-function signatures [eg where a signature is related to R(s,t) metric-spaces] have the properties of being of non-positive constant curvature, where the coefficients of the metric-functions (symmetric 2-tensors) are constants.

That is, the containment sets and "material systems" are based on (or modeled by) the simplest of the stable geometries, namely, the discrete Euclidean shapes (tori) and the discrete hyperbolic shapes (tori fitted together), where the discrete hyperbolic shapes are very geometrically stable and they possess very stable spectral properties.

One can say that these shapes are built from "cubical" simplexes (or rectangular simplexes).

Both the (system containing) metric-spaces and the "material" systems have stable shapes of various dimensions and various metric-function signatures, where material interactions are built around the structures of discrete Euclidean shapes (sort of as an extra toral component of the interacting stable discrete hyperbolic shapes), within a new dimensional-context for such material-interaction descriptions, and there are similar interaction constructs in the different dimensional levels. The size of the interacting material "from one dimensional level to the next" is determined by constant multiplicative factors (defined between dimensional levels) which are (now) called physical constants.

Furthermore, the basic quantitative basis for this description, ie the stable spectra of the discrete hyperbolic shapes, forms a finite set. The quantitative structure is, essentially: stable, quantitatively consistent, and finite.

This descriptive construct can accurately, and to sufficient precision, and with wide ranging generality, describe the stable spectral-orbital properties of material systems of all size scales, and in all dimensional levels. It is a (linear, solvable) geometric and controllable description so it is useful in regard to practical creativity.

The many-dimensions allow for new high-dimension, well organized, controllable models of complicated systems, such as life-forms. These ideas provide a "map" to help envision these geometric structures.

These new ideas are an alternative to the authoritative (and overly-domineering) math patterns used by professional math and physics people which are based on non-linearity, non-commutativity, and indefinable randomness (the elementary event spaces do not have a valid definition), where these are math-patterns, which at best, can only describe unstable, fleeting patterns, which are unrelated to practical creative development, and whose measured properties can only be related to feedback systems (whose stability depends on the range of validity of such a system's differential equation).

That is, it is a math construct which is not capable of describing the stable properties of so many fundamental (relatively) stable physical systems, eg nuclei, where within this authoritative descriptive context it is claimed that these stable fundamental physical systems are "too complicated" to describe.

There are many social commentaries, in this book, this is because such a "new context of containment, in regard to measurable descriptions," which possesses so many desirable properties, one would think that such a descriptive language should be of interest to society. But inequality, and its basis in arbitrary (and failing) authority, and the relation which this authority has to extreme violence (in maintaining its arbitrary authority, and in maintaining a social structure (as Mark Twain pointed-out) which is based on: lying, stealing, and murdering) have excluded these new ideas from being expressed within society.

People have been herded, and tricked, into wage-slavery, where deceiving people is easy with a propaganda system which allows only one authoritative voice, and that one-voice is the voice of the property owners (with the controlling stake), and the people are paralyzed by the extreme violence which upholds this social structure, where this extreme violence emanates from the justice system, and whereas the political system has been defined as "politicians being propagandists" within the propaganda system (politicians sell laws to the owners of society, for the selfish gain of the politicians, and then the politicians promote those laws on the media).

PART I

PART 1

Chapter 1

A NEW PARADIGM FOR PHYSICS

Author: Martin Concoyle Ph. D, and George P. Coatimundi
e-mail: martinconcoyle@hotmail.com

Abstract:

Currently physics says that most fundamental physical systems are too complicated to describe, eg both the general nucleus and subsystem interactions in crystals are too complicated to describe.

The currently accepted paradigm of physics is not useful. Its most usual statement about many fundamental physical system is that these systems are too complicated to describe, eg the general nucleus is too complicated to describe, and interactions between current and lattice sites in crystals (much needed to model semi-conductors so as to move to the next generational level of smaller micro-chip developments) are also too complicated to describe. Furthermore, arbitrary parameters, the measured masses of elementary particles which are included in the "calculation" of the elementary particles' (spectral) masses [Note: the standard model can essentially only calculate elementary particle cross sections and their mass spectrum], turns the standard model into pseudo-science.

The alternative: Discrete isometry subgroups, or equivalently space-forms (or "cubical' simplexes), in a dimensional hierarchy of different signature metric spaces (composed of independent dimensional levels), which must be organized around Hermitian, or quaternion spaces, due to the metric-space spin-states of both material systems and space itself. Discrete isometry subgroups create a spectral-geometric structure which exists between the material geometry of classical physics and the spectral description of quantum mechanics. The different metric spaces, ie different signatures and different dimensions, have particular physical properties associated to themselves. Putting these ideas together allows descriptions of: material, space, time, interactions, vector bundles, tensor bundles, so that differential equations can be formulated with Lie algebra valued potential 1-forms (or connections) on a principle fiber bundle. Differential equations of interacting material can be defined on sets of differential forms whose functional domains' are space-forms, so that the solution functions to the differential equations determine material geometry and spectral structures of physical systems.

If this is done, then it resolves the probability-geometry dichotomy in favor of geometry, and surprisingly, it describes life, mental activity, which are non-reducible systems, as well as material, in a context in some ways similar to the big bang singularity, but without a bang, more-or-less a steady state "big bang."

It is such a geometric theory that it allows the role of dimension to be clearly related to space-form geometry, and most differential equations are processes defined on space-forms. The hidden structure of what appears to be reducible particulate material is revealed. It is a description which is _not_ at odds with classical physics and it re-organizes quantum mechanics by giving the hidden structure of the world's microscopic systems.

It allows Einstein's statement that "God does not throw dice," to be examined in great detail. Though, because there are so many energy generating, high dimensional, "cubical" simplexes, which can "dominate through resonance," the organization of material geometry in 3-dimensional space, it is difficult to decide what "God" actually is.

These ideas can be used to model, in a mostly geometric context, and in a more precise manner than is currently possible the actual patterns observed for the:

> nucleus,
> shapes of molecules,
> crystal's current flows interacting with the crystal's spin structures,
> life (both origins and embryonic development),
> mind (memory, consciousness),
> stability of the solar system,
> [All these systems have such precise properties, over which such precise control can be affected, which the current paradigm of physics, based in probability, cannot begin to describe.]
> These above systems can now be described accurately, and the new ideas also give the proper logical basis for:
> the standard model,
> the "big bang,"
> and string theory.

Preamble

Though it is your good fortune to receive these new ideas about what is true about physical existence, and they are ideas which uphold the notion that science can develop useful truths about the external world from uncovering its hidden structure. I do not expect this communication to come to much. Sending these ideas to the institution which now upholds dogmatic interpretations (or models) of "obvious truths" as unquestionable "objective truth," is like Copernicus sending his ideas about the hidden structure of existence to the church for their approval. This is because now a days the physical science community defines a high social class, ie those with knowledge that the few powerful people of society need. The knowledge of the science community is needed in order to develop the projects that are desired by the powerful, for their own selfish gain. That is, it is believed by the powerful, that scientific knowledge allows them to get what they want, but now the knowledge needed to accomplish what is desired seems to be lacking for many highly valued projects, eg no useable model of a general nucleus, no model of electron flow in crystals interacting with the material at crystal lattice site, not to mention the origins of life. None-the-less because you are the institution with the title of science, there is a formal obligation to send these ideas to your institution for the possibility of your education.

Physics, as an institution, claims that the standard model of elementary particles, with its arbitrary parameters, is actually science. Whereas it appears to be more like a religion based in the dogmas of materialism and reductionism. It has the characteristics of a Ptolemy-like science, ie an epicycle-like mathematical patterns associated to physical data (held together by a funding process which has turned physics into bomb engineering). Namely, the epicycle of point particles rotating into other particles, in a dance about an arbitrary parameter, whose subtle value (mass) is "calculated" with such precision. It is this epicycle structure which provides the theoretical support for the obvious interpretation of micro-particle events, and that obvious interpretation is that micro-particle events are the reduced (particle) components of a material existence. This is exactly analogous to how Ptolemy's evidence showed the obvious truth that, the sun and planets rotate about the earth. In fact Ptolemy's calculations were also more precise than those of Copernicus. Precision can often be used as a means to deceive.

People of your ilk play the grand inquisitors of intellectual worth, measured from a metaphysics of ideas about material, space, time, dimension, and probability all encased in a language built out of the quantitative language of your metaphysics which is used to prove that the world, put in your terms, is consistent with all of your metaphysical assumptions. It might be noted that though Von Neumann proved hidden variables could not exist, the Bohm-Aharanov hidden variable theory has physical evidence to support it. Godel's incompleteness theorem implies that there will always be patterns that lie outside of any fixed quantitatively based language, as the patterns of Copernicus existed outside the language of Ptolemy. Godel's incompleteness theorem defines the true context of both science and mathematics, namely, that of adjusting language to find patterns of ever greater utility and range of application, ie descriptive patterns that match the external patterns of existence. The truth needs to be better approximated in terms of making a new language, built out of the universe of ideas that are already being used, leading to new ideas that can be more useful than the currently accepted ideas.

How is the institution of physics different from the church officials in the age of Copernicus? In the time of Copernicus they could smell the worthless rat that Copernicus was, because he challenged their God-given truth and hence their God-given value, as I'm sure you will detect a low value intellect who can not comprehend the physic's institution's difficult absolute truth, ie you will detect one who is outside your social class of high intellectual value. As a community physics upholds an intellectual value which is needed to perceive the absolute "objective truth" as given to "the world" by science, which also happen to serve the most powerful. This is analogous to how the church was the arbiter of truth and worth in its society, in the time of Copernicus, but always at the service of the most powerful.

The efforts of institutional physics are all shrouded in pious rigor. None-the-less the following five listed achievements of physics are either non-rigorous or they represent formalism which is not understood: (1) general relativity (ie material follows the geodesics of space, the geodesics of space are determined by the general metric of a general metric space, the general metric is a solution to the second derivative of the metric, or the Ricci curvature, which is set equal to the distribution of "energy-material" in space), (2) the H-atom (the crowning achievement of quantum mechanics), (3) the Dirac operator (a first order [linear] metric invariant differential equation (because of special relativity) should carry all the same physical information as the second order metric invariant differential equation, and be easier to use), (4) all of the standard model of elementary particles, along with (5) all the other speculative theories of modern physics (mainly string theory where point particles are given string geometry within the language of probability, this, along with big bang theory, and inflation). These five descriptive aspects of physics have either not been rigorous at all, or as in the case of the Dirac differential operator, have only added formalism to the descriptive context without a coherent story, ie spin, in a mathematical sense, was given a new formal structure with the Dirac differential operator. The following question remains unanswered, "What is being spun?" The currently accepted answer is formalism, namely, spin implies an internal state for the particle. These five examples of physical description (with the exception of general relativity and black holes) are built from a language based in probability, which is fundamentally inconsistent with geometry, (Note: geometry, and probability logically oppose one another). For example, particle collisions create an elementary geometry, ie the collision, for the standard model of elementary particles, hence the standard model is fundamentally an inconsistent theory. Furthermore, these five examples are all based on either (1) describing the obvious interpretation of the reducibility of material systems into microscopic events, and their associated spectral properties (eg the masses of elementary particles), which together suggest a probability basis for the descriptive language of physics (including the Dirac operator), or (2) (in regard to general relativity and the Dirac operator) that differential equations of physical systems can all be based in <u>either</u> a unifying principle about frames, and the local linear structures that they contain, ie all frames are equivalent and the laws of physics must be formulated in terms of tensors so they remain invariant to locally invertible transformations of a system's local coordinates, or about the order of differential equations.

Yet, in regard to general relativity, both the special relativistic principle of the constancy of the speed of light in space-time frames, and the experimental evidence that light in a gravitational field is red shifted, together imply that light does stay within a flat space-time metric space. This implies that the Schwarzschild solution cannot be correct. Whereas in either an inertial frame, or an arbitrary frame, within which material displacements take place, these displacements should be defined within a Euclidean frame (or general metric space [Riemannian manifold]). This is based on an idea of E Noether's about fundamental conservation symmetries. Likewise, the frame for the observation of light and time should be a flat space-time metric space. This sort of consistent and coherent use of both language and its logical structure has become lost in the currently accepted paradigm of physics which is full of arbitrary formalism and subsequent mathematical opportunities, ie the analogy of Ptolemy's epicycle descriptive context.

New ideas

The new ideas challenge both (1) the idea of material reducibility, by giving a "hidden" geometric structure for material which leads to both a new viewpoint about the shape of space, and to a new way in which both material and space fit into physical description,

> [though the principle of spherical symmetry (or equivalently in a practical sense, general relativity) is still applicable to generalized "Euclidean frames," ie Riemannian geometry, at least in dimensions three and less],

as well as challenging (2) the principle of materialism itself, ie material does not determine existence.

Rather, in the new theory, existence is determined by material's relations to (a) metric spaces along with the metric space's discrete isometry subgroups, ie space-forms and "cubical" simplexes, and (b) to a dimensional hierarchy of different signature metric spaces, so that (a) and (b) together, along with properties of resonance filtering between spectral "cubical" simplexes, determine existence. The viewpoint which physical description should take is not to simply emphasize the metric (as is done in general relativity and which is important, rather it should emphasize (1) the type of metric space, its dimension and metric signature, so that each of many of these distinct metric spaces in a dimensional hierarchy are associated to different distinct physical properties, eg Euclidean space associated with displacement and momentum as well as inertia, and (2) the discrete isometry subgroup, or space-form, structures of the different metric spaces. Space-forms have both geometric and spectral properties and are involved in modeling both material and space.

Furthermore, it is the contention of these new ideas that the standard model does more to clarify what a derivative is, within the context of a physical system, than general relativity. In general relativity the derivative is within a particular spatial continuum. Whereas the standard model is modeling a system at a point in which the Lie algebra of the fiber group is incorporated into the derivative is a stronger model of how differential equations and their solutions work to model physical systems than the derivative alone, so that the role of both the derivative and vector fields, which lie outside the continuum upon which the differential equation is defined. Why are local linear relations associated to quantitative models (or formulas which represent measurable physical properties) of a system so good at modeling physical systems, ie why do solutions to differential equations model physical systems so well? But also, why must differential equations acquire structures which extend outside the spatial continuum upon which the differential equation is defined.

Within the new theory, a derivative is more about attaching a physical system to space-forms of various spaces and their geometric and spectral properties than it is about a derivative attaching a physical system to either a type of function (which represents particular physical measurement) defined on its coordinates (as is done classically), or to a harmonic oscillator's vacuum properties which absorbs the differential equations extensions past the continuum defined by the containing coordinates, as is done in the standard model. In the new theory a system's local differential structure is related to its Lie algebra properties that are associated, not to a vacuum point, but rather are associated to a hidden, non-reducible (although breakable), "non-local" (but often small) geometric-spectral (space-form) structures. When the Lie algebra, is associated to transformations of a system's space-form properties, it carries the information about the incremental (or infinitesimal) rates of change of a system's properties onto the system's hidden geometric (space-form) structure.

Epicycle opportunism

Within the currently accepted paradigm of physics, the descriptive mathematics of the five above mentioned systems are essentially filled with either (1) non-rigorous mathematical opportunism, eg quasi-mathematical patterns which "work," or fit data, but the logic of a consistent language is not guiding this formalism, or (2) formalism which is not understood. It is this misunderstood formalism which provides the greatest opportunity for mathematical opportunism. Thus, caught between formalism and mathematical opportunism, the currently accepted paradigm of physics (which is fundamentally based in probability but which cannot function without geometry) is left with a language which cannot communicate any pattern that is useful. The dominating statement of the currently accepted paradigm of physics is that all systems of practical value are too complicated to describe.

It was logical consistency which Einstein was attempting to follow which gave his work so much value. In a few words he could elucidate distant horizons. It was the spirit or truth itself that he sought. Today the issue of physics authors is clearly themselves.

The standard model provides precise values for the "permutations" or "fine tuning" of the mass spectra of elementary particles about the arbitrary parameters, but no valid story as to why this should be so. Both the Fermion spin transformations and the Fermion-Boson field interactions fit into the pattern of a rotational group acting on particle states so that this model acts as an epicycle structure, but mathematical opportunism leaves the language without meaning. This language's lack of meaning is coupled to physic's basis in probability, so that when it is used to describe geometric collisions of interacting point particles at a vacuum sites [Note: a vacuum site cannot be incorporated into either a subsystem or a manifold structure for a physical system.] its logical inconsistency results in the language having virtually no physical applications other than to particle accelerators, and to the "mythological big bang." If the standard model cannot describe a general nucleus then what value does it have? The formal mathematics of the standard model is close to being correct, but what is the real basis for this formal structure? These new ideas provide an answer.

Thus talking to an uppity group of people whose absolute faith in their dogmas about physical science has greater influence on their social class than about a true description of an external world, makes one feel as Copernicus might have felt if he sought approval for his ideas from the church. Yet your institutional name is not religion or faith, rather you claim it to be science. But because I know it is easiest for a totalitarian mind set to marginalize what challenges their beliefs, it is useless to deal with you, but because you lay claim to the title science, I dutifully submit ideas which follow the true tradition of science. That is, a quantitative, verifiable description of the hidden structure of existence in a consistent, and useful language based within the ideas of geometry, is being formally provided to science. If the institution is really science, as it claims, it will embrace these new ideas. If it is no longer science it will ignore them. These new ideas form a description of the properties of the external world based in the ideas of quantity and shape and its descriptions are verifiable and they are consistent with the already measured descriptions of the world as already described by physics, but it gives the correct context for new descriptions to physical systems which have not yet been described, eg lightning, a radioactive nucleus, and embryonic development. Clear descriptions of the external world are made available to the public by this new theory, to be associated with public verifications, ie the fundamental model of science.

Modern science claims that only certain "smart" people can understand science. Modern science does not stand for understanding, rather it stands for authority, and its Ptolemaic structure of description and evidence is about faith, and I presume it will squash any perceived "lowlife" who will challenge its authority and its faith.

The institution of physical science has very few ideas that exist on a narrow range of what the fixed language allows as possibilities, and it does not allow other ideas, yet it has degenerated into a useless discipline which deals with speculation about marginal physical systems, though one must admit that the standard model associated to a gravitational singularity does still apply to bomb engineering.

If your journal rejects this paper, ie rejects these ideas, by what authority are you rejecting these ideas, which for myself and physical science, has become an, inconvenient truth? Is it rejection due to your social authority, or is it due to a lack of experts who can decide its truth, or is it that it is simply that you can discern that it is not the currently accepted paradigm, ie you can prove it "wrong" within the language and metaphysics of your own paradigm as Von Neumann proved hidden variables "wrong"?

Paper for a Science (or philosophy) (Journal)
New Paradigm for Physics
By Martin Concoyle Ph. D.
And G P Coatimundi

Short Abstract
The paradigm shift for physics

These ideas represent new patterns, for mathematics and for language, which can be used to describe the observed properties of the external world, or more generally, the patterns with which the all encompassing notion of existence can be described, ie the seen and unseen properties of existence are placed into a set of patterns which go beyond what is seen, yet it stays consistent with the observed properties of the world. These new ideas are implicated by the structure of the currently accepted paradigm of physics descriptive framework. It just so happens that the new structure, is intrinsically a part of the old description, but until now, it has "not been examined." This new structure leads to new interpretations of the currently accepted descriptive framework of physics. Namely, the old framework of: material, space, time, coordinates, metric functions (which measure lengths inside coordinate spaces), frames (coordinates with properties of motion associated to themselves), interactions, vector fields, functions (or formulas) which represent measurable properties of the described system (whose variables are the containing coordinates), derivatives, integrals, differential equations whose solutions give formulas for the functions of the measurable properties, group transformations, and invariance's for: forms (or shapes) and for lengths as well as conserved quantities. This is all placed into a principle fiber bundle whose base space is a set of metric spaces along with their new properties, and whose fiber groups are isometries, diffeomorphisms, conformal groups, and other (generalizations of these) groups. The use of this new intrinsic structure leads to new interpretations of material, space, time, and interactions, but it is not at odds with the old interpretations when the old interpretations stay within their old contexts. This is analogous to how Ptolemy's universe could be re-organized into Copernicus's solar system, eg the planets were all contained within the same space, which was the same space as the space which contained Ptolemy's planets.

In other words, the constructs of this paper should be taken seriously because they are implicit within the old paradigm's own structure.

Metric spaces are assigned physical properties, they are no longer simply Platonic forms of a mathematical language. For example inertia is assigned to the Euclidean space, while charge is assigned to space-time or hyperbolic space.

Metric spaces are characterized by their dimension and by the signature of their metric functions. A metric function is a symmetric bilinear tensor which applies to local vectors of a coordinate space so as to identify a property of length in the local coordinates. A signature of a metric function is a property of the respective dimensions of the metric space's spatial and temporal subspaces.

The natural geometric structures as well as the natural spectral structures of any metric space, ie coordinates along with a metric function, are determined by that metric space's discrete isometry subgroups or equivalently its space-forms, where an isometry group is a group defined by the property that when any group element of an isometry group acts on local coordinates, at any point in the metric space, then those transformed local coordinates leave the metric invariant. That is, when the metric function is applied to those transformed local coordinates its value stays unchanged. In other words isometries are transformations which leave the notion of length within a coordinate space (or metric space) unchanged. Often a set of isomtery transformations can be formed into a group, in the sense of algebra, and often this group is a Lie group.

Metric spaces and space-forms are put together as a dimensional hierarchy of metric spaces and space-forms, so that this dimensional hierarchy ends for hyperbolic spaces after the

10-dimensional hyperbolic space, ie an 11-dimensional space-time. This is because in hyperbolic space, space-forms no longer exist after ten dimensions.

Each different metric space can be filled with its own type of metrically flat space-forms, whose dimensions are less (usually one less) than the dimension of the metric space that the space-form is contained within. These space-forms interact with one another inside the metric space. Remember the space-forms and metric spaces are assigned to definite physical properties, eg Euclidean space-forms carry with them the property of mass, thus Euclidean space-forms represent systems or objects that have inertial properties.

This new way of organizing the properties of material and space, has all the usual properties of: material, space, time, local vector bundles, and differential equations within its descriptive structure, so that solutions to differential equations can be used to determine the properties of physical systems. However, now the solutions are differential forms whose domains are both space-forms and metric spaces. Thus it is not qualitatively any different from regular physics except it has become more constrained in its definition. This is because of its new dependence on space-forms.

It also has the advantage that the geometric and the spectral properties of a system's description now become more intuitive, ie it is no longer true that the mathematics of functions needs to be the center of one's attention when describing such physical systems, rather geometry becomes the center of one's attention. Thus, these new ideas can facilitate a wider range of creative activities than the currently accepted paradigm of physics. On the other hand the new theory is mathematical so it can be used to model systems very precisely.

Current practices in physics

What is being done now in physics is primarily mathematics, and this mathematics is based on Platonic forms and traditional ways of using language. Platonic forms will be valuable as long as they are useful. This is the ultimate conclusion of Godel's incompleteness theorem. In order to describe particular patterns, language must be adjusted so as to fit the particular patterns that are seen in the external world. Language should be adjusted rather than trying to force the observed patterns of the world into fixed way of using words, or forcing the external patterns to be described in terms of Platonic forms, so that these Platonic forms are not consistent with the external patterns that are observed.

Now a days geometries, eg Euclidean geometries, are based on definitions and traditional ways in which to use the language, and these methods are being applied to "mathematical physics" without any idea if these defined geometries are, in fact, applicable to the external world. The patterns that are being used may very well be patterns which are only applicable to abstract mathematics, and its "worship" of Platonic forms, not to the patterns of the external world itself.

For example, Euclidean space is considered a Platonic form, but some of the displacement symmetries of physical systems in space are also symmetries of Euclidean space. This suggests that physical properties be associated directly to Euclidean space. However, if this is done then it would no longer be Euclidean space. Thus, if this new pattern for Euclidean space, ie Euclidean space plus the idea of inertia being attached to the notion of Euclidean space, is more useful than the old pattern, then the patterns of "old" Euclidean space will not lead to as many useful things as the new "Euclidean space" would. Whereas the "new" patterns of Euclidean space with the property of inertia attached to itself, has more promise as a language which is capable of describing the patterns of the external world in a useful way.

So if physics is being done using the patterns of Euclidean space, or any other traditional abstract Platonic form, so that these Platonic forms are determining, "How language is to be used," then perhaps what is being described is not as valuable as descriptions based on new ways of using language. That is, altering the age-old Platonic-form notion of truth, ie not staying within a fixed traditional language, may be a better way in which to get at useful descriptions of physical systems than relying on Platonic forms and traditional language.

It seems that staying with a traditional way of using language results in complications which come about because of the attempt to place patterns (of the external world) into a language which is not consistent with those patterns. Traditional language and Platonic forms may not be consistent with the patterns that are being observed, hence a need to explore new ways of using language in order to describe physical systems.

Abstract
Broader viewpoint of the world

Physics has lost its usefulness, the standard model cannot be used to model a nucleus and it has arbitrary parameters (related to the masses of point particles) which seem to place it, as a science, in the same realm as Ptolemy's model and methods. Around the arbitrary parameters "spin" the fortunes of modern physicists.

Probability, associated to both the reducibility interpretation of the apparent particle event data obtained from micro-material systems, along with the spectral properties of micro-systems, was introduced into physics in what is otherwise a geometric description of material, and has caused the geometric language of physics to be placed in the context of probability so as to become a logically inconsistent language. It might also be noted that the obvious point particle interpretation of physical data is opposed to the Copernican tradition of finding the hidden structures of existence, particularly if a hidden structure for a point particle exists and can be described.

Godel's incompleteness theorem says that fixed languages based on mathematical patterns cannot be used to describe certain other patterns. This idea seems to be that which really underlies physic's loss of usefulness, namely, the existing patterns of the world that are being observed are outside the capacity of the currently accepted language of physics to describe. For example, the idea of a vacuum state for harmonic oscillators is a property which excludes the possibility of a physical system having a subsystem structure.

Changing the language so that metric spaces are directly associated to physical properties and by using discrete isometry subgroups (or space-forms and their equivalent "cubically" simplectic fundamental domains) to describe the hidden structure of material and its spectral properties, one finds a realm of geometry that lies between classical and quantum physics. By using this viewpoint, physical description can once again depend on geometry.

Furthermore, quantum mechanics can be re-derived from these new ideas and all properties of physical systems remain within the descriptive space of this new language. Quantum mechanics can be derived from this new intermediary geometry because of the Brownian motion that accompanies micro-system dynamics, within this new descriptive context. Quantum mechanics can be re-derived by means of the ideas of E. Nelson, who used Brownian motion of atomic systems to re-derive quantum mechanics from the perspective of the random motion of micro-systems.

The real problem with a probability basis for physical description is two fold (1) that if the measurements fit into the event space defined by a solution to the "matter-wave" equation then this is evidence that the theory is correct, but if the observed data for a particular experimental set up does not fit into the elementary event space, defined by the "wave-operator", then it is assumed that another "wave-operator" (or differential operator) actually applies to the system, thus the wrong operator has been applied to the system. Hence a probability based physical theory can never be wrong. and (2) the uses of a probability based description are very limited. A probability based theory is always correct but it is also always of limited applicability, because the probabilistic description deals with using likely-events that are outside the controllable details of a geometric context.

Consider a geometric based physical theory, when electromagnetism was described in its geometric context, then shortly afterwards the alternating current generator was invented.

We fix our attention (or perception), in three dimensions, on the inertial properties of the material geometry that exists in a Euclidean metric space. In the new theory, the massive part of a physical system is the Euclidean space-form part of this existence, yet it is the hyperbolic (or space-time) part which mostly determines a system's energy structure and stability.

In higher dimensions, new metric spaces with new metric signatures are associated to new materials, as well as their (new) material properties, and these new properties enter into the description. This may be the underlying structure which determines the so called "particle" families. It is the distinguished point of a space-form, which corresponds to the vertices of the space-form's fundamental domain, that becomes the new model for a "point particle." This gives the hidden geometry that underlies point particles.

This new basis for language can be used to provide a new way in which to both interpret and describe an entire world view. A new world view in which non-reductive space-form systems define a somewhat "self similar" structure of a space-form along with a (an apparent) metric space which together determine the composition or organization of material and space from one dimensional level to the next dimensional level. This forms a sequence of material space-forms contained within metric-space flows, and these flows are in turn defined upon higher dimensional space-forms. This sequential viewpoint of space-forms (and metric spaces) can be given perspective when it is all contained within a dimensional hierarchy of different signature metric spaces, or equivalently, a dimensional hierarchy of space-forms and/or the facial structures of a space-form's "cubical" simplexes, so that the faces facilitate interpretations of both metric-space and space-form structures which compose either a system or "all" of existence itself, depending on the dimension and containing capacity of the "cubical" simplex, ie subspace-forms exist within the various faces of the containing simplex if they satisfy certain resonance properties between themselves and/or the containing "cubical" simplex.

The space-form model is consistent with the atomic hypothesis, yet a system's description is dominated by its non-reducible hierarchical structure. Properties of dimensional independence between the faces of a simplex, for each dimensional level of the containing simplex, and the spectral resonance requirements upon which containment depends, along with the interaction processes, which (mostly) stay within each dimensional level, are clarified.

Mathematical calculations (or approximations) are provided which quickly review how space-forms can be used both to find the H-atom energy levels, and to describe how the space-form (centered) structure can be used, within a new system of interpretations so as to develop differential equations, eg spatial containment, metric space states, space-form structure, and boundary conditions, which model classical systems.

The differential equation is modified so as to give a new structure for the derivative which is intermediary between the usual derivative and the "point representation" of a system. The "point representation" of a physical system is used in the standard model of elementary particles. The currently accepted standard model of elementary particles defines a system (of interacting particles) which is centered about a "point." In turn, this solitary point defines a vacuum so that the material-particle occupancy of the Euclidean harmonic oscillator, above the "point," must always be calibrated to zero, thus defining the vacuum of that "point." On the other hand, this new model, essentially, gives both a rich spectral and geometric structure to distinguished points of space-forms, and these distinguished points of (high dimensional) space-forms can be equivalent to either vacuum points or gravitational singularity points (hence a glimpse into both the vacuum and the structure of dark matter), of the system's composing space-forms.

The big surprise that comes with these new ideas is that some odd dimensional space-forms naturally oscillate so as to generate their own energy. This can be used to give a geometric model of both life and a radioactive nucleus (a non-reductive model), but also by viewing this idea in its greater context, one sees that this energy generating space-form structure determines a living existence in which spectral centers interact with one another, and filter (through resonance) other similar (or analogous) spectral centers, all of which can be modeled as "cubical" simplexes. The different resonance's affect the type, structure, and geometry of the material space-forms contained in the metric space.

A mathematical model of a memory is also forthcoming form these ideas.

These new ideas transcend the idea of an absolute material existence as it is now understood in the physical sciences. Life and mental activity become more important than material properties because life's intent can, by spectral filtering, transition between distinct sets of material existences. Existences which are allowed (determined) by resonance between both internal (inside the observer) and external spectral structures. Namely a physical system whose guiding spectral influence is changed between two different and distinct spectral and space-form hierarchies is transformed between one material existence to another. This also provides a model of the geometric transition that takes place when a physical system changes its quantum spectral state. This also provides the idea of an initial singularity with a deeper immediacy, ie we along with everything else is still at the initial singularity.

Physics, within this new context, becomes concerned with finding the fundamental non-reductive "cubical" simplex which affects the greatest control over a system's space-form hierarchy. Lists of simplex structures, which are also fundamental domains of discrete isometry subgroups, and their spectral properties will yield much information about a system so that this spectral information can be used to effect a control over a system.

Transformations between "cubical" simplexes and their spectra become the main focus for describing physical systems, leading to descriptions of: "spectral manifolds" [ie local spectral structures, eg changing spectral faces within a system], radioactive nuclei, interactions between an electron current and lattice sites in crystals, morphogenetic fields, life, descriptions of mental activity, and intentional acts.

The ideas about the structure of material, space, and time along with their geometric and spectral descriptions are very much changed so as to conform to the properties or patterns that are actually observed.

Introduction

The currently accepted paradigm of physics is dominated by the following set of ideas or structures.

Material (scalar), space (the 3- or 4-dimensional, and unbounded set of coordinates, provide the degrees of freedom for material), time (one state), vector bundles (for the structure of differential equations), interaction potentials, [principle fiber bundles], the three primary functions used to model properties of systems in differential equations: energy function, position function, and metric function, which are used to define differential equations, which along with boundary conditions, lead to solutions, which model the dynamics or the spectra and spectral event distributions of material. The solutions to differential equations give (or lead to, as in the case of a metric-function being a solution to a differential equation) the system's properties ie event spaces, spectra, and dynamics. The entire description is dominated by the spectral properties of the energy function and its spectral expansion solutions, along with the assumption of material reducibility to particles which together lead to a probability interpretation, which models both a system's spectral properties as well as the probabilities of local particle-material events in space. This leads to the logical inconsistency that exists between geometry and probability. None-the-less all the "good" classical physics (upon which nearly all of our technology is based) is dominated by the ideas of geometry, which allows a very definite and useful description.

Particle-material events in space are an unnecessary interpretation of the observed data, much as Ptolemy's interpretation of the motions of the planets was unnecessary.

The new alternative paradigm for physics is dominated by the following set of 16 different ideas or structures. These structures are then put together as entire world views in 9 various reviews or summaries of this new set of ideas, along with its new language basis, which apply to the discipline of creating verifiable physical descriptions related to the patterns that are seen to exist out in the external world.

1. Metric spaces, signatures of the metrics, isometry groups, coordinates are generalized (or complexified) to become finite dimensional Hermitian spaces and hence the isometry groups become unitary (fiber) groups

2. Discrete isometry subgroups, lattices (on the metric space), fundamental domains, "cubical" simplexes, faces.

Note: A square on a checker board pattern (or lattice) in the 2-plane is an example of both a fundamental domain and a "cubical" simplex on a Euclidean 2-dimensional metric space.

Each different signature metric space has its own discrete isometry subgroup properties, eg lattice structures and "cubical" simplex fundamental domains.

3. Dimensional hierarchy of descriptive containment dependent on the "cubical" simplex structure of a metric space or of an over-all containment set, Note: there exist higher dimensions than 4.

4. Dimensional independence, ie the dimensional properties of the faces of a "cubical" simplex are independent of (or isolated from) the dimensional properties of the other (higher dimensional) faces or of the entire simplex, to which all the faces of the simplex are geometrically related. This is due to the properties of infinite extent hyperbolic space-forms and their relation to the physical property of light and interactions. The infinite extent hyperbolic space-forms (of one dimension less than the metric space (or metric space face)) fill a particular dimension metric space or metric space face, but these infinite extent space-forms do not extend past the fixed dimension within which they are naturally contained.

5. The way in which words are used is changed, material properties are associated directly to particular metric spaces. For example: The Platonic form of Euclidean space is changed so that Euclidean space is to be associated to:

   (a) the physical property of inertia,
   (b) the physical processes of material displacement,
   (c) the physical property (or decision) of determining frame of reference for motion, eg rotational or translational (with respect to the motion of distant stars), and
   (d) the physical properties of momentum both linear and angular momentum.

6. Interaction potentials in all faces (or metric spaces), of all dimensions, have the form $1/r$, ie a sectional curvature of a (n-1)-dimensional, interaction, Euclidean space-form. This Euclidean space-form's geometric properties are related to the separation distance of the two interacting material space-forms, so that direction of separation is a direction which is normal to the two distinct spheres that contain the two interacting "material" space-forms. This Euclidean space-form's containing n-dimensional space is determined by the normal direction, and the (n-1) directions tangent to this normal. This tangent space determines the local directions of the Euclidean space-form's local surface. Hence the (n-1) tangent directions of the space-form, along with the normal direction, determine the property of the interaction Euclidean space-form being (n-1)-dimensional so that it is contained in a n-dimensional space. The interaction potential is determined by the length, $2r$, of the normal which separates the

two interacting space-forms. Where r it will be one of (or to simplify, it can be assumed to be the maximal radii of all) the radii determined by all paths on the Euclidean torus determined by the tangent directions of the (n-1)-dimensional, interaction Euclidean space-form, at the point on the interaction Euclidean space-form determined by the normal direction connecting the two interacting ((n-1)-dimensional) material space-forms.

Actually several space-forms can interact simultaneously with a particular space-form so that each of the pair-wise interacting space-forms (one of which is always the same space-form) has its own 2-dimensional interaction Euclidean space-form structure.

In low dimensions, ie 3 and less, it is the spherical symmetry which makes the normal direction the actual direction of the interacting force which was determined by the potential, or 1/r sectional curvature of the interaction Euclidean space-form.

7. Interactions between microscopic space-forms (or atomic systems) lead to Brownian motion, a stochastic process. E Nelson used Brownian motion, or stochastic processes of atomic motions, to re-derive quantum mechanics. Hence this new set of ideas is not opposed to quantum mechanics, but quantum mechanics is usually not necessary to use, in order to describe the world's properties. If there is a more definite geometric alternative that can be used to describe a physical system, then the geometric description is usually more useful.

8. (Self similar structure) An n-dimensional space-from exists in an (n+1)-dimensional metric space, which can be viewed as an (n+1)-dimensional flow (or a face) on an (n+2)-dimensional space-form (or an (n+2)-dimensional simplex [which is a fundamental domain of a discrete isometry subgroup]).

(An (n+1)-dimensional metric space can also be viewed as an (n+1)-dimensional fundamental domain, or as an (n+1)-dimensional space-form).

9. Each structure in the previous sequence [in (8) above] of space-form structures corresponds to a face (or a set of faces, or a "cubical" simplex which itself can be a face) on a high dimensional "cubical" spectral-simplex, within which the "physical system" or space-form is contained. The high dimensional simplex represents a geometric system, that has a definite position (hence the high dimensional simplex is bounded), is also a bounded fundamental domain of a discrete isometry subgroup, which, as a space-form, is at most dimension-5. At dimension-5, the "cubical" simplex is bounded, ie compact, so that such a bounded simplex ultimately represents some "material system," so that the material subset has a definite position, and hence, a definite geometry associated to itself. Most, if not all, hyperbolic fundamental domains that are between dimensions 6 and 9 have the property that their faces extend to infinity (Theorem of Coxeter).

10. One basis for stability of space-form systems is the spectral stability of the hyperbolic space-forms. In dimension three the 2-dimensional hyperbolic space-forms form into very stable material systems.

11. The structure and properties of hyperbolic discrete subgroups at different dimensional levels: [Except for the possible exception of the "cube-like" fundamental domain]

There are no discrete hyperbolic subgroups that have dimension 10 or above, ie space-time dimension 11, and they are all infinite extent discrete hyperbolic subgroups beginning in hyperbolic dimension six, ie from dimension 6-9 all hyperbolic space-forms are infinite extent, with the possible exception of the exactly cubical fundamental domains or "cube-like" fundamental domains; as hyperbolic fundamental domains are formed by pushing together cubes (or right rectangular prisms) at their vertexes. For hyperbolic dimensions, 1-5, there exist both bounded and infinite extent (or unbounded) hyperbolic space-forms at each dimensional level within this range, ie hyperbolic dimensions, 1-5.

12. Orthogonal rotations between the faces of the "cubical" simplex are models for both the spin rotation of the spectral flows of a space-form, as well as Hermitian space rotation between the opposite metric-space states for a metric space. It is the spin rotation of the opposite metric-space states for a metric space that results in material dynamics in a metric space. For example it is the mixing (or spin rotating) of the two time states in the hyperbolic metric spaces which is needed for dynamical (or inertial) displacements to take place in Euclidean space (see below).

13. Resonance and the filtering processes between spectral-simplexes of different dimensions. This resonance or filtering process depends on the spectral sets that exist on the maximal tori of the fiber groups, where they themselves are formed through resonance with the base space's (energy generating [see 14]) spectral structure. The spectral sets on the relevant maximal torus forms the "connective medium" through which resonance's can occur throughout the full dimensional descriptive range.

14. Some odd dimensional hyperbolic space-forms, eg dimensions 3, 5, 7, etc, naturally oscillate, and in so doing generate their own energy. These space-forms become the spectral centers which determine properties of existence for the lower dimensional space-forms. Existence becomes a complicated filtering structure (by resonance's) between other (competing or existing simultaneously) energy generating spectral centers.

15. Spectral sets, on the maximal tori of the fiber group, can exist due to their resonance with the base space's energy generating spectral-simplexes. This provides a means by which to model both a memory and an intent in an energy generating (spectral) system.

16. Simplex to simplex transformations, or functions between "local" spectral structures, now become the most interesting types of transformations which can deal with, ie describe, various physical properties such as: the renormalization groups (eg thermodynamic change of phase), lightening in clouds, precession of Mercury's perihelion, stability of solutions to differential equations or the stability of physical systems, spectral manifolds or local spectral structures (new), radioactivity, the source of a star's energy generation, morphogenetic fields, life, embryonic development, the context of existence, willful acts, and mental activity, eg perception and creative thought.

All of this new structure is used to revise the language which is applied to physical description, but it also allows the same basic description as that of the usual descriptions of physics, ie as the description whose basis is given in the first paragraph of this introduction, except that the ideas about the model for both material and space are changed, though neither of these changes are that noticeable, because most material space-forms are small and metric spaces are so large.

Furthermore, the notion of probability is not needed in the same manner as the currently accepted paradigm of physics, but which none-the-less may be used. Though probability is allowed, it is no longer central, in this new paradigm for physics. That is, if the basic relations of (classical, though it could even be quantum) physics can be determined from the new principles then there is nothing excluding the use of the old mathematical techniques in the new context.

Space-forms re-define material but it does so in a way in which material depends on either its atomic structure or its classical center of mass structure, ie material can still have a center of mass, as well as most of the usual fundamental aspects of physical description, namely:

1. material (ie its center of mass), and atomic structure
2. space (though somewhat modified, for most systems that the currently accepted paradigm deals with [in which its description is detailed and mostly correct] the new viewpoint of space, for the most part, can be given the same metric-space or "symplectic-space" structure as the old paradigm, eg Hamiltonians).
3. time (has greater significance in higher dimensions, however, time only has one direction in the currently accepted paradigm), furthermore,
4. vector bundles can still be used to define either the usual old (metric invariant) differential equations or new differential equations based in space-form structures.
5. The new differential equations deal with the new aspects of a system, the new solutions identify properties of a system in terms of properties of the space-forms which compose it, and determine the boundary conditions of a differential equation and space-forms determine the system's natural processes, eg rotations between metric-space states. However, if the equations for classical physics are derived using these new ideas or methods, then classical mathematical techniques can be used.

> How do ideas enter the "market place," or enter into the public's attention,
> ie how do ideas become admissible to institutions?
> Answer: This decision is made within the social class structure
> which defines value for civilization

The issues presented in this paper are not issues of opinion and how social class allows certain opinions to be given authority, rather the issues within this paper are issues about verifiable truth, whether Ptolemy's view of authority or Copernicus's view of hidden structure should dominate the discourse of determining the truth about what exists within a verifiable context for physical description.

Professional science journals entertain outlandish speculations, eg inflationary scenarios of many big bangs and miniature black holes, and the pure mathematical speculations of string theory, all dealing with descriptions of remote physical systems which makes verification tenuous at best, while on the other hand, scientific journals do not seem to consider, very sober descriptions, which already have supporting evidence to back them up, about the hidden mathematical structure of all material and space. So that the new ideas are all described in terms of a new quantitative language for physical description. Such journals do not want to consider the hidden structures of a new (consistent) language which sheds much needed light on both the puzzles of the material patterns of existence, eg how to describe a general nucleus, as well as the structure which determines the stability of the solar system, and beyond, to descriptions of "physical" systems that intrinsically and mathematically cannot be reduced to the sum of its material components, such as life and embryonic development.

Science is supposed to be about public description and public verification, the core ideas about information within a democracy, but science has become the fundamental obscurer of knowledge. The language of science has come to be more about identifying an elite social class and the aspirations of their masters, a social class whom decides the issues of absolute "objective" truth, as opposed to being able to provide a useful truth. Both ideas about science's role in society, namely, to either be secretive or elitist, are ideas which fit nicely into a totalitarian state's interests.

Words and language impose restrictions on what can be described

In order to say something definite, words must be limited in their meaning. The more definite language of quantity and shape, ie mathematics (and physics), have limitations on the patterns that they can describe (ie the content of Godel's incompleteness theorem), simply because the words used to describe a definite pattern, need to have a definite meaning. Words that do not have a definite meaning can be associated to a lot of different patterns.

Certain patterns need to be described by words so that these words have certain properties. This means that new ways to use words need to be considered when describing new patterns because the old words were chosen because they could be used to describe certain old patterns. Thus using the old words results in a great deal of difficulty in describing the new patterns. Therefore, adhering to an old way of using words is more a function of adhering to the forces of social class, and the aspirations which were formulated in that language, than it is about finding the correct patterns of language that fit a verifiable objective truth.

Currently accepted paradigm of physics

The currently accepted paradigm of physics is that material determines existence. Physical systems are described within the context of material, space, time, vector bundles, spectral properties, differential equations, and their solution functions which represent a material system's properties such as: position, the motion and/or force field structure of a tangent bundle, spectral properties, eg energy spectra of microscopic atomic systems.

In this context material is always modeled as either a scalar or a point particle (a point with internal capacity for changes) and the 3 or 4 dimensional (local) material containing metric spaces have the properties of both being continuums and being unbounded. Material defines existence for both classical physics, where the solutions to differential equations are geometric, as well as quantum mechanical physics, where solutions to [linear] differential equations can be represented as series expansions of a function space's spectral sets. These quantum mechanical solutions describe the events of point particles in space, so that in quantum mechanics, the function spaces are interpreted as probability spaces of material point particle events in space. However, the uncertainty principle means that space must lose its property of being a continuum. Thus we see the mathematical opportunism which enters into physical description. Namely, for some questions the space is a continuum so that by using space's continuous properties, the properties that deal with the mathematical form of higher dimensional force fields can be identified, while for other questions space is not a continuum, the point-particle material composing quantum systems is allowed to "jump" from the geometry of one energy state of a quantum system to the geometry of another energy state in a discontinuous manner. In order to understand what is really true about the properties of space, the language of the physical science needs to be changed.

Putting it together (Review 1)
Physics and the spectral structures of metric spaces.

In the currently accepted paradigm of physics there is a separation of: (1) the spectral properties of function spaces, from (2) the smooth geometric properties of coordinate spaces. In the new paradigm of physics the spectral-geometric properties of metric spaces can be used to cause a re-ordering of the structure of the containing spaces of physical description. The natural geometric-spectral properties of metric spaces are contained in their discrete isometry subgroups, which reduce (or can be "moded out") to be (or to become) space-forms, the natural geometric and spectral (where spectra especially apply in the hyperbolic case) objects (or properties) of metric spaces.

This is exactly analogous to how Copernicus re-ordered the planetary geometry of Ptolemy's universe, but in this new case there is a big influence on the re-ordering of the solution space for differential equations, though the spatial and material structure of physical description is also re-ordered. The solutions of differential equations (eg matter-wave equation) are no longer considered to exist only in terms of a function space structure but rather in terms of differential forms (geometric functions) whose domain's are space-forms. This allows the interpretation of function spaces as probability spaces to come to an end.

A Dirac-like differential operator becomes central to differential equations because it naturally fits into the geometry and boundary conditions of space-forms and their metric states. The Dirac-like operator is applied to differential forms (geometric functions), which in turn, are defined on space-forms. Space-forms carry on themselves both their own geometry, along with some of the spectral properties of metric spaces, including the different states of a space-form's spectral flows ie space-forms are defined by discrete isometry subgroups, subsequently, the lower dimensional faces of a space-form's fundamental domain determine the spectral flows on these space-forms. The hyperbolic metric space's discrete groups carry the spectral properties of material systems in a stable manner.

Discrete isometry subgroups:

1. Principle fiber bundle structure, ie the space in which physical systems are best described, is composed of (1) a base space (typically a metric space, or a dimensional hierarchy of space-forms and metric spaces) and (2) a fiber group (typically a fiber group is an isometry group eg rotation Lie group, translation group, spin group, or unitary Lie group) so that the fiber group is defined over each point of the metric (or base) space. Discrete isometry subgroups emerge from fiber isometry-groups.
2. Lattices represent the geometric structure of a discrete group, and are defined on the metric space or the base space, eg a checker board pattern on a 2-dimensional Euclidean plane is a Euclidean lattice.
3. Fundamental domains are the individual regions which compose the lattice and they are "cubical" simplexes, eg one of the checker board squares for the Euclidean plane. Note: A square is a simple simplex. A "cubical" simplex has face structures of right rectangular prisms, so that the boundaries of each face are themselves faces whose dimension is one less than the region (or face) that they bound. The lowest dimensional faces of a "cubical" (or any) simplex are its set of vertices all of which become identified as a single (distinguished) point of the space-form which is formed when the simplex is "moded out."

4. A moding out process, is a process of identifying "opposite" faces with each other, and it creates the non-local geometry of space-forms in the base space, eg the checker board (or the checker board square) "modes out" to form a doughnut, or torus, in Euclidean space.
5. Space-forms are contained within a dimensional hierarchy of metric spaces (and space-forms) so that material and space have a self similar (space-form) structure as one progresses up the dimensional hierarchy.

Note: This self similar structure is a geometric model of super-symmetry, which is contingent on a viewpoint in which a metric space may be thought of as a spectral flow on a higher dimensional material space-form.

Conjecture: sparticles are so massive because they are really entire metric spaces and the sparticles associated to a field particle would exist in the dimensional level which is one dimension above the dimension of the space containing the field particles. Sparticles that exist in the same dimension as the field particles are considered to be material space-forms, and are sparticles for the next lower dimensional level.

This self similar structure of material and space is caused by material always having the structure of a space-form so as to always be a geometric shape whose dimension is one dimension less than the dimension of the metric space within which the material space-form is contained. This also implies that within each dimensional level the interaction potential always has the form, k/r. This is because material is no longer a scalar rather it has a dimensional geometric structure and an interaction potential is determined by the sectional curvature of an interaction Euclidean space-form so that the plane defining the sectional curvature, ie the 1/r potential, contains the normal that is defined in regard to the two interacting space-form's tangent spaces, so that a pair of distinct, separated tangent spaces on the two interacting material space-forms identifies their normal separation distance. It is always the inverse of this distance of spatial separation (sectional curvature) which determines the mathematical factor which applies to an interaction potential between two (or more) material space-forms. Hence, the non-existence of higher degree potentials (or force fields) is not evidence that higher dimensions do not exist. Furthermore, the faces of the containing simplex are "dimensionally independent" of one another. Where material space-forms of a particular dimension are contained in the proper dimensional faces of a "cubical" simplex.

Full picture

To gain plausibility as an alternate theory to the currently accepted paradigm, a sketch of the entire world view, that can be derived from the new alternative theory, must be given. The following description is quite robust and leads to a definite context.

New patterns, new ways to use language (summary 2)

There is a new way to look at physics. It is not a reductionist's view of existence, yet it is consistent with the atomic hypothesis. Atoms are composed of a pair of interacting, positively and negatively charged 2-dimensional hyperbolic space-forms, along with their associated, massive Euclidean space-forms. The neutrino (associated to the leptons or equivalently associated to the negatively charged hyperbolic space-form) will be an infinite extent (or semi-infinite extent) hyperbolic space-form. The nucleus along with its neutrons will be a bounded, or compact, space-form. Thus, the singularity of the mathematical formula for interactions, ke/r, does not need to be considered, because material itself is intrinsically geometric, hence non-local. Furthermore the space-forms have a distinguished point about which collisions can be centered, namely the vertexes of the "cubical" simplexes which becomes a single point of a material space-form in the "moding out" process.

It is upon the Euclidean space-forms, (1) that is carried the property of inertia (or mass), and (2) upon which we focus our attention. It is from material distributions that light emerges. Our attention's main focus is seeing and feeling material.

The charges would be 1-dimensional hyperbolic space-forms, ie circles of charge, and would occupy the stable 1-dimensional spectral flows on the 2-dimensional hyperbolic space-forms. The affect of interactions between space-forms is ultimately to "push" lower dimensional space-forms together so that they will form either higher dimensional space-forms, or dynamic systems, which will fit within the bounds or parameters of stable (space-form, or geometric-spectral) systems. The bounds and parameters of stable systems are determined by the spectral properties of the containing space-from or "cubical" simplex structures. The formation of atomic structures and orbital structures is due to interactions, but of course not all such interacting space-forms result in stable systems. Furthermore, some "stable orbits" of dynamic systems are stable because they fit into an envelope of stability defined by a containing space-form.

Hyperbolic space-forms have very stable, and somewhat limited, spectral properties. Small circles of charge fill the stable spectral flows of 2-dimensional hyperbolic space-forms. Euclidean space-forms are made stable by their resonating with the very stable hyperbolic space-form spectra. Together, the 2-dimensional: Euclidean, hyperbolic, and interaction space-forms, create physical systems in a 3-diemsnional metric space, ie a metric space that has three spatial dimensions.

Apparently a (3,2) signature metric, ie three spatial dimensions and two time dimensions, [Note: The notation means $(3,2) = R^{3,2}$, where, R, stands for the Real numbers] has no practical applications for a 3-dimensional spatial subspace unless a new material property beyond inertia and charge is introduced into a 3-dimensional spatial subspace, but even then it may still be equivalent to a (3,1) metric signature, though perhaps the idea of a new material associated to a (3,2) signature metric, explains some of the data of particle experiments.

On the other hand, Huygen's principle implies that a (4,1) signature metric space has no valid time properties, because light does not propagate as spherical shells. This is due to the existence of unbalanced space-forms, for space-forms whose (spatial) dimension is odd (see below).

Note: The space-forms contained in a hyperbolic 4-space are 3-dimensional.

However, within a (4,2) signature metric space, the second time dimension can allow for useful new properties of time, which might also be associated to new properties of material, ie a new material type, associated to higher spatial dimensional structures.

Material enters into the description within the pattern of metric space signatures, a pattern in which material existence "evolves."

Metric spaces of various dimensions and subsequent signature properties as well as the dimensional level where new materials are defined

1-dimensional metric space (mass)
Only consider the (1,0) signature
2-dimensional metric space (charge)
Consider the signatures: (2,0), (1,1)
3-dimensional metric space
Consider the signatures: (3,0), (2,1)
4-dimensional metric space (new material)
Consider the signatures: (4,0), (3,1), (2,2)
5-dimensional metric space
Consider the signatures: (5,0), (4,1), (3,2)
6-dimensional metric space (another new material)
Consider the signatures: (6,0), (5,1), (4,2), (3,3)
Etc. Where $(r, s) = R^{r,s}$, where, R, stands for the Real numbers, etc.

Note: (with the exception of the first dimension) That in the even dimensional spaces a new signature enters the list of metric spaces, thus allowing the definition of a new type of material with its new physical properties in those even dimensional metric spaces, as well as a new dimension for the temporal subspace of the metric space.

Resolution of the geometry probability dichotomy

Discrete isometry subgroups can be used to resolve the probability-geometry dichotomy of physics. The spectral structure of metric spaces allows both the geometric and spectral structures of physical systems to become consistent.

The current paradigm of physics uses the spectral structure of function spaces (as opposed to metric space spectral structures) to find and interpret solutions to differential equations. Because physics reduces systems to their micro-components, the function space is given a probability interpretation. This, along with differential equation solutions which are spectral expansions, results in the description of a material system's spectral states, to have its basis in probability. Thus, probability has become the basis for all the descriptions of physics.

Probability spaces always come with an uncertainty principle built into their structure. It is the uncertainty principle associated to probability spaces which is not compatible with geometry. Unfortunately, a physical theory must always have geometric notions associated to its description, eg interactions are always geometric, hence the language of physics becomes inconsistent. The opportunity for change is for material to have a more interesting model than either its scalar or point particle model.

A more interesting model of material may be obtained by applying discrete isometry subgroups to metric spaces in order to describe a system's spectral and geometric properties. This results in functions being defined as differential forms, ie geometric functions. Differential forms are functions whose domains are defined on the material and interaction space-form structures that exist within the (system containing) metric space, or metric-space face of the system containing simplex (ie fundamental domain of a discrete isometry group).

Flatness

A metric space has a "flat" space-form shape which is only noticed if (or barely noticed unless) it is viewed from the next higher dimensional level (or metric space), within the dimensional hierarchy of metric spaces and space-forms which compose the base space of the principle fiber bundle.

Spherical symmetry

Space-forms and metric spaces, along with differential equations, can be used in this new theory to model all of physics. It might be noted that in the new theory the low dimensional metric spaces have a slight property of spherical symmetry. This is due to an SU(2) fiber subgroup, which has the geometry of a 3-sphere, whose tangent space descends to become the base space of the principle fiber bundle. The tangent space, $T(SU(2))$, can be used to create the base (coordinate) space of lowest dimension which also contains systems which are stable, eg hyperbolic space emerges from a 2-dimensional metric space, and motion requires the structure of 2-dimensional hyperbolic space-forms, and they exist in a 3-dimensional hyperbolic metric space.

Quantum mechanics

The interactions between space-forms, at the atomic level, have a Brownian motion associated to themselves, and this stochastic motion is what E Nelson used to re-invent quantum mechanics

(QM). Hence this new viewpoint also accounts for QM. However, within this new framework probability is not needed as a fundamental axiom. This means that the spectral properties of systems as well as the geometric properties of material systems do not have to be inconsistent with one another.

This new viewpoint resolves the conflict between geometry and probability so that the description is essentially based in the language of geometry, namely space-forms.

Stability and physical description

The hyperbolic space-forms have definite, and very stable, spectra associated to themselves.

The geometric-spectral inter-relations between material, its interactions, in space, and time, along with tangent bundles and spectral properties, which all together, form the basic describing bulwark of physical systems, are "all" clarified within this new way of organizing the language of physical description. It describes the nature of physical systems and their subsystem structure, and this language is fundamentally spectral-geometric.

Applications

These new ideas provide a new "local spectral structure" applicable to the mathematical representations of: molecular shape, crystal subsystems, life, and embryonic development. It gives a clear notion about what determines the stability of physical systems. It also clarifies the renormalization group (or re-scaling group), typically used to describe phase transitions of thermal systems, in terms of space-form properties.

It clarifies: Super-symmetry because super-symmetry pertains to a self similar structure of material and space that is dependent on the existence of a sequence of a (higher) dimensional structure for the metric space and/or space-form hierarchical containment and composition properties of systems (or equivalently metric spaces) for both material and space, that form within the proper containing space for physical systems.

The mathematical structure of material and fields, and their respective symmetric and anti-symmetric properties as functions

A Fermion's spin-½ properties pertain to charged material space-forms, or their "cubical" simplicial fundamental domains. It turns out that the measure of the geometric properties of the different dimensional faces of a simplex is best done in terms of an alternating form, ie a set of anti-symmetric tensors (or operators). Spin-½ properties implies an attachment to a simplex along with a means to measure the fundamental vibrations (eigenvalues) for each different dimensional levels (or faces) of the simplex, or space-form (so that the dimensions of these faces are less than the dimension of the containing metric space). This measure of the geometric properties of the simplex's faces is an anti-symmetric function. For hyperbolic fundamental domains (or space-forms), each face, of the relevant simplex, identifies a stable spectral value, or eigenvalue, of the space-form, ie for 1-faces the measure would be length, and for 2-faces the measure would be area, . . . etc.

Bosons represent the "particles" which "supposedly" compose the fields that interact with material. Boson's integer spin properties pertain to metric space properties of material geometry. This is because the light space-forms, and other hyperbolic interaction space-forms, are infinite extent space-forms which fill the metric space (or facial subspace of the containing "cubical" simplex) and have a "wave-front" structure which propagates in a spherically symmetric geometric pattern. Metric space geometry is best described in terms of the space's metric, or measure of length, and the metric is a symmetric tensor, ie a symmetric operator or matrix. Integer spin properties of field "particles" implies being a part of a geometry which connects spatially separated material systems within a metric space. This geometry, contained within the metric space, is a property of relative separation of material, and the length functions defined on (or by) the interaction structures, ie the infinite extent space-forms, are symmetric functions.

Light's structure

Light would be a Boson, but the structure of light in a 3-dimensional metric space, is that of an infinite extent 2-dimensional hyperbolic space-form which fills the system containing 3-dimensional metric space, with the help of a "local" wave-like structure which in the metric space has the shape of a spherical shell which propagates to infinity along the infinite extent hyperbolic space-forms.

Note: In an odd dimensional spatial part of the containing metric space, the wave structure may be somewhat local or a spherical shell-like structure. Such infinite extent (light) space-forms along with their spherical shells, composed of waves, fill the 3-dimensional metric space, and are important in modeling both the dimensional independence that exists between the faces of the metric space containing "cubical" simplex as well as the mind of an observer, ie the entire region of the metric space along with the geometric shapes that the metric space allows are intrinsic to both the metric space and the observer.

The spectra of light depends on the spectral properties of the faces of the fundamental domain of the (system containing) metric space.

Atoms contained in a metric space resonate with the metric space's spectral sub-structure.

Stable planetary orbits of many body systems

Furthermore, the space-form property of a metric space allows the space-form (or metric space) to contain the earth's orbit, so that the space-form shape of the metric space allows the earth's orbit to be stable. This is a new envelope of stability imposed by the space-form structure of space. This has a very similar geometry to KAM's (Kolmogorov, Arnold, Moser) envelope of stability for planetary orbits but it is not a geometry, or stability envelope, which comes from the structure of differential equation's boundary conditions, [for KAM to work it is important that there exist no resonance's, but unfortunately the solar system has resonance's] rather the new envelope of stability is a structure of space itself.

Metric-space states and the natural unitary structure

This new theory explains why the natural fiber group for a physical system is unitary. Namely, because each metric space has two metric-space states associated to itself, and those states can be separated into the real and pure imaginary subsets of a finite dimensional "Hermitian" base space.

The two metric states of hyperbolic space are (+t) and (-t), and the (-t) state carries monopoles. The two metric states of Euclidean space are (1) the "stationary distant stars" and (2) the "rotating distant stars." The Euclidean metric state defined by the "rotating distant stars" carries anti-matter.

Spin is the rotation between the two opposite metric states, that exist as opposite spectral flows on the material space-forms or as the opposite metric-space states in metric spaces. At the level of a metric space, such rotations of states, eg states of time, are needed for the full description of the dynamic processes of material.

The new vacuum

Within this new theory, the vacuum is given a geometric-spectral structure. Namely the spectra of a high dimensional space-form at the space-form's distinguished point. The distinguished point of a space-form is formed from all the vertices of a high dimensional simplex's fundamental domain, during the "moding out" process, so that all the vertices of the fundamental domain are identified with only one point of the space-form. Thus, for example, a crystal will have a distinguished point, sort of equivalent to a vacuum point, which is the common vertex (after moding out) of a quite complicated 3-dimensional, hyperbolic, "cubical" simplex which represents the "atomic structure" in a 4-dimensional metric space, of the interacting 2-dimensional atomic components which form into the crystal (or condensed matter). Thus, the crystal will have 2-flows, and 1-flows, ie currents, which are part of the geometry and spectral structure, as well as contributing to the spin structure, of the entire crystal.

Note: the vacuum is an unnecessary construct within this new theory since neither probability nor point particle geometries exist on space-forms.

In the currently accepted paradigm, each of the different spectral functions, which are summed together to compose the series solution of a wave equation, eg the energy function, have different geometries associated to themselves. These distinct spectral functions are ultimately particle probability distributions which determine event probabilities for material particles which, because they are functions whose domain is space, have a geometric dependence. However, the solution sums all these different and distinct geometric probability distribution functions together, so that in all cases the particle events are governed by probability and randomness in selecting one geometric event distribution or another. Hence the descriptive process is no longer ruled by a language which based on geometric processes but rather it is based on random events which jump either within or between geometric event distribution functions.

Material structure

Material and space are given a self similar structure in a dimensional hierarchy of different signature metric spaces and space-form shapes.

This new theory accounts for electrons, neutrinos, protons, neutrons, light, interaction space-forms and corresponding atomic (or space-form) structures on the different dimensional levels of the base space's natural dimensional hierarchical structure. The new theory also has a finite dimensional unitary Lie group as a fiber group, because the base space is ultimately a finite dimensional "Hermitian" space, which is a necessary structure to accommodate the pairs of metric space states that each different metric space carries on itself.

Apparent particles, higher dimensions and the unitary fiber group

The results of particle experiments in particle colliders and all the diverse sets of particles that emerge from high energy collisions, might be best understood in terms of the dimensional hierarchy of space-forms, ie the (three [and/or more]) different families of such particles, and the possible breaking apart of the space-form structures of various dimensions of both material and space. The space-from structure is broken by the high energy collisions of material. These high energy collisions occur at distinguished points of space-forms. However, because the whole process is under the influence of a unitary fiber group, in both the standard mode and the new paradigm, the result is that the particles show the activity that is seen in "bubble chamber" photos, because the broken pieces of the space-forms are being put back into the structure of space-forms under the guidance of the unitary fiber groups, or their local Lie algebra transformations. However, this putting back together process rightfully belongs in a metric space with more spatial dimensions than three, hence the particle state structure (or spin, or isospin) which exceeds the usual coordinate containment of differential equations.

Envelopes of stability

The space-form shapes, which are the shapes of metric spaces, can provide an envelope of stability within which the motion of stable planetary orbits can be contained. Furthermore, faces on a "cubical" simplex which contains material subsystems identify both a metric space structure and the spectral range within which the contained lower dimension space-forms must be in resonance. That is, material space-forms that are contained in the metric space are in resonance with a spectral set defined by the containing metric space.

Space-forms are flat, hence they would not be very noticeable, thus also causing the measurements of the universe to see a flat spatial structure.

Interactions

When 2-dimensional material space-forms have a geometric relationship to one another, then other 2-diemsnional spatial space-forms naturally form in the space between the material space-forms, and in so doing create a set of 2-dimensional faces which form into an interacting system which defines a 3-dimensional space-form. This is the basis for interaction properties between material space-forms and spatial space-forms, and the subsequent sectional curvatures which dominate material interactions.

Life

This new theory gives a relatively simple model of both life and a mind. It gives a new structure to all of existence.

Some odd dimensional hyperbolic space-forms, due to electric imbalances, naturally oscillate and create their own energy. Because resonating Euclidean space-forms adhere to hyperbolic space-forms, and these adhering Euclidean space-forms also start oscillating. The energy generated by an oscillating living space-form results from a "matter anti-matter" interaction induced by the living space-form's oscillations. Note: The two metric states of hyperbolic space are (+t) and (-t), the (+t)-state carries charge and the (-t)-state carries monopoles. The two metric states of Euclidean space are the stationary distant stars and the rotating distant stars. The Euclidean metric state defined by the rotating distant stars carries anti-matter. A Euclidean space-form will follow the hyperbolic space-form's oscillation due to resonance. In Euclidean space-forms low energy oscillations push the matter and anti-matter together in order to create energy. High energy oscillations, eg as in the case of the sun, may be able to push charges and monopole-charges together in order to create energy. Thus, the sun can have stable electromagnetism structures which can suddenly change, so as to give off energy

Such oscillating space-forms are the simple forms of life.

Mind and memory

The fiber Lie group contains maximal tori, on such maximal tori there are defined the closed curves (lines) or closed surfaces (planes) etc, which will resonate with the oscillating, energy generating space-form spectra that exists on the base space. Geometric-spectral information can be stored on such a maximal torus, hence this becomes an elementary model of a memory for an elementary mind.

The structure of the stars and galaxies

A maximal torus of a fiber group of a high enough dimension could carry the spectral information of an entire galaxy. Thus, conjugating (a group operation) between maximal tori within the fiber group could be considered as a transformation between galaxies. Differentiating such a conjugation transformation in terms of one parameter (note: a model of time) can result in, "the raisins in a bread pudding" expansion of galaxies.

Science should be skeptical

None-the-less, it is not clear from the data alone that the galaxies are really moving (or space is expanding) as it is assumed they are moving (or space is expanding), ie the red shift of galaxy spectra may not be a Doppler shift or it may not result from spatial expansion. What measurement could resolve this question?

Verifiable description of existence

This new theory is a mathematical description based in principle fiber bundles, high dimensional energy generating "cubical" simplexes and the space-forms contained within the various dimensional faces of the "cubical" simplex (of the base space), and differential equations, hence predictions can be made, and it is measurable, hence it is verifiable. It transcends material existence. Existence has more to do with living intent (s) which identify resonance's on both the base space and the spectral sets of maximal tori, than it has to do with a "barren" material existence that is identified only by material properties.

Self organization and the properties of space

Self organization described in relation to the non-linearity of differential equations is limited, and would be wrong if one believed in the probability basis for existence, since then the needed stable boundary conditions for self organization would have no valid basis.

The ideas about the shape of space, as put forth by Gauss and Einstein, need to be considered in terms that allow for flatness, because evidence, as it is now analyzed in regard to our limited knowledge, suggests that the heavens are flat. Space-forms are themselves flat.

Flatness might be an equivalent idea as the idea of linearity, and thus would provide a new descriptive path to be used in order to control the physical systems such as superconductivity, as well as living systems, or any system which has this space-form flatness as part of its physical properties.

The best thing going

If one believes in the spirit, and one also believes in one's own ability to determine truth, then one must logically believe these new ideas, because there are no other descriptions of existence which are quantitative and predict properties consistent with the external world so as to provide a context for existence, which is verifiable and measurable, so that such a description of existence includes both the spirit and the (many) material world (s) that are part of existence. The material existence of the world is transcended by these new patterns for describing existence. That is, high dimensional, energy generating, "cubical" simplexes transcend the idea of material absoluteness, as they command through their spectral dominance (a dominance determined by resonance relations with the observer) the nature of the material existence that is contained within their region of influence, (or their boundaries), a region defined both by spectral and spatial bounds. That is, material existence changes its geometric and spectral properties depending on the spectral simplex structure within which it is contained.

What can be talked about, what can be described and verified

What is science? Is it a quantitative description which can be verified by experience and/or measurement? Or is it about the existence of (actual) real forms for things, such as length, time, and material-balance, and systems of certain types, as well as subsystems? Where systems are that which tie together derivatives of certain types of functions with vector bundles of the same type. Is science about the metaphysics (underlying beliefs) of a language or is it about the observed patterns of existence? Can a different language direct observations to new patterns of material that do exist but have not been described by another scientific language based upon different metaphysical basis so they have not been sought out or examined? Have external patterns been observed in the world which do not properly fit into the currently accepted language which is being used to describe these patterns?

A formula represents a quantitative property of a physical system. Is something measurable if local linear scale changes of formulas can be set equal to other (local) linear quantities of the same type, defined at the same point where the scaling relation was defined? The equality between scaling relations and local linear quantities results in what is called a differential equation. Such differential equations can be solved, by an inversion process (or inverse operator eg integration), so as to give a formula for the solution to the differential equation which represents the measurable properties of the system in the context of the formulas containing system of variables, or degrees of freedom (or dimension). If such formulas (or solutions) can be found which predict measurable properties when boundary conditions are satisfied, ie a linear differential equation's solution, then is this a verifiable description, provided there is good agreement between descriptions and measured values? Are questions about measurability and stability the same type of questions? Must there be a stable form or stable system properties with which the measurement must be in relation, or is measurement simply a local process? Does such a local process of relating a system measuring-system structure have within its small bounded region a clear notion of stability?

Can the idea of a measuring subsystem always be implemented? Can the absolute unit be reproduced and be scaled with other systems of measuring? Does this depend on a system being within a metric space?

Can a 2-dimensional shape be contained, in terms of its geometric description, within the simple ideas of lines (or line segments) and circles?

In 3-dimensions, circles could be sectional curvatures of other shapes. However, to describe geometry in 3-dimensions, does one also need planes (or planar regions) and spherical shapes, or can things be sliced and projected onto coordinates, so that lines and circles are still enough?

If existence is decided from the higher dimensions to the lower dimensions then the higher dimensional curvature may best be described in the correct dimensional context, eg the differential equation which describes the higher dimensional context may not project to lower dimensions, so that the lower dimensional differential equations, or lower dimensional space-forms give more information, ie more possible eigenvalues than the higher dimensional context allows. Thus the lower dimensional formulation may not give the full set of information about the system, namely, too much information.

Measurability in terms of probability

In the context of probability, and function spaces, a measurable, well formed, system comes from the identification of a set of "spectral" functions within a function space. This depends on finding a set of operators, ie differential operators,

> [this could imply meta-functions operating on operators, eg the Fourier transform (this becomes formalism, and it is hard to grasp, though Fourier transforms are very useful)]

so that the set of operators forms a "complete set of commuting operators," ie with respect to Lie algebra properties, which, if found, allow the diagonalization of the infinite dimensional matrix defined (usually by a Hamiltonian) on the function space, ie the complete set of commuting operators allows the finding of the set of spectral functions for the function space, which is found through the diagonalization process. Each function of this set of spectral functions corresponds to its own unique set of distinct geometries, each geometry defined by the individual spectral functions. Thus within the realm of probability and function spaces there is not a geometry, but rather a set of many geometries, so that each such geometry is associated to a particular spectral value (or eigenvalue) of the system and the geometrically dependent distributions of particle event probabilities associated to that probability distribution function.

This idea of measurement implies measurement's relation to a stable form for a system, as opposed to measurement being simply a local property of measuring subsystems at any point in space.

The geometric alternative to measurability (The differential equation defines measurability in terms of the geometry of changes of scale)

However, instead of a probability basis for physical description, there is an intermediate possibility of an existence of a metric space geometry, ie discrete groups, which allows a lot of geometric variety so that almost smooth transitions between spectral sets of functions, in function spaces, can be accomplished. That is, can the geometric properties in the coordinate space be more diverse than a sequence of spectral geometries in a function space? Such diversity is what space-forms allow. Namely, a rich geometry of space-form shapes exist in the base (or coordinate) space of a principle fiber bundle (PFB) which allows greater diversity than a probability structure allows. This is because space-forms have both geometric properties and spectral properties associated to themselves.

The fundamental geometry of metric spaces

The new idea is that the geometry of a 3-dimensional space can be represented by means of slices through both space and space-forms by planes, and projections of geometry onto coordinates in terms of lines, line segments, and circles or sectional curvatures which are defined over what are essentially flat geometric (or space-form) shapes.

Furthermore, these ideas are consistent with the atomic hypothesis, yet it is not fundamentally a reductionist theory. Thus both geometry and stable spectral structures (properties), associated to hyperbolic space-forms, can be described in a language which is logically consistent with itself.

Probability, reductionism, and the importance of geometry in making a language understandable (or useful)

The current paradigm of physics has its descriptive language based in probability. This is due to the belief that all of existence can be reduced to its elementary particles whose behavior is described in a fundamental way by probability, and then elementary particles can be built, up into complex material systems, which ultimately depend on the physical law associated to elementary particles. None-the-less physics always has material geometry as a part of the description, even if that geometry may be as simple as a collision. Hence, the descriptive language of the currently accepted paradigm of physics is logically inconsistent.

Such a reductionist view has resulted in the belief that all physical systems that have more than, say, three elementary particles are simply too complicated to describe. In other words the modern day (2006) physical theories, based in probability and reducibility, have become useless. None-the-less very useful macroscopic descriptions of material based in geometry, from 19th century physics, have been the basis for the vast majority of our present technology.

Note: Galileo's, Newton's, Maxwell's, and Boltzman's work as well as Einstein's early work, were all based in a flat metric space (velocity fields and coordinate positions), within which exist material and atoms and interactions, eg force fields and/or (statistically) atomic collisions. This was all done within a geometric context, with the exception of statistical physics, and much of classical statistical physics also depends on geometry.

Consistency

Space-forms and metric spaces along with differential equations can be used to model all of physics so that the low dimensional metric spaces have a slight property of spherical symmetry, due to an SU(2) fiber subgroup whose tangent space descends to create the base space for low dimensions.

The interactions of space-forms at the atomic level would have a Brownian motion associated to themselves, and this stochastic motion is what E Nelson used to re-invent quantum mechanics (QM). Hence this new viewpoint accounts for QM, but within this new framework probability is not needed, so the spectral properties of systems and material geometry do not have to be inconsistent with one another.

Some properties of the foundations of the new paradigm for physics (summary 3)

For the new paradigm, material properties as well as the material itself are all associated directly with metric spaces. For example, coordinate frames which determine rotational or translational inertial properties are spherical and Euclidean in nature, but the spherical metric is the Euclidean metric simply restricted to the sphere, so one can say, both rotations and translations are Euclidean in nature. On the other hand stable energetic systems (based in the [space-form] structure of charge) are associated to states of time within space-time (or hyperbolic) coordinate spaces. By using discrete isometry subgroups the metric spaces are partitioned into geometric shapes (or space-forms). Space-forms have both geometric and spectral properties, eg hyperbolic space-forms have very stable spectral properties related to energy, whereas Euclidean spatial structure determines spatial displacements related to inertial properties.

Different metric-space states, related to each metric space's associated physical properties, are built into both the (1) metric spaces and (2) the metric space's space-form objects. This is done within the context of either: (1) finite dimensional Hermitian space (where motion and force fields exist) [so that the two metric-space states can be placed into its real and "pure imaginary" subsets] or (2) the fundamental domains of the discrete isometry subgroups, which have properties of "cubical" simplexes, where the distinct metric states can be associated to the simplex's different distinct faces (or face pairs).

The geometric shapes of space-forms and their spectral properties are related to this "cubical" simplex structure. That is, space-forms, or their simplexes, of a given dimension, say, n, [existing in a particular state of a metric space (or face of a higher dimensional containing simplex)] have a dimensional hierarchy of spectral faces associated to themselves. The most significant spectra being the (n-1)-dimensional faces of the space-form's n-dimensional simplex. A simplex's different dimensional faces are associated to different levels of existence. A material space-form's spectral flows are paired off into opposite states, of the same dimension, to form the space-forms.

The opposite states are mixed (1) within a metric space so as to allow dynamics, and (2) when mixed on space-forms, this rotation or mixing of metric-space states or (n-1)-dimensional spectral flows, "spin rotates" the geometric n-dimensional "surface" of the space-form. When these metric-space states are mixed within an (n+1)-dimensional metric space it allows dynamics within the (n+1)-dimensional metric spaces, by allowing Euclidean displacements to occur with each change in time-state of the hyperbolic metric-space. Rotating the states of metric-spaces is similar to spin rotating the spectral flows of space-forms, because the spectral flows (also) represent metric-space states on a lower dimensional level, namely, the lower dimensional level whose dimension is the same dimension as the spectral flow's dimension. This is because of the self similar structure of the dimensional hierarchy of metric-spaces and space-forms which compose and contain existence. For the material space-forms, that are within the metric-space of an observer's optical perception, this rotation of states is spin.

A perceived space-form is equivalent to a concrete "cubical" simplex whose geometric measures are alternating forms. The geometric measures of this alternating form applied to a space-form's simplex, determine the spectral values or eigenvalues of the spectral flows of the space-form (or equivalently the geometrically measured values of the faces of the equivalent simplex). Material space-forms' geometric measures have the mathematical property of being anti-symmetric, hence space-forms, which provide the hidden structure of material particles, or Fermions, and as a material object are always a part of a space-form (or simplex) structure, hence they have the property of being anti-symmetric.

In a metric space there are optical, or light, space-forms which allow spatially separated material systems to be connected together, and these light (or interaction) connections are modeled as infinite extent, hyperbolic space-forms. Within the context in which optical connections exist between material space-form systems. These optical connections can be related to geometric distances by the time interval of the light signal which optically connects the two material objects. The geometric measures that exist within a metric space are lengths, which are measured by a metric 2-tensor. A metric 2-tensor has the mathematical property of being symmetric. A (hyperbolic) metric space contains many infinite extent space-forms, eg light, which, through time measurements, can be effectively used to measure lengths between material (space-form) systems. The mathematical measure of length, ie the metric, has the mathematical property of being symmetric, hence because Bosons are always a part of the propagation mechanism of an infinite extent space-form structure, by measuring the time interval between light signals, lengths between the interacting (or observed) material space-forms can be measured. Thus Bosons, which are naturally placed in relation to time by the space-time metric itself, are related to a symmetric function. "Particles" with integer spins, or equivalently infinite extent space-form that have a propagation "wave structure" associated to themselves, are related to the metric space's metric, hence are associated to symmetric functions

Infinite extent hyperbolic space-forms of dimension, n, are contained in a metric space of dimension, (n+1). This property allows the dimensions of the faces of simplexes to be independent of the dimensional properties of the over-all simplex, or of its other faces.

On a principle fiber bundle, the base space is a dimensional hierarchy of many metric spaces and their related space-form structures, while the fiber group is the set of isometry groups for each of the metric spaces, though it is possible a diffeomorphism group could also be attached to an extra added on "generalized" Euclidean space. Furthermore, there seems to be a need for a (simple) conformal group, ie multiplication of hyperbolic space-forms by conformal factors, to account for the smooth property of velocity fields. This conformal factor could also be related to the mathematical properties that determine the speed of light.

Note: Within the isometry groups, the displacement groups, acting on Euclidean space, also need to be considered.

Because 3- and 4-dimensional metric spaces are associated to pairs of metric-space states, the metric spaces are complexified so that one of the two metric-states is in a real subset, and the other metric-state is in a pure imaginary subset. Thus the metric space is turned into finite dimensional Hermitian (coordinate) spaces and the resulting fiber groups become generalized into finite dimensional unitary lie groups. Within a Hermitian space one metric-state is associated to its real part, while the other metric-state is associated to the pure imaginary part (or subset) of the Hermitian space.

However, 5-dimensional metric spaces are associated to quadruples of metric-space states, the metric spaces are quaternisized, ie the coordinates are given a "number field" (actually a division ring) of quaternions, so that one of the four metric-states is in a real subset, and the other three metric-states are in a pure quaternionic subsets. Thus the metric space is turned into finite dimensional quaternionic coordinate spaces and the resulting fiber groups become generalized into finite dimensional quaternionic-unitary lie groups. Within a quaternionic space one metric-state is associated to its real part, while the other three metric-states are associated to the pure (imaginary) quaternionic part (or subset) of the quaternionic space.

For any description of an object's position in space, x (t), such a position function can be represented on a manifold in terms of group transformations $g(t) \cdot x(0)$, where g (t) is a group element on a curve parameterized by t, and x (0) is the object's initial position. Similarly, a vector field in space can be determined by group transformations acting on initial values of fields (or potentials) over a manifold.

What about spectral transformations? For this new viewpoint, there is a simplex which represents the basic structure of the system, but there is also a set of space-form subsystems of the system, contained within the system's simplex, that are in resonance with the basic (or main) simplex structure (or spectral structure) of the system. Thus there is a new set of group transformations to consider, (1) discrete groups and their symmetries and (2) the transformations between sets of simplexes which compose (or which compete for composition properties) of a system, in a system's dimensional hierarchy, defined by either its simplex or the simplex containment structure of the system.

Can sets of vectors or tensors, eg alternating forms, be used to describe (or contain) the properties of a simplex, and hence its space-form equivalent? What is wanted is a means through which group actions on the simplexes of the base space can be identified.

The subgroup elements of the generalized unitary fiber groups (in terms of real metric signatures) include: discrete isometry subgroups, the symmetry and permutation groups of the discrete isometry subgroups, spin orthogonal rotation groups defined on the faces of space-form simplexes, unitary rotation groups acting on the real and pure imaginary orthogonal division of Hermitian spaces, Euclidean displacement groups which displace space-forms, unitary transformations acting on sets of simplexes, which can represent either a system's context, or the context of the containing simplex within which the system-containing metric space is itself contained. Then there are the real subgroup structures of this unitary group, the isometry groups, which apply to the real systems contained in particular (real) metric-space states.

This is the context in which the Lie algebra valued connection 1-form (of the principle fiber bundle) acts with its Lie algebra structure upon (1) the tangent space of the metric space's vectors, and (2) metric-space states, along with (3) the base space's potential structure of interaction space-forms. Spin operators and alternating forms apply to space-forms, while unitary transformations apply to both infinite extent hyperbolic space-forms of both the (system containing) metric space's light structure, and its interaction potential structures, so that the interaction potential is invariably a $\frac{1}{r}$, (sectional curvature) term, whose affects on displacement are related to the Euclidean metric space, and hence related to the geometric properties of a bounded 2-dimensional Euclidean space-form, so that the Euclidean space-form determines the system's interaction properties.

Note: at least 2-dimensions are needed for the Euclidean interaction space-form, because there is always an inertial part and an interaction part that exist in a metric space, so that the metric space's dominant geometric property is either length (ie relative separation) or displacements.

An interaction can be viewed from various levels of dimensional containment, in which different material space-forms are the objects of interaction, and in which different spectral containment properties can be considered. Different spectral containment structures mean that there will be different resonance properties that exist between the system and its containing simplex, eg between the system and its containing metric space, or containing metric space face. These different spectral structures can be associated with different material geometries.

(end summary 3)

The process of communicating a new paradigm for physics

If one wants to change the paradigm of physics, then first one must show reasons why the paradigm needs to be changed. The answer to this, is that the currently accepted paradigm of physics is not longer providing the world with useful descriptions, and many "practical" problems exist which need solutions and such solutions are not being provided by the current paradigm of physics. Secondly, one needs to provide the entire world view which result from the new axioms and new use of words which a new paradigm of physics must be based on, thus showing how to re-organize material, space, time, functions, vector fields, and differential equations whose solutions predict properties of systems. When presenting a new paradigm, one is not afforded the luxury nor time, of the theorem and proof structure of communication. Instead one must provide a complete world picture as seen from the new viewpoint. A new geometry, namely, existence's hidden geometric structure (because geometry provides words with definite meaning), along with its new interpretations, of how external patterns are related to the new descriptive patterns, needs to be presented within a total picture of the world. A new paradigm for science is about the context of what can be talked about, so that one might see how useful results can come from descriptions within a new context, ie new ways in which to use words, which make clear the hidden structure of existence.

Copernicus did not rely on the church's authority to develop his ideas, rather he used words in new ways which were contrary to the truth that was supported by the church's authority. In the age of Copernicus the church was the intellectual center of the society, if Copernicus wanted to challenge the truth of currently accepted paradigm then he was proposing a new paradigm for God, so such a new paradigm would be heretical knowledge, hence inquiries by the church into his wickedness and his demise would commence.

In this age the intellectual centers are universities, which are divided into departments, and within universities it is the science departments, and the relation of their knowledge to how corporations control the world's material, which results in their being the bearers of an absolute "objective truth." It is not any different to challenge the authority of civilization's science institutions now, than it was in the time of Copernicus to challenge the church's authority, because in this new case one is interfering with the power of the institution of science, (but really the power of those who control civilization) to determine truth, so if a new paradigm is claimed, then it is easiest for the institution's of science to simply ignore the new ideas, or marginalize them, by saying that the new ideas are not science.

It is not likely that the new ideas will determine a basis for new investment.

Certainly Copernicus's ideas, in his age, were not religion, though physical science has now become a religion in this age.

The religion of materialism and money and power, is the deep seated desire to control material and money, and subsequently everything else. It is justified in terms of competition and survival of the fittest, ie a modern form of social Darwinism.

The unchallenged metaphysics of physics

Here are some metaphysical questions that deal with the foundations of physics. (1) Are coordinates unbounded? (No, though they can be thought of as sort of unbounded, but their main (or dominant) properties are determined by their bounded structure) (2) Should physics be based on the language of probability? (No, nothing definite can be described using the language of probability) (3) Is material a scalar quantity? (No, though it is sometimes convenient to use such a model for material, but material is always a space-form) (4) Is material ultimately reducible to elementary point particles that transforms into other types of particles? (No, material has the structure of a space-form) (5) Do elementary particles have a hidden geometry? (Yes) (6) Are there Platonic forms, eg plane Euclidean geometry, (No, all language is about finding useful and [relatively] truthful descriptions about existence), Euclidean geometry is also about inertial displacements or (7) Are patterns related to truth more through their usefulness than through particular ways of using words and pictures (or geometries)? (Yes, this is what Godel's incompleteness theorem ultimate implies for a quantitatively based language) (8) Is space determined by (the distribution of) material? (No, material and space are the same thing but separated from each other by one dimension, on the other hand material in a metric space necessitates the formation of interaction space-forms which along with low dimensional spherical symmetry affects the local structure of space, also see 10) or (9) Is material dynamics determined by the shape of space? (Yes, but not through general relativity, with the exception of the existence of spherical symmetry up to spatial dimension three, rather spatial properties are determined by flat space-form geometries) or (10) Are material and space different dimensional levels of a similar geometric structure? (Yes) (11) Does time have a direction? (Yes, its direction determines hyperbolic metric-space states, I. Prigogine has interpreted experimental evidence that "the metric-space, time-state within which we exist has a direction) (12) Are the laws of physics the same everywhere? (No, physics changes as the dimensional level changes, ie each face, of a system containing simplex, corresponds to a new metric space which may contain different physical properties and have different physical constants) (13) Do the fundamental constants remain the same everywhere? (No, it is likely that different galaxies, or different governing [or containing] simplexes have different fundamental constants) (14) Does the language of science need to be based on geometry in order for it to be useful? (Yes) (15) What symmetries are (the most) important in physics? (translational, rotational, (ie spatial displacements), time, (ie related to a system's energy properties), metric invariance, the differential applied to differential forms, (eg symplectic symmetries) and tensor invariance to the fiber group transformations (associated to a particular metric space, namely, a generalized Euclidean space), Simplex invariance's and transformations)

How the currently accepted paradigm answers these same metaphysical questions

The faith of the currently accepted paradigm is that (1) is yes (2) is yes (3-4) both yes (5) is no, the only interpretation of localized events that is allowed, is that they are particle interactions. This is consistent with probability being the basis for physical description. However, particles do transform relative to their isospin vector, a vector whose components are different particle states, ie different elementary particles. (6) is yes (7) the structure of material and space and its laws are all well known to physics which for the most part the new paradigm is not at odds with, but the currently accepted paradigm requires that these laws be "absolute truths" and not properties of a useful way in which to use language, hence no. (8) is yes (9) is yes, through the "laws" of general relativity (10) is no (11) this is not known within the currently accepted paradigm of physics, but the answer is mostly likely yes, as I. Prigogine has shown that time, in the space-time or hyperbolic space that we experience, has a direction. (12) is yes (13) is yes (14) is no (15) (material determines existence and its apparent properties, and material can be reduced to its elementary particle constituents, material (½-integer spin) particles are Fermions and field (integer spin) particles are Bosons, physics depends on a probability basis for material (particle) description in which interactions are based on particle collisions, when particles collide they define field interactions and these field interactions take place as collisions between Fermions and Bosons at the site of a vacuum.)

Interpretations of data, along with the words that are used and the structure of the language that is used to describe the patterns of physical data, as well as the underlying beliefs (or metaphysics) create a viewpoint which can trap a belief system. But science must be consistent with external patterns of the world, if it is to be useful. Very authoritative statements can be made and proved within such a word-interpretation-metaphysics system, yet such statements may not be backed up by the patterns of the "external world" that are observed. None-the-less the patterns observed are put through an interpretive system within which it is not easy to extricate oneself. Ptolemy's system, though rather complicated, could "predict" planetary events though it could not guide a space-ship, unless perhaps if one used the principles of general relativity in Ptolemy's frame, similarly particle physics can "predict" the particle events of particle accelerators though it cannot describe a model for a general nucleus.

Note: Though von Neumann proved that hidden variable cannot exist, based on the probability structure of quantum mechanics (within that way to structure language), none-the-less the Bohm-Aronahov effect is a hidden variable theory and has data to support it.

Godel's incompleteness theorem's relation to science and math

Within the institutions themselves, with their knowledgeable people "open" to new ideas, language should be an important issue, because Godel's incompleteness theorem can be interpreted to be a technical statement about mathematical language, so that it applies to any language based upon the patterns of quantity and shape. Within this context the incompleteness theorem says that many patterns can exist outside any such fixed (quantitative) language. This could be interpreted to mean that the patterns that are observed in the world could be patterns that are outside the grasp of physics' (or even mathematics' fixed pattern of language based on fixed axioms) fixed quantitative language's capacity to describe these (such) observed patterns. Interpretations of Godel's incompleteness theorem of this sort are seldom made.

Rather academic interpretations tend to focus on technical aspects of the incompleteness theorem. Though the above interpretation of Godel's incompleteness theorem implies there is always a need for the revision of a discipline's fixed (quantitative) language structure that is used for quantitatively based descriptions. There do exist examples of the language of the current paradigm of physics having difficulty describing the patterns that are being seen. For example, a model of a general nucleus has not been forth coming from the standard model of elementary particles. Yet, of what value is the theory of elementary particles if it cannot create a model of the nucleus?

[This has become part of some of the problems that the current paradigm of physics is experiencing.]

Perhaps, amongst other things, the idea of reductionism needs to be modified.

How to communicate a new paradigm

If one challenges the current paradigm, one needs to explain the changes in the structure of language which need to be considered and then compare those changes to the language and assumptions and interpretations that are already being used in the current paradigm. One must first provide a new language, and then say how these changes in language cause changes in the mathematical context of the descriptive structure and still account for the patterns that are observed, eg metric spaces in the old paradigm become a dimensional hierarchy of metric spaces in the new paradigm, so how can there be higher dimensions when we continue, as always, to only experience the dimensions that we do?

Then one must show that the new constructs, ie new ways to use language, apply to all aspects of the real physical world. And create the context within which all the different aspects can be re-interpreted. A lot of this would be mathematical theorems, however a new paradigm must provide an entire world view, so instead of presenting a long string of individual theorems, which sustain a world view though this is often difficult to discern, the entire world context must be made clear and the key properties which are central to the proofs of theorems must be quickly reviewed as one provides this (quick) world view.

A natural communication barrier

Thus there exists a communication barrier. Institutions only trust new ideas about science that come from their own ranks, and these new ideas must be (very eloquently) expressed in the language that they know and with which they are familiar.

Just as Copernicus presented an entire picture of many planets in orbit about the sun, a similar complete picture of another way in which to use words in order to describe nearly all of existence has been presented to the world. In a corresponding way, though Ptolemy's system (or the currently accepted paradigm of physics) seemed to be more precise than the new Copernican system (or the new alternative ideas), it was the Copernican system (and these new alternative ideas) which led (which will eventually lead) to many more interesting ideas and greater development than Ptolemy's (or the currently accepted paradigm's) ideas. The currently accepted paradigm for physics practices the art of deception through appealing to the idea of the "precision" of their theory.

Say the same thing in different ways (summary 4)

One way to penetrate this communication barrier is to try to communicate in as many ways as possible. In this age it is convenient for big new ideas to be condensed so that their basic structure and content are readily available in a few short paragraphs.

For this new paradigm this could be done in many ways, one such way is as follows:

A dimensional hierarchy of different signature, [ie signature of a metric is related to the dimensions of two subspaces which compose a metric space, that is, a metric space is divided up into a spatial subspace and a temporal subspace] metric spaces exists, so that each metric space is associated to a particular set of physical properties (eg Euclidean space is associated to inertia, displacement, and momentum, while space-time (or hyperbolic space) is associated to charge, time and energy) and within this context it is the spectral (or natural geometric) properties of the metric spaces that one must consider. [Note: We have taught ourselves to pay primary attention on the inertial aspect of existence in three spatial dimensions, eg we do not "see" light until it interacts with matter.]

The natural spectral (or geometric) properties of metric spaces are contained in the discrete isometry subgroup structures of the metric spaces. Such discrete isometry group properties correspond to lattices defined on the metric space, along with the lattice's "cubical" simplicial fundamental domains, which can be "moded out" to form the natural metric space shapes of space-forms.

Both metric spaces and spectral flows, defined on space-forms, have distinct physical states associated to themselves. An example of such metric space states would be, on a space-time (or hyperbolic) metric space where there exist both states of positive time (which contain charge), and states of negative time (which contain monopoles). These two states of time can be kept separate from each other by using complex coordinates, by means of the real and pure imaginary parts (or subsets) of Hermitian space. Such metric space (or spectral flow) states are associated to the properties of a rotation (or mixing) of states within a Hermitian coordinate structure, and its distinct real and pure imaginary subsets. In turn, the mixing of states is needed to describe the dynamics of material. Material dynamical processes require mixtures, of both positive and negative time metric-space states, which have the needed properties which determine displacements of inertial, or Euclidean, space-forms.

Surprisingly, some of the hyperbolic space-form structures, that have an odd dimensional spatial subspace, can start oscillating in a natural manner. That is, the hyperbolic space-forms have both a stable geometry and stable spectral structure, but the spectral structure has a ridged geometry associated to itself so that if the spectral flows are all filled with charge, then there will exist a geometric charge imbalance and oscillation will begin. Because Euclidean space-forms are in resonance with hyperbolic space-forms the Euclidean space-form, associated to the oscillating hyperbolic space-form, will also begin to oscillate. This oscillation can cause the flows of opposite metric states to come (be pushed) into contact with each other, in turn, causing matter and anti-matter metric space states of Euclidean space to come into contact, and thus create energy. An example of such an energy generating system would be a radioactive nucleus.

Pause
The Lao Tzu says abolish current knowledge and the subject (civilization) will return to order.

Many well educated people will stop reading here (or perhaps have already stopped), but the knowledge that such people might assume to exist in the standard model is only a formal and very primitive form of knowledge, eg arbitrary parameters, and the standard model cannot be put together to create a model of a general nucleus. That is, even though the standard model of elementary particles is formal and very complicated, the information value of the standard model, when judged from afar, is actually very little. The currently accepted idea of anti-matter is simply a formal aspect of a mathematical structure (built from symmetry properties of the differential equation within an isospin context which in turn was modeled on the idea of spin). Both spin and isospin symmetries are outside the natural symmetries of differential equations based in coordinates. Though spin comes from the square root of the Laplace operator, the purpose of such a spin operator is not clear until it is applied to space-forms and the state structure associated to the space-form's spectral flows. The form of the differential equation in the currently accepted paradigm of physics has no logical reason for its structure. It comes from the formal attempt to follow data. It has more the structure of a Ptolemaic system, namely, that of following data with epicycles, than it has the structure of a usefully developable and consistent geometric description such as that which is associated to Copernicus. The currently accepted paradigm does not come from logical properties of a descriptive language, which is based on a global geometric picture, rather it is full of many examples of formal mathematical opportunism which has no consistent logical core. The idea that an interaction can be modeled at a point with Lie algebra elements and spin representations modeling different interaction types, without a geometric context (the context is the energy spectra), is tenuous at best. This is analogous to Ptolemy's model: The arbitrary parameters and the particle collisions of "massive" particles are analogous to the orbits of the sun and planets about the earth while the rotation of particle states is analogous to the epicycles of the planets.

Why does the standard model give precise permutations to mass measurements?

One will find that the mathematical structure of the new viewpoint is logical, and it also happens to be quite similar to the mathematical structure of the standard model. It also contains many of the key properties of string theory. It is the fact that this new math structure is close to that of the standard model that explains why the standard model seems to do well in determining some of the precise permutations of mass properties of particles, eg their mass corrections. Because the standard model is dependent on arbitrary parameters makes the standard model a joke in terms of it being a scientific model, since after all, the standard model is describing the spectrum of the elementary particle's masses. In the new set of ideas, particles do have a hidden geometric structure, but this geometric structure is quite different from string theory's geometries, though these new ideas might also (most likely do) contain the proper logical foundations for string theory.

Return to descriptive summary

This construction of a space-form structure for existence allows for a description of the universe which is both steady-state and expanding, so that we still are exactly at the (initial) point, which is a point that is analogous to the idea of an initial singularity point of the "big bang," but this new singularity point would not be associated to gravitational singularity. This new idea depends on the notion that, stars have spectral sets, and the spectrum of an entire set of stars which compose a galaxy naturally fit onto some particular maximal torus of the fiber group of the principle fiber bundle, within which a spectral description of other (competing) spectral centers which process each other's spectra by resonance or by spectral filtering of the stars' properties of material and space are contained. It is the time differentiation of the conjugation transformation of these maximal tori, ie transforming between spectral sets of galaxies, which results in the "raisin in muffin" expansion of galaxies.

Space-forms similar to material space-forms but different (energy generating space-forms)

Some space-forms naturally oscillate and generate their own energy. Such energy generating structures exist in a dimensional hierarchy of spectral structures. This causes existence to be a complicated set of inter-relations between spectral centers which can actively, spectrally filter both themselves and other spectral centers.

These new ideas encompass the structures of both material and life, so that life depends on non-reducible energy generating structures that are simple mathematical patterns.

This new description contains the correct part of the old description, ie This new structure absorbs the old descriptive structure

These new ideas have essentially the same structure as regular physics. Namely, they depend on ideas about material, time, space, vector bundles, differential equations, spectra, etc, which are all used to describe the properties of existence. That is, solutions to differential equations still provide the means by which to quantitatively model the world's physical properties. However, now space and its properties depend on discrete isometry subgroups, hence differential equations are defined on differential forms whose domains depend on the space-forms which compose the system. It is the differential forms, defined on space-forms, upon which the derivative operator acts. It is interaction space-forms which determine the form of the differential equations which, as derivative operators, act on differential forms that, in turn, are defined on space-form objects. The space-form properties allow vector fields (to define differential equations), function states, eg metric-space states (eg spin states), Lie algebra valued potential 1-forms (which move the system along the space-form changes), a very constrained yet very real dimensional hierarchy of "cubical" simplexes (which allows new degrees of freedom for differential equations), and a spectral range (associated to the relevant [or immediate] "cubical" simplex structure) that defines stability of physical systems. It is the space-form objects upon which the spectral properties of physical systems depend, as opposed to spectral properties depending on the properties of function spaces. Solution to differential equations are either differential forms which represent the space-forms themselves, or are the force fields of interaction space-forms which affect the motion of the centers of masses of space-forms in a metric space, the metric space within which the dynamical or interacting system is defined.

In the currently accepted paradigm of physics, it is the dependence of spectral properties on function spaces along with the idea of material reductionism, based on the interpretation of local "particle-like" events in space, as in fact being particles, which leads to function spaces being given a probability interpretation. As a result of the particle interpretation, probability has come to form the basis for the descriptive language of currently accepted physics. This leads to the dichotomy of geometry-probability, ie all probability spaces have an uncertainty principle associated to themselves, and this uncertainty principle destroys any relation between geometry and its containing spatial continuum. This would imply that solutions to differential equations should not be used to determine geometric properties.

This new description avoids this logical inconsistency.

Furthermore, in reference to an old problem of Einstein, these new ideas create a context within which Einstein's statement that "God does not throw dice," can now be explored in great detail. However, there are so many high dimensional, energy generating "cubical" simplexes, whose power over the material existence of a 3-dimensional space is unfathomable, that what God might be, is difficult to decide.

Obscuring practices of science

Conjecture: When science no longer is providing practical knowledge about the world, it starts proving complicated theorems whose properties make the discipline ever more irrelevant to the practical problems of the physical world, and thus ever more difficult to challenge. Thus the need for the presentation of the new paradigm's entire world view, because of the great many individual theorems needed for this new paradigm, if all these theorems were introduced individually then their complete relation to the world would get lost in the shuffle of so many theorems. The rigor of the day, ie in this age, would require that Copernicus complete all of Kepler's and Galileo's and Newton's work, before his ideas would be considered for a "peer review" journal. Are "peers," scientists or are they the faithful? The aggressive competitors, obedient to the dogma, and diligent helpers of the faith, keeping the unworthy heretics at bay which are referred to as the ignorant (savages). The faithful are willing to exterminate all (ideas) but theirs.

Some authors to reference

Most of the math is standard math except for the several new mathematical ideas, but the ideas about discrete isometry subgroups can be found amongst many sources, but also in the books by: C L Siegel, E B Vinberg, K Katock, J Humphries, W Thurston, H Coxeter, and J Wolfe.

Another attempt at communicating new ideas and comparing them with the old ideas: (summary 5)

In the currently accepted paradigm, material determines existence, it determines its own spatial degrees of freedom and it determines the degrees of freedom for material type (including isospin degrees of freedom). Coordinates form an unbounded continuum of either 3 dimensions (mechanics and Newtonian gravity, Hamiltonian mechanics, and quantum mechanics) or 4 dimensions (electromagnetism, special relativity and the Lagrangian density part of the standard model). Material itself is modeled as a scalar quantity (on a spatial continuum) or as a point particle (and their event probabilities) defining an isospin vector, and the fields of a material's geometric distribution, or particles which collide with one another at a point to, determine interactions and/or changes in motion or a manifold's geometry determined in terms of a local continuum. In quantum mechanics a point particle of material is modeled as a probability event in space, a point particle which also has scalar properties but the property of a spatial continuum is lost. Thus the current paradigm is logically inconsistent.

One of the main arguments against the existence of higher dimensions is the nature of the inverse square force field and it necessarily having to change in higher dimensions. This argument depends on material being a scalar, so that material objects have a center of mass, and space being a continuum upon which Stokes' theorem can be applied to differential forms of material interaction fields (or potentials) and/or material densities.

The mathematical formulation of the standard model of elementary particles is that a unitary fiber group is defined over a Euclidean harmonic oscillator space so that the isospin-vector wave-function of Fermions that occupy the harmonic oscillator is integrated by (or acted on by) an exponentiated Lagrangian density (based on a Dirac operator, which is based on a Fermion's vector field) which is acted on through an interaction term, eA, where, A, determines a unitary interaction field, or Lie algebra element, over a metric space, ie it determines a Boson, so that the Lagrangian density is used to measure the harmonic oscillator's particle occupancy, while at the same time determining the Fermion (vector) [material or particle] interactions with fields (Lie algebra matrices) at the site of a vacuum.

Note: isospin properties are determined by particle properties defined on a tangent bundle, ie over coordinates upon which a matrix group acts, so this implies that there exists a geometry on these coordinates that accounts for the isospin degrees of freedom. However, in the standard model, the idea of isospin depends on dealing with the elementary particle properties, hence it supposes a reductionist viewpoint of material. The geometric systems that such an isospin geometric structure would compose, depend on: (1) a Lagrangian density, which is an energy operator (ie energy in the context of Fourier transforms), which in turn implies the geometry of both motion and fields, and (2) probability. However, geometry and probability are inconsistent with one another. This happens to be the logically inconsistent context of string theory. Hence the need for a new logical foundation for string theory.

Electromagnetism and special relativity deal with charges, their dynamics, and the metric invariance of electromagnetism in space-time (or equivalently hyperbolic 3-space). When the coordinates of space-time are differentiated in terms of time, the spatial parts become associated with momentum, while the time component, which now has mass as a scalar factor, is associated to energy, this leads to the idea that mass equals energy.

The Laplacian in a general metric gives Ric + Second Order Differential Operator, the 2nd term, whose Fourier transform is related to kinetic energy, where Ric is the Ricci curvature tensor. This almost has the structure of general relativity. The above ideas about material determining existence seem to be challenged by the ideas of general relativity, Ric=kT, where T is energy density, an equation whose solution is a general metric, with which the shape of space is found. The shape of space (or the general metric) can be used to determine the motion of test particles, but even in this example it is material distribution which determines most of T and hence the shape of space. Furthermore, it must be noted that general relativity has only been solved (in practice) for Ric=0, a 1-body problem (one material body) in which the one material body's metric field defines spherical symmetry. The problem of finding the general metric for, Ric=0, may result in a metric which has singularities, for certain physical values that are constants (variables) of a solution which is a Riemann manifold which is a spherically symmetric solution. The solution of course is associated to the system's metric.

However, is all this much different from the formulation in terms of, V(r)=k/r, potentials in K+V=E, where K=kinetic energy, and E is the total energy? General relativity descends into a lot a difficult formalism for a very limited payoff. The payoff being mostly the precession of the perihelion of Mercury, as its other "successes" can have other interpretations, eg the red shift of light in a gravitational field can be interpreted to mean that light stays within its flat metric space structure, so Schwarzschild's solution is suspect.

The precession of the perihelion of Mercury can be obtained with the "new ideas" in which potentials [including many new potentials for new materials, though perhaps not relevant to the dynamics of Mercury's orbit] are the usual method for determining dynamics. However, in the new theory, it might be noted that for low dimensional spaces, the Euclidean part has the property of spherical symmetry.

In the new theory there is also material, space, position functions (ie relative separation functions), energy functions, time, tangent bundles, spectra, and differential equations, ie the usual context of physics, so that electromagnetism is associated to space-time, and both Newtonian gravity and Hamiltonian mechanics are associated to 3-dimensional Euclidean space. Furthermore, there are fundamental symmetries associated to coordinates, such as (1) displacement symmetries which are associated to properties of both being in a particular metric space, in this case Euclidean space, and momentum, and (2) time symmetries which are associated to properties of both energy and space-time.

Thus for the new theory there are many similarities with the currently accepted paradigm, and a few crucial differences, but the fundamental context remains in many ways the same as the currently accepted paradigm of physics. This is similar to how Copernicus's ideas were in the same context as Ptolemy's ideas, but with some crucial differences.

Some mathematical contexts for physical description of the new paradigm: (summary 6)

1. Dimensional hierarchy of different signature metric spaces
2. Independence of dimension at each dimensional level
3. What are particles? What is their hidden geometric structure? Answer: this is found in the geometry of the discrete isometry subgroups of the different metric spaces.
4. Metric spaces, distinguished by their dimension and (metric) signature, can be associated to definite physical properties, eg material-type, displacement, time . . . etc
5. What are the different states of matter? eg matter and anti-matter, charge and monopole
6. Where are these different states of material located? Either in Hermitian space or on space-forms whose spectral flows are associated to different metric states.
7. What is the shape of space? Answer: it has the shape of space-forms of different dimensions, or space-forms and faces of a "cubical" simplex which is perceived as material space-forms within a Euclidean (or inertial) metric space.
8. What is the structure of material? A space-form geometry, or a lower dimensional "cubical" simplex than the dimension of its containing metric space.
9. Are material and space analogous structures at different dimensions? Yes
   Is this the idea behind supersymmetry? Yes
10. How is one dimensional level is related to the next? The highest dimension that is considered would be a metric space, and the lower dimensional space-forms are material systems. Through resonance's [between higher dimensional spectral "cubical" simplexes and lower dimensional material space-forms (or lower dimensional "cubical" simplexes) contained in the "cubical" simplex] the higher dimensional "cubical" simplexes can affect the lower dimensional "cubical" simplex (or material) structures. The lower dimensional "cubical" simplex (or material) structures are contained within a higher dimensional "cubical" simplex. The higher dimensional "cubical" simplex can model the containing metric space, or the containing metric space can be one of the faces of the higher dimensional "cubical" simplex. Note: The geometric distribution of material in space can change in going from one "cubical" simplicial face (which is determining containment of material space-forms) to another.
11. What is the geometric structure of material, metric spaces, and interactions? Answer: "Cubical" simplexes and the subsequent material space-forms that they contain determine the geometry of material in space. Material space-forms induce the formation of interaction space-forms which either cause either material to change and move or stable space-form systems to from. These interactions are contained in a higher dimensional "cubical" simplex whose spectra control the stability of the formed material systems.
12. What is the nature of interactions, (space-forms mediating or connecting together other space-forms) and thus, what is the definition of a system? Interactions are determined by space-forms which form between the interacting material space-forms, within the containing

metric space, thus forming interaction space-forms. Interactions take place within a context of both space-form geometry and space-form states as well as metric-space states. A system can be related to a "cubical" simplex (or its space-from structure) of some dimension, along with all the resonant space-forms that the system simplex contains (within itself) and the metric space simplex contains, so that the system's simplex is resonant with the metric space's spectral properties, which are determined by the metric space's "cubical" simplex. Thus, the interacting material system and the metric space together allow a system to form, but again, the new formed system's stability depends on resonance with its containing "cubical" simplex.

The simple mathematical structure of the description of a physical system

1. A formula representing a quantitative property of the system is assumed, ie a number of a certain type, so that this numerical value is associated to a formula in terms of certain variables eg containing coordinates. The formula could be for position, vector field, energy, or for a general metric, or in the currently accepted paradigm of physics the occupancy number of a Euclidean harmonic oscillator, or wave function defined on a sea of such Euclidean harmonic oscillators, in Euclidean space.

Note: In the new context a formula for a differential form could represent a space-form of the system, or field properties of a metric space, or properties of spectral containment within a "cubical" simplex, or properties of spectral (hierarchical) organization, ie the geometric and spectral structures of space-forms and "cubical" simplexes.

2. A (partial) derivative of this function defines a new, linear, quantitative type, which in turn, defines a corresponding tangent bundle for the system, ie a bundle of vectors that are of the same type as the derivative of the assumed formula upon which the derivative acts.
3. The derivative and the correct tangent bundle, along with boundary conditions, allow a differential equation for the system to be defined. It is the solutions to linear differential equations which are stable and allow control over the system based on its description, or interpretation of such solutions.
4. The solution of the (linear) differential equation gives the formula for that property of the system, which was referenced in (1) of this paragraph, so that the formula, expressing a property of a system, extends into the neighborhood, and often times, extends out past the neighborhood of the point where the derivative was defined.
5. The principle fiber bundle structure of the differential equation has a base space which is the domain for the functions in the function space upon which a derivative is defined, and a fiber group which can either transform the base space's local coordinates, or it can transform the vectors and/or the differential equation of the system's various tangent bundles.

6. The form of the interaction process and the interaction potential should be based in geometry, within an inertial frame or within a context of flatness. In this new theory, interactions result from sectional curvatures of Euclidean space-forms that both (1) form the geometry of the interaction, and (2) are in resonance with hyperbolic space-forms, which are all part of the system's interaction structure.

In general, a derivative (or a connection) is often given the form of a partial derivative with either a potential energy term (which is a Lie algebra valued differential form) or a partial derivative plus a term which is a Lie algebra element of the fiber group with a potential energy as a factor. The group elements, or their associated Lie algebra elements, are transformations of either local coordinate or vector bundle.

If the potential has a vector, or state, structure, eg a particle state structure, of say, spin, (for the vector formula of the potential) which is acted upon by a Dirac-like differential operator, then this requires that the connection depend on the Lie algebra of the fiber group. The vector structure of the formula, for the system in question, ie a formula for either position or for energy, acted on by a Dirac operator reflects the state structure of the type of quantity, that the formula represents. The formula must fit into the metric space structures, eg either metric states or local vector fields, which contain the system's description (in terms of that formula) so that the formula depends on variables which fit into the metric space, eg a hyperbolic or a Euclidean or a Hermitian space.

The basic changes in language usage of the new paradigm

1. Certain materials and certain material properties are associated to metric spaces of certain dimension and which have certain metric signatures, eg inertia and momentum are associated to Euclidean space (for all dimensions).

Material is no longer simply a scalar value or a point particle associated to various states of particle-ness. Particles are mental constructs based on the obvious interpretations of existence of micro-events, but material really has a hidden geometric structure. Namely, material has a space-form structure. This is similar to how Copernicus provided the hidden structure of planetary existence.

Note: Though von Neumann proved that hidden variable cannot exist, based on the probability structure of quantum mechanics (within that way to structure language), none-the-less the Bohm-Aronahov effect is a hidden variable theory and has data to support it.

2. The spectral-geometric properties of materials are associated to space-forms. The natural geometric and spectral structures of metric spaces are determined by that metric space's discrete isometry subgroups and their subsequent "cubical" simplex and space-form structures.

Hyperbolic metric spaces have very stable spectral structures.

Space-forms are consistent with the atomic hypothesis, and they also provide a means through which the non-reducible properties of systems can be described. Thus space-forms have an affect on the geometric properties of material systems due to their own shape. However, it is the very stable spectral properties of the hyperbolic discrete groups which have a large control over the system's organization and over its properties through the effects of spectral selection based on spectral resonance of the material space-forms which compose the system and subsequently how they are organized within the controlling spectral hierarchy of the containing 'cubical" simplex (see (3) below).

Such non-reducible forms are required for describing the properties of: nuclei, atoms, molecules, molecular shape, crystals, macroscopic systems (they are part of the re-normalization group structure of thermal systems that are undergoing changes of phase), life, planets, stars, and also metric spaces (solar system stability is a result of the space-form metric space structures that surround the planets in a manner which allows for stable planetary orbits).

3. Space is given the structure of a dimensional hierarchy of space-forms (or "cubical" simplexes) and metric spaces. The metric spaces of each dimensional level are dimensionally independent of the other dimensional levels.

This is due to the fact that metric spaces are faces of a high dimensional "cubical" simplex, and each of those faces are contained within an infinite extent metric space. These infinite extent metric spaces which contain "cubical" faces of the simplex are determined by an infinite extent hyperbolic space-forms that are contained in the metric space so that these infinite extent space-forms are one dimension less that the metric space's dimension (the infinite extent space-forms are one dimension less than the metric space's dimension) and they which represent both light and interaction space-forms, which mediate between the material space-forms' interactions, that take place with the metric space. Thus, the infinite extent space-forms fit into, or contain, the dimensional subspace determined by the face of the "cubical" simplex, ie a face which determines the metric space.

We can only perceive the things in the sub-space determined by those infinite extent light space-forms.

This is similar to the dimensional independence (or the topological closed-ness [or open-closed topology] of the metric-space subspace) of branes, that are defined in string theory.

In particle physics some of the patterns described in the standard model of elementary particles, exist in the laboratory and some do not, ie virtual particles. When they do not, it is claimed that such patterns are virtual, and then their non-existence is ignored.

4. Metric spaces have their own shape. In the dimensional hierarchy defined by a high dimensional "cubical" simplex, which is a representation of a space-form, the various lower dimensional

metric spaces or faces of the higher dimensional "cubical" simplex, can be interpreted to be spectral flows of higher dimensional space-forms. When viewed from a vantage point that is two dimensions above the given metric space's dimension, a metric space appears to become a spectral flow on a material space-form. Metric spaces, within which Bosons are defined, ie the infinite extent hyperbolic space-forms which model light and interactions, become, in the viewpoint of two dimensions above the given metric space's dimension, part of the spectral flows (or spectral states) of a Fermionic, material space-form, upon which spin is defined. This is the natural (geometric) super-symmetry structure of the dimensional hierarchy of the base space of the principle fiber bundle. The dimensional space that the Bosons define is identified with the material space-forms, of Fermions, whose dimension is two dimensions greater than the dimension of the metric space that the bosons define.

5. Time is associated to the time subspace of a metric space of a given signature, ie the signature depends on the dimension of the time subspace for the metric space, namely the signature of space-time, or equivalently hyperbolic space, is determined by its 3-spatial dimensions and 1-time dimension subspaces. Time is associated to hyperbolic metric-space states, (or equivalently, in a self similar sense) time directions (which identify hyperbolic metric-space states) are associated to certain spectral flows, which represent the metric-space, on a higher dimensional space-form that fits into a hyperbolic metric space.

Time is also related to Huygen's principle. Huygen's principle deals with wave propagation troubles in even dimensional spatial subspaces of metric spaces. In metric spaces with even dimensional spatial subspaces, light waves do not propagate as spherical shells, rather the light signal fills the entire sphere as the sphere expands due to light propagation.

The signature of a metric divides a metric space into two subspaces, namely, a spatial subspace and a temporal subspace. Time is also associated to the possible definition of new material in higher dimensional metric spaces which have new signature types that do not exist on the lower dimensions.

Conjecture: The higher temporal dimensions of metric spaces allow the field, of the new metric space's new material which can interact within one of the (other) time dimensions of the metric space, so as to no longer be ruled by Huygen's principle in the even dimensional spatial subspaces of metric spaces. However, a time direction is, thus effectively, associated to a spatial subset of a hyperbolic space.

Paradigm shift

These ideas about material properties being associated to different signature metric spaces, combined with the idea that the basic shapes and spectra in a metric space are related to its discrete isometry subgroups, together compose the paradigm shift.

Equivalently the basic spectral structures of a metric space are determined by its discrete isometry subgroups, ie space-forms of the discrete group's "cubical" simplicial fundamental domains, defined by the discrete isometry subgroup's lattice structure, that exists down on the metric space (or base space), so that the space-forms have the properties of material in the metric space and several metric spaces, eg Euclidean and hyperbolic, each with pairs of opposite metric-space states associated to themselves, are needed to contain the description of a physical system.

6. Interactions are determined by both infinite extent hyperbolic space-forms, as well as their associated (by resonance) bounded Euclidean space-forms. The bounded Euclidean interaction space-forms are connected to the interacting material space-forms, and maintain a resonant stability with the spectrally stable hyperbolic space-forms which compose the system. It is the sectional curvature of the bounded interaction Euclidean space-forms which determine how the shape of space affect material interactions, which in turn, allows interacting material space-forms to cause either changes in the motions of material space-forms or to form into physical systems composed of space-forms. Namely, the sectional curvature [of the slicing plane that contains the normal to the two (or more) interacting space-forms] determines a $\frac{1}{r}$, potential factor for the interacting materials.

Note: In the dimensional spaces three and less, there exists a natural spherical symmetry This spherical symmetry causes interactions in low dimensions (three and less) to identify an interacting force along the normal direction that exists between the two interacting space-forms. The spherical symmetry exists because in order for material to move in the low dimensions the space must be at least of dimension three, and within the fiber group, which ultimately determines existence, a 3-dimensional space would be determined by SU(2) or T(SU(2)), where SU(2) is isomorphic to a 3-sphere.

On a space-form, a metric is defined, and this metric is always of the same form, eg for any hyperbolic tangent space of a hyperbolic space-form, either light always has the same speed, or the metric always has the same form. This is what it means to say the metric space is flat. However, when a material space-form is within the natural metric space which contains it (a metric space of one higher dimension than the material space-form), then the space-form's shape can be sliced with a (dimensionally local) plane so that the slice can contain a sectional curvature. This sectional curvature is connected to both interacting material space-forms so as to cause changes in motion of the material space-forms that are contained within the metric space.

Concrete examples of using the new mathematical patterns

When the context is finally presented in terms of the new language, then examples of how this new context relates to familiar problems of physics. For example, the H-atom and finding a "local spectral structure" which can be used to describe how a molecule, of one particular chemical composition, can have several molecular shapes depending on the properties of the "local spectral structure."

Splitting away from traditional physics

This paper presents a new paradigm for physics. It changes the reductionist, materialistic model over to a non-reductionist model in which spectral filtering centers "compete" with a multitude of spectral structures for resonance filtering processes. This implies a substantial split with the traditional, "material defines existence," viewpoint. In this new theory higher dimensional "cubical" simplexes which generate their own energy define existence by means of resonance filtering.

The relatively complete summary of a picture of the world based on this new paradigm of physics (summary 7)

The currently accepted paradigm of physics is materialism, ie material determines existence. Material determines space and field structure and it seems to have a property of inertial motion whose changes are determined by the vector fields (or the force fields) of the system's material geometry. This is a true statement for classical physics, where the solutions to differential equations are either geometries or classical waves. However, in quantum physics microscopic material events determine the elementary event spaces for (linear) matter-wave differential equations whose probability distribution solutions are series expansions on a function space's spectral sets. A quantum mechanical solution describes material reduced to particle events in space, thus causing the function space to be interpreted as a probability space. The reductionism hypothesis for material, as well as the spectral expansion of solutions to linear matter-wave equations, are both necessary conditions for the function space to have a probability interpretation. Unfortunately, probability and geometry are logically incompatible with one another.

Just as Copernicus presented a new paradigm in which the geometry of Ptolemy's universe was changed so as to reveal the hidden geometric structure of that universe. Namely the earth goes around the sun, [instead of the obvious fact that the sun clearly goes around the earth, such as can be witnessed everyday] so too this new paradigm of physics stays in the same universe, or has the same context as modern physics, but it reveals the hidden structure of that universe.

In order to avoid requiring that the description of physical systems be in terms of probability, then consider the spectral properties (or natural geometric shapes) of metric spaces, ie the discrete isometry subgroups (or space-forms) of metric spaces.

This new paradigm contends that existence is contained within a dimensional hierarchy of different signature metric spaces along with their discrete isometry subgroup (geometric-spectral) structures. Such discrete group structures identify "cubical" simplexes which when moded out determine space-form geometries. It might be noted that the hyperbolic space-forms have very stable spectral properties. The faces of the simplexes determine spectral flows on (the hyperbolic) space-forms and these flows are associated to different metric-space states, eg (+t)-states and (-t)-states. The property of spin is the continual discrete rotations that occur between the two metric states of space-form flows.

Instead of modeling material as a scalar (or a point-like event), attach physical meaning to the different signature metric spaces and their space-forms and consider the shapes which are defined from discrete isometry subgroups which determine space-forms as the new model for material. For example, 3-dimensional Euclidean space is associated to mass (or inertia), displacements, and momentum, while space-time (or hyperbolic-3-space) is associated to charge, time, and energy.

However, our senses consistently "see" only the inertial (or massive) part of this existence, so as to visually be in 3-dimensional spatial subspace, for example, we do not see light except when it is related to or interacts with inertial material.

When differentiating with respect to some given instant of time, discrete rotations can be used to identify all the metric states which the different state spectral flows determine. This allows these states' relation to time differentiation to be well defined. The spin rotation, of the flows of time states, also sweeps out about the "rotation axis" the space-form's surface geometry. On metric spaces this spin rotation of metric space states also takes place, and in this case, it is needed so that inertial material can be displaced at microscopic time intervals, so as to allow dynamical properties for massive material. This is needed because Euclidean space does not have intrinsic properties of time of its own, so that displacements in Euclidean space occur after each discrete time interval that is defined by spin rotation of time states on hyperbolic metric-space states.

Considering the dimensional properties of the stable hyperbolic space-forms:

The 1-dimensional space-forms, ie circles, which are charges so that these charges interact with one another through a space-form mechanism of both infinite extent hyperbolic space-forms and finite extent Euclidean space-forms, so as to form 2-dimensional hyperbolic space-forms. That is, if the boundary, or initial, conditions for the differential equation which models the system allow the stable solution to be a 2-dimensional hyperbolic space-form. Although there can be free 1-dimensional material particles, these particles want to stay on the natural space-form structure where they naturally interact with one another, ie the 2-dimensional space-forms. Such 2-dimensional hyperbolic space-forms would include compact electron and proton interactions which form both compact neutrons and combinations of neutrons and protons, ie compact nuclei. Similarly there are infinite extent (or very large, ie the neutrino having a small mass) space-forms of electron-proton interactions which form neutrinos and combinations of neutrinos and electrons, ie the electron cloud, about the nucleus. It might be noted that energy goes into the infinite extent of the space-form. Such interactions between charges take place in a base space of a principle fiber bundle whose fiber group is a set of finite dimensional isometry groups or more generally a set of finite dimensional Unitary Lie groups. Such interactions could be modeled using a Dirac-like operator, [because the Dirac differential operator naturally accounts for the metric-state vector structure, due to the different metric-space states which are needed to compose the system] and the space-form vector potentials (or fields) may be summed together to form a differential equation defined on differential forms whose domains are space-forms. Hence the solution will be a differential form defined on a space-form. It must be noted that now charge occupation is within the 2-dimensional hyperbolic space-forms as opposed to their being within Euclidean harmonic oscillators.

It might also be noted that when individual electrons are isolated they do not remain still, rather they maintain an orbital structure consistent with the structure of a 2-dimensional hyperbolic space-form (Dehmelt). This has further implications about the structure of space and its relation to the Fourier transformation. The Fourier transform might be said to search for the periodic structure of a function, due to the $e^{i\omega x}$, ie exp(iwx), [or $e^{i\omega t}$, ie exp(iwt)] factor applied to its integral measure, dx [or dt]. That such a search for the periodic structure of functions is so useful makes sense if the space is a dimensional hierarchy of space-forms as opposed to the structure of space being a homogeneous unbounded continuum.

The 2-dimensional hyperbolic space-forms can also interact with one another, in a similar manner as the 1-dimensional hyperbolic space-forms interacted with one another, in a process which begins with the formation in space of 2-dimensional interaction space-forms which interconnect the interacting materials. This process initiates the formation of stable 3-dimensional space-forms or: (1) atoms (composed of interacting nucleus and electron cloud [Note: This atomic structure for physical systems is repeated on each dimensional level]), (2) molecules, and (3) crystals.

In a 3-dimensional space, one would see the 2-dimensional surfaces or faces of the 3-dimensional space-form which projects down into the 3-dimensional metric space. This is because a 3-dimensional space-form is contained within a 4-dimensional metric-space.

The compact 2-dimensional hyperbolic positively charged space-form would be the nucleus of an atom, while the electron cloud is built from either pairs of infinite extent and compact space-forms (or possibly from semi-compact or semi-infinite extent space-forms) which carry negative charges. The compact, eg nuclear, structures allow such material space-forms to have a particular location in space, which ultimately allows for the geometric shapes of material.

Euclidean space-forms adhere to (or form around) the hyperbolic space-forms through resonance so as to conform spectrally to the spectra of the hyperbolic space-forms, which have very stable spectral properties, thus the hyperbolic space-forms and their spectral properties are the cause for the adherence.

The following is an example of using space-form techniques to answer questions about the mathematical techniques used in the new theory to answer the "old" question about the H-atom's energy spectra.

Outline of H-atom atomic levels

Some approximate methods within the new context of space-forms are used to derive the fundamental energy relation for the H-atom. It might be noted that the H-atom is a stable space-form so its dynamics would only be relate to its process of formation, this description is about its stable properties.

To find the energy levels of the H-atom depends on solutions to an energy equation The differential operators, in the old context, are applied to a space-form function which is related to energy because it is a hyperbolic space-form and through the (energy) differential operator. The differential or energy operator applied to the space-form function gives the value for the H-atom's energy. The Laplacian plus the potential can be Fourier transformed so as to become an energy equation. Thus, in the new context, the eigenvalue equation for the energy Laplacian can be applied to a space-form function representing the compact part of the electron cloud and/or interaction space-form, in order to estimate the energy levels, or spectral flow structure, of a space-form. The potential is given by the simple relation, k/r, where, r, is the distance between the nucleus and the electron cloud. The expression k/r, is the sectional curvature of the interaction Euclidean space-form. The Laplacian can be thought of as

$$(K + V(r))f(r,\omega) = Ef(r,\omega)$$

where K is the second order partial derivative terms (or kinetic energy term), and E is the energy eigenvalue for the space-form system, whereas $f(r,\omega)$, is a differential form which represents the stable space-form representing either the electron cloud's interaction with the nucleus, or its stable spectral flows (see below).

Assuming a fundamental domain for the electron space-form to be built up of a simple square-like lattice (though still hyperbolic), then the higher energy levels will correspond to the spectral length of an electron current-flow on the electron space-form winding around, n, of the holes of the space-form's spatial hole-like structures.

(sectional curvature)=(constant)/(length of current-flow of the electron space-form)

These orbits, winding around holes of a space-form, are assumed to be essentially circular current flows (because of its square-like fundamental domain), resulting in,

[(sectional curvature)=(constant)/(r)], for each hole, where r is the radius of one of the circles which defines one electron current on the compact part of the electron (or interaction) space-form. One might assume that this would be the distance from the nucleus to the electron space-form. The constant would be related to the space-form's physical properties.

The compact part of the electron cloud's spectral flows can be thought of as a set of circles whose centers lie on a line, or half line, so that one center is at a distance, r/2, from the end of the half line, then the centers of the other circles are placed at, [r/2], 3r/2, 5r/2, 7r/2, . . . along the half line. These circles can be thought of as formed as a result of slicing the compact electron cloud space-form by a plane containing the half line. Each circle has the same diameter, namely r. The first circle will be filled with current and have the length kr. When the second circle is also filled with current the total length will be,

$$kr + kr = 2kr$$

etc . . . so the current length of, n, such filled circles would be, *nkr*. This geometry is consistent with the geometric properties of hyperbolic space-forms. The space-form's spectral flow can be represented as a simple circle in terms of the function,

$$f(r,\omega) = nkr(\sin\omega + \cos\omega)$$

so that the circle is translated or repeated n times, and the potential would be represented as (k'/nr) where k' is of the order 10^{-11}. When the cylindrical (or polar) Laplacian term is applied to the function $f(r,\omega)$, where r is constant, one gets $-nkr(\sin\omega+\cos\omega)\left(\dfrac{1}{(nr)^2}\right)$, and this is summed with the potential term, $V(r)\cdot f(r,\omega)$, so that $f(r,\omega)$ can be divided out, leaving the kinetic energy term, K, and the potential term, $V(r) = \left(\dfrac{k'}{nr}\right)$, where k' is of the order $10^{-11}\dfrac{F}{m}$, and r being of the order of $10^{-11}m$, which gives the dominant term as being of the form,

$$\left(\dfrac{k''}{nr}\right)^2 = E$$

where k'', is a constant, gives a first order approximation to the energy levels of the H-atom.

The mathematical processes described here are very much like Bohr's original arguments. Thus using Sommerfeld's elliptical orbit approximations as better representations of the space-form orbits now has a more formal underlying reason as to why they gave such good approximations to real H-atom energy levels.

Using the Bohr radius for, r, a suitable constant, k'', can be selected to give exactly the energy eigenvalues that are obtained from the radial part of the Schrodinger equation for the H-atom.

However, now the eigenvalues are obtained without the embarrassing diverging radial solution of the Schrodinger equation, and its miraculous truncation properties, to obtain the experimentally determined spectra of the H-atom, the first process by which infinities were swept away in the application of quantum mechanics.

The radial solution of equation for the H-atom no longer diverges, because there is no radial equation. This is because the context is now a compact space-form and its closed current (or state) flows. None the less, because there exists a residual amount of spherical symmetry for low dimensional metric spaces, thus in Euclidean space spherical harmonics are still applicable but now there is a bound on the system's radius, ie the radius of the minimal sphere which contains the H-atom's hyperbolic space-form.

This short review has revised a similar discussion that exists in the literature.

Back to summary

Existence in a dimensional hierarchy of metric spaces and space-forms, can be described in terms of everything ultimately being contained in a higher dimensional "cubical" simplex or a system of high dimensional "cubical" simplexes, so that for each face there exists a set of infinite extent light and interaction space-forms whose dimensional properties conform to the dimension of the face within which they originate. Thus the geometric structure of light contributes to the property that our perception is limited to a face of a certain dimension, within which is confined the material we see, ie the structure of light defines the (same) face within which we confine our attention. Furthermore, interactions are also confined to this same metric space which is determined by the face from which originated the infinite extent, hyperbolic, interaction space-forms upon which interactions between charged hyperbolic space-forms depend for their existence.

Spin is the property which rotates pairs of opposite (metric) state spectral flows into one another. Ultimately this gives a "dynamic" to the space-form's geometric shape, organizes the differentiation process at an instant, and allows dynamics to exist in a metric space, ie a face (of a particular metric-space state) on a simplex, or a metric space, which is filled with material space-forms.

Thus material is the space-form quantization of space, and spin rotated mixtures of spectral flows in opposite metric space states in hyperbolic space (or as a subset in Hermitian space), with the help of the speed of light (or an analogous constant on higher dimensional hyperbolic metric spaces), is a quantization of time.

At each dimensional level, of hyperbolic space, there are positive and negative charges which can interact to form 2-dimensional hyperbolic space-forms for both neutrons and neutrinos. Such 2-dimensional hyperbolic space-forms would include compact electron and proton interactions which form both compact neutrons, along with combinations of neutrons and protons, ie compact nuclei. On the other hand, neutrinos, of the external (negatively charged) electron cloud, form into infinite extent (or "nearly" infinite extent) space-forms. The structure of an inner compact part of the space-form allows for the existence of definite positions for material and this allows material to have a geometric structure. This property of compactness, and hence the existence of material geometry, continues up the dimensional ladder until the dimensional level where a space-form can no longer have the property of being compact. The last compact hyperbolic space-form has dimension five.

The outer infinite extent (or nearly infinite extent), electron cloud, part of the atomic space-form structure facilitates interactions between a particular atomic space-form and the surrounding physical systems. Note: In higher dimensions, new materials are introduced as space-forms of different signature metric spaces that exist in the higher dimensions are also introduced.

Thus one sees that this new theory models material in space and time, along with a system's vector bundle structure of motion and fields, as well as metric space states, which together allow the differential equations of physical systems to be defined. These systems can have both geometric and spectral properties as they do in both classical and quantum physics. However, the derivative is now defined on differential forms (geometric functions) whose domains and boundaries are the space-forms of the system. Note: In the H-atom the Laplacian was applied to a 1-form representing an energy 1-flow on a 2-dimensional space-form.

The universe of the new paradigm of physics is the same universe as the currently accepted paradigm of physics, namely, material, space, time, vector bundles, and differential equations and their solutions can be used to describe the geometry, the spectra, and the dynamics of material. However, now there exists a new set of spectral-geometric shapes upon which systems and their material properties depend.

Outline of how space-forms can be used to describe macroscopic physical systems

The goal here is to show that from the principles that deal with space-forms, [for example, the natural differential operators to use, an interaction's relation to space-form properties, and the new models of "ordinary" physical properties, eg potentials and motions] the basic properties (or formulas) of the interacting materials of systems of classical physics can also be obtained.

There are no principles in the new theory which prohibit modeling a system, composed of large material objects, in terms of the object being at a point, ie at its center of mass, along with various functions, derivatives, and vector fields so as to find the usual formula by means of solutions to differential equation which the currently accepted paradigm of physics uses. However, by examining classical systems in terms of space-forms one sees how derivatives are related to both metric-space states and connection Lie algebra transformations. The tradition of the old paradigm is to represent a property of a system with a formula and differentiate it so as to relate this formula to the system's local linear structure, thus allowing the derivative and its inverse, or integral, to be the defining ideas underlying the process of measuring and describing physical systems. These new ideas focus on space-form material spectra and geometric properties along with a notion of a containing spectra (of geometric origins) which can organize the physical systems of material geometry and material spectra that do exist.

Dynamics in this new theory, when focusing on the system's space-form properties, are a bit complicated. Inertial dynamics is caused by spatial displacement in Euclidean space which takes place in each of the subsequent time intervals of a spin rotation of metric-space states. There is a process of "stable" inertia, and a process for "inertial change" which are very closely related (or very similar) to one another. The two processes of (1) maintaining motion and (2) of changing motion, depend (in three spatial dimensions) on the sectional curvature of a 2-dimensional Euclidean space-form. Both of the two slicing planes of this Euclidean space-form (needed to determine this Euclidean space-form's two sectional curvatures) contain the normal direction which defines the distance, r_1, between the interacting material space-forms, ie the normal to both of the interacting space-form surfaces. This normal direction, in turn, identifies a tangent space to the normal direction which is located at one space-form or the other. It is two orthogonal directions in the tangent space along with the normal direction that determines the two slicing planes. One of the two slicing planes determines r_1 while the other slicing plane determines r_2. The radii r_1 and r_2 determine the two radii of an Euclidean 2-torus, and these radii define the two potentials needed for the (entire) interaction, ie (1) and (2) in this paragraph.

There are two parts to the interaction: (1) the direct interaction part for either masses or charges (or other interacting materials) related to r_1, and (2) the inertial interaction between the two "virtual charges" related to r_2, ie the virtual charge in the (+t)-metric-state, and the virtual monopole currents in the (-t)-metric-state, (monopoles form into current-flows on space-forms contained in the metric state, (-t), which, in turn, generate an electric field due to these monopole currents). Thus r_2 is related to a pair of "virtual" hyperbolic space-forms and their interacting "virtual" charges. These two states of time are parts of the dynamical system structure needed for motion within both of the two metric state structures (of metric spaces) and which necessitate the use of a finite dimensional Hermitian space in which one metric state is in the real part, and the other metric state is in the pure imaginary part of the Hermitian space.

Consider a non-charged dynamic system such as two interacting masses, m_1 and m_2, so that $m_1 > m_2$. The frame will be the center of mass frame, which will essentially be the center of mass of m_1. The mass m_2 is moving with respect to the center of mass frame, hence it has associated to itself an additional hyperbolic space-form, because of its relative motion, and this relative motion is associated with r_2 of the Euclidean inertial space-form, and hence, the inertia of this relative motion.

Note: This extra dynamic hyperbolic space-form structure also accounts for its relativistic mass increase.

This dynamic hyperbolic space-form has a (virtual) charge, q_1, on itself in the (+t)-metric-space state, so that this (virtual) charge is interacting with a "charge" (or monopole current), q_2, on a second hyperbolic space-form in the (-t)-metric-space state. The normal direction connecting the (+t)-space-form to the (-t)-space-form, projected into the real part of Hermitian space [which would be a direction consistent with the tangent direction of r_2 on the real (or (+t)-space) Euclidean interaction space-form (in the simplest case)], defines the direction of inertial motion in the real metric space. The distance r_2 separating the (+t)- and (-t)-interacting (virtually charged) space-forms (in Hermitian space) is determined, as mentioned above, by one of the two minimal, yet independent (or orthogonal in the simplest case), sectional curvatures of the interaction, 2-dimensional Euclidean space-form in the traditional coordinate spaces used to describe material with three spatial dimensions.

Note: The charge, in the dynamical (+t)-state has a mirror charge, q_2, in the dynamical (-t)-state which is actually a monopole current on a (-t)-space-form which generates the equivalent electric field of an equal "mirror" charge, q_2, so that, $q_1 = q_2 = q$.

Thus, there are two aspects to Euclidean dynamical displacement, (1) the $\frac{km_1}{r_1}$, interaction potential part, whose (gradient) direction of interaction is determined by the normal direction between the two interacting material space-forms, ie this directional property of the system is due to the slight spherical symmetry of the underlying low dimensional space, and (2) the

$$\frac{k'q_1 q_2}{m_2 r_2^2},$$

force, affecting the inertial part of m_2's motion, where $q_1 = q_2 = q$.

Note: either a non-rational slope, ie an infinite length curve on the Euclidean checker board lattice (or Euclidean torus) can correspond to r_2 being associated to a direction of motion which is in the same direction as r_1 [in this case the inertial property of motion has two terms: a rational term giving inertia's tangent direction, and an non-rational term giving inertia's radial magnitude], or two perpendicular inertial Euclidean space-forms can be part of the dynamical process so that the inertial part of the motion can have both a tangential part and a radial part to its motion.

The displacement distance for each spin rotation of the two time states is, $d = \frac{1}{2} ta^2$, or velocity changes, $v = at$, where a, is a constant acceleration, which is constant because in each time interval the force fields are essentially constant, and where, $t = \frac{(atomi - radius)}{c}$, where c Is the speed of light, and t, is the time interval for each metric state spin rotation (this "short" time interval allows changes in atomic energy states to initiate light waves within atoms). For each displacement in this time interval, the value of the constant acceleration, a, is (the sum of) either

(1) $a = \frac{km_1}{r_1^2}$ and/or

(2) $a = \dfrac{kq_1 q_2}{m_2 r_2^2}$,

depending on whether it is (1) the interacting materials or (2) the inertial properties of the moving material, respectively, which are being described.

The context for the Dirac-like eigenvalue equation is that of many metric states for a system's properties, ie functional values of system properties can be in different metric states, can exist on either space-forms or within the containing metric space [for the system mentioned in the above couple of paragraphs]. This set of space-forms, metric spaces, and metric states along with a "good" selection of coordinate directions upon which the Dirac differential operator is defined, are the context within which the differential equations for the system are defined.

What is central to the definition of such a differential equation for a macroscopic system are the square root properties of the Dirac operator, which mix the different metric states of a system's physical properties, along with an interaction connection term which is added to the differential. Namely, both the energy eigenvalue E, and the potential V (r), or the connection 1-form A, are square roots due to a difference in squares factorization of the generalized Laplace operator. Letting, γ, be the Dirac matrix, and, *, the differential "adjoint," (which applies to differential forms) there is the factorization:

$$(\gamma\partial - iA^*)(\gamma\partial^* + iA) = \partial\partial^* - i\gamma\partial A + i\gamma\partial^* A^* + AA^*,$$

where the last equality is true either because $i\gamma(\partial A)^* = i\gamma(\partial^* A^*) = i\gamma(\partial A)$, in a 4-dimensional (flat) space-time, ie because a 2-forms' differential-adjoints are also 2-forms in a 4-dimensional flat metric space, or $i\gamma(\partial A)^* = i\gamma(\partial^* A^*) = 0$, because dA is the curvature and on a flat metric space the curvature is zero, ie $dA = 0$. Thus $(\gamma\partial - iA^*)(\gamma\partial^* + iA) = \partial\partial^* + AA^*$, and when applied to a differential form, f, whose domain is a space-form then $(\partial\partial^* + AA^*)f(r,\omega) = Ef(r,\omega)$. The connection 1-forms, A, and the energy eigenvalues, E, act like square roots in the square root Dirac operator. This generalized Laplace differential operator needs to be of the form,

$$(\partial\partial^* + V(r_1) + V(r_2))f(r,\omega) = Ef(r,\omega), \text{ or } (dd^* + A(r_1) + A(r_2))f(r,\omega) = Ef(r,\omega),$$

so that ([dd*(of a differential form defined on a space-form domain)]=0), ie dd*f=0, for a short time interval (see below) related to the spin rotation between metric states within the metric space.

This would allow one to convert the inertial interaction field of

$$(2) \ \frac{kq_1q_2}{m_2r_2^2}, \text{ into a kinetic energy term.}$$

Namely,

$$v = at = \frac{kq_1q_2}{m_2r_2^2}t = \frac{kq_1q_2}{m_2r_2^2}\left(\frac{atomic-radius}{c}\right) = \frac{kq_1q_2}{m_2r_2^2} \cdot 10^{-18} = 10^{-29}\left[\frac{q}{r_2}\right]^2,$$

so if q is about 1 coulomb, and m_2, is about 1 kg, so if v is about 1 m/s then r is about 10^{-14}m, or about the size of a nucleus, which, it might be noted, is also carrying a very large charge,

ie 1 coulomb. So as to form the kinetic energy formula, $\frac{1}{2}mv^2 = T$.

The kinetic energy, T, can be found from the r_2, radius of the Euclidean inertial space-form, so that the system has an energy (or eigenvalue) structure of the usual classical system,

ie T + V (r) = E.

Thus in order to set up the equations for the system in the context of space-forms one begins with the Dirac operator and considers one of the positive energy eigenvalue equations, along with a "good" selection of coordinate directions upon which the Dirac differential operator is defined in regard to the space-form's geometry, one gets

$$(\partial_x + A(x))f_-(r,\omega) + (A(r_1) - \partial_z + A(z))f_+(r,\omega) = E_+f_+(r,\omega),$$

Where both A(z), and A(r)=A(r_1), determine the radial motion, where A(r) is the spherically symmetric interaction, and A(z) is the inertial property in the same radial direction. Note: That the gradient of A(r) is directed along, r, the direction connecting the two interacting space-forms, is a property of spherical symmetry in 3-dimensions and less. Before the displacement, which takes place after the time interval for the rotation of the state of time, it is assumed that the space-form structure at the instant of the displacement is constant, hence all df=0.

[In other words, the function, f, if viewed as an eigenfunction on a 2-dimensional space-form, ie f represents a spectral flow on the space-form, then it can be modeled as a uniformly circular orbit (or winds about circular orbits) on a hyperbolic space-form, and for the instant before the "spin rotation time interval," assume that there are no coordinate changes, ie f is evaluated at fixed values of the coordinates, hence df=0, for the instant before the "spin-rotation time-interval" within which displacements occur.]

Furthermore, it is also assumed that [f(+) / f(-)]=1, thus leaving either an energy equation or a square root energy equation, ie square root potentials and the square root energy eigenvalue. For the energy eigenvalue equation, consider the energy relation

$$(A(x) - A(z)) + A(r_1) = E_+$$

but $(A(x) - A(z))$ Are both related to inertia,
hence they determine the kinetic energy, T,
and A(r)=V(r),

so this gives the usual energy relation of macroscopic physics. However, now all of this was derived from the space-form properties of the system for a new context in which the differential operator is strongly associated to its natural Lie algebra transformation relations, associated to the Lie algebra valued connection 1-forms, within the geometry of the space-forms and their different metric state flows and processes associated to these states, such as these states spin rotation, which all together compose the system.

This viewpoint applies to both macro- and micro-systems in which the system depends primarily on the Euclidean space-from to determine the interaction structure. This short review has revised a similar discussion that exists in the literature.

Reviewing: Two material objects, one large mass and one small mass, which could be modeled in terms of their center of mass frame, so as to be separated by a distance, r. The distance r/2 represents the radius of an Euclidean interaction space-form which touches the two separate points of the two separate and interacting material space-forms. Thus, 2/r represents the sectional curvature of the Euclidean space-form which determines the interaction between the two masses, as well as the lighter masses inertia. Namely, the inertial properties of the system are described above with respect to the sectional curvature properties of the Euclidean interaction space-form's pair of sectional curvatures.

Seeking the correct mathematical interpretation

In searching for the correct mathematical formulation of physical systems being defined in the context of space-forms, one needs the correct space-form structure, the correct rotation of metric-space states for both the space-forms, and the metric space itself, as well as the correct Lie algebra transformation properties eg inertial displacements.

It is assumed, that during the time interval between metric-state jump rotations (or spectral flow state jump-rotations) of the time period of a discrete spin rotation, the geometry of the space-form structure is fixed, ie its formula is evaluated at a specific values of the containing coordinates, so the formula's derivatives are zero, ie $\partial_x f = \partial_y f = \partial_z f = \partial_t f = 0$. However, at the end of the time interval of the spin rotation between states, the Lie algebra's "jump" transformations occur so that the system's geometry is adjusted and/or changed to conform to the properties that the fields and interactions require. This information is carried on the Euclidean space-form along with the information needed for determining a conformal factor of a standard "unit" hyperbolic space-form.

Differential operators I

In order to find the differential equation which can be used to model a physical system one needs to know the space-form context of the system. A differential, acting on co-homological forms, defined on a manifold in the base space of a principle fiber bundle, has a Lie-algebra-valued 1-form added to itself, so as to account for actions which are contained in the "horizontal" space (related to the base space) of the principle fiber bundle's tangent space, so that these Lie algebra actions have factors which are related to sectional curvatures. These Lie algebra transformations can be used to change the geometry of the interacting space-forms in a time interval, related to a spin transformation cycle of the metric-space states, by transforming the space-forms in their metric spaces.

The structure of the interaction potential in the base space (which contains the space-forms which in turn compose the physical system) is related to the principle fiber bundle's fiber group, primarily by means of a discrete isometry subgroup, ie space-forms. However, there are also the conformal groups which are applied to the system's vector bundles, as well as the possibility of applying the diffeomorphism group so as to relate the geometry of the system's space-form structure to a general Riemannian metric, ie a Riemannian manifold. However, it is logically easier to think of the system's space-form geometry, with its mixing of states and its jump displacement adjustments in terms of changes of geometry caused by Lie algebra transformations. It is energy which holds a system together, and energy is a property of the very stable hyperbolic space-forms. It is natural to consider the metric invariant Laplace operator as applicable to either the system's energy properties or its geometry.

An ordinary guitar string's eigenvalues are proportional to the guitar string's length. The eigenvalues of a 2-dimensional space-form system are related to the spectral lengths of the space-form's natural (1-dimensional) spectral flows, eg hyperbolic space-forms. Each of the spectral lengths, or spectral flows, of a space-form are in one metric state or another. The eigenvalues of a hyperbolic space-form will be energy, in which case the Laplacian will be the "space-time" Laplacian, ie the wave-operator.

A system is composed of many space-forms and many metric-space states. Using the Dirac operator results in a system of differential equations acting on both positive time and negative time solution states (or matter and anti-matter solution states in Euclidean space) so that these separate solutions will have positive and negative energy (momentum) eigenvalues, respectively. However, in the Dirac operator both the positive and negative function states as well as their positive and negative eigenvalues are all mixed together.

That is, within a system represented as a space-form, the states of the space-form are rotated by the differential operator, this can be done by the spin matrices of the Dirac operator, though other rotations might be more appropriate. Which leads to the question. Should the rotation of states on a space-form have the same form as that of the Dirac operator?

Note: The Dirac operator is first order. Hence, if it is used in a linear differential equation, then that differential equation should (will) have solutions. Thus there will be a relation between the structure of a differential equation based on the Dirac operator and the stability of the system modeled by such a differential equation and this system's consistency with the spectra which allow stable physical systems.

In this new theory, a function representing a system's energy has different forms or different function components depending on the function's metric state. The function can be defined as a vector which carries those different function components of the different metric states, by means of a state vector. Thus, not only should the derivative have potential terms introduced into it structure, but also its vector states need to be separated and then related to the system's differential equation. [Note: The state-vector property of the function naturally causes the Lie algebra to become relevant to the system's properties, both to alter space-form structure and to inter-relate a system's natural metric-space states.] Because the metric states are used in a mixed form in order to allow dynamics to take place, the components of the differential equation need to be mixed or rotated amongst themselves. This could be done by a unitary matrix acting on the vector components of the differential equation. The Dirac operator mixes the components of the state-vector function and the terms of the differential equation together. Thus the Dirac operator is used, though it may not be the best choice for a differential equation defined on space-forms. Thus the new theory is barrowing some of the mathematical structure of the currently accepted paradigm so as to use it in a new context of space-forms.

Connections on the principle fiber bundle are Lie algebra valued differential 1-forms, which are added to a differential (or Dirac) operator, so that the Lie algebra acts on differential forms (or state vectors), whose domains are a set of many space-forms.

If A is a connection 1-form then it has the property dA=F where F is the force-field 2-form, and d*F=m, where, here, m, is the material source of the force field, ie scalar or vector quantity type (or tensor). The sectional curvature is related to a spectral curve on a space-form and its radius of curvature. The curve is contained in (or projected into) a 2-dimensional plane, and it spatially connects interacting space-forms so as to change the nature of space, eg a geodesic is formed which is different from the flat coordinate structure. This property of spatial deformation causes one of the interacting material space-form's motion to change. In 3-dimensions this change is spherically symmetric.

Though [grad (V(r))]=F, where F is a spherically symmetric force field, for the scalar potential V(r)=k / r. This is not necessarily the "usual" relation between curvatures or potential 1-forms and force fields that are contained in a metric space of arbitrary dimension.

In electromagnetism the force field's source is the vector current, j, whose vector properties affect the properties of the force, eg its directional affects. Thus in electromagnetism there exists a force field and a Lorentz force, so that the force field's affects of the magnetic field are not spherically symmetric.

In a metric space with four spatial dimensions the analog of sectional curvature is surface areas of bounded 3-dimensional space-forms. This can be related to the mean curvature, essentially determined by the two minimal sectional curvatures, k(1), and, k(2), [the eigenvalues of the (locally) diagonalized curvature tensor (is diagonalizable since Ricci is symmetric)] of the space-form, so as to get a similar term as that given in the Lorentz force as (v x B), where v is a test particle's motion, and B is the magnetic field, but now it is (r x B), where r is the position vector and B is no longer the magnetic field, rather it is a force field related to one of the two "minimal" sectional curvatures of a 2-surface, which, of course (being related to circles), can be given distinct directions. In metric spaces, with higher dimensional spatial subspaces, the force field caused by the interaction space-form's curvature properties is no longer necessarily spherically symmetric, but rather it could be tangential to the interaction space-form's normal direction, where the normal direction is determined by the separation distance of the two interacting space-forms.

What form of the potential should be used for the connection form?

In a metric space, with a 3-dimensional spatial subspace, interactions between a pair of 2-dimensional material space-forms depends on there being a bounded, 2-dimensional, interaction Euclidean space-form whose geometric structure depends on the (system) connecting (or geometric) properties (including direction of interaction) of the two space-forms. The connecting (or geometric) properties are relative separation and relative velocities of the two interacting space-forms. For the interacting space-forms there is a separation face (or simplex) which naturally forms between the two space-forms, so that one of the 1-flows of the Euclidean interaction space-forms is related to the separation distance, while the other is a relative velocity that is in the frame of the velocity of the previous time interval that is caused by the inertial interaction potential. The relative velocity of the two interacting space-forms is also carried by a "virtual" inertial hyperbolic space-form's conformal factor. The connecting direction is the direction which is normal to both interacting "virtual" space-forms, so that the two space-forms can be thought of as being contained in two separate spheres. This normal can be used to identify a 2-dimensional tangent space on each of the interacting material space-forms. The length of the normal, the normal direction, and selecting a pair of particular directions tangent to the normal, which are to be associated with the interacting system's properties of motion, can be done so that one of the particular direction's in the tangent space determines (part of) an inertial radius, and the other tangent direction determines the plane of the Euclidean space-form's (curvature) circle which is also the plane of the radius which determines the interaction potential, so that together, ie inertia and interaction, this information determines the 2-dimensional Euclidean interaction space-form. Note: It is the spherical symmetry of three dimensions which allows the gradient of the potential to determine a radial direction of inertial change, and this is caused by the interaction term's spherical symmetry, which in turn is due to a 3-diemsnional space's spherical symmetry.

The two radii, of the Euclidean interaction space-form, determine a pair of potentials which determine the system's relative motion. The potentials determined by the Euclidean torus's two radii are changing, and this change can be identified, or caused, by a Lie algebra transformation. Namely, Euclidean coordinate displacements, associated to the potential factor(s) of the Lie algebra term in the connection of the principle fiber bundle.

The two potentials, related to inertia and interaction, means that the Lie algebra matrix form should be associated to two potential factors (in the sense of tensors), namely the interaction factor on the interaction terms along with a conformal factor multiplying the "inertial" hyperbolic space-form. The connection is given by

$$dg + Adg \cdot dx,$$

where d is the differential operator which either acts on differential forms or acts on state-vectors which are functions which define differential 1-forms, and where g is a fiber group element (hence dg is a Lie algebra element), and where x is a base space element, and A is the pair of independent potential interaction (tensor) factors which are determined by the pair of interaction space-form's sectional curvatures.

On a manifold one wants to differentiate but one wants to differentiate by $(d + dg + Adg \cdot dx)$, in a manner which stays parallel with the base space, so that the connection operator can be used to transform the space-form structures of the system by an infinitesimal amount as part of a matrix transformation Lie algebra element. This Lie algebra transformation allows the measured changes of the derivative to take place in the base space.

That is, the measurable changes of the space-form's position or energy with respect to the containing space-form's coordinate or energy structure are due to Lie algebra transformations. That is, the measured changes of the function type, f, (that the derivative acts on), which result from coordinate changes, are caused by Lie algebra transformations, including changes in relative position. Changes in relative position of a system's interacting space-form result in changes to the system's interaction potential term. The infinitesimal Lie algebra transformations cause changes to the interaction and material space-form properties within the base space manifold. These include changes to position, changes to space-forms, such as a space-form's spectral state (eg the solution function is a space-form) as well as changes to potential terms in the differential equation, where potential terms are also determined from space-form properties, and changes in relative motion, due to the changes in the conformal factor applied to the "inertial" (or virtual) hyperbolic space-form. The infinitesimal Lie algebra transformations change the properties of the space-forms and their relations within the coordinates (and other space-forms which compose the system and) which contain its description.

The inertial properties of one of the interacting space-forms are represented (and also determined) by an "inertial" hyperbolic space-form related to the system's inertial Euclidean space-form's displacement properties (that are associated to Euclidean space), because the motion of the either interacting system's or inertial systems depends on a hyperbolic inertial space-form. The inertial properties of the system thus, also depend on an infinitesimal conformal factor which operates on (or multiplies) "inertial" hyperbolic space-forms.

A conformal factor is needed because the "inertial" hyperbolic space-form cannot adjust to any fundamental domain whose 1-faces are any length that would be convenient for the system's properties of motion. The inability of the hyperbolic space-forms to do this is because of the rigidity of the hyperbolic space-form. Therefore a conformal group can cause the hyperbolic "inertial" space-form (or velocity space-form) to have values of the velocity that are within a continuum, if the "inertial" hyperbolic space-form can be multiplied by a conformal factor whose values are taken from the real line (or complex plane). Thus, there are three groups associated to the principle fiber bundle: isometry groups, diffeomorphism groups, ie for an extra added general Euclidean or Riemannian metric space (Note: This is not very well established, but the hypothesis of the shape of space affecting motion is so compelling.) and the conformal group; which is associated to smooth transitions of velocities in a hyperbolic metric space, or on the (hyperbolic) tangent space (where velocity exists), defined on the base space of the principle fiber bundle.

Hyperbolic space-forms define the stable basis for material systems in terms of their electromagnetic properties, ie essentially all electromagnetism is determined by the 1-form, A, which is the solution to the non-homogeneous wave equation, $d \cdot d^* A = j$, that the current vector, j, exists on the stable hyperbolic space-forms, which represents charged material. The potential, A, exists within the containing metric space. However, because Euclidean space only has physical properties associated to material displacement, this means that Euclidean space has no property of time associated to itself, which in turn, means that the local instantaneous property of motion (a vector) must be related to the hyperbolic space. Thus, motion is attached to material space-forms by attaching, to a Euclidean (inertial or massive) space-form, a hyperbolic space-form which can be used to define a tangent space to itself, upon which the local velocity (or acceleration) vector depends, ie including time dependent changes in position and motion. Thus the attached "inertial" hyperbolic space-form is associated to a system's tangent bundle. The vectors within such tangent bundles can be given the properties of being continuous by allowing that hyperbolic space-form to be associated to a conformal scaling factor transformation, ie to a conformal scaling (or multiplication) group.

Conformal factors

The velocity is, $v = v(r_1) + v(r_2)$, where $v(r_2)$ is, $[V(r_2) \text{ (perpendicular)} + v(r_2) \text{ (radial)}]$. The conformal factor which represents an "infinitesimal change" in velocity is a function of the material space-form's instantaneous (observed) velocity, v (r_2), along with its change in velocity due to the interaction,

$$v(r_1) = \left(\frac{km_1}{r_1^2}\right)t.$$

This formula can be used to define a conformal factor which changes the sectional curvature of the "inertial" hyperbolic space-form of the velocity tangent space of the space-form. This conformal factor should also be related to the inertial Euclidean space-form's r_2 radius. Translation changes the interaction Euclidean space-form radii, r_1, so that both r_2, and r_1 can be changed for each time interval $t = \frac{(atomi-radius)}{c}$. The conformal factor should also have the mathematical structure which allows a natural bound for the addition of hyperbolic velocity vectors at a point in space.

The conformal factor is related to the speed of light as a natural bound for this metric space

The "inertial" hyperbolic space-form has an "inertial" bound (or limit) caused by a conformal factor which approaches infinity as the space-form's "inertial" velocity approaches the speed of light, ie is equivalent to infinite speed. The form of a metric dependent factor emerges from the metric, ie (*idx, idy, idz, cdt*), so that differentiating by time in the formal metric structure leads to $\left(1 - \frac{v^2}{c^2}\right)$.

An "inertial" hyperbolic space-form is associated to the inertial properties of the system, ie a space-form in relative motion, whose (real) interaction and inertial Euclidean space-form radius is related to the relative velocity, v, where $v = [v(r_2) + v(a(r_1))]$, and r_2, is the Euclidean inertial radius, and $v(a(r_1)) = \left(\frac{km_1}{r_1^2}\right) \cdot t$, is the change in velocity due to the fixed acceleration (for spin time interval, t) associated to the value of the potential radius, r_1. The addition of velocity is with respect to hyperbolic velocity vectors in a hyperbolic metric space. This represents the new velocity after the time interval needed to rotate between opposite metric states, so that the time interval of the spin rotation is given approximately by $t = \dfrac{(atomi-radius)}{c}$.

Differential operator II

The Euclidean inertial space-form radius, r_2, determines the radius of the "inertial" hyperbolic space-form.

The derivative acts on both the function which represents the systems properties, as well as the manifold structures of the base space, M, that are involved in the local derivative of (f, M), so that the system is contained in a principle fiber bundle,

$$(d + dg + Adg \cdot dx)(f, M) = F(k, x, f, dgM).$$

where k is a constant, eg an eigenvalue, and F(,) is some general function, and *dgM*, is a Lie algebra element acting on the manifold, M.

Because in a principle fiber bundle this is defined on a base space, within which metric-space states are contained in the real and pure imaginary subsets of a Hermitian structure for space, and because these states are rotated or mixed to allow for dynamical displacements, it is necessary that this differential operator act on the function, which represents a measurable property of the system, by a rotation operator which mixes together the different states that exist on both the space-forms and the metric-spaces. This is much as the Dirac operator rotates angular momentum properties of states of micro-systems.

However, in metric spaces with higher dimensional spatial subspaces than three, translational inertial properties may have more degrees of freedom than a simple vector relation, eg tensor relations, or new independent material and its independent motion. That is, there can exist different (new) materials, in spaces with more than three dimensions, that move with different motions (or move in different directions).

The connection differential operator has the form:

$(d + dg + Adg \cdot dx)(f, M) = F(k, x, f, dgM)$, ie a principle fiber bundle operator.

[The Dirac differential operator has the form $(\gamma(\partial) + A)f = Ef$, or $\gamma(\partial + A)f = Ef$,

within a flat metric space, where, γ, is the Dirac spin Matrix]

This principle fiber bundle operator acts on f, and f defines a vector of eigenfunctions that exist for the different states of a space-form and/or within the (Hermitian) metric space of the base space. The vector, f, has components which represent the same function on different metric-space states, where each metric space has two states which are distinguished on the subsets of Hermitian space.

The function, f, if viewed as an eigenfunction on a 2-dimensional space-form, ie f represents a spectral flow on the space-form, (then it) can be modeled as a uniformly circular orbit (or winds about circular orbits) on a hyperbolic space-form, so that for the instant before the local Lie algebra space-form transformations which occur after the "spin rotation time interval," assume that there are no coordinate changes, ie f is evaluated at fixed values of the coordinates, hence df=0, for the instant before the Lie algebra displacements actually do occur. That is, displacements of the system that are associated to the "spin rotation time interval, ie a spin rotation, between opposite metric states," occur after the time interval, and are caused by Lie algebra transformations.

The usual electromagnetic vector potential, A, within an atom, can be obtained from the laws of electromagnetism applied to the nuclear space-form's currents flows. The vector potential can be evaluated at a point along a current flow of an electron cloud's compact space-form part. The formula for the potential, A, has a (common) factor of $1/r$, which is the sectional curvature of the Euclidean interaction space-form involved in the nucleus-electron interaction in terms of interacting space-forms within any particular dimension metric space.

The energy eigenfunctions of a hyperbolic differential operator are determined by the sum of the potential and kinetic energy terms related to the solution function, so that, these values are (or can be) determined by Lie algebra transformations applied to the two radii of the Euclidean interaction torus and its associated "inertial" hyperbolic space-form. In order for this to work the potential differential form, A, multiplies the Lie algebra element, dg, giving Adg. The term, Adg, can act so as to cause displacements on the two radii of the Euclidean torus, that are contained in M. This Lie algebra action causes changes in the two radii by incremental amounts. It is also related to the value of the conformal factor which, in turn, is applied (by multiplication) to the "inertial" hyperbolic space-form.

If the function, f, represents one of the properties, though with multiple states, of the system's manifold structure on the base space, then one might want to ask, "Are such Lie algebra actions on M equivalent to actions on f?" One would want to believe that such a formulation in terms of, f, would be a very good way in which to represent the properties of the physical system made of space-forms. In fact, if the system is a set of interacting 2-dimensional space-forms, these space-forms can be thought of as composing the faces of a 3-dimensional hyperbolic space-form. Thus, f, in its stable energetic representation can represent a 3-dimensional hyperbolic space-form whose spectra are 2-dimensional flows on the 3-dimensional space-form, whose stability is determined by the spectral set of its 4-dimensional containing metric space. It is the 2-dimensional spectral flows that exist by projection down into the 3-dimensional space, the space in which we experience material geometry.

There are three types of systems that a function, f, can represent, the (1) stable space-forms which are compact and (2) the dynamic systems that can be stable and either compact or unbounded but which can be identified as systems which can be reduced into their components, ie a stable bounded system, eg elliptic (bounded), a free system is parabolic (semi-bounded so its bounded-ness depends on values of key physical parameters, eg initial conditions) and hyperbolic (free, material systems approach one another but do not bind together), and (3) systems that do result from interactions but which are not stable, though they may be bounded as the example of the irrational sloped curve on the Euclidean checker-board lattice in the plane. The instability of such systems is because the interaction geometry and/or its boundary conditions do not fit into the spectral structures that the containing simplex allows.

The electron cloud is an example of system which is accessible to substantial changes of its components, eg component and/or spectral-state re-arrangements. This ease of accessibility to change is due to its infinite extent space-form structure. Thus the electron cloud is readily available for interactive changes, whereas the compact nucleus is not so susceptible to changes.

Dynamical systems' descriptions seem to depend on the Lie algebra transforming the space-form structure of the system's space-form properties, eg transforming the two radii of the 2-dimensional, interacting, Euclidean, space-from, ie Euclidean torus, as well as the "inertial" hyperbolic space-form related to velocity fields. If the dynamic system is stable, then the transforming Lie algebra elements would need to be "trapped" within some region of the Lie group (about the identity element), ie within a fundamental domain of one of its a discrete groups, or within a finite number of periods of the identity's fundamental domain, so that the Lie algebra elements form a repeating cycle, eg acting on an interacting, inertial, Euclidean space-form.

Furthermore, the entire context of this description is discrete isometry subgroups in which the isometry group is a Lie group, so Lie group transformations, or their local linear Lie algebra maps should have some strong relations to space-form's discrete group properties.

Ultimately this strong connection seems to be the spectral set, or spectral-simplex, of the entire discrete group context, which is relevant to a stable system's spectral and non-reducible geometric (and subsequently spectral) properties.

The positive charge structure of an atomic system, on any dimensional level, is associated to stable, compact space-form structures (at least up to and including space-forms of dimension 5), while the negative charge space-form structure is infinite extent (or very large range), and thus negative charge space-form subsystems readily allow changes and transitions of its space-form structure. Note: compact hyperbolic space-forms only exist in hyperbolic metric spaces which are dimension five or less. All the six and higher dimensional hyperbolic space-forms are infinite extent. The infinite extent space-form structure is readily accessible to becoming a part of, or connecting with, either other space-forms or dynamical systems. Thus stable and bounded 3-dimensional space-forms represent the nuclear (or positive charge) part of the 3-dimensional hyperbolic space-form system structure. This is analogous to the 2-dimensional hyperbolic systems which have both compact and unbounded parts of its atomic system space-form structure.

When an atomic structure changes its "quantum" level, light is exchanged. However, in this new viewpoint each atomic energy level, or spectral flow, is actually a metric space of a particular dimension, eg 2-dimensional space-forms have 1-dimensional flows so that each 1-flow is a 1-dimensional metric space. Hence a spectral flow has the defining properties of a metric space, ie its metric-space states, its dimension, and its spectral subsystem structure, along with the "fundamental eigenvalue" of the spectral flow, itself, which represents the metric space. Light carries the fundamental energy spectral characteristic of the particular dimension metric space, which it defines. The light applied to the particular dimension spectral level (or metric space) causes the charge contained in the one metric space to jump to the metric space which the light identifies as its characteristic spectra ie the fundamental eigenvalue of the metric space's spectral flow.

The two atomic levels between which an atomic transition of spectral levels might take place are each metric spaces in their own right and have different fundamental physical constants, eg different speeds of light associated to the metric space which they are, hence they would have different fundamental laws of physics associated to themselves.

Dynamic systems are part of the electron cloud type systems, which none-the-less can be quite stable. Dynamic systems can also be stable because they can be within a bounded "nuclear type" space-form, such as being within the bounded part of the containing metric space. It is this property, of a planet being contained within its containing metric space's bounded space-from structure, which allows the many bodied solar system to be so stable.

Spectral transformations between "cubical" simplexes

The most important transformations for physical systems are those transformations between spectral sets, eg between two different "cubical" simplexes or between different faces, of either the same dimension or of different dimension, on the same simplex.

The spectra of discrete groups can be listed. Understanding the hierarchical spectral structure of a simplex might be the most useful type of information about physical systems.

Conformal factor and inertia
The conformal factor of hyperbolic metric spaces and inertia

Remembering the discussion about Euclidean interaction space-forms;
Regarding the dynamical properties of material, there are two aspects of motion:

(1) the inertial part, or conserved constant motion, where $v(r_2) = \frac{kq_1 q_2}{m_2 r_2^2} \cdot t$, so that r_2, is one of the two radii that define the Euclidean inertial space-form, and $v(r_2)$, is the conserved constant motion, which, in turn, is related to a conformal factor which multiplies each of the pair of simplexes, of the minimal uniform hyperbolic space-forms, in opposite metric-space states, which are associated to a relatively fixed (or conserved) velocity of a moving space-form, and

(2) a change in velocity of an object due to the spatial interaction of either mass or charge (in 3-dimensions), ie the gradient of the sectional curvature (or potential energy) of the interacting Euclidean space-forms causing changes in the inertial velocity by the amount

$$v(r_1) = \frac{km_1}{r_1^2} \cdot t,$$

in one time interval, t, where $t = \frac{(atomic-radius)}{c}$.

Hyperbolic space-forms have spectral flows which identify both a property of time and a current of charge. Hyperbolic space-forms can exist everywhere in space. A hyperbolic space-form, multiplied by a conformal factor, so as to be associated with a velocity tangent space (or bundle) over a coordinate point in the base space, is an "inertial" hyperbolic space-form. An "inertial" hyperbolic space-form can be thought of as a minimal, uniform, ie a hyperbolic fundamental domain based on "squares," space-form defined within a unit circle (or sphere) so that the limit approaching the inside boundary of the unit circle (or unit sphere) represents the unattainable points at infinity. The conformal factor is related to the "inertial," hyperbolic space-form and the velocity vector field at each point of the system's space. The velocity vector field in space depends on both the structure of the "inertial" hyperbolic space-forms and the inertial radius of the Euclidean interaction space-form. The Euclidean space-forms inertial radius follows the radius of the "inertial" hyperbolic space-form, while on the other hand the "inertial" hyperbolic space-form follows the geometric displacements that occur on the Euclidean space due to the displacement Lie algebra transformations after each time interval determined by spin rotation of states, ie $t = \frac{(atomic-radius)}{c}$. The conformal factor adjusts the vector bundle's velocity vector values in the interaction processes so that the velocity vectors belong to a continuum. The velocity which defines the hyperbolic metric space that we experience is, c, the speed of light. Thus a "flow" of "charge" on the metric space will have velocity, "c," but the dynamic processes between the material space-forms that are contained within the metric space can have relative velocities which take on values from a continuum. These velocities can form into a velocity field continuum which defines a vector bundle that is contained in (or defined over) the coordinate metric space. For example, On a 2-dimensional flow the spectral 1-flows break free of their ridged eigenvalue positions that they hold on 2-dimensional space-forms and move through the new 2-dimensional metric space. This results in a "net" of 1-flows which, together, define a 2-flow current. Analogously "nets" of 2-flows form into a 3-flow current, where in our metric space one might conjecture that the 2-flows defined by the earth, and perhaps the earth's orbital motion which together define the 3-flow current that our metric space has within its containing 4-dimensional space-form. Though such earthly 2-flows define the current path, the 3-flow current itself may none-the-less have a current speed much faster than the earth's orbital period.

Momentum and its changes for a material space-form are determined by an "inertial" hyperbolic space-form's conformal factor. This conformal factor allows velocity to have the property of being continuous in its vector bundle. There is an effective bound to motion caused by a velocity bound which identifies a metric space's current flow. This metric space current flow is (approximately) equivalent to defining a metric-space bound at infinity (or connection to infinity) for the interacting systems, eg velocities (or "change of state" energy signals), that the metric-space contains. It is the "change of state" energy signals, or light, that defines a current-flow on a 3-flow that our metric space defines

The effective (or natural) boundary of the metric space is the natural space-form boundary (or facial boundary) of its containing space-form structure (or "cubical" simplex structure), or spectral flow structure (if it is bounded) if the metric space is viewed as a spectral flow on a space-form. The boundaries of a metric space are well beyond the size of the material space-forms that are contained within the metric space. A hyperbolic metric space's bound on velocity is related to the size of the metric space's space-form, and if that size is large then this causes the material space-form processes, ie the atomic processes, to occur at a fast rate so that the incremental changes in a system's geometry are small.

The hyperbolic metric space's bound is effectively equivalent to (or a natural approximation to) infinity.

Hyperbolic metric space's contain material space-forms which carry on themselves (or their tangent spaces) the property of velocity. The motion of material space-forms also depends on the Euclidean space's interaction space-form. In regard to the magnitude of the velocity vectors in a hyperbolic tangent bundle, the formula for velocity is given by,

$$v = v(r_2) + v(r_1) = v(r_2)\left[1 + \left(\frac{km_1m_2}{q_1q_2}\right)\left(\frac{r_2^2}{r_1^2}\right)\right],$$

so that the expression $\left[1 + \left(\frac{km_1m_2}{q_1q_2}\right)\left(\frac{r_2^2}{r_1^2}\right)\right]$, is the conformal factor which changes the value of an object's velocity with respect to an "inertial" hyperbolic space-form, where r_1, r_2, and m_1 are all Euclidean space quantities (or properties). Note: The value for, v, must be less than or equal to the speed of light, c.

One can find the extreme relations between these values when one sets the velocity, v, equal to c. Solving for r_1 and r_2, gives:

$$r_2 = r_1 \left[\frac{k_e}{G}\right]^{\frac{1}{2}} \left[\frac{q_1q_2}{m_1m_2}\right]^{\frac{1}{2}} \left[\frac{c}{v(r_2)} - 1\right]^{\frac{1}{2}},$$

The factor $\left[\frac{k_e}{G}\right]^{\frac{1}{2}}$, is of the size of 3. If the charges are the size of one coulomb, and m_1 is the size of the earth's mass, and m_2 is 1 Kg, then $\left[\frac{q_1 q_2}{m_1 m_2}\right]^{\frac{1}{2}}$, is the size of 10^{-12}. The ratio $\frac{r_2}{r_1}$, depends on $v(r_2)$, so one can consider the bounding values for this equation.

Thus, as $v(r_2)$ (the inertial velocity) approaches c, in the r_2 equation, then r_2 approaches 0, hence $v(r_2)$ appears to approach infinity. On-the-other-hand, as $v(r_2)$ approaches c, in the r_1 equation, then r_1 approaches infinity, ie the system separates quickly.

The conformal factor for the hyperbolic space-form, which affects inertia, is related to the speed of light and the size of the metric space's natural bounding faces.

The unit circle (or unit spherical boundary) is usually thought of as the unreachable boundary of the metric space, but it can now be thought of as corresponding to the spatial (or geometric) measure, from the center of a metric space's space-form to its outer boundary. That is, the radius of a metric space's space-form's outer or bounding circle (or ribbon [or band], eg a sphere), ie the radius of the containing sphere of the metric space's space-form (so the containing sphere is tangent to the space-form). In a 2-dimensional torus there two such radii, in this case consider the smaller (or minor) radius of the two Euclidean space-form radii. The smaller radius of the two radii, is the boundary of the metric space's defining space-form. This boundary, provides the main structure which contributes to the stability of a planet's orbit. The bounded region (of the smaller of the Euclidean space-form's two radii) provides a region of stability for the planet's orbit. This region is full of a "current" which flows within the toral ring about the sun. This flow is characterized by the flow of light which moves between the earth's orbit which forms a net of 2-flows and the metric space's outer face.

The radius of this outer circle (or band), for this metric-space's space-form, is to become the defining "infinite distance" (or unit sphere) for the boundary of a hyperbolic metric space.

The spatial measure of this radius (the length of the radius) divided by a time interval of some significant space-form spectral process, which is defined by an interaction process such as perception (perception seems to be related to the heart-beat), which might be estimated to be a time measure of about one second in duration, so that these two fundamental measures, of length (of the ("traditional") smaller radii of a toral-like geometry) and time interval (needed to have a perception), identify the relation that Euclidean space (and its displacement distance) has to hyperbolic space (and its time) and that relation is the speed of light, c. The speed of light, c, relates displacement, of Euclidean space, to either the energy or time, of hyperbolic space.

Relative to the (our) metric space's boundary, the next lower geometric scale which dominates the geometry within the metric space is the earth (where perception occurs), then the next lower size scale would be the earth's geological features, followed by other macroscopic systems, along with the general crystal so as to include living systems, and then finally the atomic scale for the size of material systems. An atom's physical processes, eg changes in energy levels, are defined by (or must take place within) a time interval which is the atom's radius divided by the speed of light, c, which gives the time interval, t, within which atomic interactions must take place, and this also defines time intervals for dynamical processes, eg displacements, in the metric space are determined within this time interval.

Note: a metric space is defined by a face of a "cubical" simplex along with the infinite extent hyperbolic light and interaction space-forms which define the metric space as a certain dimensional coordinate space. Though a metric space can have a natural boundary, its infinite extent, light space-form structure, in fact, defines the metric space as unbounded and infinite extent, though the practical applications of its infinite extent properties are "not very important," though they exist. However, objects which enter from outside the metric space's natural bound may have [are at the risk of their also experiencing] new properties that result from their existence outside the metric space's natural bound, yet within the bounds or distance range of the next containing face in the simplex. For example, these new properties might cause there to exist slightly greater values of gravity or cause high energy interactions, gamma rays.

It is with respect to the metric-space's scale of size, eg the bounding (minor) radius of the metric space's spectral flow structure, that inertial motion is perceived. The inertial properties of an object's motion, caused by a constant acceleration due to constant field properties to exist within the time interval of, $t = \frac{(atomi - radius)}{c}$, so that within this time interval, changes of state occur for the metric space. In this case the constant acceleration would be as follows:

ie (1) $a = \frac{km_1}{r_1^2}$ and/or

(2) $a = \frac{kq_1q_2}{m_2 r_2^2}$,

the dynamic properties of inertia, ie (2), and inertial change due to (1). These changes need to be (relatively) smooth in regard to changes of the metric space's material space-form processes, eg changes in motion and changes in geometry, so that such changes occur in small increments that are related to changes in (metric) state, ie related to the time interval, t=[atomic radius]/c.

The inertial part of an object's motion is dependent on hyperbolic space-forms, and the process of determining the inertial vectors, eg velocity, and its changes, ie acceleration, by means of the local tangent geometry of the hyperbolic space-form [so that this geometry is associated to an object's motion in space] needs to be smoothed out. This smoothing can be accomplished by multiplying a uniform, hyperbolic space-form within a hyperbolic metric space that has a "unit" boundary by a conformal factor. The mathematical expression

$$v = v(r_2) + v(r_1) = v(r_2)\left[1 + \left(\frac{km_1 m_2}{q_1 q_2}\right)\left(\frac{r_2^2}{r_1^2}\right)\right],$$

gives $\left[1+\left(\dfrac{km_1m_2}{q_1q_2}\right)\left(\dfrac{r_2^2}{r_1^2}\right)\right]$,

as the conformal factor for the "inertial" hyperbolic space-form.

The metric invariant processes of the metric space, eg electromagnetic structures, along with the dynamic process of adding velocities, are related to the speed of light with which the conformal factor must be made consistent. This was demonstrated above.

Geometric changes in a metric space take place in time increments determined by the time interval, $t = \dfrac{(atomi-radius)}{c}$. If the boundary of the metric space determines one boundary or scale of size, then the smallest size of a uniform hyperbolic space-forms defines the other bounding property of (small) sizes, in a metric space structure within which inertial motion is determined. It is the capacity to move between these two bounds on size, within the time interval in which a defining (thought) process takes place, that allows geometric models of existence (or perception) to take shape.

The conformally determined, uniform, hyperbolic space-form is related to velocity through the conformal factor. The "inertial" hyperbolic space-form determines the inertial properties of the moving object to which it is attached, in the moving object's (or space-form's) tangential space. It does this in conjunction with Euclidean interaction space-form radial properties, (a) Inertia, which is determined by "virtual" charge amounts on hyperbolic space-forms and separation distances, and (b) Inertial changes, which is determined by separation distances of interacting material (Euclidean and/or hyperbolic).

It is also this conformal hyperbolic "inertial" space-form and its associated Euclidean space-form (or associated radius of the interaction Euclidean space-form) which applies to the "wave properties" of material, eg,

h (1/ t) = E, (hyperbolic space-form) and
h (1/[wave-length]) = p, (Euclidean space-form)

where Planck's constant, h, gives the scale conversion of sectional curvatures of both the conformal hyperbolic space-form and its associated Euclidean space-form to the physical properties that are associated to such coordinate symmetries eg energy and momentum of the space-forms of the respective metric spaces, ie hyperbolic and Euclidean.

Sectional curvature of the conformal space-forms are related to both momentum and energy, as well as the gradient of the Euclidean interaction space-form's sectional curvature also being related to forces.

The sectional curvature of the conformal hyperbolic space-form, 1/ t, is directly related to the conformal factor of the uniform minimal "inertial" hyperbolic space-form. When considering the formula, [displacement] / t = v, is $t = \frac{(atomi-radius)}{c}$, or is, t, related to the conformal factor?

(answer) Apparently the "inertial" hyperbolic space-form multiplied by the conformal factor results in 1/t for the "inertial" hyperbolic space-form's sectional curvature so that,

$$[displacement] / t = v.$$

Does the form: Closed curve integral of, $\nabla\left(\frac{k}{r}\right) = 0$, ie $\oint \nabla\left(\frac{k}{r}\right) dr = 0$, where $\nabla\left(\frac{k}{r}\right)$, is Grad (k / r), imply that $\left(\frac{k}{r}\right)$, is a potential energy?

Conjecture: Conformal models of high dimensional simplexes, whose bounding distances may be great, may be used to store information, about "material" geometry in a metric space, within the spectral record of such a geometric existence on a maximal torus within a fiber group, so that it can have resonance with a corresponding conformal spectral structure within a spectral structure of a living being. That is, the external simplex in question, when represented conformally is a (or almost a) face which is related, through spectral resonance, to the composition of the internal structure of the living entity.

Brownian motion stochastic processes and the re-derivation of quantum mechanics (back to summary 7)

At the atomic level, interactions cause Brownian motion, which is a stochastic process. E. Nelson used the properties of stochastic dynamics (or Brownian motion of microscopic particles) to re-derive quantum mechanics. Hence these new ideas are not opposed to quantum mechanics, However, these new ideas do not depend on probability for their logical basis, rather they depend on the hidden geometry of material and space for their logical basis. That is, probability is a derived property of these new geometric-spectral ideas.

Space-forms that are similar to material space-forms, but they generate their own energy

It turns out that some of the odd dimensional hyperbolic space-forms (dimensions 3 and above) are spectrally and geometrically stable, furthermore, spectral flows on such hyperbolic space-forms have a very ridged geometry associated to themselves, so that if those flows are filled with charge, then the charge would form an electromagnetic charge imbalance because of the geometry of the space-form's spectral flows. The electromagnetic forces caused by this charge imbalance results in the space-form oscillating, and this oscillating process creates energy.

Energy is created because the electromagnetic forces push on both the charge filled spectral flows of the hyperbolic space-form, as well as the inertial spectral flows of the accompanying inertial (Euclidean) space-form. Thus, the charge filled spectral flows are pushed into one another. The spectral flows on the space-form which are pushed together are of opposite metric states (the electromagnetic force which causes the oscillation is due to charge imbalance). The Euclidean space-form associated to the hyperbolic space-forms due to resonance will also oscillate The inertial (or Euclidean) space-form has less energy separating its two, opposite state, metric-state flows, hence when the inertial flows of the Euclidean space-form are pushed into one another, then the matter and anti-matter associated to these flows are also pushed together. This, in turn, causes energy to be released. This energy goes into the resonating energy of the system which the oscillating space-form defines.

Such an energy generating oscillating space-form system would form a model for a living system.

Such an energy generating oscillating space-form system would also form a model for the radioactive nucleus. The nucleus is a set of electrons and protons on 2-dimensional hyperbolic space-forms so that these 2-dimensional charged space-forms are interacting with one another, thus forming a 3-dimensional compact hyperbolic space-form composed of neutrons and protons. Some of the 3-dimensional hyperbolic space-forms have charge imbalances, thus allowing oscillation and energy generation and, hence also allowing radioactive nuclei.

Where does a star's energy come from? It comes for the same mechanism upon which both the radioactive nucleus and life depend for their energy. Namely, odd-dimensional, oscillating, energy generating hyperbolic space-forms. If life is (or we are) elaborate 3-dimensional space-forms so that our stability, as well as the energy from which our action comes, depends on the hyperbolic space-forms of which we are composed, then 3-dimensions is the least dimensional space-form structure into which life can be formed, because that is the first dimensional level within which oscillating, energy generating, hyperbolic space-forms exist.

Furthermore, the sun is not only the energy source upon which life on earth depends, another energy source would be the spectral simplex within which the earth, we know, is contained. The sun is an oscillating, energy generating space-form, hence it is a hyperbolic space-form of an odd dimension and therefore, because it contains the spectral structure within which the earth is contained, its lowest dimension would be five.

The sun's magnetic structure is a natural consequence of its hyperbolic space-form structure with its charge filled spectral flows or currents. Its large, energy generating, space-form oscillations can, it is speculated, have enough energy to cause the sun's hyperbolic space-form structure to break, ie disintegrating the spectral structure of that space-form. This is possible because of the large amounts of material along with the sun's materials that exist in the opposite (metric-space) states are being pushed together to generate the sun's great energy. That is, both matter and anti-matter as well as charge and the material of its opposite state, namely, monopole material are being pushed together and turning to energy. Hence cataclysmic changes in the electromagnetism properties of the sun result from the energy generation of the sun's high dimensional oscillating hyperbolic space-form. This provides a more definitive pattern in which to model the sun's electromagnetic properties, than the models of the sun that the currently accepted paradigm is providing.

Resonance

A resonance structure, that exists inside the fiber group's maximal tori, can allow different faces of the oscillating (or energy generating) space-form (or simplex) to resonate with different energies. Thus, energy generating spectral sets on the fiber group can direct the spectral energy by resonance, toward different faces of a spectral simplex (by means of a filtering process on the spectral sets of the fiber group). This structure is the mathematical basis for a local spectral space which is capable of varying its spectral structure, eg a morphogenetic field. It also allows intent to enter the description of existence. It is a simple model of a living system with its own energy and intent.

Thus there are some space-forms which are stable, and they form material, while there are other space-forms which oscillate and generate their own energy. The stable material space-forms depend for their existence on their resonance properties with respect to an energy generating spectral structure. The ones that can generate their own energy can form into living systems (or space-forms). A high dimensional oscillating simplex can control, through resonance, the faces that energy is directed towards. That is, such an energy generating structure can decide which of its faces is to be given the most (resonant) energy of the simplex, and as a result control the space-forms, their currents and motions that are contained within the simplex itself. Thus the energy generating space-forms can control what material structures exist (or are perceived) by controlling the spectral basis of metric spaces through both resonance and the channeling of the oscillating space-forms spectral energy to particular faces of the space-form's "cubical" simplex structure, ie different spectral basis will have different material geometries associated to themselves.

For example, a molecule can have different shapes at different times depending on the spectra that is influencing it. Certain (or some) of an energy generating space-form's faces can be energized, while other such faces are not as energized. One set of spectral properties can be associated to one molecular shape, while the other spectral set can be related to the same molecule having a different shape. The capacity of spectral structure selection exists on higher dimensional simplexes, and can be used as the basis for a (or can be described in terms of a) "local spectral structure," ie a spectral manifold, which can be applied to the spectral sets, and hence it can be applied to describe the different shapes that the same protein has, when it exists within a certain spatial region about a cell. Thus, when one spectral face in the spectral manifold of the cell is energized then a particular protein can take on one shape, while on the other hand, when another spectral face of the cell's spectral simplex is energized then the same protein can take on a different shape, thus allowing the same protein to take on different shapes depending on the faces to which the (protein containing) oscillating spectral system sends its energy. That is, a cell would have influences over such various changeable spectral properties within a certain region about the cell.

Low dimension spherical symmetry

From dimension 1 to 3, the system containing base space has the property of being slightly spherically symmetric. This spherical symmetry influences the local system's properties of interaction. Such spherical symmetry is needed because in order to have dynamics in low dimensions there must exist at least a 2-dimensional hyperbolic space-form to identify motion. The geometric form of such a space-form exists in a 3-dimensional hyperbolic metric space. In order for this to be possible, the base space is formed from the subgroup pieces of the finite dimensional unitary fiber group which have the fewest dimensions. The smallest dimension subgroup, which would have 3-dimensions (the number of dimensions needed to contain the 2-dimensional dynamic hyperbolic space-forms) is the SU(2) subgroup. SU(2) is isomorphic to a 3-sphere. Thus the base space becomes T(SU(2)), the tangent space of a 3-sphere, so that spherical symmetry becomes a property of local interaction properties of physical systems.

Thus spherical symmetry is allowed along with the singularities of the spherical symmetry's equations, in Euclidean space. Note: A spherical metric is a Euclidean metric restricted to a sphere. However, even if singularities exist and have the force to crush the low dimensional space-forms, there will be some dimension at which a space-form is still formed. Such a space-form would contain a gravitational singularity, and it would also determine the shape of space (and the spectral properties of material) within which that singularity, eg a black hole, would interact. Thus high dimensional simplexes are the geometry of dark matter.

It might be noted that simplexes also determine the spectral properties of the vacuum. However, the idea of a vacuum is not needed in this new viewpoint. This is a very good thing since the vacuum has only been a problem, or a limiting idea, or a limiting way in which to use language, in describing physical systems so that those systems can be controlled. The idea of a vacuum actually interferes with the ability to control a physical system which has been described (one has been trying to describe). This is because there cannot exist either a subspace structure or a manifold structure for a vacuum point because a system built around one vacuum point cannot interact with another system built around another vacuum point. The vacuum is a relic of the old paradigm and it is not too relevant to this new, fundamentally geometric way of using language.

The idea of the shape of space, depending on material-energy distributions, determining material dynamics is being altered with these ideas. That light's speed remains constant in an accelerating frame implies that light consistently stays in a space that has a constant and unchanging metric structure. This means that the Schwarzschild solution cannot be correct, and the residual spherical symmetry of space needs to be incorporated into a Euclidean (or spherical) space frame and Euclidean space's property of allowing action at a distance for inertial systems.

Note: This makes more sense in terms of a black hole having a gravitational field, even though the (assumed) gravitons cannot escape its pull. The so called existence of a black hole's gravitational property, ie caused by time dilation, this time dilation is a property of a region of space (inside the black hole) where it is not clear if the mathematical methods that are used to demonstrate that time dilation occurs, actually still have meaning, (or if those methods which describe time dilation still apply in that region of "space").

The precession of Mercury's perihelion can be accounted for in terms of space's higher dimensional space-form structure (see The Authority of Material vs. The Spirit, www.trafford.com/05-3038). Thus, it is space-form structures, predominantly interaction space-form structures and their sectional curvature properties, which are to be used to determine the Ricci curvature tensor, which in turn, determines a general metric, which in turn, determines geodesics and hence material dynamics, in such a generalized Euclidean space or Riemannian manifold.

However, now the shape of space is determined by the geometry of space-forms, and can be used to solve more than the one-body spherically symmetric problem, ie Ric=0, which is solved in the old paradigm in an assumed spherically symmetric space. This one spherically symmetric problem is nearly the entire content of general relativity. However, in the new theory, the spatial and geometric context of both a black hole and "dark matter," can be modeled. Indeed the "event horizon" of a black hole may be the surface of a higher dimensional space-form, which will have it own intrinsic spectral structure, projected down into 3-space.

Do material's properties of relative position require reference frames of coordinate motions relative to the distant stars? Yes. Perhaps a set of material space-forms determine the geometric properties of space, (and that geometry determines how material moves. Namely, material has a shape and it is that shape which causes other space-forms to be formed in the space between the interacting material space-forms, so that the connecting shapes determine a spatial context so as to cause material motions.

Within this space-form context, motion itself is associated with space-form structures. However, the original material shape may be formed as a result of its resonance with a (or set of) containing spectral structure which is also a space-form shape. Are material, space, and time separate constructs or are they the same thing, emerging as relatively different dimensional levels of a "cubical" spectral simplex (or from several "cubical" spectral simplex structures) which (together) determine the spectral properties of what exists. Material and space shapes determine shape and motion in a metric space, but that metric-space is part of a material shape in a higher dimensional metric space or higher dimensional "cubical" simplex, in which the metric space identifies a material (or current) flow with respect to higher dimensional faces, or higher dimensional space-forms.

It appears that the Lie algebra elements transform the simplex structure. General relativity insists on a fixed dimensional structure for inertial material. General relativity seems to confuse the issue of whether material properties determine space, or whether space determines material properties. The theory of material (energy) and space determining each other, ie G=kT, has not proven to be useful. On the other hand these new ideas point to an intermediate viewpoint, namely, that space and material are closely linked in a self similar structure, an idea related to supersymmetry.

The constants of physical law

There are many high dimensional space-forms, and many high dimensional oscillating (or energy generating) space-forms (or simplexes) each with its own unique spectral structure. This means that the constants of physical law can change from one such high dimensional simplex to another. That is, the sun defines a high dimensional, energy generating simplex (or complex of simplexes) which have a dominating influence on the physical systems that are contained within the energy generating "cubical" simplex of the sun, so that for the contained space-form's the sun's spectra affects both their spectra and their geometry. The sun's dominating "cubical" spectral simplex also influences the constants of physical law, which are determined by the sun's high dimensional "cubical" simplex structure so as to affect the differential equations of systems that are contained within its high dimensional, oscillating spectral simplex. It may be assumed that the earth fits into the spectral structure of the sun, though the fit may not be exact.

An example of how physical constants can be related to space-form geometry, note that when a natural value for a time interval might be determined for say, a non-reducible system. This natural time interval can be taken to be the interaction time (or time for a cycle) of such a non-reducible system. For example the average rate of heartbeat of an animal. Then the speed of light can be determined by such a natural time interval for the given system related to the average radius of that system's space-form. For the case in which the space-form for the system is the metric space itself, then the moon might define the outer edge of the planet earth's metric space space-form structure. Hence, the distance from the earth to the moon would define a natural distance in the space-form structure of this metric space existence so that the 1.25 seconds needed for light to get to the moon may be the needed short time interval within which a light signal must be processed for the special interaction (or process) to take place within the subsystem, ie an animal's perception, in the metric space. Perhaps that special process is a heart beat or some mental process or perceptive process. Hence the metric space and the time interval needed to complete a process within a system contained within the metric space, can be used to identify the physical constant, c, the speed of light.

The assumption that the laws of physics are the same throughout the universe is an assumption upon which one should be skeptical, especially if those laws do not predict dark matter. On the other hand these new ideas do predict both higher dimensional space and material (which interacts as a 1/r potential) this higher dimensional structure could be the structure of dark matter. These new ideas give a lot of geometric and spectral structure to dark matter, upon which one can build ideas about how to describe it.

Mental activity

Mental activity can be modeled in terms of a memory structure that exists on a fiber group of a principle fiber bundle. Within a finite dimensional unitary Lie (fiber) group there exist a set of maximal tori, ie a higher dimensional doughnut or Euclidean space-forms, so that the oscillating spectrum of a spectral simplex in the base space can create a resonating spectral set on the maximal torus of the fiber group. This spectral set on the maximal torus is filled with the spectra of virtual material, eg charge, related to discrete isometry subgroups of the fiber group. Such a spectral set on a maximal torus, or one of its subsets, can be used as memories in the sense that the subset can be given more energy through resonance with either the energy generating set which composes a living system (eg memory) or an energy generating simplex outside the living system (eg senses). The spectral set in the fiber group can act as a common medium so as to determine resonance between spectral sets, and it can also determine which part of a spectral set gets the energy of resonance channeled to it by a filtering mechanism. The many worlds just described as a natural part of existence, now has a very accessible quantitative structure through which such worlds can be examined in quantitative detail.

Geometric information, both of one's experience and one's senses, may be stored on a maximal torus as spectral simplexes. Thus an oscillating, energy generating space-form (or simplex) has a memory mechanism associated to itself, found in the fiber group, and this memory mechanism can also be used to single out particular spectral sets. These particular spectral subsets (obtained from memory) can be used to describe both the spectrum and geometry of intent, ie describing how one is able to move one's body. The ability to use certain spectral sets in certain ways, within a context of knowledge and memory, all modeled in terms of space-forms, spectral simplexes, and hence, also modeling: material, space, time, tangent bundles, fiber groups, and differential equations.

Multiple levels to existence

The intent of a life form can determine its own development, especially if the life from has knowledge about both the higher dimensional spectral simplex structure of which it is composed and of which its surrounding is composed. Existence is about an intricate filtering system of high dimensional oscillating simplexes, each determining existence as well as a perception of existence through resonance and interpretation. Existence is a complex of inter-related spectral filtering centers.

In this new theory, the Dirac differential operator can be applied to space-forms and simplexes in order to model both classical and micro-systems, as well as modeling systems beyond these classifications, by means of differential equations and their boundary conditions. Space-form geometry can be used to model crystals, and oscillating 3-dimensional space-forms can be used to model a radioactive nucleus, as well as life.

Note: the standard model does not seem to be able to describe a general nucleus in terms of high speed particle collisions.

Living systems and their memory structures also seem to be modeled by this new way of using discrete isometry subgroups and the fiber group's maximal tori all existing within a principle fiber bundle.

The new mathematical structure has some similarities to both the standard model and string theory. In this new theory the base space of a principle fiber bundle is a dimensional hierarchy of different signature metric spaces and their subsequent discrete group simplexes which together form the domain for differential forms (geometric functions) upon which higher dimensional Dirac operators (or an analogous differential operator) act. The fiber group is (can be) a set of generalized unitary groups, eg $U(n,m)$, and defined on the base space is the measure (or inverting operator) which is an integral (homology) acting on differential forms (co-homology) which is easily defined on space-forms, differential forms which are (usually) solutions to metric invariant differential equations which originate from differential connections.

The standard model as well as string theory have many strong relations to this new set of ideas.

Apparently the standard model can find the precise adjustments to an experimentally determined arbitrary parameter because its mathematical structure is a good an approximation to this logically more consistent new mathematical structure where the harmonic oscillators of the old theory have become the space-forms so that their spectral flows can be occupied by "charges."

Whereas string theory stops at space-time dimension 11, this new theory also stops at space-time dimension 11, but this new theory stops at dimension 11, because this is a property of discrete isometry subgroups of the hyperbolic spaces, ie a generalization of space-times. That is, hyperbolic discrete groups (or reflection groups) do not exist beyond an 11-dimensional space-time metric space, this is a theorem of Coxeter's about hyperbolic reflection groups, ie discrete isometry subgroups. String theory also defines "branes," eg a face on a simplex could be an example of a "brane." The "branes" of string theory can be immersed in higher dimensional spaces, but inside the "brane" the dimensional properties are determined by the "brane," not the higher dimensional containing space.

In the new theory, faces within simplexes also have this property of dimensional independence. (end summary 7)

Modern Science

First it is not clear that people, especially institutional scientists, know what science is. Fundamentally it is seen as the absolute authority, derived from the idea of materialism, ie material defines all of existence, and secondly it is seen as a descriptive language based in mathematics that has data that supports it. This second definition, one might suppose, is the most elemental definition of science, the one most followed by institutional scientists, most notably S. Hawking, but this definition qualifies Ptolemy as a scientist, which would seem to be correct. However, this could create discontent because most people analyze science in terms of its authority, namely, Copernicus being right and Ptolemy being wrong. That is, people interpret science in terms of authority and absolute truths. Science seems to want authority but it does not want to be responsible for its dogma, nor for how science has degenerated into uselessness. The scientist in most cases simply wants their social rank. With its absolute dogma of materialism, science usurps the authority of religion. Science is right, religion is wrong, which would <u>not</u> be true except that religion falters, because religion is itself materialistic rather than spiritualistic. Hence science can be considered to be more honest than religion. If we actually loved (or tolerated) both our neighbor and our enemy, the world might be dominated by democratic socialism as opposed to totalitarianism. Our totalitarian system leaves civilization with Nietzsche's master liars grubbing for control over material at the expense of the slaves.

In fact it is Ptolemy's idea of science, adjusting outlandish dogmas with epicycles in order to obtain corroborating physical evidence, which also dominates the modern idea of science and its overly weak probabilistic description. The result is modern science's has so many similarities to Ptolemy's scientific ways, that it is quite striking.

But the problems of science are really about the limitation of words. Namely, a language's inability to contain all patterns in its descriptive range. Thus when a new observation of a new pattern, that exists in the external world, often this new pattern cannot be incorporated into the fixed language which science is using. This is also why a probability based theory is so enticing, its nature is to set up sufficiently encompassing elementary event spaces in order to use such event spaces to chase data. Understanding Godel's incompleteness theorem results in the acceptance of a need by science to keep changing its basis in words or language.

Though science is about always being willing to alter one's language in order to incorporate the new patterns seen, it is also about trying to determine the "hidden structure of existence," so as to put this new structure (dealing with existence's mysteries) into words so as to be consistent with what is observed. Subsequently, science gains from the development of such new ideas.

This is what the example of Copernicus really means. The ideas of Copernicus led to development, the ideas of Ptolemy did not. Copernicus could not submit his ideas to the intellectual authoritative institution of his day for approval, he was describing a hidden structure that the authorities of his day did not believe. That is, existence is always a mystery open to new ways in which to view and describe its properties, and new viewpoints will always be opposed by authority.

Godel's incompleteness theorem revisited

When one uses language based in ideas about quantity and shape so that the words have definite, limited, meanings, then there will exist patterns which such a language cannot describe. This is the case of the quantity and shape based language of modern physics because some of the patterns of the external world are patterns which the language of modern physics cannot describe, and such cases of language's limitations are what Godel's incompleteness theorem deals with.

For example the pattern of a non-reducible material system, eg perhaps a living system, is a pattern that the language of modern physics cannot now describe because a living system might be non-reducible. Such speculative possibility, eg that living systems are non-reducible systems, can now be described with great detail leading to new measurable properties and experimental activity based within these new ideas.

A new way of using words and language
(summary 8)

The new theory changes the structure of words, and by changing the emphasis of what mathematical patterns are fundamental to the organization of physical description. Instead of letting material define existence, eg how material is modeled (as a scalar), its degrees of freedom (3 or 4), and the properties of its containing space, such as it being an unbounded coordinate space continuum, rather it is the metric spaces of different signature metrics, and of different dimensions, which are associated to particular physical properties. This is motivated by the properties of symmetries and invariance's which <u>exist for the metric spaces</u>, eg displacement . . . momentum . . . inertia . . . . Euclidean space, eg time . . . energy . . . . charge . . . hyperbolic space, etc. <u>and the idea</u>, to look at each metric space's natural geometric shapes or (in hyperbolic space) equivalently examining the metric space's natural spectral properties. Such metric space natural geometric or spectral properties are determined by their discrete isometry subgroup properties or their "cubical" simplicial fundamental domains associated to their discrete group lattices, so that when the fundamental domain is "moded out" it leads to space-form shapes.

The properties of material, space, time, interaction, vector bundles, spectra, and differential equations are all still a part of this basis for description. Essentially what changes the most in physical description is that: (1) metric spaces are associated directly to physical properties including material properties (2) this is done through the use of discrete isometry subgroups or space-forms which also provides micro-particles with a "hidden" geometric structure, (3) the containing space is a dimensional hierarchy of different signature metric spaces and (subsequently) space-forms, and (4) spectral properties of physical systems are no longer dependent on the properties of function spaces and their probabilistic interpretation, instead differential equations are defined on differential forms ie geometric functions, whose domain is the space-form structure of space. Space-forms provide a model for both geometric and spectral properties of material systems.

The implications of the new theory

What results from such a new way of trying to describe the properties of the world? One finds the hidden geometric structure through which one can explore the idea of Einstein that "God does not throw dice." However, there are so many high dimensional, energy generating "cubical" simplexes, whose power over the material existence of a 3-dimensional space is unfathomable, that what God might be, is difficult to decide.

Quantum mechanics can be derived from the natural spectral geometry of space-forms thus the probability patterns of the world are derived from geometry.

It gives the natural non-reductive structures for: nuclei, atoms, molecular shape, crystals, solar system stability, living systems, mental systems, and it provides the structure for a steady state universe which is none-the-less expanding.

It gives the spectral structure of a vacuum, though in this new way of viewing things the vacuum becomes irrelevant, it gives the higher dimensional structure of dark matter, and it gives the higher dimensional geometry of a black hole's singularity. It provides for a new local spectral structure for "new" spectral manifolds, and it gives the "true" mathematical structure for describing geometric and spectral systems in a consistent language. This mathematical structure has similarities to the standard model's mathematical structure. It is these similarities which explains why the standard model leads to its "apparently valid" quantitative properties, but with not much breadth of application. The standard model is not actually doing physical science, where science is defined as using clear mathematical models in a clear, consistent language with which to describe and predict properties of physical systems. For example the mass spectrum of elementary particles form the arbitrary parameters of the standard model "theory." Thus the standard model's main function should be to wonder why it works at all.

On the other hand the new ideas describe the families of particles in terms of the various types of space-forms, eg different signature metric spaces, in a dimensional hierarchy. The new structure of the principle fiber bundle has a set of guiding principles based within properties of space-forms, which logically determine its structure, instead of the mathematical opportunism that dominates the standard model. The vertices of the simplexes, which define space-forms, are all identified with each other in the "moding out" process, which creates the space-form geometries, so that the vertices form into a distinguished point on the space-form. This distinguished point associated to space-forms, is the hidden structure of charged elementary particles. This is because collisions between hyperbolic space-forms focus on (or are led to) this distinguished point, thus they appear to be particles.

It is the stable spectral properties of hyperbolic space-forms that allow this discrete group model of physical systems to work so well, and allow physical systems to have the stability that they do have.

A dimensional hierarchy of different signature metric spaces [and their associated space-forms, which can all be defined on high dimensional "cubical" simplexes] forms into a self similar pattern of material, space, time, interactions, and the spectral properties of stable systems. Within a dimensional hierarchy of different signature metric spaces, and their associated space-forms, metric spaces turn into a spectral flows on higher dimensional material space-forms as one goes up the dimensional ladder.

Differential equations are defined on differential forms which, in turn, are defined on a domain of space-forms and metric spaces whose boundaries are space-forms. Thus the universe, of the new paradigm of physics, is the same universe as the old paradigm of physics, in which material and its properties have been adjusted so as to be within a context created by the spectral-geometric properties of discrete isometry subgroups, which none the less can include a metric space as the base space, so that this would be consistent with the currently accepted paradigm of physics. However, the new paradigm is defined on a more restrictive geometric setting, as opposed to a probabilistic setting of the currently accepted paradigm. The description of the world has been turned around by means of new ways to use language. Furthermore, the axiomatic structure of the mathematics is changed, ie metric spaces are associated to physical properties, ie Platonic forms are changed. For example, Euclidean space is now Euclidean space as well as the space which carries the physical property of inertia.

When material interacts, it is the sectional curvature of a bounded Euclidean space-form [which is defined so that one dimension of the slicing plane (for the sectional curvature) is along the normal line (to the two enclosing spheres of the two interacting material space-forms)]

which connects the two interacting material space-forms

(each of which has been enclosed in a sphere so that a normal direction between the two interacting space-forms is easy to define).

Because material has a geometric structure this model works for each dimension within the dimensional ladder (or containing dimensional hierarchy). This implies that material interactions in each dimensional level are determined by a sectional curvature, of a 1/r potential. Though subsequent force field directions, which result from the potential are dependent on the dimension at which the system exists.

The space-form structure of the metric space within which our experience is contained, is also the structure that allows our planet to remain in a stable orbit. This is because of the self similar structure of the dimensional hierarchy of space-forms and metric spaces, that exists on the base space of the principle fiber bundle, has the property that our metric space is actually (or can be seen as) a spectral flow on a hyperbolic space-form structure (in higher dimensions). The earth's orbit is contained in this spectral flow of a higher dimensional space-form.

Hyperbolic space-forms can be infinite extent, and this provides a new way in which to deal with the idea of infinity in a mathematical way, which is also very geometric.

New ways in which to define what a stable solution to a differential equation is, can be determined.

The renormalization group, that is related to thermodynamic phase changes, is also related to the properties of the system's space-form structure, so that this space-form structure can be used to describe the system's distinct states which identify a jump in space-form structures to be describable in a simple manner, ie a "quantum jump" which changes macroscopic geometry.

A new idea about a "local spectral structure" can be defined and applied to the new idea of a "spectral-geometric manifold" and this can be used to describe both the shapes of molecules, along with embryonic development.

Energy generating space-forms

Some odd dimensional hyperbolic space-forms are spectrally and geometrically stable yet they may have a charge imbalance due to the ridged spectral structure of hyperbolic space-forms, this charge imbalance can cause the space-form to oscillate and hence push opposite metric state spectral flows together, eg matter and anti-matter come in contact with one another thus causing energy to be generated. This is a simple model of a living system.

Such energy generating systems control the spectral structures that are contained within their boundaries, and in a context of equal or higher dimensional simplexes which oscillate, this causes existence to become a competing set of spectral sets which can filter the spectra associated to other spectral centers or spectral simplexes. It is this which causes a shift from a material centered description of the external world to become a description centered on the intent of a living spectral simplex.

Inter-species communication becomes more about the real question of life's creative relation to all of existence, and away from ideas about selfish exploitation.

(end summary 8)

Entire world view
A new paradigm for physics requires that its entire world view be presented

When describing a new paradigm of physics based on new axioms, and new ways of using language, one must present the new language and then one must proceed to describe an entire world structure. It is not enough to present single theorems, instead the world constructed must fit together so well that important technical properties of the world structure must be presented in a convincing manner, quite simply and directly.

Similarities between this new theory and string theory

The new ideas give a geometric model of super symmetry, ie metric spaces (Bosons: Infinite extent space-forms which define metric spaces of particular dimensions within which geometry can be identified) turn into spectral flows on material space-forms (Fermions: Up to dimension six can be associated to compact material space-forms, through atomic system structure, so that this material associated compactness implies a material geometry), as one climbs up the ladder of the dimensional hierarchy of metric spaces.

The new ideas also have many similarities to string theory. The new theory has both 1-strings which represent charges, and n-branes, which in the new theory are faces on higher dimensional "cubical" simplexes of fundamental domains of discrete isometry subgroups, so that these faces have a (similar) property of dimensional independence (a closed topology) as do branes in string theory. The relationship between the different dimensional structures is better organized on "cubical" simplexes than with strings and branes. The new description has a natural cut-off at space-time dimension 11, as does string theory, due to the properties of hyperbolic discrete groups, but the different dimensional levels of the new theory can be related to space-forms (or their simplexes) for both micro- and macro-sized systems. Physical systems are organized around non-reducible spectral structures, namely, the (higher dimensional) space-forms of discrete isometry subgroups.

Note: an 11-dimensional space-time is equivalent to a 10-dimensional hyperbolic space.

Furthermore, complete models of simple living systems and associated mental systems come straight from the new mathematical structure of the new theory.

History of physical science and Nietzsche's ideas about civilization

The history of science has its beginning with rebellious individuals, Copernicus proposing a new way in which to re-order Ptolemy's geometry of the universe, used a new language with which to talk about the universe, as well as a new geometry which he used to reveal the hidden structure of the universe. The solar system's simple properties were modeled mathematically by Kepler. Galileo developed the basics of dynamics and found important supporting data for Copernicus. Newton and Liebnitz synthesized this with a philosophical discussion about the containing coordinates and presenting the differential equation. The differential equation is about the changes of scale of certain functions which represent properties of the system (or formulas which depend on the containing coordinates) which, in turn, are related by equality to the properties of one of the system's tangent bundles whose components are of the correct type.

The degrees of freedom of such a description were decided by experience, though this experience was not the experience of Buddha and his ability to "see" the world's true reality. The equality relations of both derivatives and (linear) quantitative types results in the definition of a differential equations. Differential equations are defined on function spaces, ie the differential operator takes functions to functions. Classical physical systems are characterized by solutions or formulas which depend on coordinates, and give a geometry of the system within the system's containing coordinates. The rules for creating differential equations were called the laws of physics, and the solutions to these differential equations can be used to predict outcomes of experimental set ups. These rules for the most part depend on metric invariance and the geometric properties of differential forms, which are the mathematical basics of Gravity, Electromagnetism and Thermodynamics.

Such knowledge became both practically and economically useful, thus science was institutionalized. Within this context, thermodynamics was developed. Thermodynamics, along with the atomic hypothesis, led to statistical physics. The properties of charge were developed by Faraday and Maxwell. Light had unusual properties related to its velocity and its containing metric space. The quantum properties of both light and material were "discovered," and Einstein stepped in to solidify the atomic hypothesis, organize the probabilistic and quantum behavior of light, and identify that the speed of light can be used to define a particular metric space, thus expanding the discussion of the properties of the containing metric space which Newton and Liebnitz began. However, Einstein's general relativity has not been as productive. General relativity is difficult to understand as well as use, and this has led to an avenue for mathematical opportunism and for obscuring science. Because it has only been applied to the one-body problem, its claims about the nature of material and space are dubious.

In trying to understand the spectral properties of micro- or atomic systems deBroglie proposed the wave idea of matter which allowed headway to be made into a form of an answer, but it was only Einstein who was able to allow this idea of deBroglie's to enter into institutional physics.

The institutionalizing of science has created the master-slave relation, dependent on the interpretation of "ancient" physics texts (and now peer reviewed journals), a socialization process described by Nietzsche for its admitted institutional members. The slave-like members aspire to write an "important" paper which will elevate them to the "top," thus ideas tend to be conservative and narrow, as opposed to science's illustrious beginnings, in which a whole language was discarded, thus leading to new developments in (of) knowledge.

Problems with the currently accepted paradigm for physics

It is with some reluctance that I write paper for a science journal, since being peer reviewed, the servants of the journal will only deal with ideas that already have a history of well known papers behind them.

The ideas of this paper are not conservative ideas they are "Copernican" in their scope, thus not admit-able into institutional physics, but these radical new ideas are now necessary because of the dysfunctional way in which science has developed over the last 70 years. The last 70 years in the physical sciences has been characterized by institutional interests in lasers, radioactivity, particle physics, atomic physics, solid state physics and cosmology. But the institutional theme which has integrated these interests has been the funding of bomb engineering, hence modern physics more than anything else is bomb engineering. The idea behind lasers was presented in 1905 by Einstein, the Schrodinger equation led to an imperfect model of the H-atom which had some limited value in chemistry. Quantum mechanics mostly fails at pin-pointing the causes for molecular shape, and the properties of heavier atoms tend to be different from the properties of the H-atom. Though the H-atom model gives some guidance for the two top rows of the Periodic Table of elements and the three boundary columns. Quantum mechanics works for simple systems whose boundaries could be modeled as infinite potential barriers, and this has led to some understanding of the spectral properties of crystals and to ideas such as tunneling etc, but most of the emphasis for physical science in the last 70 years has been on radioactivity and particle physics; Dirac operators, spin, isospin, unitary transformations, and a sea of harmonic oscillators are used to model all of Euclidean space, but beginning with a single harmonic oscillator's zero occupancy, ie the vacuum, and ending with particle perturbation interactions at the vacuum site, whose calculated values are used to make the value of arbitrary parameters (the laboratory measured masses of elementary particles) precise. This bazaar mathematical model led to some correlation between itself and: particle collision events, beta decay, and some understanding of fusion of elementary nuclei was imperfectly developed, but a model of a general nucleus has not been accomplished. Hence radioactivity of a general radioactive nucleus has yet to be properly described. These ideas about the standard model were applied to superconductivity, but high temperature superconductivity has exceeded the critical temperature of that model and it has never been recovered. This means crystals are not understood with either quantum mechanics or the standard model. In cosmology the red shift of distant galaxies led to the big bang, winning out over the steady state idea of the universe, but this new set of ideas allows a steady state in which galaxies also have the property of their motions causing them to drift apart, but now without needing a "big bang" assumption. In fact, the new theory says that all the galaxies can be modeled as if they are still all together, that is, matter is still at the initial starting point, sort of like an initial singularity, but an explosive model associated to this idea is not needed.

Right now it is claimed that the standard model applies to particle collider physics, and the big bang. It seems to have failed as a physical model otherwise, and the standard model applied to the initial singularity is really death star bomb engineering, hence actively funded.

It is clear a non-reductionist model for physical theory is needed to pull away from the failures of modern institutional physics.

Institutional science has become the authoritative replacement of religion, and that religion is based on the dogma of materialism, ie material determines existence. Namely, there are either 3 or 4 degrees of freedom for material depending on whether the material is inertial or the inertia of charged material, that is, unless the many other degrees of freedom that may exist are curled up into microscopic geometric structures.

What underlies this dogma is the belief that space can only exist if material and its properties exist. The properties of material as seen by the currently accepted paradigm are that it has a scalar structure, it has spectral properties, it is reducible to elementary particles, so that the space that it defines is either a <u>continuum</u>,

> [so that Stoke's theorem can be applied to a point of material which defines spherical symmetry in a higher dimensional space to determine its field, so as to result in a force field of higher degree than (negative) two if higher dimensions were present]

<u>or</u> a space which has a chaotic foam-like structure when material particles are defined over a vacuum. Thus, modern physics insists on material having a fixed set of degrees of freedom and that it reduces to particles, so that microscopic material particles depend on the rules of probability, but then its argument that higher dimensions cannot exist, is suspect.

Another short summary (summary 9)

In reference to the new ideas, instead of letting material define existence, let existence be defined in terms of the different signature metric spaces (which together compose a dimensional hierarchy) being associated to different physical properties, eg Euclidean space is associated with inertia and momentum, while space-time (which is isomorphically equivalent to hyperbolic space) is associated to charge and energy-time. Then consider the spectral properties, or natural geometric properties, of metric spaces ie their discrete isometry subgroups or "cubical" simplexes, or space-forms. Differential forms can be defined on these space-forms, as well as being defined on "cubical" simplexes, and can be used to describe the spectral properties of material systems, as opposed to using the spectral properties of function spaces which are traditionally used to describe the spectral properties of material systems. Thus in the new theory the differential equations which define material systems would be defined on the differential forms whose domain is the set of space-forms that together compose the system.

Dimensional independence of faces

In reference to the new ideas, the dimensions of the faces of a higher dimensional hyperbolic "cubical" simplex can be made independent of the rest of the dimensional structure of the simplex, if each face has within itself (and beyond its bounds) an infinite extent simplex (or space-form) which models light [as well as the structure of material interactions] within the face (or metric space) so that this model of light exactly fills the spatial dimension of the face and extends this space, with its fixed facial dimension, out to infinity.

This property makes the dimensionally independent properties of these faces similar to the dimensionally independent properties of branes which have been immersed in higher dimensional spaces, so that the branes still retain their dimensional properties independent of the space within which the branes are immersed.

Surprisingly such a shift in viewpoint allows one to examine the idea of Einstein's statement that "God does not play dice," in quite a bit of detail. This is because this new context resolves the probability-geometry dichotomy that exists in the currently accepted paradigm of physics, in favor of geometry. However, there are so many high dimensional, energy generating "cubical" simplexes, whose power over the material existence of a 3-dimensional space is unfathomable, that what God might be, is difficult to decide.

That is, a metric space's space-forms determine the primitive geometry of that metric space. Space-forms can be used to model material, space, time and interactions. Interactions depend on the sectional curvatures of the (if needed higher dimensional) Euclidean space-forms that connect two (or more) interacting material space-forms. The interactions between microscopic atomic space-forms result in Brownian motion. Brownian motion can be modeled mathematically as a stochastic process, and it is the properties of the stochastic dynamics of microscopic material which E. Nelson used to re-derive quantum mechanics. Thus geometry is primary and probability is derived.

19th century physics vs. 20th century physics

The reason science has become so dogmatic has to do with how it is used in society, as well as the way it has become dependent on a fixed language based on the truth of "Platonic forms," and a language which also suffers from its own built-in probability-geometry inconsistency. Geometry is the basis for a very useful science up to the 19th century. Probability is the basis for the description of material system's particulate and spectral properties. It is believed that material can always be reduced to elementary particles, so that their spatial event properties are defined by probability. Thus, material is reducible to particles and it can have no hidden structure. In the probability based language of physics, the differential wave equations which determine the spectral properties of any particular system has not been that easy to come by,

> [with the exception of a few systems (1) infinite (wall) potentials, and (2) the H-atom (though the radial part of the solution of the H-atom's differential equation mathematically diverges, none-the-less its subjective truncation has led to the much heralded success of quantum mechanics), and (3) Einstein's 1905 laser model]

and when the "correct" form for the wave equation is found, eg for particle physics, they are so complicated that they have a very narrow breadth of application. For example, the standard model seems to only apply to particles in particle accelerators and the religion of the "big bang." The big bang is not science since it cannot be scientifically confirmed that the explanation of the spectral properties of distant galaxies actually do have the properties that it is claimed that they do have. Is the spectrum of these distant galaxies really what we see? Are they moving? Are they still and space expanding? On the other hand, the semi-empirical, Functional Density Method for finding ground states of quantum systems seems to be the natural way to apply the probability structure of quantum mechanical operators.

As for the standard model, it has not resulted in a model of a general nucleus. Thus, since it is believed that the nucleus is composed of violently colliding particles, if the standard model cannot be used to develop the theory of a general nucleus, what can it be used for?

It certainly has not been able to predict the mass of the Higg's particle, in fact arbitrary parameters, ie the experimentally determined masses of elementary particles, used when "calculating" the elementary particle's mass, seem to indicate that something about the mathematical structure of the standard model is correct, but as a scientific theory the standard model is sort of "a nothing." This is because it is supposed to be a theory about the spectral properties of elementary particles, ie their spectral masses, but it can only precisely get the perturbations of a particle's mass. This is interesting, but it does not seem to lead to anything useful.

The derivative process and physical science

The standard model is sort of an extreme example of the differentiation process, whereas a function's derivative, when placed in a context of tangent bundles, results in differential equations whose solutions model properties of systems, on the other hand, a vacuum point, defined by a Euclidean harmonic oscillator, is modeled as an isospin vector, acted on a by a measure, which counts the particle occupancy of the harmonic oscillator, as the particle types appear and disappear. So that the measure is a group element found by exponentiating a Poincare invariant Dirac-like differential operator (whose representation depends on the particle being modeled) acting on isospin functions, which have a unitary gauge group, ie the unitary Lie algebra elements which model the field interactions between particles, which is applied to material particles by one interaction term in the Dirac-like differential operator, so that all properties of the universe, applied to a particular system's wave function, defined on a sea of harmonic oscillators which fill Euclidean space, can be modeled at the system's vacuum point.

This also means that the system can have no subsystem, or manifold structure, so that the context of the universe being modeled is contained in either an isospin wave function defined on the sea of harmonic oscillators which fill all of Euclidean space, or the gauge elements which interact with the system's isospin vectors. How this is done is not clear, and the whys and the wherefores of this extremely eclectic mathematical structure, just described, seems to be more related to mathematical opportunism than it has to do with what words mean. Furthermore, the soft results that are obtained, or the limitations of its applicability, reflect this weakness. Thus modern science has come to have the same structure of the physical description and verification as Ptolemy's system, where the epicycle structures are whatever mathematical opportunity presents itself in the process of fitting data.

None-the-less when this context is altered by introducing a containing space which is a high dimensional "cubical" simplex, the interactions and non-reducible system relations can be made clear.

Further problems with modern physics

It was believed the standard model could be used to describe superconductivity, but high temperature superconductivity exceeded the theory's critical temperature and it has never been recovered. As microchip components shrink to the point where classical physics needs to be abandoned, in order to design such ever smaller microchips, unfortunately the standard model is providing no model as to how electrical currents in crystals interact with their electromagnetic surroundings, and subsystem structures simply do not exist in the language of the standard model. Penrose is trying to incorporate the standard model into his model of a mind but the coherence of an extended system that can be obtained at a single point depends on the annihilation and creation operator wave that is defined on the set of harmonic oscillators which fill all of space, and/or depends on the material-field interactions that are modeled by the Lie fiber group and the Dirac-differential-operator-matrix related material vector. A correlation of such disparate and independent structures seems to be more a function of luck, if it ever really exists, rather than due to a validity of description.

The sun was supposed to generate energy based on the standard model, but the neutrinos are not present in the predicted amounts, the experimental data was collected and it was hypothesized that if the neutrinos have a small mass then they can oscillate between each other so as to fit the data. Note: probability theories always chase data, they can only predict if "the correct" operators can be guessed in advance, but this is too difficult so, in effect, they never predict anything. However, if the neutrinos oscillate between neutrino states amongst themselves, ie between the different families of leptons, then more interactions amongst these other families should be present, ie more Bosons associated to the other particle families would be present, and so by Einstein's laser (Boson) formula more of these other family Bosons should also be detectable. Are they?

The same molecule can have different shapes at different times. Is there a general model of molecular shape? No, the probabilistic theory is to cumbersome and too dependent on data to be able to be of much use in describing such systems. Apparently living systems use the properties of different shaped molecules with precision. How can such systems be described with the same type of efficient precision?

The most definitive statement that modern physics now makes is that, "Virtually all physical systems of practical interest, are too complicated, and hence, too difficult to describe," too difficult to describe, at least in terms of mathematical patterns which allow predictive control.

The epicycle technique of mathematical opportunism applied to particle and other system behaviors has led to a descriptive wasteland for modern physical science.

Then there is the matter of the stability of the solar system. Geological evidence suggests that the solar system has been stable for at least a billion years, but all the differential equations which model the solar system are non-linear equations hence the solar system should be chaotic, but the evidence suggests otherwise. Why is the solar system so stable? This is a question for which physics again has no answers.

Consider general relativity. When light is emitted from a star with a large mass the gravitational field causes the light to red shift, ie no change its velocity. This means that the metric for space-time consistently keeps its (flat) electromagnetism space-time metric structure. This means that The Schwarzschild's solution does not make any sense. Thus, general curvature, such as spherical symmetry, must be modeled on Euclidean space, or more generally on a Riemannian manifold, plus time.

It is claimed that in general relativity that space determines material dynamic properties but in order for this to be credible more than a 1-body problem, ie Ric=0, needs to be solved so as to provide useful results.

In modern physics, energy emerges either from the systems composed of matter or from the vacuum or the big bang. Life generates (or uses) its own energy, which is used by mental activity, and so does the radioactive nucleus, as do the stars. A radioactive nucleus generates energy, but without a model of a general nucleus there is no convincing explanation (or description) of why some nuclei generate their own energy. It has already been shown that it might be a good idea to be skeptical of the current explanation of how a star generates its own energy, one should also be somewhat skeptical of neutrino theory.

On the other hand, if an alternative model of material existence had a mathematical structure that was similar to the standard model's math structure then it would explain why the standard model can approximate some measurable consequences about the particle mass spectrum that it does approximate so well. Perhaps the standard model's math is close, but not correct, and perhaps the idea of reductionism is also wrong.

It should be noted that this paper presents the fundamental principles of a new idea about how to describe the properties of physical systems and the way these principles can be applied to both create a context of physical description and then to give actual examples of actual physical descriptions which lend to one's understanding. These new ideas provide a description which is verifiable, and some of the descriptions deal with very complicated physical systems, ie systems which no one has known how to describe.

Not for destructive purposes

These new ideas cannot be used for selfish or destructive purposes. This includes both systems of destruction, as well as the destructiveness of holding science up as a discipline which only a few "high caliber" people can master. That is, no destructive weapons and no brow beating based on special knowledge. Knowledge is about man's relation to infinity, not about hoarding money and material for power, in the material world of "civilization."

It would be in the best interest of physics to publish this paper in a widely read physics journal.

CHAPTER 2

REVIEW I

Summary of the language of physics and math and their relation to an alternative descriptive math language based on metric-invariant discrete-groups and many-dimensions.

Classical physics is simple, the laws of classical physics instruct one as to how a differential equation of a physical system can be determined. The (position and motion) properties of (or within) a classical physical system (or a material object) can be determined from the solution functions of the system's differential equation which identifies its (or its components') inertial properties so that these inertial (or position and motion properties of objects or components) are associated to the system's geometric field properties.

The more complicated math techniques applied to classically defined systems such as function space techniques applied to electromagnetic or sound wave equations (properties) and to fixed types of ridged charge distributions (or geometries), eg polar, quadra-poles, and multi-polar charge distributions . . . , used in order to solve "difficult to solve" classical differential equations or their subsequent boundary (or initial) conditions . . . , work very well when such techniques stay within the classical context and within the limited (well defined) properties of the differential equations, and when the properties being described are physical.

Furthermore, when the classical differential equations are linear, or both linear and (geometrically) separable then the solutions are stable, definitive, and very controllable, where control is effected by means of controlling the system's initial and/or boundary conditions.

The relation of "simplex geometries to the structure of differential-forms" results in the existence of a simple process which leads from defining the differential equations of fields through the exterior derivative applied to differential-forms to solving classical field equations (when these equations are linear and separable) within the context of the fundamental theorem of calculus (the integral of a function defined on a region in space is equal to its integral on the region's boundary). The idea of inertia can also be related to the geometry of cubical simplexes, eg the edges of the cubes, (and both field and inertial properties can be related to a new space-form construction which can also be related to energy and interactions) are all math structures which are consistent with the "cubical" simplex structure which can be used as a model for both material and space, as well as material's force fields, where both material, its force fields, and space can all be modeled as space-forms, where the material space-forms and the spatial space-forms and (the relatively newly defined) interaction space-forms are all related to different dimensions, so that these geometric structures as well as boundary (or containment, or resonance) conditions, are related to <u>both</u>, (1) (primarily) the spectral properties of higher dimensional space-forms, or as space-forms in a multi-dimensional context, so that this multi-dimensional space-form model of physical systems, is (can be made) consistent with both classical physics as well as the observed properties of quantum systems, <u>and</u> to (2) the geometric properties of the local coordinate transformations of the fiber group of the metric-space (or base space of a principle fiber bundle, a math construct which works well for defining, and refining, a physical system's differential equation as well as its geometric properties which are related to "cubical" simplexes).

The problem with classical law is that (1) it is not consistent with a bounded charged dynamic (micro) system, eg the bounding forces (for such a bounded system) must cause the charged system to dissipate energy (by emitting electromagnetic radiation), and (2) the vast majority of classical differential equations for classical systems are non-linear and thus they are equations which are, for the most part, non-solvable, and their subsequent numeric solutions are chaotic and uncontrollable, and thus have very limited descriptive value.

That is, physical systems which are characterized (or defined by) non-linear differential equations provide information which is very limited in regard to such a description's usefulness, in terms of control over the physical system which is being described. That is, the Newtonian equations for the solar system are non-linear, and this suggests a chaotic solar system, yet apparently the solar system has been stable for billions of years. Note: General relativity is essentially a non-linear description and thus does nothing to resolve the problem of understanding solar-system stability, one of the main (fundamental) problems of physics.

On the other hand, the (relatively) new descriptive languages of quantum physics, particle physics, (both of which are linear) and general relativity (which is non-linear) . . . , where (all of these descriptions) are subsequently related to string-theory and other types of abstract "physical" theories through their principles (or laws, eg Feynman diagrams of particle-state transformations) . . . , are creating a language structure which is more general . . . , ie claims to be relevant to a wider range of physical systems so as to be able to incorporate more details of each system in the description . . . , than . . . (it appears, or it seems that) . . . the language of classical physics is capable of describing such details (as these (relatively) new languages are trying to describe).

It is well known that the math techniques of classical physics cannot be used to describe quantum systems. It appears that the interactions and motions of quantum systems appear to be fundamentally random, but to assume this random property of microscopic material is (as being) fundamental, results in releasing from the mathematical structure of description "the capacity to describe the (more) fundamental stable, definitive structural properties of micro-physical systems" where the stable, definitive spectral properties of quantum systems are fundamental to the properties of quantum systems.

That is, a description of a quantum system must model this very stable and definitive set of spectral properties so that at the same time this relatively-new descriptive language (about one-hundred years old) is based on randomness. However, more fundamental than their random properties, quantum systems are also extremely stable physical systems which have definitive spectral properties (along with being fundamentally random). This balancing act between the two extremes of randomness and stable definiteness has not succeeded, the stable definitive structures of general quantum systems go without verifiable (or measurable) descriptions. The spectral function structure of function spaces has not resolved this dilemma, the spectral properties of quantum systems are not being found by operating on function spaces with sets of operators, which are supposed to resolve a set of spectral functions for a function space.

Furthermore, some of the properties of these languages of physical description are inconsistent with the math structure of the most useful descriptive language of physical systems, namely the language of classical physics. A glaring example of this is the "uncertainty principle" of quantum physics, which is not consistent with the assumed continuity of classical physical properties defined in space and time in regard to the definition of a classical system's differential equation. This "uncertainty principle" might be based on a mathematical structure which does not allow for a proper (or correct) descriptive language which would allow a physicist to describe both the fundamental stable, definitive spectral properties of very small physical systems (or quantum systems) as well as its random properties (this is what the function space model of quantum systems is trying to do).

That is, describing stable, definitive spectral properties of general physical quantum-systems based on random processes has not yet been accomplished.

For example, (1) the high atomic number, of say 5 or greater, atoms go without adequate description, (2) the general nuclei of say the helium atom (and higher atomic numbers) all go without any description based on fundamental physical laws of particle physics, (3) furthermore, crystals go without adequate descriptions, eg BCS predicts a critical temperature which has been exceeded by the new high temperature superconductors.

Furthermore, macroscopically, the stability of the solar system, apparently for billions of years, has yet to be described based on physical law, either classical or general relativistic.

At every size scale, except the size scale of classical physics, (both quantum and astronomical systems) the vast majority of the fundamental systems within these size scales go without adequate (or without useful) description.

The descriptive language of general relativity is geometric and (like classical physics) it describes properties which are assumed to be continuous, but it is defined over a greater (more general) set of "system containing" frames (or moving coordinate systems) which are transformed by a diffeomorphism (locally invertible) fiber group (of local coordinate transformations (transformed either between frames or as dynamic coordinate changes)), as opposed to the more restrictive classical isometry (both locally invertible and metric invariant) classical fiber group for local coordinate transformations.

However, this construction (or set of physical laws), which is supposed to be capable of describing a wider set of very detailed systems, ie quantum physics, particle physics and general relativity, has not lived up to its promise. In particular, general relativity leads to a set of non-linear differential equations, which means that the detail (which the "new" language is "supposedly" capable of describing) is not resolvable, and subsequently systems more general than . . . , one-body systems which can only have spherical symmetry . . . , are indescribable (are not describable) in the "more general" mathematical descriptive language of general relativity.

The descriptive language of quantum physics is based on the random particle-spectral events (in space and time) of quantum systems, where the spectral (aspects of the) events are contained in (or define) an elementary event space, and (so that) the elementary event space is modeled as the eigenvalues of a (very general) function space (which can be used to model an extremely large set of: either geometries, or waves, or elementary spectral events), so that the properties of the function space are constrained to be related to the measurable (or observed) properties of the (quantum) system, where these measurable properties are in turn, modeled (ie modeled on the function space) by (as) operators which act on the function space of the quantum system, where sets (complete sets) of (commuting) operators (which represent measurable properties) model the (complete set of the) stable definitive system's measurable properties, and subsequently this math structure is supposed to identify the complete set of observable (measurable) stable definitive events in a quantum system's elementary event space, and this is supposed to be done by identifying a subset of spectral functions, whose energy eigenvalues are supposed to exactly match the energy eigenvalues of the observed quantum system which is being described with these mathematical methods, but this has not been accomplished in general.

That is, this ideal has only been accomplished, in an approximate way, for one actual (observable) physical quantum system, namely the H-atom.

The spectral functions found by the above methods of quantum physics are to be adjusted by placing them in a perturbation structure determined by the Lie algebra valued 1-form applied to this function through a measure mediated by the changes of particle-state caused by the Lie algebra particle-state transformations during particle collisions which leave the wave function invariant because they are unitary Lie algebras, where unitary invariance identifies the energy invariance of the quantum description of the quantum wave-function (or the quantum wave-equation, or the solutions of the quantum wave-equation).

Thus, we have the questions:
Field particles colliding with material particles whose particle-state is changing results in what forces, and how are these forces related to a physical system in a general manner?
When do charged particles emit other (or electromagnetic) field particles, eg photons?
Answer: Only during collisions? or Are they emitted always in random directions?
If the charged particles are emitting field particles always in random directions, thus identifying spherical symmetric force-fields for all elementary particles in all dimensions, then this would always cause random changes in the charged particle's kinetic energy, as well as always randomly changing its momentum.
Changes are resulting from "what set of forces?" ie random particle collisions which have no meaningful direction and no interaction structure other than that of random collisions between different types of particles whose dynamical directions are always random? and
To which wave-system do these random forces apply?
The random set of the global wave-functions which are related to (or apply to) a single point in space?

That is, the mathematical construction of forces applying to a quantum wave-function (or quantum wave-system (or quantum wave-equation)) do not have enough structure to have any meaning, or any geometry, and they describe only random interaction processes defined on a random set of descriptive functions. It describes nothing which has order, and it is not clear what the order of the usual (classical) mechanics of a particle collision would be for elementary particles. Thus, the orbital properties of nuclei and atoms etc cannot be determined through the math techniques of either quantum physics or particle physics.

nor can the orbital-spectral properties of quantum systems be determined from the math structures of particle physics.

Random particle collisions associated to random particle-state changes induced by Lie algebra transformations during collisions so as to identify quantum interactions, without any valid definition of a quantum system context, ie no valid wave-function for the system to begin with, so that random collision interactions are applied to an unknown global wave-function (because it is global, and it is unknown) which is defined only at a single point in space, and this point is applicable to a whole set of random wave-functions, and thus this is a descriptive language for quantum systems which has little prospects of describing such a quantum system that has any order of its own, nor is it a description of a perturbation process for a system with some already identified order since it is not clear what global wave function is defined at a solitary point. This leads one to ask "What is the correct order (or correct wave-function)?" so as to subsequently have the correct order (determined from the perturbations). Answering such a question seems to not be a possibility within this type of descriptive language (of both particle physics and quantum physics).

Complex coordinates, algebra, algebraic geometry, and unitary related geometries are used in an ad hoc manner within an imaginary particle-state, particle collision representation of quantum interactions within an external context which has no discernable context, ie at a single point (where) any wave-function might be represented. That is, the perturbed quantum wave-function is being described within a random context as to what the wave-function might be, yet it is a space whose context is that of detectable particle collision patterns as observed within particle accelerators. Within this context, "holes in the Feynman diagrams of these particle collisions" and detectable particle mass and life-time are related to (pieced together so as to form) extremely small geometric patterns (in the hidden space of particle existence) which are "curled-up" so that the (mostly) unstable particles can be represented as vibrating 1-dimensional geometric-curves within this hidden space, ie piecing together an assumed materialism (the assumed God (of physical science)) on a mentally imaginary (or hidden) space of complex-vectors of particle families and particle-states so that these geometries (Feynman diagrams) and constrained vibrations are consistent with the patterns (upon which it is claimed material existence depends) of collisions of field particles, where particles are represented as complex vectors transformed by the fiber group, U(1) x SU(2) x SU(3), as well as such equations (or solution wave-functions) being consistent with rotational symmetry in either 3-space and time or in 4-dimensional space-time, ie the assumed spherical symmetry of the Lagrangian density operator.

But it is a model only consistent with probabilities (or cross-sections) of particle collisions in a system going from order to disorder, eg a violent nuclear reaction, which is based on the rates of random particle collisions of particular (but random) particle types, within the context of a random but global wave-functions (of an assumed random set of ordered quantum systems) to be determined at the point-like site of the particle collisions.

But the collisions from random directions of random types of particles from random particle families related to random wave-functions cannot (has not) been put together so as to describe stable definitive orbital properties of observed quantum systems, neither nuclei, nor atoms, nor crystals.

These ideas have no practical value except within the model of an energy releasing nuclear reactions that take place within a bomb.

Important social comment

The ideas of quantum physics, particle physics, and general relativity, only identify a theology of material deism, whose existence is unseen except by the expert physicists who accept the dogmatic descriptive language which describes nothing of any practical value except within the destructive context of an exploding bomb, ie a God whose only influence is to terrorize others into learning modern physics which has essentially become useless on any practical level other than (if even there) bomb engineering. That is, it is all about "authoritative dominance" and the relation that such dominance has to violence and coercion. This has the same structure as the authority of authoritative dogmatic truth in the time of Copernicus. Today's (2010) authoritative dogmatic physics has the same value in truth as authoritative dogmatic religion, because it is only a Platonic truth (a truth based on accepting a set of assumptions) whose claim to experimental verification is exactly analogous to Ptolemy's claim to experimental verification of his model of an assumed material existence.

This religion of materialism is based on random material particle collisions and an assumed spherical symmetry associated to random (field-)particle emissions from a reaction's (or its material) center.

It is a descriptive language which is irrelevant to the development of useful knowledge about existence. That is, it is irrelevant to knowledge which can allow wider ranges of human creativity, than the idea of dominance, its authoritative dogmas and the threat of violent destruction.

Based on random material-spectral events in space and time, and spherical symmetry which is defined in only one potential energy function which has any basis in reality, along with unitary invariance related to either energy conservation or to random particle-state changes when particles randomly collide, has been the basis for a religion which has been built, where the main focus of the religion is the actual authorities (the persons, the personalities) who are claimed to have the proper "intelligence" to discern truth for the rest of us in society, but really these highly intelligent people (who discern truth for the rest of us) are tricked by the military industry, and thus in reality these great intellects are people whose personality traits are (mostly) aggressive, competitive, yet obedient (personalities). The "great physicists" are people who covet their social position of being the intellectually dominant people upon the earth.

But "intelligence" needs to identify a real truth. Unfortunately, in quantum physics the claim about truth, is that a function space represents both all random possibilities for the quantum system and it represents all stable, definitive measurable eigenvalues of the quantum system and both these possibilities and measurable properties have an order based on measurable properties such as motion, and the geometry of a potential energy function, as well as the random particle collisions, and algebraic properties which are represented as differential and symmetric (or invariant) operators which act on the quantum system's function space. Nonetheless this mathematical process (or mathematical model) does not work for general quantum systems.

The truth identified by the intellectually dominant people of society seems to not be useful, its main use is similar to the use of religious truths, it is a dogma which keeps them in their social positions. Measurable properties represented as differential operators and symmetries are not sufficient (in order) to determine the order on the assumed random structure of (micro) material spectral-particle events in space. The function spaces and the "quantum mathematical processes" are not capable of the describing the observed order which is possessed by the quantum system. The observed order of physical quantum systems is more sophisticated than the invariant properties of geometric operators within a metric-space with too little structure, where that limited descriptive structure is based on a model of "materialism of random colliding elementary particles." The model of material properties being determined by random colliding elementary particles has too little structure to be able to describe the observed stable, definitive spectral properties of quantum systems.

The language the, institutionally appointed geniuses of physics and math, use does not work, and they (the geniuses) are too "stupid," or too obedient, to figure out this fact for themselves.

Furthermore they guard their authority with great ferocity, always assuming that any other articulated description of existence to be something coming from an inferior source of "intelligence." If intelligence is the ability to discern truth, then the "intelligence" that really exists within our society is an ability to discern authority, and to be subservient and obedient to that authority.

Only Einstein considered ideas from other people during the 20th century, all the other physicists (to this day) are "too important" and "too authoritative" to consider ideas from other people who do not already agree with their dogmas (this is the main point of the peer review process, the maintenance of dogma), ie the thing the scientists most covet is their (intellectual) dominance, which they are allowed to possess through the power of the truly dominant military-business interests of society.

The truth that these "divine geniuses" espouse is their authoritative theology based on . . . :

(1) materialism,
(2) the fundamental randomness of material existence confined to space-time (or to space and time), and
(3) the spherical symmetry of space caused by the spherical symmetry of the random process of material emitting field particles,

. . . . and this truth is neither widely applicable (at a practical level) and it is not useful for describing any of the many fundamental physical systems which go without a useful description.

It is authoritative and irrelevant to everything except bomb engineering.

That is, technology is still driven by classical physics and the kitchen or foundry science of chemistry, the science of physics has nothing useful to say in regard to practical technical development, it can only describe initial gravitational singularities of a 1-body spherically symmetric system, and the random properties of particle collisions, the two main properties related to bomb engineering.

These intellectually dominant geniuses are trapped in a bag from which they are incapable of thinking their way out.

These people are supposed to be the most highly educated group of people in society and they should the greatest proponents of the Socratic method and the wide-open set of possibilities free inquiry might be able to uncover, but instead they are the most dogmatic and authoritative set of people in society. They are sure that they have proved their intellectual superiority over everyone else by means of the successfully competing within their overly authoritative educational institutions, institutions which narrow their range of considerations at the behest of selfish commercial interests. They are more authoritatively dogmatic than the ancient religious authorities of the past, and this has resulted in a narrowing of technical development because a narrow dogmatic view of knowledge also narrows the range of creative possibilities, and commercial interests want to be in control of all creative efforts within society.

It is a natural property of language to have limitations built in to its range of possible description, where these limitations are related to the limits of description associated with any fixed set of assumptions and definitions, so that if the language stays fixed and dogmatic, and people confine their thinking to remain within such a fixed, (eventually) useless language, then (they) that language will begin to be able to describe only illusions (as opposed to useful descriptions), and upon these illusions a theology (based on absolute truths) and personality cult (the cult of the high intellect) can be built, and this is what has happened.

In physical descriptions of quantum systems physical interactions are not determined by large scale (large size) geometric properties but rather by random collisions of a set of both stable and unstable particles so that large scale geometric properties are (supposed to be) recovered by (through) [or related to] high energy (or large size) averages of large numbers of quantum particles (or quantum subsystems), where geometric properties are again supposed to dominate the descriptive properties.

However, it is unfortunate, but one can define a fundamental problem with physical description, "that quantum description cannot describe the stable definitive spectral properties of fundamental quantum systems such as atomic systems with high atomic number, and general relativity cannot describe anything but a 1-body system with spherical symmetry."

Thus the descriptions of random particle collisions under extreme spherical symmetry (systems associated to gravitational singularities) are only related to a model of an exploding bomb, and the peer reviewed ideas of science (of physics) have little to do with being skeptical of authority and challenging the authoritative truth of military engineering, and instead of adhering to peer-review, to instead try to find a useful widely applicable truth, which goes beyond bomb and military engineering.

That is, physics is basically bomb engineering, the subject of most significance to the military businesses in the US.

If one analyzes its descriptive structures, ie its descriptive language, and its descriptive capacities to be able to describe fundamental physical systems then one cannot take modern physics (2010) seriously as "real" science. Where "real" science should be honestly trying to describe the observed properties of physical systems.

This means that true science should be about suggesting (or providing) new alternative descriptive language for physical systems which can be successfully be applied to the properties of physical systems which have yet to be (have been) described in a useful way.

Complicated abstractness does not make for a valid descriptive language. That is, both mathematical truths and religious truths are Platonic truths, ie truth based on agreement about what words mean and what is true, that is, the agreement about the truth of the axioms or assumptions upon which a descriptive language is built. Descriptions based in a mathematical language should be measurably verifiable, yet Ptolemy's descriptions where mathematical and were also measurably verified, in a similar way in which today's descriptions based on physics are measurably verified, but they have a very limited range of application and they are related to only a small set of technical innovations when compared with the set of technical developments directly associated to the descriptions obtained from classical physics. One can very reasonably contend that it is the practical usefulness of abstract (or measurable) descriptions which give the descriptive language its relation to truth, not its abstractness nor its complication nor its relation to Platonic truth, since Platonic truths can in a practical sense be very much related to illusions.

For example, consider the question: How many angels fit on the tip of a pin? Although this might be an interesting challenge to understand within the Platonic truth of a religious dogma, but it has yet to be associated to a practical widely applicable truth, other than the truth that "By means of violence and coercion people can be forced to agree with the dogmas of Platonic truths." Similarly, in mathematics, the questions which come from considerations of the Platonic truths (of mathematics) might have interest within the realm of the authoritative dogma of a mathematical truth, but this does not mean that these complicated abstract Platonic truths have any practical value. Thus, Platonic mathematical truths can be as limited in their usefulness as is the language of religion, which is also based on the dogmas of authoritative (or absolute) truths, or truth based on assumption.

This problem characterizes all descriptive truths. Language has limitations as to its range of description and its descriptions should be believed only in so far as the description is practically useful and widely applicable, ie useful in regard to creativity or in regard to creative acts, eg either in regard to controlling material (or in regard to creating beyond the material realm, eg in higher dimensions that the dimensions identified by material (or by a belief in materialism)).

Back to physical description

To go on with a much more interesting discussion of physical science than can be obtained from our overly dominated media and education systems.

Should (such a more interesting) physical description capture the general geometric properties of physical systems within the coordinate space, ie general relativity, or should function spaces be used to model the random, yet definitive spectral properties of material (quantum) systems?

Are there other mathematical models (or descriptions) of existence in general, where the properties of the described existence are about both stable material geometries, such as stable planetary orbits, as well as descriptions of the stable, definitive spectral-geometric structures of quantum systems and a description within which the apparent randomness of quantum systems can also be accounted?

Answer: Yes, there are other mathematical models or other ways to organize a descriptive language in a mathematical manner which is capable of describing both stable planetary orbits as well as stable atomic orbital-spectral structures, and stable nuclear orbital-spectral structures. This new descriptive mathematical language has the structure of a principle fiber bundle which has a base space and a fiber group, the base space is the system containing coordinate space which has geometric properties which can be described based on metric-invariance (or Hermitian invariance) and the fiber groups are finite dimensional isometry (or unitary) Lie groups, ie a mathematical context where linear differential equations are both solvable and useful, ie controllable. Furthermore the base space can have many dimensional levels, it can be based on real numbers or complex numbers, the metric-functions can have many different metric-function signatures, and both material and space can both be modeled as discrete isometry (or unitary) groups (or space-forms). The domain spaces of function spaces can be defined to conform to this new domain space based on space-forms, thus limiting the possible things a function space can describe (or which it is capable of describing), but perhaps most importantly, the description can be made to conform to the stable definitive spectral-orbital properties of quantum systems. That is, the spectral properties of quantum systems which are observed can be described in a more useful way in this new descriptive language. The apparent randomness of quantum systems can also be accounted for in the new descriptive structure by (still) allowing a function space to be defined on the structure of the new domain space, ie the same domain space for each of the functions in the function space.

One could say that the idea of describing material or energetic (or other) existence in a context of:

(1) metric invariance, and
(2) many dimensions, and
(3) many signature metric-functions,
(4) where the coordinates can be either real numbers or complex numbers, and
(5) there exist pairs of metric-space-states for each dimensional metric-space,

.... is a statement about (that) the geometric domain spaces upon which either differential equations for physical systems are defined (eg classical, or wave equations, or space-form equations), or it is the domain space upon which functions in a function space are defined . . . ,

> (6) ... so that this base space in both cases has the properties of (or is controlled by) metric-invariance as well as being modeled as the discrete isometry subgroups of the isometry fiber groups.... (or more generally, modeled as discrete unitary subgroups, in the complex-coordinate context where pairs of metric-space-states are defined and identified, ie a Hermitian invariant context [or the context of unitary fiber groups]), ... and their subsequent space-form geometric structures, so that ....
> (7) .... each (individual) dimension metric-space structure is part of (or contained in) a ridged higher dimensional space-form structure, ie the metric-space is a flow on a higher dimensional space-form, at least up to a 10-dimensional hyperbolic (or Euclidean) space.

It should be noted that either diffeomorphism geometry or overly simplified flat geometry, upon which (quantum) function spaces are defined, are not useful mathematical structures. These simple (fairly un-useful) metric-space structures associated to the currently accepted dogma of physical description are either spherically symmetric, with only one-body in a fixed lower dimension (3-dimensional Euclidean) manifold (diffeomorphism), or they are the simple metric-invariant (and flat and extend out to infinity) metric-spaces within which classical physics is defined.

It should be noted that a stable quantum system . . . .

> (no other quantum systems are actually defined [where quantum physics is based on finding the spectral functions of a quantum system's function space])

.... is (also) invariant to unitary transformations, where invariance to unitary transformations is associated to the idea of conservation of energy, (and the domain space for all the functions in the function space is metric invariant).

The domain space for the Schrodinger equation is 3-space and time, where 3-space is Euclidean 3-space, and it has an isometry group of SO(3), while the domain space for the Dirac equation is space-time, which has an isometry group of,
SO(3,1) = SL(2,C) = SU(2) x i SU(2) = SO(3) x i SO(3). (Note: The = sign denotes equivalences)

It needs to be noted that the wave properties, or sets of elementary events, and/or the geometric properties of "general quantum systems" cannot be described within the descriptive language of quantum physics. Only the H-atom has an approximately valid description within the language of quantum physics, otherwise quantum physics fails to describe in a valid (or in a useful) way any other fundamental, stable, definitive quantum system.

Either based on an all encompassing elementary event space . . . , or the function space model upon which a quantum system is modeled, it should be noted that all of the observed details of a quantum system are, in fact, observed on the domain space, and that (furthermore) these observed details should be describable within the "all encompassing" context of a function space. But descriptions of these observed quantum properties based on the principles of quantum physics are not being provided by quantum physics.

Perhaps it is the overly simple domain space (upon which the functions in the function space are defined) which causes the function space to be so incapable of describing the relevant details of a general quantum system, and perhaps it is that "not all the details" of quantum systems should be described, rather perhaps there is a simple skeleton of very ordered mathematical structures, forming a skeleton of space-form structures within the domain space, upon which the description of both stability and the definitive eigenvalue structures of quantum systems should be attempted.

Perhaps currently the language of function spaces is too broad and currently the domain space (upon which the functions are defined) too limited, in order to allow for good descriptions of quantum systems.

That is, though the function space as well as the descriptive language of probability should essentially be "all-inclusive" (or "all-encompassing") the random-quantum based language cannot describe the properties of general quantum systems.

Apparently (or perhaps) the "overly simple" domain metric-space is the main obstacle to the descriptions of general quantum systems. That is, perhaps there are more physical constants (or conformal factors) similar to Planck's constant, h, associated to both quantum systems and to other spectral structures in a higher dimensional containing space, so that these higher dimensions have (hidden) geometric properties which are not curled into microscopic geometry . . . .

(or not curled into geometry of extremely small size, where these small curled-up geometric properties are irrelevant to the properties of material geometry (as in string theory where the small curled-up geometry is distinct from the space which contains material))

. . . . but rather the higher dimensional geometry is formed into space-form geometries of various different size-scales, both microscopic and macroscopic (thus a need for conformal factors between different dimensional subspaces and/or different dimensional levels). A function space with functions whose domain space is restricted to (such) a new multi-dimensional geometric structure, will in turn, also have a more restricted set of properties to which the functions . . . , which compose the (such a) function space . . . , must conform.

That is, the descriptive language of quantum physics cannot describe all the (spectral) details of general quantum systems that are observed on the domain space. Most notably it cannot identify the spectral properties of general quantum systems. Though in Euclidean 3-space there is spherical symmetry and the isometry group of rotations, SO(3) . . . , which is associated to Euclidean 3-space (along with its group representations [in Hermitian space]) . . . , seem to be an important (and prevalent) set of properties of most confined quantum systems which are composed of charges (and thus with inertial properties in 3-space).

In the context of unitary invariance and hidden particle-state spaces . . . .

(or in string-theory, where higher dimensions are curled into tiny shapes [or curled into microscopic geometry which is irrelevant to the properties of material geometry])

. . . . the descriptive language of quantum physics (apparently) is trying to find a context where quantum physics actually works, but instead the descriptive language is creating a description with two distinct contexts: (1) the set of material existence and (2) the set of hidden dimensions (where strings can "vibrate.")

However, the new language of string theory (and other such abstract languages are) is not capable of identifying (useful) relationships between these two distinct sets. Thus string-theory creates a Platonic truth into which a great deal of energy can be focused, but it is a Platonic truth which has no useful relation to the world, useful in the practical sense, and thus it is the same as religion which is being supported by the "state," ie it (the religious Platonic truth of string theory) is being supported by the US government.

A metric-invariant metric-space can be a space-form or a flow on a higher dimensional space-form, where a flow is equivalent to a face on the "cubical" simplex. Material modeled as a space-form can be contained in metric-spaces, modeled as space-forms whose dimension is "one dimension" greater than the dimension of the material space-form which is being contained in the metric-space.

It is conjectured that material (modeled as space-forms) is usually contained in flows (or faces of the equivalent "cubical" simplex structure) of a highest dimension containing space-form structure of the over-all containing space-form system, so that each flow can contain material space-forms of a lower dimension than the flow's dimension.

However, any such "over-all" high dimension containing space can have any dimension greater than 1-dimension.

A material space-form element contained in a space-form flow will be in the metric-space state of the flow within which the material space-form is actually in.

If the greater (dimension) containing metric-space (ie the flow or face) is a complex-coordinate space then both metric-space states can be contained in the complex coordinate space. In such a complex-containing space material will always be in a mixture of metric-space states, where the two "pure" metric-space states are in the real and pure imaginary subsets (of the complex coordinate space) respectively. A mixture of metric-space states . . . , distinguished by the complex structure (orthogonal structure) of the complex coordinates . . . , is induced by the "spin-rotation" of the metric-space states which make it appear that a material's containing metric-space state is always changing to its opposite state. A material system must exist in such a mixture of metric-space states because opposite metric-space states are needed for the material dynamic process related to spatial displacements (either rotations or translations).

It is within this complex-coordinate structure that the Dirac operator and its further relation to Lie algebra valued 1-forms, where the Lie algebra elements locally displace (in space) a material's coordinate positions, are naturally contained (or defined). This is the correct material-geometric and material-metric-space-state interpretation in which to determine (or in which to understand) what derivatives and function components (of the metric-space states) mean.

Material space-form spin-rotation of metric-space states are the source of spin-½ (metric-space) state rotations, while such spin-rotation of states of macroscopic metric-spaces leaves such spin-rotations of metric-spaces undetectable, so that rotations within "relatively macroscopic metric-spaces" (ie the metric-spaces within which an observer is contained) are integer (or geometric) rotations.

Thus spin rotations can be placed within a context of higher dimensions where there are many flows and thus there are many metric-space states within which spin-rotations . . . , placed within the context of complex numbers (or within the "number-field" of quaternions) . . . , can be interpreted and understood.

When material space-forms are contained in a metric-space whose structure is rigidly associated to a higher dimensional space-form structure so that metric-spaces are flows within this ridged high dimensional space-form structure [with its different dimensional flow structures (or faces on a high dimensional "cubical" simplex structure)] then each dimensional level (or each dimension metric-space) is geometrically independent of (or hidden from) other dimensional levels or other dimensional metric-spaces (in the sense of space-form interactions). That is, the geometric structures, and geometric properties of higher dimensional metric-spaces are hidden from lower dimensional metric-space structures, ie space-forms in one level metric-space do not interact, in a geometric way, with material space-forms of other dimensional levels. However, there do exist resonances between equivalent dimension subspaces of higher and lower dimension space-forms, or between material space-forms and space-form flows of different dimensions that are associated to different subspace structures in a high dimensional containment context.

Lie algebra transformations of local coordinates determine the structure of material interactions based on its relation to both the tangent structure of interaction space-forms and the material space-form's tangent structures.

The existence of conformal factors and their relation to the relative sizes of physical systems

Consider the ratios of orbiting systems. For example, the radius of the sun to the radius of the earth's orbit about the sun and compare this to the ratio of the radius of the nucleus to the radius of the atom.

(Sun's radius / Earth's orbital radius) = (10^8 / 10^{11}) = 10^{-3},

(Radius of nucleus / Size of atom) = (10^{-14} / 10^{-11}) = 10^{-3}.

This suggests that a fixed conformal factor is relating the 3-dimensional solar system size scales to the 2-dimensional atomic-nuclear size scales. What is this conformal factor? The speed of light multiplied by Planck's constant is (c h) = 10^{-25} Jm [J is Joules] (Note: J can be related to k/r on space-forms, for some constant k, thus Jm=km/m=k). Consider (Radius of nucleus / Sun's radius) = (10^{-15} / 10^9) = 10^{-24}. Some of the discrepancies of these numbers are related (or attributed) to, "How to actually determine the true radial measures of these systems?" In which case, there would be four unknowns and only three simple equations. Thus another equation would be needed (perhaps the equation J=k/r), and (thus) this leads to the question as to "Which fundamental constants to consider? Eg permeability, permittivity, and/or the gravitational constant, all constants which give differential equations the correct scale properties." or "Perhaps mass should be considered so that a relation between mass and the sectional curvature of space-forms, 1/r, are related." The Boltzmann constant may also be of relevance, since h is related to properties of probability as is the Boltzmann constant.

The question is, "Does this relation (the existence) of a conformal factor confine the nucleus and the atom to their relative sizes within their containing metric-space, or does each subsystem (the nucleus and the atom) have its own conformal factor, each having comparable orders of magnitude, which determine their size scales, in the containing metric-space.

The main question in regard to the dynamics of material interactions is: What (type of) geometric properties do unitary transformations identify? This is because it is the local coordinate transformations on the tangent structures of interaction space-forms by the fiber groups as well as the factors determined by the 1-form potential functions which determine the direction and magnitude of spatial displacements for each time interval of the interaction system.

Can unitary invariance (which is associated to energy conservation) only be modeled on a base space which has coordinates whose components are complex numbers? Yes. (?)

Unitary invariance is equivalent (or analogous) to metric-invariance (or transformations on real coordinates which are rotations, such as in relation to the metric-invariant Euclidean spaces), unitary invariance of complex-coordinates leaves the Hermitian-form . . . (analogous to both the metric function of real space [and to the inner product (or scalar product) on real spaces]) . . . . invariant.

The invariance of the Hermitian-form on finite dimensional complex-coordinate spaces leaves angles and lengths invariant, ie the Hermitian-form is also an inner product as is (or can be) the metric-function and its close relation to an inner product on a Euclidean space.

For example, (can) a unitary transformation of local coordinates, [can] be associated to a rotational transformation on the real part of the complex coordinates, as well as another rotational transformation on the pure imaginary part of the complex coordinates, ie in coordinate transformations of these types on complex coordinates where it is desired that such unitary transformations are the same as a pair of independent Euclidean 3-spaces, so that this identifies (can identify) a geometric context within a (3-dimensional) complex coordinate space (?)[.] How close to such a form of pairs of independent rotations can unitary complex coordinate transformations come? Such a set of independent pairs of transformation conditions are best associated to a 4-dimensional Euclidean space whose isometry group $SO(4) = SO(3) \times SO(3)$.

How should (are) these two independent sets, ie real and pure-imaginary sets, be related? Answer: Each subset contains within itself one of the two different metric-space states, ie this is true for different dimensions and for different metric-function signatures (ie a metric-function signature is the dimensional of the spatial subspace minus the dimension of the temporal subspace of the total dimension metric-space).

Note: New time dimensions are added as new material types can be defined in the higher dimensional metric spaces.

Two types of material exist in both 2-dimensional and 3-dimensional metric spaces, namely charge in R(2,1) and R(3,1) and inertia (or mass) in R(2,0) and R(3,0) spaces, with Euclidean and the usual space-time metric-function structures. While for 4-spatial dimensions there can be the signatures R(4,0), R(4,1), and R(4,2), where R(4,2) is related to a new material type which exists in a metric-space which has 4-spatial dimensions.

Is there a geometric (or algebraic) correspondence (or relation) between the real and pure imaginary subsets of the complex coordinate system, or are they independent sets (spaces) as in the case of Euclidean 4-space and its isometry group, SO(4)=SU(2) x SU(2)=SO(3) x SO(3)?

Quantum physics and simultaneously particle physics have a function space which has a domain space which are either Euclidean 3-space and time (Schrodinger) with isometry group SO(3)=SU(2), or space-time with isometry group SL(2,C)=SU(2) x i SU(2)=SO(3) x i SO(3).

How is time to be understood? Answer: As being related to "energy and (+t) and (-t)," or as a living mind and its intent.

On space-time, SO(3), can be related to the electric field, E, while, i SO(3), can be related to the magnetic field (or some such approximate field, eg magnetic induction field), B, where in the electromagnetic 2-form (in space-time) the B-field has three components of time. That is, each of the coordinate transformation sets, SO(3) = SU(2), and, i SO(3) = i SU(2), are equivalent to rotations on Euclidean 3-space, and inertial 3-rotations always have a local axis of rotation (identified with an angular momentum [or moment of inertia] axis) thus the isometry group for space-time can be identified with pairs of [local] Lie algebra elements, each associated to the Lie group SO(3), that are related to particular rotation axes.

E and B fields can either be rotating in regard to electromagnetic waves (where SO(3) is the rotation group), or in regard to identifying a 2-surface of the electromagnetic force-fields which are projected down from a 3-space-form (contained in 4-space) into a 2-surface (or 2-space-form) contained in 3-space,
 .... or identified with spin-rotations of metric-space states on material 2-surfaces, ie 2-dimensional material space-forms.

The light in a metric space with three spatial dimensions should be related to the 2-dimensional boundary properties of the metric-space's space-form structure, or 3-space boundary properties of a 4-space-form structure, etc.

Aside

Note: Rotations of Euclidean 3-space have 2-dimensional representations with (local) vector components in the complex numbers.

These 2-dimensional complex number representations take the linear wave-equations . . . , either Maxwell or Dirac . . . , to linear 1st order differential equations, which always have solutions, but it is a unitary representation.

Are solutions to SU(2)-invariant linear 1st order differential equations, such as the Dirac equation, the 2-space-forms, which are represented in complex coordinates (and thus would be 1-space-forms) with properties of "spin rotation of (metric-space) state" associated to themselves, ie spin invariance.

Does this mean (projecting down to) 2-dimensional complex surfaces? Answer: Apparently, yes.

Summary

Classical mechanics is based on the laws of inertia and the laws of differential-forms.

For example, the exterior derivative applied to differential 0-forms . . . , which model inertial physical systems within the context of a three spatial dimensions and time . . . , so as to define a material-distribution's gravitational field, while classical thermodynamics is based on the two (or three) laws of thermal physics and the application of exterior derivatives to differential forms which represent thermal functions, eg energy function, defined on the thermal coordinates, and classical electromagnetism is based on both inertial laws and the exterior derivative applied to differential 1-forms and 2-forms which represent both potential vector functions and tensor electromagnetic fields defined on space-time, in order to define charge and charged current and its field structures, which in turn, are related to inertial forces.

These math structures can all be fit into a multi-dimensional, metric-invariant, multi-metric-function-signature, and a multi-metric-space-state structures, which represent spaces (or material) are filled with space-forms, as well as material, space, and material-interactions being organized by space-forms. The properties of these space-forms can all be determined by applying the exterior derivative to differential forms, which represent space-forms, in each different: dimension, signature, state, context. All the descriptions of physical systems either classical or quantum or otherwise can be fit into such a: dimension, signature, state and space-form (or differential form) context within (or upon) which an exterior derivative can act on a physical system's (or a space-form's) differential forms, which represent both physical properties and properties of space-forms, in a context in which the dimensional-spectral-containment sets are fundamental to determining a (useful) description of systems and their place within existence, so that within such a context the stable definitiveness of both classical and quantum systems (and beyond) can be determined (or described) for systems of any dimension and of any size scale.

Consider how the exterior derivative is used in physical description of gravity (3-spatial dimensions) and electromagnetism (4-spatial and temporal dimensions, ie 3-spatial dimensions and 1-temporal-dimension). In the description of gravity a scalar potential energy is defined, ie a 0-form, the exterior derivative is defined on this scalar gravitational energy function so as to obtain the gravitational field, in three-dimensions the gravitational field 1-form is isomorphic to a 2-form upon which the exterior derivative can be defined so as to create a 3-form in a 3-dimensional containing metric space where a 3-form is a volume form so that this 3-form represents, and is set equal to, an inertial material density function defined on 3-space. Thus the Laplace operator is set equal to a mass density function defined on the 3-dimensional metric space. This equation, along with the assumption of spherical symmetry for the field geometry of a single particle of mass, leads to Newton's theory of gravity, where the potential energy scalar function is determined by a line integral on the gravitational field.

Consider the use of the exterior derivative in the description of electromagnetic properties in space-time. The potential energy function for electromagnetism is a 1-form where the time dimension is associated to the energy of a charge distribution in 3-space at a particular instant of time, associated to the electric field, (where again a "stationary" point charge is associated to an electric field which has the geometry of spherical symmetry) and the other terms of the 4-dimensional potential energy 1-form are associated to the energy of the magnetic and electric fields due to charged currents, ie moving charges, and time changes in the electric field. The exterior derivative is applied to this potential energy 1-form so as to create the electromagnetic field tensor (or electromagnetic field 2-form). There are two sets of equations associated to the electromagnetic field, one set associated to the exterior derivative applied to the electromagnetic field 2-form so as to get zero, and the other set is obtained from the "adjoint" (or dual) of the exterior derivative applied to the electromagnetic 2-form so as to get a 3-form which represents the charge and current density in 3-space for an instant of time (or in space-time). This set of linear differential equations can be integrated for separable geometries to find the potential energy 1-form, which is determined from the geometry of the current and charge distribution in space-time. Basically this potential 1-form can only be found for charges which fit into a set of separable coordinates or for ridged multi-pole charge distributions where properties of spherical symmetry can be used to solve the geometry of the electromagnetic field (or electric field). Because of the extra dimension of time, two independent vector fields, ie for the electric field and for the magnetic field, can be defined within the 2-form electromagnetic field so that these two independent vector fields can be geometrically orthogonal to one another in 3-space. Is this process similar to finding a 3-dimensional interaction space-form in hyperbolic 4-space?

Consider an electromagnetic field defined on 2-spatial dimensions and 1-time-dimension. In this context "stationary" point charges can still have circular symmetry, based on the geometry of the fiber group SO(2), and due to the extra time dimension the electric and magnetic fields can still be defined so as to be independent and possibly orthogonal to one another within the 2-space. If the 2-space is "bent" into the shape (or form) of a space-form then light (the electromagnetic field in the R(2,1)) can be trapped so as to stay within the 2-dimensional space-form. Thus space-forms with only light defined within themselves might only have the conserved property of mass whereas space-forms with charge or charged current confined to themselves also have the property of charge and current associated to themselves.

An alternative mathematical structure not based in materialism

If one considers how the properties of space-forms are (or can be) determined within (or on) a general n-dimensional Euclidean metric-space, ie a metric-invariant coordinate space, (also, or and) where metric-spaces whose metric-functions have different signatures, that is, coordinate space with different dimension temporal subspaces as well as different spatial dimensions which when summed together determine the total dimension of the coordinate (or metric) space, are essentially equivalent to a particular "total dimension" for a given the metric-space, regardless of the metric-function's signature, but in metric-spaces with different metric-function signatures there will be different interpretations of how to understand (or interpret) the dimensions of the temporal subspace for the different dimension differential-forms in the metric-space with a given total dimension.

For Euclidean space of metric-space dimension, n, the space-forms contained in such a given dimension metric-space will be of dimension (n-1), and they will have spectral properties characterized by the (n-2)-flows on their (n-1)-dimension space-form geometric-spectral structures.

The same dimensional relations also exist for discrete hyperbolic shapes.

Thus to determine the spectral properties of an (n-1)-dimension space-form one could consider the exterior derivative of the (n-2)-flows of the space-form so as to form an (n-1)-differential form which in turn represents the (n-1)-dimension space-form. However, the (n-1)-differential forms are dimensionally isomorphic to the 1-forms of the same n-dimensional metric-space, and the (n-2)-dimensional flows of the space-form which are represented as (n-2)-differential forms are dimensionally isomorphic to the 2-differential forms of the same n-dimensional metric-space, thus the geometric-spectral properties of an (n-1)-space-form can be modeled as 1-forms upon which an exterior derivative can be defined (or can act) so as to identify 2-forms which represent the (n-2)-flows of the given space-form. Thus space-forms and the geometric properties of fields, dependent on both the geometric distribution of material and the dynamic distribution of material, within the metric-space are exactly analogous mathematical structures within a metric-invariant metric-space which is the containing space for the material and its motions of physical systems.

The geometric structure of space-forms result from the process of "moding-out" or making "opposite" faces (of the space-form's "cubical" simplex fundamental domain) equivalent to one another. That is, the geometric properties of a space-form are directly associated to the facial properties of the space-form's "cubical" simplex fundamental domain structure, thus they are properties of the space-form's (or equivalently the given dimension metric-space's) subspace geometry based on adjacent (or bounding) facial properties of the "cubical" simplex of the space-form's "fundamental domain," and thus they are exactly analogous to the geometric measures which differential-forms represent on adjacent dimension subspaces of the given metric-space of the particular dimension.

The multi-metric-space-state properties of physical systems, where these metric-space states are associated to types of material or their characterizing physical properties, where these metric-space states are in turn, associated to subsets of a complex coordinate structure, and thus the physical description of metric-spaces is based within complex coordinates, and thus in order to represent curvature as a Lie algebra valued 1-form then within the similar metric-invariant context (of the real spaces physical systems) the Lie algebra is (would be) unitary, (as opposed to it being a Lie algebra of a diffeomorphism group). (and) so that This "flat" Hermitian-form space (or Hermitian invariant space, or metric-function structure) makes complete sense within this new higher dimensional Hermitian invariant (similar to metric-invariant) containment space (or structure), and thus physical description is not related to the general-metric structures of general relativity.

The fact that micro-interactions between quantum systems results in a local dynamic structure exactly like Brownian motion and thus (because of the descriptions of E Nelson, Princeton) this allows the new "multi-dimensional, space-form structure (or model) for existence" to be related to an apparent harmonic structure (or) of (apparent) wave-like probabilities, as in the case of quantum physics, but this wave-function or function space basis for a derived wave-function, or derived probability based description, must be defined on a domain space which is the new "multi-dimensional space-form structure, or model of existence."

The new "multi-dimensional space-form structure or model of existence" provides a model for both material and metric spaces, ie material and metric spaces both have space-form shapes (or equivalent) but they exist at different dimensional levels, where one space-form structure is the containment structure (or containing metric-space) and the lower dimensional space-form structure is the material which is contained in the metric-space (also modeled as a space-form) in a many dimensional context for containment.

That is, the space-form structure for material and space in a multi-dimensional, and multi-metric-space-state, and multi-material-type structures, is fundamental to the stable, definitive properties of spectral and geometric systems and properties which characterize material systems, which have been observed. The apparent probability based structures can be described by adding a function space structure upon (or onto) the multi-dimensional, space-form structures, so that the multi-dimensional, space-form structures are (or identify) the domain space for the function space. Thus, in many ways the function space structure is superfluous to the important multi-dimensional, space-form properties of physical systems and their descriptions. That is, the function space, or random, properties of the descriptions of physical systems are often unimportant details of the existing material and spatial systems (of existence).

The conclusion is that "differential equations," both classical and in quantum physics, are too general for the set of properties which must be identified in a descriptive language which is describing the stable definitive properties of physical existence, and the further existence of many dimensions within which (or of) space-form geometries and space-form spectral properties can be identified, eg resonances of stable space-form spectral-geometric properties that exist between different dimensional levels.

These stable properties are all related to the separable geometric properties of space-forms (and the fiber bundle descriptive mathematical structure), and thus the linear and separable geometries of "differential equation models" of physical systems, be it classical equations or be it the unitary invariant energy operators of the Schrodinger equations, (separable geometries of space-forms) become the fundamental, skeletal, structure upon which useful and widely applicable describable properties of physical systems emanate.

Trying to describe physical systems in greater detail, with either probability or with general coordinates, results in a descriptive structure of mostly unimportant details which essentially result in identifying the unstable properties (or unstable details) of those descriptive mathematical structures which are not fundamental to physical description.

Chapter 3

REVIEW II

Review of both math and science

A short review of a new direction that mathematics can take and explore, whose pay-off could be a new set of possibilities for technical development, both physical and biological, and a new model of mental structure, (as well as a new context for human creativity).

What is the context of "scientific" description? Precise descriptions have limitations as to what descriptive patterns can emerge from a set of assumptions and a precise language (upon which the description language depends).

What should the context of a creatively useful (or "scientific") description be? Should it be about existence (in general), or about material, or about life?

What is life?

One answer is that:
"Life is about knowledge (truth about existence) and the useful relation that knowledge has to creativity, thus measurably verifiable descriptions which are widely applicable and very useful in regard to creative processes may be regarded as being true," because of their usefulness in regard to creativity. Should all creative efforts within society be organized to maintain and increase the power of a few individuals as is the basic belief of capitalism, where acquiring property is the only rule? or Should all individuals have an equal right to create and to know (so that property belongs to everyone equally)?

Furthermore, creativity needs to be placed in a context of equality, since creativity cannot be judged based on markets. For example, the ideas that Copernicus published in his revolutionary book could not be the basis for a "best seller" when he actually published them. Something less than 60 books published by Copernicus were ever bought. Should people create for themselves, based on their selfless desire to create, where they believe that their creative efforts are helpful to others (or to society), or should they "work for others" so that they are wage-slaves who are only allowed to be involved in the creative efforts of others, ie the few who control property and money (or who control the economy).

Working cooperatively with others, where "all of the group" owns the creative effort in an equal way, is certainly a valid way in which to create, but a limit as to the (market) size that such a working group can be, must exist. A market can only be free if everyone in a society are equal.

or another answer is that:

Both material and space are space-forms which are removed from one another by one-dimension (see below), and life is also a space-form, where a "living space-form" has the fundamental properties of being both of odd-dimensional and it has an odd number of flows associated to itself.

This odd-number of flows, when occupied by "charge" are unstable because of the space-form's geometric shape (or geometric properties), so that the unstable space-form begins to oscillate on its own, where the oscillation generates energy for the "living" space-form.

Mind is modeled as a set of spectral resonances contained on the different maximal tori contained in the finite dimensional unitary fiber groups. One can move between maximal tori by group conjugation.

Then this leads to the question, "What is the true creative context of life?" Is it about material creations? or Can it be about art, or Can it even be something beyond material existence, based on a greater "set structure" for existence beyond the idea of materialism, but which contains the descriptions of material properties? What is life, eg what is the dimension of its space-form structure, and how can it affect the structure of existence itself?

Should people be interested in a mathematical representation of existence which describes both material properties and the fundamental properties of life so that both descriptions are contained in the same mathematical structure (see above)?

There are descriptions of material existence which are very useful and widely applicable, namely, the descriptions of classical physics. Classical physics is based on defining material in the descriptive language of (or systems as) differential equations, and the coordinate context of material containment (is) in a metric-invariant context, of either 3-Euclidean space or space-time, and in 3-spatial dimensions (namely, in Euclidean 3-space) there is an observed spherical symmetry for both gravitational and (much of) electric field material interactions. Interactions are defined in terms of two types of differential equations which define: (1) inertia determined by force fields, and (2) the force fields related to material geometry.

However, fundamental micro-systems are stable, bounded and definitive in their spectral-energy properties, where classical physics cannot describe the properties of these quantum (or small scale) systems, and their descriptions (or properties) appear to depend (or be based) fundamentally on probability (or function spaces) and sets of measurable operators which act on the function space to determine the function space's (or the quantum system's) spectral set.

But this language, based on probability, is not capable of describing the stable definitive properties of general quantum systems.

Quantum physics can only describe the H-atom, and it describes, in a qualitative way, the quantum properties such as tunneling and (small scale) material's wave-like properties. Quantum physics also describes the two "types of material," being either Fermions or Bosons, whose distinction is related to how these material types fit into the stable states of quantum systems.

In regard to technical development, quantum physics has only led to: lasers, electron microscopes, and tunneling devices (described in an inexact qualitative way).

On the other hand quantum properties can be used in highly controlled classical systems, but such uses do not depend for their usefulness on their quantum descriptions, but the usefulness of these systems depends only on the classical systems to which they couple.

Particle physics is only relatable to the probability of particle collisions which only seem to model bomb explosions, ie a system which starts in an ordered state and ends in a very disordered state.

That is, particle physics has no relation to the descriptions of fundamental, well defined, stable, definitive quantum systems. That is, particle physics is not capable of describing the stable orbital properties of general nuclei. Thus, it is hard to identify the value of particle physics as a valid description of physical systems.

Supposedly the unitary invariant wave-function of any given quantum system, "which is almost known," can be used to adjusted such a wave-function's properties by reducing the global wave-function to its value at a single point in space, and by means of unitary invariance and Lie algebra valued 1-form models of a potential function used to model particle collision interactions at the given point, and subsequently adjust the wave-function's properties, but it is not clear which wave-function to which the interaction particle-collisions at that single point are to be actually applied, so as to make an adjustment to the wave-function.

General relativity is only about a one-body problem with spherical symmetry whose local coordinates have a space-time structure (or should it really be a Euclidean space structure?), ie it has a very limited range of description, and its equations are virtually always non-linear, and thus it is nearly always describing a chaotic system, and thus it has virtually no uses for describing physical systems which are stable, such as the solar system.

Are there other mathematical structures which can be used to describe the stable definitive properties of quantum systems, as well as difficult macroscopic problems such as the stability of the solar system (apparently stable for billions of years), so that the descriptions are widely applicable and lead to more creative uses of the measurably verifiable descriptions, such as being related to having greater control over microscopic (or quantum) material properties (as well as very large systems) and so that the description leads to a viewpoint of existence which is not based on the idea of materialism.

There are such mathematical structures.

These new mathematical structures, at least newly applied to describing material existence, are based on:

(1) Metric-invariance
(2) Many dimensions beyond 3-spatial dimensions,
(3) Discrete isometry subgroups, where isometry groups are the natural fiber groups of local coordinate transformations which are metric-invariant. Discrete isometry subgroups are also called space-forms. Space-forms are geometrically separable (all coordinate directions are locally orthogonal to one another at each point in space, eg a cylinder). Furthermore, hyperbolic space-forms are very stable with definitive spectral properties.

Space-forms are based on the geometric properties of "cubical" simplex structures, the same structures upon which differential forms are most naturally defined, where multi-variable calculus also depends on differential forms and exterior derivatives.

(4) (both many dimensions and) On each dimensional level there can be a variety of metric-function signatures which one can consider so that different signature metric-functions on the same dimension metric-space are related to both fundamental physical properties and to different (new) material types (such as material which is "living).
(5) Physical properties associated to particular metric-invariant spaces can be related to pairs of metric-space states, ie a physical property and its "opposite," eg time.
(6) Many metric-space states can be considered, thus leading to natural coordinate systems based not on real numbers (as metric-spaces require), but now based on complex numbers (or complex coordinates) as well as Hermitian forms, analogous to both metric-functions and inner products, and subsequently these complex coordinates have a relation to unitary fiber groups.

A space-form (or function [ie a differential-form], or physical system) can be in many various types of states dependent on metric-function signatures, where each metric-space is dependent on both metric-space "states" (associated to both fundamental physical properties and space-form interactions) as well as "spectral states" associated to a new construction related to spectral-dimensional-set containment properties. Can such set-containment be described, controlled, and mixed in a controlled manner?

This type of complex coordinate is the structure upon which the Dirac operator (the square root of the Laplace operator) is naturally defined.

(7) One may define function spaces on such a space-form domain space (though this might be found to be a complication of structure which is not all that useful).
(8) The model of interactions based on space-forms leads to microscopic Brownian motion, and (thus) through relations identified by E Nelson (Princeton, 1957), microscopic interactions seem to identify the same random structure of quantum physics, but the basis for physical description is no longer fundamentally based on probability but rather it is based on the geometry of space-forms which can be used to model the definitive, stable orbital properties of: nuclei, and high atomic number atoms, as well as well defined stable orbital properties upon which superconductivity in crystals depends.
(9) This space-form structure of many metric-space states, defined on complex-coordinates, leads to unitary fiber groups, perhaps the only valid physical pattern which the particle collision experiments (of particle physics) identify.

Claiming that diffeomorphisms are the proper fiber groups needed to describe the general frames (ie properties of moving coordinates) of physical descriptions (in regard to classical [or general relativistic] systems) has not led to any useful information, nonetheless "gravitational singularities" are claimed to be important (the only other system, besides bomb explosions, which particle physics is claimed to be able to describe), but these ideas have no useful value and seem to only be related to the theology of materialism, the idea of which (physical) science cannot let go, even though the theories based on materialism: quantum physics, particle physics, string theory, and general relativity have not shown to be able to describe the observed properties of existence, ie the stable, definitive orbital properties at all size scales. Classical physics is based on materialism and it is useful, but it clearly has its own limitations as to what it can describe (neither quantum physics nor stable solar systems).

Though it seems that curvature (and its related diffeomorphism local (general) coordinate (or local frame) transformation group) should be related to the ideas of: inertia, force fields, and differential 2-forms on a metric-invariant containing coordinate spaces, because it is in this context where the ideas of forces and interactions do work (in both widely applicable and very useful ways), but such a relation has not been made in a useful manner.

The actual structure of a physical system's differential equation is a square-root structure (or Dirac operator) which, if it is linear and defined on a separable geometry, then it would be solvable and controllable through initial (or boundary) conditions, but because of the square-root structure (of a metric-invariant differential operator) the coordinate base space (in the principle fiber bundle structure of the descriptive containing space) will be a complex-coordinate space. That is, the coordinates of a metric-space which has metric-space states are complex coordinates and these coordinates will contain the description of the system as a mixture of time-states and/or a mixture of spatial displacement states, ie rotational and translational inertial frames, and in this structure the "description containing space (ie a principle fiber bundle)" is a Hermitian-invariant base space and a unitary fiber group, and thus can be called a unitary-Hermitian-invariant (or unitary-flat) space.

There is a . . . "square of the (square-root) (kinetic motion) term plus a square-root (potential energy) term" . . . . type of structure for the real energy function and the real geometric structure of the metric-function of real metric-space states.

The square-root representation of a geometric term, ie the potential energy term, is a Lie algebra valued 1-form (potential term). That is, the real issue of physical description is position-geometry and motion-frames but in real coordinates geometry and motion combine as a sum of a square of a square-root of motion and a square-root of the squared metric-function.

Force fields are about the change of motion of an object with time (or some parameter) . . .

> [or about minimizing action eg integral of pdx, ie (resulting in) a 2nd order differential equation which can be linear on a metric-invariant domain space]

or

Forces are about following geodesics whose definition depends on determining the general coordinate space's metric-function, which is found by solving a 2nd order curvature differential equation, which is always non-linear, and hence its solutions are nearly always related to chaotic patterns, and thus they are descriptions of limited usefulness.

or

Force is about "how spatial displacements change with time," thus it is the Lie algebra valued 1-form . . . , where the 1-form's value determines a factor which affects how spatial displacements change with time, ... which locally transforms coordinate directions in a manner which is consistent with "how spatial displacements change with time."

When the local coordinate changes are done in a unitary context this determines spatial displacements [due to geometric properties (the factor determined by the 1-form)] in the context of opposite metric-space states.

Higher dimensional forces are not determined by spherical symmetry as is the case in 3-dimensional Euclidean space, where inertia is defined, but rather by space-form geometry of both spatial and material space-forms (or faces) which compose a 3-dimensional interaction space-form (or "cubical" simplex, where 2-dimensional faces represent both spatial and material properties of the interaction), and the directions of the spatial displacements are being determined by the local coordinate transformations of the Lie algebra values, or elements, (of the Lie algebra valued 1-form) which act on both the spatial and material tangent structures of the interaction space-form.

Thus the geometry of interaction depends on:

(1) the tangent structure,
(2) the relation that the Lie transformations (or spatial displacements) have to these tangent structures, and
(3) included in this descriptive structure is the geometry of the Lie group, eg the Lie group SO(3) has the geometry of a 3-dimensional sphere, and this geometry (of the Lie group) affects how the Lie algebra elements act locally on the tangent structures of the interaction space-forms.

Note: Most interaction structures and/or geometries of interaction in higher dimensions are tangent in their direction and not radial, in regard to a spherical geometry [of the Lie group local coordinate transformations (or Lie algebra local coordinate transformations)].

A material (space-form) interaction results in spatial displacements, which can be associated to displacements for each time cycle . . .

> [where a time cycle is defined by a return to the original direction of time when the flows of different metric-space time-states (in a space-time metric-space) are spin-rotated],

. . . . and in the complex coordinate space (or two-state-space [two metric-space states]) context, the displacements are in opposite spatial directions (in the two distinct metric-space states [or metric-space subsets]).

(10) A metric-invariant metric-space can, itself, be a space-form, or a flow on a higher dimensional space-form, where a flow is equivalently a face on the "cubical" simplex associated to the

higher dimensional space-form. Material modeled as space-forms can be contained in metric-spaces (also modeled as space-forms) whose dimension is one dimension greater than the dimension of the material space-form, which is contained in the metric-space.

It is conjectured that material (modeled as space-forms) is usually contained in flows (or faces of the equivalent "cubical" simplex higher dimensional structure) of some "highest dimension" containing space-form structure (of the over-all containing "space-form-metric-space-space-form" system, so that each flow can contain material space-forms of a lower dimension than the flow's dimension.

However, any such "over-all" high dimension containing space can have any dimension greater than 1-dimension. That is 2-dimensional space-forms can represent a 2-dimensional metric-space which can contain 1-dimensional space-forms, which represents material (contained in a 2-dimensional metric-space).

When material space-forms are contained in a metric-space whose structure (as a flow [or as a face]) is rigidly associated to a higher dimensional space-form structure so that metric-spaces are flows within this ridged high dimensional space-form structure [with its different dimensional flow structures (or faces on a high dimensional "cubical" simplex structure)], then each dimensional level (or each dimension metric-space) is geometrically independent of (or hidden from) other dimensional levels (or other dimensional metric-spaces) in the sense of space-form interactions between material space-forms of different dimensions, ie the different dimension material space-forms are contained in different dimension metric-spaces. That is, the geometric structures, and geometric properties of higher dimensional metric-spaces [within a high dimensional space-form metric-space structure] are hidden from lower dimensional metric-space structures, ie space-forms in one level metric-space do not interact, in a geometric way, with material space-forms of other dimensional levels. However, there do exist resonances between (equivalent dimension subspaces [or different facial structures of the same dimension] of) higher and lower dimension space-forms, or between material space-forms and space-form flows of different dimensions (that are associated to different subspace structures), where the flows exist on a ridged fixed dimension over-all containing space-form.

However there can be many higher dimensional space-forms of various dimensions (contained in the over-all containing space-form) which can simultaneously contain the material space-forms of (or contained in) the lower dimension metric-spaces of the high dimensional "complex."

This determines a new problem of "set containment" based on both dimension and spectral structures, so that such properties of "set containment" are (or seems to be) relevant to determining material spectral properties which resonate with the full containment set, where each set which contains some given material system (of some given dimension) is modeled as a space-form structure, each of which can be of various dimensional values (ie each containing set [or space-form] can be of various dimensional values).

(11) Resonances exist between material space-form (or quantum) systems and their higher dimensional containing metric-spaces (which are also modeled as space-forms). Material is modeled as space-form which is contained within metric-spaces which is also modeled as a space-form, and by means of this structure the spectral properties of material space-form (quantum) systems can be more precisely described in the language of space-form physics, than their spectral properties can be described in the math structures of "quantum physics" (where quantum physics cannot describe the stable definitive spectral properties of quantum systems).

(12) There can be many high-dimension space-form structures into which a low dimension material space-form can be simultaneously contained. This identifies a new set of properties within which set-containment can be defined.

Within how many "states" can a representation of a function (or of a system, or of a space-form) be defined?

(a) Metric-space states, and
(b) Spectral containment states, based on being simultaneously be contained in various sets of higher dimensional space-forms, ie higher dimension than the material space-forms.

The spectral properties of the entire over-all high dimension containment structure and all of the component space-forms (of various dimensions) that it contains must all conform the spectral properties of the highest dimension over-all containing space-form.

(13) It is difficult to understand higher dimensional geometry and discrete isometry subgroups are consistent with the fundamental properties of geometry (as classified through topology) and these discrete isometry subgroups give a blueprint with which one may try to both follow into higher dimensions and use to interpret higher dimensions.

Constraints on the dimensions and geometric properties of higher dimensional space-forms have been determined by Coxeter, especially for hyperbolic space-forms.

In 3-dimensional Euclidean space it is the geometry of the group, SO(3), through which the local coordinate transformations are related to the tangent interaction structure (of interaction space-forms) upon which the local Lie algebra elements act in a spherically symmetric manner.

To outline the new mathematics:

(1) set containment based on dimension and spectra of space-form's as well as sets of conformal factors between (space-form) sets of different dimensions.
(2) calculus (forces and/or curvature) is related to the tangent (or facial) structures of the interaction space-form's associated "cubical" simplexes, and their relation to local Lie algebra spatial displacements determines interaction dynamics. "Cubical" simplexes and

local Lie algebra coordinate transformations (derived from Lie algebra valued [potential] 1-forms) is also the math structure of force fields as well as the structure of curvature, which is related to the Lie algebra valued 1-form connections (where Lie algebra actions on local coordinates are affected by the conformal factors associated to the 1-form evaluation on the local coordinates), within a metric-invariant structure of space-form geometries which have non-zero sectional curvatures, where curvature is determined by the commutation of two Lie algebra elements defined at different points in the limit as those points approach one another on the space-form (or within the metric-space). This naturally fits into the structure of a Dirac-type (square root of Laplace-like) operator, resulting in a 1st order differential operator, which acts on differential forms, but it does so in a manner which mixes the metric-space states in the complex coordinate structure.

(3) 2-forms (or 2nd order operators) . . . , curvature . . . , Thurston's topological classification of geometry which fundamentally depends (in regard to metric-invariant spaces) on space-forms and discrete isometry subgroups, where a space-form's properties can be determined based on solutions to differential equations defined on 2-form models of a space-form's spectral flows.

Quantum physics can have its function spaces defined on the above mentioned space-form domain structure with its various types of "set-containment" properties. But details of random properties of systems are relatively unimportant in relation to the space-form structure of:

1. material,
2. space, and
3. interactions

and all of these space-form entities can have a relation to the resonances that can exist between different dimension (or different subspace) spectral properties (in relation to the over-all containment set).

Perhaps it is of value to identify important subset structures of the over-all containment set, of various dimension space-forms, and use function spaces to study (or piece together) such important subset structures.

Classical physics is the same in space-form physics, but it can be represented in the language of space-forms, but this is usually unnecessary.

General relativity must be framed in the new metric-invariant space-form context, a containment set where curvature is not usually zero but the math is based on both isometry groups and unitary groups.

Conclusion

In an abstract sense a scientific description (or scientific theory) is a measurable, widely applicable, and very useful description in a language based on well identified assumptions and a language in which words have definite and limited meanings.

It is clear that a geometric model of existence would be widely applicable and very useful.

It has been found that probability is not a good basis for describing (material) existence, because probability, by itself, has not been able to describe the observed order which material quantum systems possess, eg particle physics cannot describe the stable orbital properties of general nuclei etc. The only "ordered systems" which are descriptions of quantum physics are very simple systems which have well defined geometric properties as part of their potential energy term, and these few successes are mostly unspecified qualitative quantum properties, eg tunneling based on the vague notion of a potential energy "wall" of finite height.

It is observed that material systems have stable, definitive properties, and a descriptive language based on the probability distributions of random events has not shown itself to be a valid basis for describing these stable, definitive physical properties.

Thus the issue of scientific description, or physical theory, must revolve around geometry.

Geometry has two main parts the descriptions of general coordinates with a fiber group of diffeomorphisms or descriptions of geometrically separable coordinates with isometry (or unitary) fiber groups, where the geometrically separable coordinates (space-forms) are derived from discrete isometry (or unitary) subgroups.

It is clear that diffeomorphism fiber groups do not lead to descriptions that have any order (other than the 1-body system with spherical symmetry) and thus diffeomorphism geometry is too complicated to be of any useful value in describing existence, where the (apparent) stability of the many bodied solar system . . . . (and thus it should be a non-linear, chaotic system) . . . goes without a valid description.

Interactions between microscopic systems leads to Brownian motion and E Nelson, Princeton, 1957, showed such Brownian motion was equivalent to quantum randomness and thus discrete isometry subgroups are consistent with the apparent randomness that quantum systems seem to possess.

Using metric-invariance, discrete isometry (or unitary) subgroups, many dimensional levels, and the existence of spectral resonances between dimensional (or subspace) levels, leads to simple geometric ideas about set containment in regard to a physical system's spectral structure, with a built-in well defined stable geometric structure associated to spectra and resonances within the containing space. Conformal factors between dimensional levels play an important part in this new descriptive language. These (new) conformal factors would be similar to the fundamental constants, such as Planck's constant and the gravitational constant etc, which can be interpreted to be just that, ie they are part of the set of conformal factors between dimensional levels.

PART II

Chapter 4

TOPOLOGY AND GEOMETRY

Math is about identifying types of measures, eg local measures of change of a measurable quantity, and it is about identifying the correct process or equation for determining a measurement of a particular measuring type. It is about applying measurements to shapes and patterns so that the measurements correspond to experience. Math is not so much about details, rather detailed measurements (of a pattern) are the result of refinements which occur after the proper measuring relations (or processes) are determined.

Unfortunately, much of the work in math is focused on refining techniques so as to identify details for measuring processes which have not corresponded to experience in a convincing way, eg quantum physics, particle physics, general relativity, string theory, etc. That is, using function spaces to identify unknown spectra in a mathematical structure of probability and discreteness (quantum physics), and forcing low dimension random local unitary transformation patterns onto this random structure (particle physics), and using non-linearity in a fundamental way to uncover ordered structures in regard to gravitational forces (general relativity), as well as other related physical models, are not describing the observed ordered structure of the world in useful ways.

Much of the way in which math and science is used in society is associated with industrial interests, and the biggest industrial interest in the US is militarism. This way of focusing on the math techniques . . . of quantum physics, particle physics, general relativity, string theory . . . is associated to a "public relations" ploy [as is the search for the "dark matter" particle] for the purpose of associating militarism to science and math. These math techniques of quantum physics, particle physics, general relativity, string theory, either do not identify order, rather they imply randomness and more instability, or they do not describe "anything at all" (as is the case of string theory) and they lead one to the conclusion that "the stable, definitive spectral-orbital properties of material systems at all size scales are properties which are within in physical systems which are simply too complicated to describe." Subsequently, the stable fundamental spectral-orbital systems of the: nucleus, the general atom, crystals, and the stability of the solar system go without useful descriptions in regard to their stable properties.

The one assumption that is common all of these descriptive languages is the assumption of materialism, ie that existence is determined by material, yet particle physics and string theory are identifying dimensions beyond the idea of materialism, yet these subjects (or these descriptive languages) are forced into a math structure which is consistent with materialism.

Math descriptions

In quantitative description one (often) has a containing space and a set of measurable properties which can be identified within (or based on) the containing space, the measurable properties can be represented as functions of the variables which identify the containing space (or the function's domain space) and together these properties and measures (functions and coordinates) identify a (physical or geometric) system which has measurable properties. If one can control (based on the system's "set-up" conditions) and predict the measurable properties from a solution to a differential equation then one can use this information within a further context for creative purposes.

> Are containing spaces composed of sets of space and time, where each dimensional type (space or time) can be associated with subspaces of the (over-all) containing space . . . ,

or

> . . . . are they random events modeled as function spaces upon which the actions of differentiable and other (infinite-dimensional matrix) operators . . . (such as operators based on multiplication in regard to the variables [or different dimensions] of the functions' domain space) . . . are defined, wherein the containing space (of the measurable properties of such a descriptive language based on probability) is an elementary event space of the spectral values of the function space and the probabilities of these elementary events are associated to the spectral decomposition of the function space . . . (or its spectral function covering set [of functions]) . . . and where the spectral values are associated to a (quantum) physical system, which is characterized by its stable, definitive distinguishable spectral properties associated to (or observed as) (random) point-like spectral events in space and time, ie events of spectral-values observed as point-events on the domain space?

Quantum physics is plagued with singularities, divergences, and subjective interpretations concerning probability, and quantum physic's subsequent uncertainty principle, in regard to singularities and divergences, and interpretations about the relation that quantum physics has to classical physics, where it is claimed that at high energy (or at high numbers of components) for a quantum system then the quantum math for that physical system changes to classical math, whereas it should be noted that classical physics depends on continuity and causes, ie forces, while quantum physics is discrete and based on probability and thus they are irreconcilable math structures, though they have some common ground in statistical physics.

Are the spectra of quantum systems all about:

(1) (1/r)—geometry (or spherical symmetry) of point particle energy terms (in wave-equations),
(2) unitary-invariance of the wave-equation (or the energy operator),
(3) the geometry of point-particle collisions,
(4) random unitary changes in a point-particle internal structure,
(5) rotational symmetry,
(6) spin,
(7) the Slater determinant and estimates of point-charge distributions in a general atom,
(8) no topological obstructions for spin, and
(9) randomness
?
(No.) (though (5) and (6) have some legitimacy)

Can quantum physics claim to have meaningful geometric bounds, which introduce particular spectral properties into "the quantum system which is being described," due to an assumption of the quantum system's function space being L^2 bounded? (No.)

The conservation of angular momentum in a spherically symmetric geometry seems to account for some spectral properties in some simple systems where such symmetry might exist, but the radial equation (for quantum systems assumed to have both the property of spherical symmetry and conserved angular momentum) cannot identify the fundamental spectral properties of general quantum (or general atomic) systems. Rather the solution to the radial equation diverges, unless subjective adjustments are made to the math structures.

This is what the main subject of quantum and particle physics is about subjective adjustments, such as renormalization, to "cover-up" divergences due to the math singularities of point charges. When Dehmelt isolated and cooled the electron the electron did not stand-still, rather it formed into a relatively stable orbit. This orbit was claimed to be a natural solution of the Dirac equation, but when Dirac solved the Dirac equation for an isolated electron the solution did not imply a fundamental lowest energy orbital structure for that isolated electron.

That is, one has a set of authoritative statements about what is true (this authority frames an academic contest) and then one has lying and conniving to up-hold such an authoritative structure which cannot be used to describe the observed, and fundamentally stable properties of the material world.

Conjecture: A descriptive language which is based on randomness (or equivalently, based on probability) cannot be used to describe the ordered properties of existence, ie probability is a descriptive language which cannot account for the stable definitive properties of quantum systems, and general relativity, a description of existence which is based on non-linear differential equations, cannot describe the (apparent) stability of the solar system.

Note: Though an assumption of god is the basis for a Platonic truth, as math theorems are Platonic truths, the practical (or creative) usefulness of a belief in god seems to be (at best) very limited. It is only the wide applicability and great usefulness which distinguishes a Platonic truth from being either a useful truth (or useful pattern of description) or an illusion. Furthermore, most mathematical Platonic truths are illusions, they have very little bearing in useful (or real) creativity, this is true most often because mathematical truths become too complicated in relation to the assumptions upon which they are built, or the assumptions have an esoteric quality, such as the axiom of choice, which implies a special choice or a jump to continue an idea which has naturally concluded. For example, the structure of the real numbers (whose structure depends on the axiom of choice) is too complicated for them to be useful (see below *).

> Neither
> The descriptive language of Darwin's evolution . . . ,

based on random mutation and the random properties for an organism trying to survive (or live) in an environment (or an eco-system) in which natural selection takes place, ie survival of the fittest,

> Nor
> The languages of either quantum physics or particle physics are . . .
> . . . . capable of deducing . . . , or describing . . . , or identifying . . . .
> . . . that from which the order (of such well organized systems) comes,

[where the order comes from either a sequence of events . . . , or a principle of randomness . . . , or a cause (such as geometric properties)].

This failing of descriptive languages based on probability to describe the observed order of either living systems

or

> how the many living systems change "over time" toward life forms of very ordered complexity, or more refined (or highly developed) properties of different (but specific) organs which are common to most animals

or

> .... in regard to the physical world, probability based descriptive languages have failed to describe the stable definitive order of physical systems eg nuclei, atoms etc.

That is the order of the world cannot be described using the language of probability, though properties can be identified and counted so that, if these distinguishable properties are stable (so that an elementary event space can be identified) then probabilities can be determined for these stable, identifiable properties. But the causes or processes through which these stable properties emerge from existence cannot be described by a language based on probability or a descriptive language whose fundamental structure is random such as solutions to non-linear differential equations.

Furthermore, random descriptions depend on the existence of an elementary event space of stable, definitive events. However, neither quantum physics nor (biological) evolution have such elementary event spaces associated to themselves based on the properties derived from their probability based descriptive languages. There are no elementary event sets, which these descriptive languages (based on probability) provide, so that these provided elementary event spaces (or spectral sets) match the observed spectral properties (or elementary event sets) of the physical systems. That is, there is no complete set of stable, definitive elementary events of a quantum system or an evolutionary sequence which can be deduced form the laws of either evolution or quantum physics.

Back to math descriptions, and the descriptions of classical physics

However, if the descriptions of measurable properties of physical systems are based on the derivative, ie based on solution functions to differential equations, as are both classical and quantum physics (but quantum physics is more about operators acting on function spaces where many of these operators depend on derivatives), then the properties which identify a physical system must be measurable properties, ie a solution function to a linear differential equation, which are consistent with the "geometric" (and any other) measures, such as spectral measures of the system, and these function values must be consistent with the coordinate measures on the function's domain space.

The derivative is about y=mx, or an equivalent many-variable tangent structure used for relating uniform stable measuring sets, ie the function's values and the domain's values. If y is not consistent with the measures of x then the derivative is not giving a valid linear relation for measuring between the function's values (or the measurable property) and its domain or containing coordinate space, in which case the solution formulas (to differential equations) are not valid as identifying properties of a (relatively) stable system, where the formula is expressed with the variables of the coordinates. If the derivative is not a valid local linear measure then this results in non-linearity and chaos, though the chaos might have critical points which identify limit cycles, but a semi-ordered math structure, eg a limit cycle, of a solution function to a non-linear differential equation is only of use if the system has continual feedback, so as to adjust the system to ensure that the system reaches the desired result, eg driving a system toward a limit cycle.

A function's local linear approximation by means of a derivative is based on an elementary idea that two measuring sets (of different number types) need to have two stable (but different) uniform units of measurement which need to be related to each other through a (simple) linear relation, ie y=mx, so that the two different measuring sets can be related to one another in a consistent way in a small local region about a particular point where the derivative is defined.

The derivative is a tangent structure which is supposed to identify a linear relation between two different measuring sets, so that (in classical physics) a physical system's differential equation can be defined in relation to, a locally measurable property being set equal to an equivalent geometric property.

There are three aspects of the local measuring process:

(1) is identifying the "set structure" of both the function's values and the (its) domain's values, through which the property of closeness (or localness) can be identified, ie the properties of continuity (the set structure of closeness with respect to both the function's values and its domain values), and
(2) the local tangent (or local linear) structure of the function's (or measurable property's) graph, in relation to both the function's values and the system's (or geometry's) . . . [upon which function's (or measurable property's) are defined] . . . . containing coordinates (or the function's domain space). This local linear structure, for the function, is the derivative and its associated (nearly inverse) integral operator, and
(3) using these local linear measures of the function's values in relation to the local geometric measures on the containing coordinates, where the local geometric coordinate measures actually measure the geometric properties in relation to tangents of the coordinate curves within the containing coordinate space (or the domain space of the function).

The property of continuity allows one to define "local measures" in regard to function values and their domain sets, ie it allows derivatives and integrals to be defined. If these local measures, eg the derivative, can be equated to another equivalent quantitative pattern then a differential equation is defined for a function which makes the equation true, ie for a new measurable property (ie the solution function to the differential equation). If a differential equation is linear then the equation allows for a local linear approximation for the solution function.

The point of a linear approximation to a function . . . , ie the tangent structure to the function's graph, of the measured values of the function . . . , (or to a function's value at a given point in the domain) is to approximate the function's value with a linear relation between partial derivatives (of the function) and a geometric (or coordinate) measure, in relation to the geometric measures on the domain space.

In this context the function's values must be consistent with the measured geometric properties on the function's domain space (or system (or shape) containing coordinate space), or equivalently, (the function's values must be consistent with) the measurable (geometric) properties of the coordinate variables (in which the [solution] function's values are expressed as a formula of the coordinate variables).

If there is no consistency between the function's values and the coordinate space measures, then the fundamental numerical values (or arithmetic properties) of the function and the numerical values of the domain space do not match-up so that the rational number structure of the function's values and the rational number structure of the domain space cannot be placed in correspondence with one another.

Thus, new function values can emerge which violate the property of closeness, ie discontinuities created by inconsistent rational number structures between the two sets. That is, new types of rational number values come-forth (or pop-up) with small changes in domain values which are not consistent with the properties of continuity.

Note*: One only needs to be concerned with the rational numbers since G Chaitin has identified the property that a general real-number requires an infinite amount of information to identify. Thus, the real-numbers are not practical quantities to consider, and properties identified so as to be related specifically to real-numbers (not finite rational approximations) seem to have no information which is useful, and thus are as good as illusions.

Claim: This non-continuous change in a function's values in relation to the domain variables is what the property of chaos . . . (of a solution function to a non-linear differential equation [see below]) . . . . actually is (ie a numerical property, of inconsistent jumps in value (inconsistent with the local linear approximation of the function), which is the cause of chaos).

The differential equations of classical physical systems are about one type (or set) of local geometric measures (of material) which are set equal to an equivalent (but different) type of geometric property (which is represented in a different mathematical form), so that the values of this other geometric property can be identified at the point (or points) where the local geometric measures of the system are being measured.

In classical physics, this is modeled as "a geometric representation of force is set equal to a mass's acceleration at the same point in space and time."

In geometry once a metric-function is known for a manifold then one can find a differential-integral equation whose solution, associated to a path (or sets of paths of equal length), minimizes (or optimizes) the measure of distance between any two points on the manifold.

If a solution function of a physical system's differential equation is to be useful then the differential equation needs to be linear (to preserve the function's quantitative structure (or its uniform unit)) and the system's coordinates are to be both separable and their metric-function is to be metric-invariant (so that the coordinate values are consistent with the uniform-unit structural basis of the coordinates' geometric measures), where separable means that the tangent directions of all the coordinates are always perpendicular to one another at each point in the coordinate space, eg the sphere, the cylinder, and rectangular coordinates.

However, if all of these properties: metric-invariant, separable, linear, are not the properties of a system's differential equation . . . [or a local tangent relation (on a solution function's graph) between a system's measurable properties and its coordinates] . . . then non-linearity becomes a property of such a system's differential equation, in which case changing boundary conditions no longer identify linear relations between a solution function's values (with different initial conditions), this is because the derivative no longer identify a valid linear relation between the function's values and its domain values.

Because of the property of non-linearity, new quantitative relations between the two sets (the function's values and the domain space) enter the description without (apparent) cause, which leads to chaotic (quantitative) properties, such as chaotic motions. That is, non-linear differential equations will not be useful at identifying "how order enters the material world" because non-linear relations do not identify valid quantitative relations between a function and its domain space, and chaos is a result of the non-existence of an actual quantitative relation between the two quantitative types being linearly compared by the differential equation.

This means that the coordinates for the physical system must be geometrically separable. That is, the tangent directions of all the coordinates are always perpendicular to one another at each point in the coordinate space, eg the sphere, the cylinder, etc.

If the coordinate measures of geometry of the system's description are not consistent with the geometric measures of the system-containing coordinates (or the function's domain space) then the measurements of the properties, ie the (solution) function's values, do not fit into the quantitative containing space in a consistent way. That is, the variables associated to both the coordinates', and the function's values . . . (based on a formula which is dependent on the coordinate variables) . . . are not consistently relatable measuring values, if the coordinate measures of geometry of the system's description are not consistent with the geometric measures of the system-containing coordinates. For the coordinate measures of the geometry of the system's description to be consistent with the geometric measures of the system-containing coordinates, then the coordinates must be separable (and metric-invariant, and the function be a solution function to a linear differential equation).

Such a measuring inconsistency between a function and its domain is the (a) property of non-linearity.

In the viewpoint of classical physics, measurable properties (or solution functions to the physical system's differential equation) are related to geometric properties within the containing space so as to identify a differential equation either for the system or for the geometry itself.

How are these geometric properties manifested?

(1) Are these geometric properties manifested as material, which in turn, identifies a shape within the containing space?
(2) Are geometric properties associated to abstract shapes within the containing space, ie shapes identified by abstract functions?
(3) Is the abstract shape the actual containing space or the domain space for the function or equivalently for the measurable property defined on the shape, ie a manifold?

What is the math structure within which these geometric properties are measured, eg coordinates and their geometric measures?

If the geometric properties are material shapes then coordinates whose shape is consistent with (or follows) the material shape can be used, in which case tangent directions to the coordinates, along with the use of differential forms, can be used to identify geometric measures on the shape, but the differential equations which identify a measurable property, ie a solution function to the differential equation, must be linear. If the natural curved coordinates for the material system are separable then it is possible for the system's differential equation to be linear.

A manifold of "tangential" dimension, n, can always be immersed in a containing (spatial) space of dimension, 2n, or less (ie down to dimension n+1), within which the geometric properties can be measured with the math structures of a higher dimensional coordinate space and then these measures are dimensionally projected down to the point of tangency on the manifold.

Thus, a manifold containing space is the same as a coordinate containing space in regard to how (or the structure of the) geometric measures are made on the shape (or manifold) in the containing coordinate (spatial) space of the manifold's local coordinate structure, where for a local coordinate structure modeled as R(s,t) the full dimension of such a manifold's local coordinate structure, R(s,t) . . . , where R denotes the real numbers, and where s is the dimension of the spatial subspace, and t is the dimension of the time subspace,

.... so that the full dimension is, n = s + t.

That is, local geometric measures in coordinate patches on a manifold can be used to make the local measurements of the system's (or geometry's) properties, but there are new dimensional considerations on the local coordinate patches, ie one must remember to project back to the manifold's tangent space.

Non-linear differential equations emerge from an inability of the function's values to be consistent with the measurable values on the domain set. That is, local linear approximations of a function's value can be made which are consistent with the function values of the (linear) differential equation.

The property of non-linearity is brought about due to:

(1) the differential equation is not linear and thus the function's values can not be identified with a uniform unit (of the function's values) in local approximations, and
(2) the fact that a metric-function can be represented as a matrix which is always symmetric, and thus there is always a coordinate system within which the matrix of the metric-function is diagonal. However, general coordinates are not always (usually not) separable and these coordinates are not consistent with the local measurable values of the metric-function. Thus, the differential equations defined for such a system which is defined on general coordinates are non-linear differential equations, and thus (in general) their solution functions will be chaotic,
(3) the fact that, if a coordinate system has the same diagonal structure for the metric-function everywhere so that the coefficients of the diagonal terms of the metric-function's matrix are constants, then it is a metric-invariant space. If the diagonal values of the metric function are themselves functions so that these diagonal values change from point to point in the coordinate space then this metric-space will have curvature then the coordinate measures in the local approximation of a function's values at a point will be functions so that the local approximation of a function's value will not be linear thus the function's values and the coordinate values will not be consistent with one another. Subsequently, the differential equations defined in these coordinates will be non-linear.

or using more refined (but standard) ideas for a strong statement,

If the diagonal values of the metric function are themselves functions so that these diagonal values change from point to point in the coordinate space, then this metric-space will have curvature and this implies if one uses coordinates which diagonalize the metric-function locally [or the local geodesic coordinates which can be obtained from the metric found by solving the Ricci curvature tensor set equal to zero] then in these coordinates a differential equation will be non-linear, which in turn implies chaotic dynamics, or chaotic interactions.

Thus the values of a metric-function in a metric-invariant space whose coordinates (for a system's differential equation) are geometrically separable, ie they diagonalize the matrix of the metric-function, then the values of the function and the values of the coordinate space will be consistent if both the coordinates are metric-invariant and the system's differential equation is also linear.

Note: A linear differential equation preserves the structure of the numerical values of the function's values.

Thus, numerical consistency (a linear differential equation) and the property of consistency of the function's values with the geometric measures of the domain space (geometrically separable) and that the coordinate space be metric-invariant, are all three properties which are necessary for a quantitative description so that the function's values and the domains geometric measures are consistent.

It also turns-out that the properties associated to linear differential equations lead to the control of such a system by controlling its initial conditions or its boundary conditions, where the different controllable conditions affect the solution functions values (at the point of differentiation) only through the relation of a linear function.

Note: There are metric-functions for separable coordinates whose diagonal elements change with position . . . , eg the metric-function of a sphere,

Note: The one problem in general relativity which has a solution, ie the one-body problem with spherical symmetry,

> . . . . such metric-functions are also related to non-linear differential equations and subsequently they are also (usually) related to chaos.

The Ricci tensor is symmetric . . . , due to a general metric-function being C^2 . . . , the second partials are continuous . . . , thus there is always a coordinate system in which the Ricci tensor is symmetric, thus the diagonal Ricci tensor has diagonal elements which are functions, not constants.

If the Ricci tensor has only constants on its diagonal then the space would have constant curvature, spaces with constant curvature are topologically equivalent to space-form geometries. That is, otherwise the metric-function would be metric-invariant.

Chaos (whose cause is identified above) is a condition (of a property of randomness) in which the elementary event space (of chaotic events) cannot be identified, and thus it is outside the capability of probability to identify probabilities of these random chaotic events.

Main problem of physical description

The problem of physical description which quantitative description is most in need of describing is to find the math structure (or proper measuring context) of the stable, definitive spectral-orbital properties which material systems form at all size scales: . . . . nuclear, atomic, molecular, crystalline, planetary orbits, and the structure of galaxies (the motions of stars in galaxies and the shapes of galaxies).

Conjecture: Stable, definitive spectral-orbital properties at all size scales is more related to the controllable, predictable, useful properties of the stable, separable geometries of controllable systems of classical physics, than it is to probability, function spaces, or than it is to non-linearity.

Non-linearity of a differential equation means that local geometric measures are not compatible with the (its) solution function (or not compatible with the measurable property which the solution function represents).

Conjecture: The non-linearity of a differential equation implies that a separable geometry does not fit into either the physical system, or the geometry, and (Conjecture): such a physical system (or such a geometry) is not stable.

Claim: The problem of describing the proper context for the stability of physical systems is to be solved (in relation to describing measurable properties which lead to [control] or to stable definiteness) in the structures of metric-invariance, linear differential equations and geometrically separable.

The place in mathematics where these three properties (the structures of metric-invariance, linear differential equations, and geometrically separable) are prevalent is in relation to discrete isomtery (or unitary) subgroups of any dimension and of any metric-function signature, where the signature of a metric-invariant metric-function on R(s,t), where s+t=n, and the signature is s-t.

Note: R(s,t) could be C(s,t) where C denotes the complex numbers (with a Hermitian structure defined on these complex coordinates). The complex coordinates C(s,t) are related to the shapes of discrete unitary subgroups (or complex space-forms).

These three properties (ie the structures of metric-invariance, linear differential equations, and geometrically separable) must be satisfied if the main property concerning the stability of physical systems is to find its proper descriptive structure (or proper measuring context), but implied in this descriptive structure, based on solutions to differential equations, is the idea of continuity.

What relation do discrete isometry subgroups have to the properties of continuity?

Continuity establishes them as having a very important relation to physical description and to the fundamental properties of geometry.

What are the main properties of continuity?

Materialism is the idea that all of existence is defined by material existence. If one considers the density function of material, the relative amount of material, ie (material measure) / (volume measure), in a small region about any point in space, then the property that "this density function is continuous" is a necessary implication of materialism, otherwise there would be a further (unseen) context within which material exists, ie an unseen context beyond the three-dimensions within which material can move.

Note: Both particle physics and string-theory claim that there is an "unseen" context for existence but particle physics only follows data, as Ptolemy's epicycle structures followed data, with very limited uses of the particle collision description (claimed to be important in bomb engineering) and string theory has no useful results. Thus, these are both descriptive languages of abstract speculations, whose abstract basis may be as illusionary as a general "real number" is an illusion.

A geometric property which is conserved under continuity (or continuous deformations) is the property of the existence of holes in the space, ie not simply connected [or not higher dimensional homotopy analogs of the simply connected idea of continuously deforming a closed curve in a space],

> (a space is simply connected if any closed curve can be deformed to a point, ie the space has no holes which obstruct deformation to a point).

Both "material is conserved" and "the hole-structure of the space is conserved" under continuity. Another conserved physical property is the total energy of existence, ie the "total composition" of existence is invariant to translations in time.

This leads to the evidence that during some high energy collisions of small particles (or small material systems) the particles change to energy, thus challenging the idea of conservation of material, but these types of events are observed under exceptional conditions, and they are marginally measurable events, usually dealing with "particles" which are unstable in the first place, ie they do not naturally belong in the space within which their instability was observed (in a small time interval) in the first place.

Atoms break apart under high energy collisions, and atoms can emit light (if not in its ground energy state) and thus lose energy.

Continuity preserves the number of holes of a shape. Does it also preserve the quantitative content of the property which a space-form represents and which are contained within the hole preserving structure, ie the conservation of material and energy? That is, are the "conservation laws" a result of a multi-dimensional space-form structure as the correct model for the over-all containment space of all existence, and thus the geometric structure of the hole is preserved and the quantitative content of the property which a space-form represents . . . . and which are contained within the hole preserving structure . . . . is also preserved, because each different dimension (general) face of a high dimension space-form simplex also identifies a geometric hole structure . . . , by means of identifying congruent opposite faces (of a general dimension) with one another so as to form a geometric shape (the space-form shape) by an equivalence topology. The conservation of a hole structure also conserves both material and/(or its) energy.

It can be conjectured that: The main measurable geometric property, or the stable distinguishable property of shape, is the number of holes the shape has.

The main (associated) geometric property of discrete isometry (or unitary) subgroups (or space-forms) is their property of defining holes in space, and the natural coordinates of these shapes are separable.

One can assume that a multi-dimensional structure for existence is determined by space-forms of various dimensions, so that each dimensional level is geometrically independent of the other dimensional levels, within a high dimension space-form containing space structure. Furthermore, each dimensional level contains independent material space-forms whose dimension is (usually) of one-dimension less than the dimension of their containing metric-space (level).

Space-forms are related to "cubical" simplexes and thus the different dimensional levels of existence are related to the different (but particular) dimensional faces of the over-all containing space-form, or containing "cubical" simplex (of high-dimension).

Interacting material space-forms [contained in a particular dimensional level (or contained in a particular dimension face)] along with their spatial separations together form an interaction space-form, which is the same dimension as the material containing metric-space's dimension, but as a (moded-out) space-form it would be contained in the next higher dimensional metric-space.

This interaction space-form can be interpreted to have the structure of a (classical) system's differential equation.

That is, a differential equation for a physical system is (fundamentally) identifying an interaction space-form structure.

Thus, systems which have non-linear relations which define their differential equations, ie the material geometry of such a non-linear system is not separable yet the interaction space-forms are still space-form structures, but the dynamics which the interaction space-form causes are chaotic, nonetheless the interaction space-form can have holes which can become associated to the solution function (or dynamic) properties, ie the cause of the existence of limit cycles in the (qualitative properties of) solutions to non-linear differential equations.

Note: Usually the more detailed structure of a geometric shape (different from its hole structure) is related to general curvilinear coordinates, thus the geometric measures on the curvilinear coordinates are not consistent with the natural (separable) geometric measures on the metric-invariant metric-space, and thus it is a non-linear property, and thus not stable. A description of detail which is non-linear is a description (of detail) which has very little useful value.

Nonetheless, the main properties of physical systems are their conserved and/or stable properties.

Note: The many component statistical structure of interactions between many small material space-forms results in each small space-form following a pattern of Brownian motion, and this property can be used to build a math structure equivalent to the random, wave-function, properties of quantum physics. (this relation was described by E Nelson, Princeton, 1957, 1967 (?)). That is, space-form dynamics can provide a cause as to why quantum events appear fundamentally random so that this cause is statistical in relation to the large numbers of many small scale interactions, where the interactions are fundamentally geometric, not probabilistic.

When the geometry of an interaction is (or can) changed into a separable geometry then the interaction structure can become stable, or by resonance the interaction can result in a new stable space-form system, not simply a stable dynamic system, which might eventually destabilize due to friction, or due to other interacting space-forms.

Conjecture: The stable solar system is stable because it has entered into a stable, space-form, structure which is in resonance with some corresponding space-form geometry in the greater containing space.

Non-linearity does not describe systems which are stable.

On the other hand, metric-invariant, separable coordinates associated to a system's measurable properties which identify linear differential equations will be the context in which the ordered measurable properties of the: nucleus, the atom, the molecule, the crystal, the planetary systems, and the structure of galaxies can be described, so that the properties of these systems will define stable and definitive spectral-orbital properties.

It might be noted that continuity is related to conserved physical properties and conserved geometric properties, namely (for geometry), the number of holes that a geometric shape has will be preserved for continuous deformations of shape.

In metric-invariant spaces discrete isometry (or unitary) subgroups (or space-forms) identify separable geometries characterized by the number of holes which they contain on their space-form shapes.

The proper context within which to describe the fundamental stable, definitive spectral-orbital properties of physical systems without paying attention to irrelevant details is that of space-forms in a multi-dimensional context.

Note: The differential equations which define space-forms are linear (see Riemannian Geometry, L Eisenhart).

Furthermore, differential equations are best placed in a context of "interaction" space-forms which inter-relate three different dimensional levels of the multi-dimensional containment set, so as to identify both the local dynamic properties (the edges of the interaction space-form's simplex structure) and their relation to "global" geometric properties of the interaction (the spatial 2-faces of the interaction space-form's simplex structure).

Hyperbolic space-forms have stable definitive spectral-orbital properties for any size scale. Hyperbolic space-forms are geometrically separable yet they have a limited number of geodesics. (Why?)

In a multi-dimensional containing space, whose geometry is based on a high dimensional space-form shape, there is geometric independence between dimensional levels, yet for a space-form contained within a particular dimensional level, there is spectral dependence on the spectral-geometric properties of higher dimensional space-forms (or their different subspaces of the same dimension) contained in the over-all high dimensional containing space-form geometry. Thus, there is a dimensional-spectral property which relates, by spectral resonance, the spectral properties of (different) subspaces, of other higher dimensional levels, to the spectral properties of space-forms which are contained in a (or any) given dimensional level of this containing set.

Material interactions within a particular dimensional level are modeled as interaction space-forms, so that classical differential equations (or any [physical] differential equations) are best interpreted in this context.

Thus, the descriptive context, for the descriptions of the properties of existence, is a "multi-dimensional"—"space-form"—differential-equation—"interaction-resonance" context. This is a context which transcends materialism; and life, intent, mind, and morphogenetic fields (needed to describe embryonic development) can all be modeled as space-form structures, or can be modeled in regard to space-form properties (or space-form relations) in this new context.

If this knowledge is correct (and it is) then this will open-up a new context (within the properties of existence) for creativity.

** To get an idea about what the media presents as math, both professional and popular, one can read an interview with a Fields-prize winner, ie an intellectual oligarch of our society, in the magazine Discover to get an idea. It is the ideas of society's intellectual oligarchs which get expressed in the professional journals of our media-education system. Some mathematicians became intellectual oligarchs by aggressively (or blindly) competing in a very narrow context, extending the traditional dogmas which have come to define the professional mathematician, essentially a dogma defined for them by the military industry.

Mathematicians claim that non-linearity is fundamental to describing the properties of the world . . .

> (apparently he believes there are non-linear properties that cause order to emerge out of a world based on the random events of the non-linear math structures,
> [Can order emerge out of ever greater levels of randomness?])

. . . . but since solutions to non-linear differential equations are chaotic they cannot be pieced together in a manifold structure to get a valid picture of the manifold's geometry. Thus, (it is claimed) topology is needed to identify the over-all topologically conserved properties of the manifold, but the geometric properties of the manifold must emerge from (local) geometric properties themselves and subsequently pieced together, but (because of chaos) they cannot be consistently pieced together, so "From where (what process) are the geometric properties identified by topology emerging?"

Mathematicians spend much time working on string theory and general relativity, where they claim all the properties they identify (or prove) are rigorously true. However, a rigorous truth, ie a truth based on assumptions, such as Ptolemy's epicycles (which results in a never-ending correction of the assumption so as to fit data) can lead to (or be based on) illusions.

Note: This never-ending correction of the assumption so as to fit data is the state in which current science and math are now in.

Later in the article, the mathematician claims "It is difficult to imagine that such deep structure (the math of string theory and general relativity) corresponds to nothing. Everything fundamental in math has ultimately had meaning in the physical world." Apparently, unless it does not, epicycles seemed to be a very deep math structure.

Nothing useful or even relevant has come from the descriptive languages of particle-physics, general relativity, or string-theory (except particle-physic's relation to bomb engineering) etc.

Mathematicians have the markings of a self-centered, autistic types (with all the accolades a society can give to its loyal, devoted experts) who is more pleased to deal with a bunch of memorized assumptions and its contest-like relation to social success, subsequently his interests follow the math related to bomb-engineering (where probability of particle collisions is central to describing a bomb's properties, ie the narrow place of application for particle-physics and string theory etc) and military engineering interests of the military industry, than he is capable of seeing what his social position represents . . .

> (namely, complete support of and devotion to the military industry, which is what science and math became after WW II, where one of the chief architects in this academic institutional processes of change was E Teller, and Teller's [and the NSF's] vision of domination over what they consider to be its intellectual aristocrats, though elite science was already well formed in the nineteenth century, where Faraday held [for a long time] a low position in the arbitrarily hierarchical science community of his day)

. . . and even mathematicians being able to analyze the validity of non-linear differential equations on the simple level of "Do their solutions even represent valid quantitative relations with their domain spaces?" Where the obvious answer is that, they do not.

The US "press" as well as the education system represent important parts of the country's public relations institutions which turns science and math into propaganda. Academic math and science builds highly rigorous mental structures which seem to have either no uses or very limited uses.

That is, what the academic scientists and mathematicians describe is either irrelevant or a description of an illusion.

Authority and high social rank have been re-instituted as the correct representation of truth in the US society, just as the authority of religion used to represent absolute truths in the days of Copernicus.

Chapter 5

UNITARY TRANSFORMATIONS:

Relativity, Quantum Mechanics, and the Observer
or
Unitary Transforms and Geometric Changes in Metric Spaces
(or Changes Between Metric Spaces)
or
Practical Teleportation and Astral Travel

By George P Coatimundi

Relativity, Quantum Mechanics, and the Observer
or Unitary Transforms and Geometric Changes in Metric Spaces
(or Changes Between Metric Spaces) or
Practical Teleportation and Astral Travel
By George P Coatimundi

Abstract

The ideas of this paper and the two books "A New Copernican Revolution," and "The Authority of Material vs. The Spirit," mostly come from the idea that the spectra of physical systems can be determined from a more constrained viewpoint than the viewpoint of function spaces. Namely, the spectral properties of physical systems come from the spectral properties of metric spaces, ie discrete isometry subgroups. Discrete groups are also represented in terms of reflection groups and E Cartan formally invented the idea of spin from considerations of reflection groups.

These spectral structures, which when "moded out" form space-forms, are very geometric, easily pertaining to atoms, crystals, dynamics, the solar system, the structure of stars and the structure of galaxies. Except for the Euclidean space discrete group, or space-form, the properties of the spectral structure of metric spaces with metrics that have other signatures (and other physical properties associated to themselves) are hidden from perception, including most of the properties of the space-time metric space, also known as hyperbolic space.

In this paper, the spectral-space-form structure of an observer is modeled and their relation to star travel elucidated. General relativity is examined and often found to be improperly developed.

The solar system's stability depends on its discrete isometry subgroup structure. Hence travel to particular regions of the universe can be described in terms of spectral transitions from one spectral structure, of one planetary system about a star, to another such spectral (or orbital) structure. The Euclidean part of a physical system's interaction structure is accessible to action at a distance. Our perception is dominated by our sense perception of the Euclidean part of physical systems, yet systems must be composed of various fixed metric spaces with a variety of spectral properties, including spin and the spectral properties of light. Remember we see only where light interacts with matter, we do not see the propagating wave itself.

The dynamics of space-forms and their spectra can be related to the covariant derivative.

This theory is consistent with both the relativity theories, ie special and the general theories.

Quantum mechanics can be derived from the Brownian motion (or stochastic processes), which this new theory describes, and this was done in an abstract way by E Nelson and his students. Yet this theory is opposed to the philosophy of quantum mechanics, in particular, the quantum mechanical claim that all physical theory is (or that all physical description must be) based in probability. This is because the spectral properties of metric spaces can be substituted for the spectral properties of function spaces.

This new theory is consistent with dynamics and extends and clarifies the notions of integrability and stability of physical systems. It also supports the atomic hypothesis, though in a limited sense. This is because it is not a reductionist theory, yet at each of its dimensional levels there exist natural (space-form) components upon which physical systems are built at that level. Both atomic stability as well as planetary orbit stability result from the natural space-form structure of material systems at each (different) dimensional level. Existence depends for its properties on many levels (or dimensions of metric spaces) so that these metric spaces have various properties depending on the signatures of the different metric spaces.

This new theory clarifies the re-normalization group which is used to describe sudden transitions between new spectral-geometric structures for physical systems.

Furthermore, this new theory re-arranges all notions about the shape of space, and it questions how general relativity frames physical problems, eg using all of space to define a one body problem while using a questionable form for a general metric. General relativity is framed within a fixed dimensional, unbounded containing space. This is another underlying assumption that goes unchallenged, and the reason for this, is that such an idea about space can be used in conjunction with a particular interpretation of differential equations in which a scalar model of macroscopic material is used, so that when these ideas are combined together they necessitate a particular form of spherical symmetry for the geometric structure of the field structure for material. These false arguments are also used to infer that space can only have dimension three, otherwise higher (inverse) degree forces dependent on material separation distance that are involved in material interactions would be observed. However, there are other mathematical patterns which can be used which can supercede this dogmatic notion of the absoluteness of space and how material must be modeled (essentially as a scalar quantity) in a differential equation in order to fit into this space.

Introduction

Civilization has developed a totalitarian viewpoint of science. The leaders want only ideas that they already possess. There can be no new ideas, only variations on the old theme. Thus scientists have reduced themselves (or uplifted themselves depending on one's servile viewpoint, [eg a few important scientists desiring greater compensation]) to be the servants of the few owners (in civilization) of all material. Science is now driven more by authority than by data and reason. It is those that head the institutions of science that uphold the old theme, and in doing so support the interests of the few owners of all civilization eg star wars becomes more important than cheap clean energy or sanitary conditions for all.

John Horgan, in his 1994 book, "The End of Science," laments the ineffectiveness of modern science and how it has found its refuge in doing what he calls "lyrical science," the art of describing physical systems that are so remote from experience so as to make their physical properties (save a few) to exist outside the boundaries of scientific verifiability, eg cosmology, the standard model, and string theory. The major problem with the standard model is that it is not an actual well defined structure, either in terms of mathematics or in terms of an understandable language. It is not a mathematical structure, and it can really only "predict" cross sections of colliding particles. However, the "authorities" say that many of these particles which are colliding do not really exist, because they are part of a probability framework. The standard model makes other "predictions" (particles' masses) based on arbitrary parameters, ie experimental values (the measured masses) of the "predicted" values, so that such "predictions" do not seem to be legitimate "scientific" predictions. The foundational bedrock of the standard model are not well defined nor understandable patterns and descriptions, but rather the authority of its practitioners. Is this science or religion?

And finally the conclusion of the standard model is that all interesting physical systems, like a nucleus, are simply too difficult to describe. In other words, modern scientific theories do not describe anything that is of practical interest.

How flowing electrons might interact with spin sub-systems inside a crystal is far too complicated a system for the standard model to describe, in fact it is beyond what the language of the standard model allows. The standard model can only model global (or universal) systems of harmonic oscillators in which all (important) interactions of the system can only take place at a single point. The only thing that the standard model predicts are cross sections of particle collisions, ie probabilities of random particle events

The standard model only describes the limited attributes of elementary particle collision events within particle accelerator instruments.

Science is about a description of the patterns of the external world, so that the description actually predicts measurable aspects of the world, thus leading to the description being of practical value. In other words science should be a description that uncovers a method of control over the external patterns so these descriptions can be coupled to in a classical manner and thus have a substantial number of practical uses. However, it is questionable what the pattern basis for the standard model actually is, the evidence about which the standard model is designed, comes from particle accelerator data which has the mark of irreproducible particle tracks in "bubble chamber" photos. The set of tracks in particle track photos is continually differing from photo to photo, and this indicates the data is coming from a non-linear structure (or pattern). The point of non-linearity is that it is quite possible that one can measure it all one wants and still not identify a pattern that is of any value, or which leads to a description which allows "predictions," ie allows control. Indeed a spectral expansion may not be valid, mathematical patterns are not really identified, renormalization is not mathematically valid, arbitrary parameters makes it suspicious that the "theory" of the standard model is even science, mass remains an invalid idea (2006), the unitary structure still seems to be a lucky guess, and the zero point or vacuum for the single point which represents the system has a fictitious nature, ie systems cannot have subsystem structure. The only place where the standard model is in its element, is in the mythological theory of the "big bang."

New ideas

If one wants to go to new places with ideas, then one must really believe in oneself. Philosophically there are many ideas that one can consider. Science today wants to only deal with the philosophy of materialism based in a logically inconsistent language which depends on probability and a very constrained notion of space and how space can be structured. Yet despite its probability basis today science has very geometric models, eg interactions are modeled as particle collisions, in a probabilistic context, string theory gives these "particles" a string-like geometry, yet still in a probability based language.

Institutions of civilization fix their attention on only a few ideas upon which the leaders have fixed their attention, the "the leaders" being the few owners of material and hence rulers of civilization. It is upon only a few ideas that the owners apparently have placed their bets, and those interests are dominated by ideas about military weapons.

It is creating at a level at which "the forces of one's life are most real" which moves one along. To hedge one's bet and to believe in what others do (or think), as opposed to what oneself does, will make one lose one's way.

In review what is being done and what things people are attempting to do in physical science, but failing, do not be mindful of their claims, that they are about to get it right.

Here is a review of modern physical science's lack of models for:

The nucleus, there exists no model of the nucleus,

No valid or inexpensive control of (or clean) energy generating (nuclear) systems,

Superconductivity, no valid or understandable model of (high temperature) superconductivity,

No scientific progress at controlling the micro-flow structure of the micro-chip, ie no useable model of the interactions of an electron current with the spin subsystem in a crystal lattice,

No ideas about why the solar system is so stable,

No valid model of the energy source of the sun, (How do oscillating neutrinos affect the model of nuclear interactions?)

No understanding of star motion in galaxies, or galaxy motion in the universe. (no understanding of dark matter)

Yet, none the less, there are only authoritative viewpoints about how the professional journals are willing to discuss these topics, only refinements on "lyrical science" are allowed by the professionals, ie those whose "caliber" (or worth) as a person is evaluated so carefully by corporations, so that these "highly developed" people are known to be able to function as corporate handymen. [genetically superior servants of the corporations]

Virtually all the ideas now being discussed in the currently accepted paradigm of physical science are baroque (overly complicated) and useless. This indeed is totalitarianism at its worse.

This is authoritarianism at its worst because science is supposed to be an effort that depends on both public description and public verification. However, it is clear no one understands anything about what science is even trying to describe today (2006). It simply provides a contest (often about "unification" of the laws of physical science) for the clash of "dominant" personalities, ie those obedient to the authoritative law of the ancient "science text." (see below)

How convincing are their claims of truth? What are their claims? Namely, how useful is the description in terms of control and/or guidance? Science has become useless and it follows technology, mostly making claims about immeasurable, hence un-verifiable, systems, ie systems too remote or too small to be measured, ie science remains in the realms of "lyrical science.".

The institutional scientists seem smug in their proof that every fundamental physical system is simply too complicated to describe. But science is supposed to predict and that means to lead. The language of modern science is self contradictory, thus resulting in the state of affairs in science in which all fundamental systems are too complicated to describe.

There do not exist descriptions of either the origins of life, or embryonic development, or mind, or a living organism; yet, these just named systems, are stable and/or controllable, hence either they are simple linear (apparently not) or simple flat systems (the alternative) and not reducible.

Understanding the secrets of existence must mean that hidden underlying structures do exist and need to be described. This is what science is about, and that is exactly what modern science now excludes. In modern physical science micro-events that take place at some point can only be collisions of either a point particle or some localized geometry, ie a string. In either case it is a micro-geometry within a probability based theory, ie either a contradiction or at best an inconsistent use of language.

On the one-hand there is science, and on the other-hand there are the wants of the few who control the material world and its money. Then one has the scientist-authority who wants power, which he thinks is determined within the value structure of other people's judgment. Namely, the judgment of those who control money. It is the few who control the material world and its money who also control where the attention of science is held and how science is used. They do this in concert with ambitious scientists.

Thus, science should be anonymous, and it is technology which should be the natural place where political bodies should make decisions. How technology should be used, is an issue of politics, where politics may be defined as the social power to do what one wants, or to do certain things, eg decide what is made, who uses it, etc.

However, the context of existence is not materialism as science suggests. Science has been pushed into this stance, by social forces, not by its true nature. Science is about using language in new ways in order to find the hidden structure of existence. It will be about truth in the relative context of its language, and because of this, it must be anonymous. Otherwise it will be owned by those who "own" the world's material. In the hands of those who control, and defend their property, science reduces to a convenient list of those well versed in the language of technology, ie a list of handymen.

Science is about adjusting language in order to describe and/or predict a set of patterns that are found by looking at the data of measurements that are related to all of existence. This can only be done if the descriptions are clear and all aspects of the data collection and interpretation are made transparent. The rule for valid science should be: If it is too complicated for an average 12 year old to understand then it does not qualify as science. This means that the fundamental ideas about math, eg the idea of differential equations, need to be brought into elementary education.

Thus the question is, "Who are these authorities to determine the course of science either by 'peer review' or by economic interest?" the answer is that these things are being determined by the authority of society, ie the few rich, and this is made easy by listing the people who use science to work on technology.

If science does not satisfy the criterion of clarity and open-ness for explaining what it is doing so that its descriptions possess clarity, and the ideas are anonymous, then the ideals of freedom and liberty are lost. People are free only if they are also equal.

The ideas of science have been brought within a world social order which ultimately is based on the worship of personality cult. It is clear that we have sunk into a totalitarian social order. Language has become controlled by the few. The authorities (important personalities) that are now defining science, are upholding a mumbo-jumbo of self centered gibberish called scientific authority. These personalities are toads to the few, and thus to the economic and coercive "masters." It is these "masters" who in effect represent the sensual interests [violence, breeding, emotionalism, and self righteousness] of civilization. It is the few who own and control civilization and the bad communication systems, or more appropriately the self serving communication systems, which institute the sensual and the other destructive forces of selfishness on civilization.

In a word, the authorities, who represent the cult of personality, that are upholding (or own) both civilization and modern science (2006) are imbeciles.

Descartes pointed out long ago that learning other people's knowledge is impossible, and all knowledge is really generated by self examination and one's own integrity. Hence the communication level of technical issues needs to be at a level of a common denominator, namely that of interested 8 to 12 year olds. Science is about uncovering the hidden structure that underlies the world that we exist within.

Hiding this scientific information, by instituting an authoritarian educational system and government, though profitable to the few, is detrimental to the vast majority.

The idea of an ancient "Text" of science, as developed around the ideas of Nietzsche

Of course those hardened to the rigors of a functioning personality cult will chuckle at all of this, since they are in awe of the vast intellectual power of their favorite scientific hero, but the question is: "Of what utilitarian value is today's physical science?"

The answer seems to be that essentially modern science is incapable of describing the outputs of its own instruments. Today technology is essentially being driven by 19th century science because modern science does not have the capacity to describe any fundamental stable physical system, from the nucleus to the atoms, to crystals, to the stability of planetary orbits, to stars and galaxies.

Nietzsche's theory of a "will to power," in which power is all based on fraud and lies, that are all derived from ancient absolute texts, is amply supported by the ancient "Text of Science," elucidated above and below, and by the actions and beliefs of the personalities filling the institutions of modern science.

Science has also been caught up in Nietzsche's notion, that the root of the lies and deception upon which an individual's "will to power" is based, rests upon the "(old) text." In science this "(old) text" was written by Copernicus, Kepler, and Galileo whose combined efforts began a new way in which to use language in order to describe the hidden structure of the world. The world, in their day, was dominated by the authority of its centers of knowledge, ie religious institutions. In our day science usurped the authority of religious institutions and has come to be controlled by corporations (of which the catholic church may be the first best example of a corporation), but this power is not used in the name of science, ie science is about questioning current knowledge, rather this power is used in the name of authority and power itself.

The new language of Copernicus was further developed mathematically by Newton and Leibniz, so as to begin the technical structure of the differential equation and its relation to physical description. The differential equation along with materialism, is the core of the "scientific text." The "text" of (old) science is the idea of materialism and the differential equation. These ideas (in the text) include the coordinate context of metric space frames, which are of a given a fixed dimension and these coordinates are assumed to be unbounded. Thus differential equations which include terms that have material as a scalar factor, also have certain aspects of their boundary conditions pre-assumed, determined by assumptions about the containing coordinate system.

The hidden structure of the world, described by Copernicus, became illuminated by solutions to differential equations. Then the idea developed that the form of the differential equation has become consistent with the appearance of a system's local properties. Thus the law of the scientist's "text" became that, "Local properties are to be interpreted in ways that are consistent with what they appear to be, and that the existence of material defines an unbounded metric space of a fixed dimension." It is this unbounded metric space defined by the existence of material that has become the context within which differential equations are defined, differential equations which include terms that have material which is modeled as numerical factors.

The properties of material, ie solutions to differential equations, which describe material positions in a coordinate space, depend on the containing metric space's dimension and its unbounded properties along with the scalar properties of material. The differential equation re-instituted the idea that "local" appearances can be trusted, that is, local appearances can be interpreted to be, what they appear to be.

Einstein contributed to this "text" of science, in that he emphasized the central position of the metric space (or its metric), both its dimension and its transformation properties in physical description, ie invariance of the form of differential equations which model physical systems, are (both) very important.

Though quantum mechanics commands that probability be the basis of physical description, its origins are really simply the extension of the "text" in which solutions to linear differential equations can be represented in terms of sums of spectral functions of function spaces. The derivative is an operator, ie a general function, that is defined on function spaces, ie the (derivative's) general function's domain set is a function space. Though wave equations and heat equations were already using spectral expansions as solutions a new differential equation was needed for matter waves. Thus, for quantum mechanics, it was also needed to extend the list of differential equations that are applicable to physical systems to include deBroglie's matter waves, ie Schrodinger's wave equation. In quantum mechanics each spectral function, of a spectral expansion, represents a quantum (or discrete) property of the system (often a harmonic oscillator), so that together the spectral functions determine a probability for finding material events. Localized "small" events, eg discrete light spectra events or discrete (small) energy changes, became interpreted in a manner consistent with differential equation formulation. Namely, these local events are what they appear to be. (This interpretation of local events has become a big thorn in the side of science). Thus the only really new things for science that comes from quantum mechanics was Schrodinger's equation and subsequently, that the function space was re-interpreted as a probability space. This probability interpretation is something that mathematics and physics already had done for certain of its problems, namely the physical systems that were composed of many atomic particles, as well as solutions to wave equations could be interpreted in terms of probability.

This has become the essential "text" of the institution of physical science: materialism, differential equations, an unbounded metric space with a fixed dimension, a probability basis for language which describe physical systems, material usually modeled in a metric space as a numerical scalar, and specialized (face value) interpretations of local events. This is the textual basis upon which physical scientist's "will to power" now depends.

Real science

The real science of any age are the attempts to use a new language in order to describe the hidden structures of the world. It is important that the description gives predictions so as to lead the mind into the hidden realms of the world.

An inability to actually predict anything, with any reliability, when no data has been collected is the reason why probability is not of value for a scientific description. The clinging to the fixed structures of old science, ie clinging to its ancient text, is the result of the engineering successes that the old science had, along with a desire to keep the old structure is an attempt to extend that engineering tradition.

Using differential equations to model material contained within an unbounded metric space of a fixed dimension leads to the conclusion that material cannot have the properties that it does have if it existed in higher dimensions. Hence higher dimensions cannot exist. However, this is to be true only if one believes in, and stays within, the parameters of language as put down by science's "ancient text."

Likewise, if the description of material is based in probability, then one concludes that hidden variables (or hidden geometry) cannot affect material properties. Rather, only the "natural" geometry of material are what determine the probability patterns of material (micro) events. That is, the obvious interpretation of local properties, namely, the geometry of micro-particle collisions. Thus language and underlying assumptions of material differential equations, unbounded-ness, fixed dimension, and the "natural" probability basis for the solutions to differential equations all keep material and its properties constrained to certain patterns, ie the discipline of the "official" fixed language.

Or vise-versa, fixing material and differential equations, in turn, requires unbounded, fixed dimension metric spaces, and the local interpretation of micro-material requires probability.

However, it is not clear that such patterns as required by fixed assumptions about language and the patterns this language, and its assumed context, can describe are actually the patterns that are being observed and measured in the world.

A differential equation has two types of solutions

The differential equation has two types of solutions: (1) integrable solutions which give the system a geometry structure, and (2) spectral expansions which are interpreted in terms of probability.

Probability and geometry are logically inconsistent ideas. Spectral expansions depend on the properties of function spaces, however, perhaps there should be another source for spectra.

The alternative is the spectral structures of metric spaces, ie discrete isometry subgroups.

Perhaps the unbounded coordinate structure of single metric spaces of some fixed dimension is wrong too.

Perhaps Einstein did not go far enough in stating the true importance of a metric space in physical description.

Existence and metric spaces

The issue of one's place in existence. How one wants to use one's life force in one's life. The need to know what really exists and its hidden structure. These are questions a person who wants to know about the world asks, ie these are the basic questions of science, and they are the basic questions of religion.

For most people, the "real world" has become a world of money, personality cult, and the decision about what institutional discipline to which one wants to belong. However, this is a fixed viewpoint of the world and it is the opposite of science.

New basis for the description of physical systems

The spectral structure of metric spaces can be used in place of the spectral properties of function spaces in order to allow both integrable, or geometric, and spectral expansion, or series, (or probability based) solutions of differential equations to be logically consistent with one another.

The spectral structure of metric spaces are the discrete isometry subgroups of a metric space and the subsequent space-form structure of these discrete groups. Space-forms allow material to progress beyond the scalar models of material, they allow point interactions to have some hidden structure, and they combine spin spectra and metric space states together in particular in the context of Dirac operators.

Discrete isometry subgroups are defined over metric spaces, and they identify a lattice along with a fundamental domain, ie a simplex on a metric space. When the lattice (or fundamental domain) is moded out, it [the moded out lattice] provides the shape of a space-form upon which both spectral properties and geometric properties join together in a consistent manner.

The spectra of the hyperbolic space-forms are very closely associated to the face structure of its fundamental domain, and these spectra are very stable.

The hyperbolic space-forms have the properties of having very limited and very stable spectral properties. Hyperbolic space-forms can be either compact, ie bounded and with a boundary, or they can be infinite extent.

We are attached to the infinity (or apparent unbounded-ness) of a particular metric space through lines of the world, ie infinite extent interaction space-forms, eg light. Though some of the light and interaction space-forms do extend to infinity they are mostly dependent on, or the result of the spectral properties of the metric space within which they are (mostly) contained. Namely, in their spectral relation, eg their colors, to the metric space's space-form boundaries which are, in fact, sub-faces of a simplex structure which, in turn, form the basis of the spectral structure of the space-forms and material that are contained in the metric space.

There are many different metric spaces which can both contain and be used to compose systems. There is a dimensional hierarchy of self similar metric structures and space-forms upon which one can define the following fundamental aspects (attributes) of a physical system: metric spaces, material, and interactions, forming self similar patterns of space, material and interactions, as one moves from dimensional level to dimensional level.

Infinite extent space-forms are connected to properties of a system which are all distinguished by metric space properties such as: dimension, signature, and metric space state, as well as material, its metric space properties and spectral types. Infinite extent space-forms identify the field interaction structures of a metric space, eg electromagnetic (E&M) structure (or light structure). The E&M theory fits into a space-time or equivalently a hyperbolic metric space. For each metric space spatial sub-space dimension, there exist different material properties which are associated to each different signature metric spaces that exist at that level. For example, for three spatial dimensions there are two distinct signatures: Euclidean space and hyperbolic space (or [equivalently] space-time). Furthermore, each such different metric space has a different physical property associated to itself. We are (or existence is) composed of those different materials and properties distinguished by a set of different metric spaces, which are themselves distinguished by signature and dimension, and can be partitioned into spectra due to their discrete group structures. Spin ½ structures and their associated geometries, eg space-forms which define both material (lower dimensional than the metric space) and dynamics of material (higher dimensional than the metric space) [though we consider dynamics to be part of an integer spin structure], are also part of our diverse composition.

Our perception is based on both perceiving (at each dimensional level) the Euclidean space-form shapes that exist at that level, and perception is affected by a spectral filtration process within a structure of spectral set containment, eg a structure of set size, and set dimension. That is, the spectral filtration process depends on the structure of various types of spectral simplexes. In such a view, we are composed of a large spectral-simplex set (or high dimensional space-form) which can be connected with (or made to be aware of) different metric spaces and their geometries. The world we (most of us) experience is but one of these many metric spaces that we are attached to. But it does not need to be this way, we can perceive higher dimensional structures, along with the many different metric spaces of the same dimension, ie other spaces within which we are capable of perceiving.

The observer

At what dimensional level do we really connect to existence? What is our awareness at that level (dimension)? Within our knowledge (of existence, its parts and its various relations) what do we intend to create? These are fundamental questions of science. Whose answers can depend on the language one uses, or the "text" one reads from.

Indeed the material side of life is a limited attribute of one's over-all existence. The structure within which people act and intend, affects both their belief and context of reality. The point of connection to a metric space, within which we often feel trapped, is related to both an infinite extent space-form connection to that metric space (which dominates the structure of our minds) and our education.

Basic education

Limited and fixed knowledge and intention within this context, results in selfish limited action and false issues of value.

The role of civilization is relatively minor, it is about allowing the material side of human life to be easily dealt with, irregardless of people's beliefs, and this material side of life is to be taken care of in as harmonious a relation of mankind to the earth as possible, eg its material capacities and its natural abundance. The choice between material sensuality along with a (false) sense of importance (materialism), as opposed to life's creative relation to existence (the spirit), should be easily understood and distinguished.

Unfortunately, the minor connection to materialism that human life has, is turned into a major point of worship. This is done by stirring up people's desire for the sensual and by institutional religions which are based in materialism but claim to be connected to the creative side. Unfortunately the creativity of civilization is based in selfish interest and it is miss-named the spiritual side. Helping people with their material needs, and helping people become successful is considered the mark of goodness, but underlying all of this do-good-ness are always selfish or petty motives. Helping people deal with their material needs in a harmonious manner in which those being helped are not saddled with an indebtedness to those doing the material providing does in fact define good but these are issues that should not require a deep control by one set of people over another set of people, ie the whole act should be harmonious and without much energy drain and without a lot of interference within the interaction of giving a receiving. Civilization turns all of life into a veiled form of materialism and subsequently into petty issues of self importance built around ideas of absolutes. Actions based in selfishness (childishness) have bad out comes in this process of allowing material needs of people to be satisfied, however, satisfying people's material needs should be done in a context of selflessness (without self). Harmonious relations with earth are more important than profit, and success should always be defined by an individual in terms of their creativity, which is not motivated by (or based on) selfish gain.

Educationally, the baroque technical structures that allow limited applications need to be distilled down into the logical organizing structure which is easy to understand, and which symbolically contains the greatest range of useful descriptive structures.

Fraud, science and civilization

In fact the scientific process is too hidden and so hyped that it is not clear that experimentally determined beliefs are in fact experimentally valid. R Milikan got the Nobel prize for only presenting the data that supported his belief that an electron has a fixed quantum of charge associated to itself, while throwing away the "bad" data.

Science is about uncovering the hidden structure of the world.

In terms of science's attempt to uncover the hidden patterns of existence; any mathematical structure which gives a verifiable description of an external world and which has all four properties of being: (1) quantitative and (2) geometric and (3) spectral and (4) its language is logically consistent, needs to be taken seriously.

Mathematics is key to understanding the problems that modern science is having, or that modern science deals with.

Mathematics

Math is about size, ie quantity (a well defined unit, along with a counting process), and shape (geometry). Quantity, ie numbers (integers and their relative sizes along with more precise fractions and mixed numbers only to mention the difficult structure of non-rational numbers), fits into measurable (geometric) patterns, eg metric spaces, scale changes (or rates of change) and formulas for "newly" defined quantities.

Quantities which are defined as formulas are always put in relation to certain "fixed" quantities, (ie systematic allowed changes of scale), that is measurable quantities, whose measurement can be made with rods (length), pendulum swings (time), balances (material), or with formulas whose variables represent these measurable quantities. Many quantities (or sets of quantities) determine containing spaces, ie metric spaces, or thermal (differential-form) quantitative coordinate systems, which can have many dimensions.

Differential equations =systems.

Local scale relations (or rates of change) set equal to the same type of other local linear quantitative properties of the system, at any particular point of the system, allow differential equations to be defined. The structure of differential equations which represent physical systems are based on what are called physical laws.

The solutions of these differential equations are either integrable (geometric) or spectral expansions (probability). Most solutions to differential equations are solutions to differential equations which have the property of being linear, ie all solutions that one can find to the differential equation can be summed together to form another solution. Though a few integrable differential equations are non-linear (most are linear) none-the-less all solutions based on spectral expansion are solutions to linear differential equations.

Properties related to the form of the differential equations, and the affect this form has on their method of solution, are:

(1) Coordinate orientation,
(2) The shape of the coordinate system, and
(3) The symmetries and invariance's of the system, these properties affect the form of differential equations and their solutions.

Invertability: coordinate transformations, or local coordinate transformations, eg matrices acting on vectors (or n-tuples), must be invertible. Local orthogonal coordinates allow the local linear coordinate transformation structures to be invertible, ie allow local matrices, to have the same dimension as the coordinate space itself, eg n, as opposed to having a dimension of, n x n. This is called diagonalization of the coordinate transformation matrix. The same idea of diagonalized infinite dimension matrices, that act on infinite dimensional function spaces, allows the definition of a function space's spectral sets. The idea of orthogonalality allows the idea of local independence, which, in turn, is related to the idea of a minimal set of containing coordinates, and in the infinite dimensional case, to a minimal spectral set with which to represent the system's properties, ie a spectral set with which one can represent the solution to the system's differential equation.

Consistency of form of the (mathematical) measuring process for length through out the coordinate system (isometries) implies metric invariance's. Metric invariance's that are of a certain type, in a certain type of metric space imply properties of conservation, eg rotations (rotation momentum), translations (linear momentum), time (energy). Consistency of measuring scale is like identifying a unit for a set of quantities of a certain type. If in a containing fixed metric space the unit of measurement is always changing, as in a general metric space, then this change of metric, or measurable scale, must not be detectable by using material instruments of measuring (as mentioned above), where the changes in the metric at different positions of the system would only be detectable if the general metric space were to be embedded in the higher dimensional frame of a fixed metric space.

Global, invertible measuring processes, on a "large" set of the domain within which the coordinate transformations are invertible (eg isometries) vs. local invertible coordinate transformation (or measuring processes, eg diffeomorphisms) identify the difference between an inertial frame and a general frame of reference, within which a physical system can be measured. The laws of physics are supposed to be tensor differential equations and hence invariant with respect to the more general diffeomorphism transformations of coordinates (or reference frames). Thus solutions to a system's formulation as a (tensor) differential equation will be equivalent when transformed between reference frames, equivalent in the sense that in a higher dimensional reference frame, in which a fixed metric exists, the dynamics of the two distinct reference frames for the same system, ie the same form tensor differential equation, can both be mapped into the single higher dimensional reference frame (within which they can both be embedded) and both maps (of the dynamics) will be exactly the same, ie the exact same curve and the exact same motion.

Spectral properties:

Most solutions to differential equations are obtained based on the idea of using the solution's (or its containing space, ie context) spectral properties, ie spectral expansions and spectral transformations, eg the Fourier Transform. Note: These techniques imply that the differential equation is linear, ie sums of all solutions are also solutions to the same linear differential equation.

Note: sometimes such spectral expansion techniques are claimed to be non-linear in nature. This confusion might be a result of the logical inconsistency which the language has built into itself. Indeed the ideas changed in this manner will be more difficult to manage mathematically than the transformation ideas of general relativity. I would have difficulty trusting any conclusions beyond local relations that this method might claim.

> The ideas about a function's spectral properties:
> (fixed ideas) function spaces
> (alternative) metric space spectral properties, ie discrete isometry subgroups.

This paper assumes that the spectral structure of the world is best described in terms of the spectral structure of metric spaces and their space-forms, as opposed to the spectral structures that are described within the framework of a function space.

Outline of paper's main ideas

Special relativity requires that there exist an observer whose sense perceptions and appearances, ie his mental constructs aided by perceived data from his senses, are affected by his relative motion.

Quantum mechanics requires that an observer be connected to the space where measuring takes place, so that how the observer, and perhaps his instruments, interact with the system affects the system's properties.

It should be noted that a physical description which requires containment within a function space dependent probability structure, is far too general (or too broad), so as to make the description essentially useless, as has become the case with modern physical science.

General relativity: though the general transformations within the locally invertible transformation group suggests that the metric space structure within which we exist should allow frames where the speed of light is not the fixed value that we know, none the less, the evidence that gravity causes either a red shift in light or a curvature in its path rather than a change in its speed implies that the speed of light remains constant within the metric space from which we draw our experience. This, in turn, implies that the electromagnetic (E&M) metric space structure is fixed. This, in turn, means that the Schwartzchild solution cannot be valid because the space-time metric space is fixed. Hence it is the Euclidean spherical symmetry plus time which should be used to form a general metric that should be used to represent system's curvature (or metric) defined in terms of general relativity.

This also means that general relativity does not give any valid prediction of the precession of Mercury's perihelion because such a prediction was based on a Schwartzchild solution. However, a prediction of the precession of Mercury's perihelion can be found in the book, "The Authority of Material vs. The Spirit," based on the new alternative principles of higher dimensions, metric spaces naturally associated to spectral structures intrinsic to the metric space, a space within which differential equations can be defined and solved, so that the solutions are dominated by the geometry of space-forms, as opposed to being dominated by properties of function spaces.

The E&M field can be associated to hyperbolic (equivalent to space-time) space-forms which have an infinite extent, which in turn can be associated to the construction of a mind, (see below) in a very strong way.

Thus the observer's mind needs to be (and can be) within the model of existence, ie the ingredients are: The containing metric space and its natural spectral structure, along with the maximal torus in the fiber group, along with a model of the observers themselves.

The world's essential material aspects, ie its geometry, its spectra, and its changes, are carried by light in a metric space with a fixed metric.

Spherical symmetry is a property of Euclidean space, but not a natural property of the space-time (or equivalently hyperbolic space) structure of E&M. If this is true, then the E&M changes that occur in the physical world must result from flat-shapes that exist within a set of metric spaces.

Materialism myths

People believe that life and mind have a basis in materialism. There is no scientific evidence that has any validity which demonstrates this idea. This idea is based on the prejudice that material defines existence. There is no valid theory based in materialism which describes life's origins, nor embryonic development, nor is there any theory that the mind is of a material origin, other than pre-conceived ideas. So there is no basis for these viewpoints. To tell someone otherwise, is like telling someone that the earth really goes around the sun, but not long ago, it was clear from observation, that the sun goes around the earth. Likewise, the obvious "observation" made today is that life and mind must be made of material as we already know it to be. In fact the pre-conceived ideas about the properties of life and mind being based in materialism are not scientific ideas because science is about the hidden structures of existence which are to actually be described and verified.

In fact mind does depend on spectral structures in a very strong way, and spectra are important in understanding material properties. So the new alternative ideas are based in some ideas everyone is familiar with, just as everyone, in the age of Copernicus, was familiar with the idea that some space contained both the sun and the earth, it was simply not clear, at that time, how this space was organized, and it seems to be organized in a way in which appearances and pre-judgments are wrong.

Return to outline

The fact that the E&M structure stays in a fixed metric space implies that the mind is a stronger component in determining physical existence, than the affect that the distribution of a system's material in a containing metric space has in determining the hyperbolic metric space's metric. The issue becomes that of determining all the properties of the system containing metric spaces. That is, several metric spaces simultaneously contain a physical system. A metric space's local structure is deformed by both the space-forms that it contains, and the underlying spherical symmetry (for dimension <4) that emerges from the low dimensional group structure properties of a principle fiber bundle and manifests on Euclidean space.

The observer is best represented as a complex, high dimensional, hyperbolic spectral-geometric simplex that has many faces (of the same dimension), which may be associated to metric spaces, within which the observer can perceive. An observer perceives within an external metric space structure that is in resonance with one of the observer's internal faces of her simplex structure.

Regular dynamics can be related to the derivative or connection (covariant derivative) which is naturally placed in a fiber bundle with a unitary (or spin) fiber group. As for properties of E&M and hyperbolic space consider $dA = \gamma \cdot j$, where, γ, is a spin factor related to the square root of the generalized Laplacian (or wave equation), and this is an equation defined on the differential forms that are naturally defined on the faces of the material and interaction space-forms' fundamental domains, and results in a Lie algebra element because the geometry of the solution depends on the structure of space-forms in the containing metric space, so that these space-forms determine how the local Lie algebra elements act (or operate) on the metric space's coordinates for the material (or space-form) system.

Change in a function, $f(x)$, with respect to its coordinates, x, ie the derivative, d, must be accompanied by a change in the coordinates themselves, ie the infinitesimal coordinate change, A, where A is a Lie algebra element of the principle fiber bundle's fiber group. Thus the derivative, D, becomes $D(f(x))=(d+A)f(x)$. However, the change in coordinates is not intrinsic to the coordinates themselves, but rather due to space-form shapes that form within the metric space. These space-form shapes form between interacting material space-forms that are contained in the metric space, and they determine the local nature of how the system's coordinates change, ie they determine A.

Local spectral changes are related to nearby simplex changes within a complex number framework, so that the simplex changes can be related to unitary fiber group transformations and subsequent connections (or covariant derivatives).

The observer, who is represented by a larger simplex than the metric space containing simplex, is also part of this model. The observer is simultaneously connected (related) to several different but equivalent metric space structures, so that his perception is exclusively contained in one of these equivalent metric spaces.

Within this more general containment model of existence the metric space of physical containment can change depending on what happens within the observer's simplex structure. The observer's simplex is a larger containing simplex than the metric space simplex. Furthermore, the sub-spectral-set resonating properties so as to determine within which metric space an observer's perception will take place, depend on the observer's will. This leads to a slightly more general connection. This connection, or derivative (covariant derivative), analogous to examining the structural changes of the local affects caused by applying transformation groups to change either (1) the spectral structure of the metric space or (2) the geometry within the metric space (or coordinate) structure. In case (1) the local affects point to the higher dimensional faces (which represent different metric spaces) in the observer's containing simplex, as opposed to the spectral faces of the metric space's spectral properties and shape. In the second case the metric space connection properties are contained within the metric space itself, while in the first case the metric space can be changed.

The over-all affect of this newly based physical description, ie based on the spectral properties of metric spaces, eg discrete isometry subgroups, is to make the description much more geometric and intuitive. Furthermore the observer is tied into the description of interactions within a new way of organizing a principle fiber bundle upon which differential equations can still be defined.

Review

It has been the power of differential equations and their solutions which has made the descriptions of physical science so effective. This effectiveness has recently been lost, this is mostly due to the desire to hold on to a reductionist view of materialism that is within a probabilistic framework. This is caused by the belief which the nature of interpretation of description that differential equations instill, namely, that local structure can be interpreted in terms of its appearance, with an underlying assumption that there exist both a global spatial (or coordinate) structure of fixed dimension and unbounded containing coordinates.

Advantages of non-reductionism

This new model is geometric and this fact gives new power to its descriptions, and it gives a new vantage point to a non-reductionist viewpoint of physical systems, which enhances its geometric properties. Non-reductionist models of life and mind enter the descriptive context in a strong way.

Now the observer is modeled as a spectral simplex and has a certain set of faces which represent different metric space structures, ie and these metric spaces not only have a metric structure but they also have a spectral structure. However, the faces on this simplex may either resonate with, or interact with, other high dimensional simplexes. This would allow physical connections, by a path determined by resonance's (or interactions), to many metric space structures.

Furthermore, the (ordinary) observer is contained in the earth's large high dimensional simplex, hence other metric space structures can be obtained by conjugating upon the earth's simplex structure.

Advantages of the new model of science in terms of new geometric descriptions

This new, non-reductionist, way of describing physical systems also allows a geometric description of: the nucleus, atoms, molecular shape, crystals, dynamic systems, that is, it is consistent with both the old descriptions and the usual dynamics of physical systems, yet it gives greater structure so that it can explain the stability of both the nucleus and the solar system, and it shows that the motions of stars and galaxies are associated to both higher dimensional space-form structures and to group conjugations. This would mean that stars and galaxies are outside the metric-space-mind structure of a containing coordinate system that in the currently accepted paradigm are assumed to be both unbounded and the same in all directions.

Note: The metric structure does have a huge influence on the properties of a mind.

The world model when the world is interpreted in terms of discrete isometry subgroups

Interaction-motion with respect to space-forms and metric space states.
   A 2-dimensional light space-form identifies a metric space state. The 2-dimensional light space-form does this by forming into an infinite extent space-form so that its infinite reach extends to and past the boundaries of the metric space state within which an observer experiences the material geometry of that metric space state. These 2-dimensional states (or spectral flows) are defined on 3-dimensional space-forms (of infinite extent) which are geometrically contained in a 4-dimensional metric space. Within the 3-dimensional space-form the particular 2-dimensional spectral flows of light (which may be interpreted as space-forms), which fill the 3-dimensional metric space state that we experience, are rotated into one another.
   This can be done by multiplying the spectral flows by the complex number, i, ie rotated into the opposite metric space states of one another by a 90-degree rotation. This occurs within the 3-dimensional space-forms upon which 2-dimensional spectral flows are contained. This rotation of the 2-dimensional infinite extent spectral flows is seen in a 3-dimensional metric space state, in terms of an oscillation between metric space states, so that material motion within metric states can occur. The positive and negative time states of material geometry guide each other by giving the dynamic system its geometric incremental order with each new incremental change of time, ie each new spin rotation period.
   The idea that time has an intrinsic direction to itself has scientific evidence in its support. Namely, I. Prigogine has claimed to have proved that the time, we know from our experience, has an intrinsic direction to itself, eg the direction of moving material.
   Consider the metric states (that we typically experience) as being within a 3-dimensional spatial part of a simplex (or space-form) structure. That is a 3-dimensional spatial part is one state

of a metric space (one part of a pair of metric space states). Each different metric space state is either purely real numbers or purely complex numbers. The moving, compact, part of a material space-form, contained in a metric space state, oscillates back and forth between a metric time state and its opposite. This allows Euclidean displacement in time. Motion needs both Euclidean and hyperbolic metric spaces, along with a pair of hyperbolic metric space states of opposite times each of which identifies the motion's geometric order.

The infinite extent light and/or interaction space-forms fill the metric space's state and identify that metric space state's material geometry. Each material object is bathed in light, and the light records that material objects geometric relations to the space-form boundary of the metric space, through the color that its lines to infinity have. These colors depict the size of objects, their masses, their positions, and even their relative separations that exist in the (Euclidean) metric space. The colors of the infinite extent, light and hyperbolic interaction, space-forms can all be translated into spectral information depicting material geometry. Geometry can be represented by vectors, hence identify properties of right rectangular prisms (blocks) in Euclidean space.

Space-forms and metric states (super-symmetry)

Lower and higher dimensional metric states define:

(1) on lower dimensions; spectral flows on space-forms and
(2) on higher dimensions metric spaces; metric-space-states.

In both cases the metric space states identify a complex number-system structure.

This complex structure, associated to spectral flows of opposite metric space states, influences:

(1) Either the material geometry of the one metric space state [opposite state spectral flows on space-forms define a complex number system] ie the space-form within the metric space (state),
(2) Or the material dynamics of the one metric space state that the complex numbers define.

All the coordinate transformations of physical systems thus properly belong with unitary (or spin) transformation groups, ie Lie groups.

Interactions and motion

Interactions (usually) imply i4nfinite extent hyperbolic space-forms connecting two interacting material space-forms (or perhaps other spectral structures), which implies metric space (state) geometry which the interacting (or interconnecting) light identifies, in turn, this implies an integer spin structure, ie the complex structure is outside the metric space state, ie the metric space state appears real but the motion of material implies oscillations between metric states, hence the micro-structure of dynamics is a Hermitian space or a unitary Lie group.

The existence of material space-forms imply compact "sub-space-forms," ie the "nuclear" part of a pair of interacting space-forms [that are each one dimension less than the dimension of the containing metric space]. A compact space-form implies that the material space-form has a position and a geometric shape within a containing metric space. A material space-form (usually) has a spin ½ property. Such a material space-form contributes to the metric space material geometry, ie the individual space-form and its spin ½ rotation of its metric space states (of lower dimension) (or its spectral flows) are within a containing metric space state. The individual material space-forms join together (just as atoms do) to form material objects. The space-forms "in three spatial dimensions" are the atoms. However, now there are several parts to them, two separate metric space parts, hyperbolic and Euclidean and each metric space part, in turn, also has its own two parts, ie the two metric space states, upon which the atomic system depends. Furthermore, the atom is composed of two interacting and stable space-forms the nuclear compact part and the semi-infinite electron-cloud part.

Infinite extent interaction space-forms connect material to the boundary of the next dimensional level so as to allow motion.

In regard to 3-spatial dimensions, mass and its relative position-momentum properties are contained in Euclidean space, while charge and its incremental time-energy properties are contained in hyperbolic space.

Life and mind

What dimensional level is life connected to? Answer: At least dimension five but probably dimension seven. The fundamental simplex of a living system must have an odd dimension since space-forms that have an odd dimension can oscillate and generate their own energy.

A mind deals with (1) the set of spectral structures on a maximal torus, that is contained within the fiber group, a spectral set on a maximal torus reflects, through resonance, (2) the spectral geometric properties of the base space of the principle fiber bundle. The spectral set on the maximal torus is resonating with the original spectral set of the material geometry in the metric space. That is, the original spectral set is (usually) the metric space state, on the base space of a principle fiber bundle (PFB).

The infinite extent light space-forms connect to all the geometric "surfaces" of the material contained in the metric space state, including the boundary of the metric space. Hence both the light and interaction space-forms can carry the entire geometry of the metric space on the total set of all their infinite extent space-form arms so that these arms carry colors on themselves which relate material positions in the metric space to the metric space's boundary.

Many of these infinite extent lines can converge on a living system's distinguished point and there re-create the spectral structure of the material geometry of the metric space onto a Euclidean torus (maximal torus), that exists in a the fiber group and this is done by means of the colors that the lines of light carry on themselves.

Mind can resonate with many spectral structures, ie many geometries.

Rule 1: It is the intent or interaction which must change the structure of spectral-geometric set containment.

This rule must be correct, particularly if the model of life contains many different metric space structures inside itself, so that it can perceive within each of these metric space structures. That is, the world itself is held in place by the intent of living energy.

Rule 2: Systems with a great amount of stability and control must be either linear or flat. This is an extension of the idea that stability deals with either linear or integrable differential equations.

E&M properties and fixed metric spaces

The light and interaction, infinite extent space-forms can carry the geometric information of both the material and spectral distribution that exists within the metric space. These light and interaction space-forms extend to infinity. Thus, geometry depends on being in some fixed metric space which has space-forms that have the property of being infinite extent. That is, not all the space-forms of the given metric space need to be infinite extent but some of them need to be.

The electromagnetic field properties of the world are described in terms of force fields, whose properties are determined by both relative positions between charges (the electric field) and relative motions of charged material (the magnetic field). The E&M potential, A, can be used to define E&M fields in terms of a connection (or covariant derivative). Within this new space-form context it might be that the magnetic field is caused by oscillations into the opposite metric state of the space-time metric space.

The force field is not seen, yet its predictions (solutions to differential equations) give answers which are fairly close to what the measured data is. The force field depends on solutions of the potential, A, to a non-homogeneous wave equation that is defined on the metric space in which the field is contained, a metric space whose metric depends on a fixed value for the speed of light.

The observer

The observer and his mental apparatus, ie metric spaces and their material geometries along with a spectral set in the observer's fiber group within the maximal torus, so that the spectral set in the maximal torus gives a representation of the material geometry of the (containing) metric space, and this is done mostly through the metric space properties of light. That is, the metric space and the light it contains are a central part of the structure of a mind.

Both the observer's mind and intent determine and model existence. An existence which includes the observer's structure as a system and the containing metric space's material geometries (and subsequent mental geometries) which are ever changing.

Light and other infinite extent hyperbolic space-forms contribute to mental awareness. Light essentially brings all the space-form spectral geometry of a metric space to the spectral set of a maximal torus that exists within the (observer's) fiber group. The observer also has a separate distinguished point where the observer makes decisions about either what metric space within which the observer will perceive or what changes to make in the existing geometry of the metric space within which the observer is actually choosing to perceive.

The simplex which models the observer has a number of different spectral faces of 3-dimensions which the observer can choose between. Each of these 3-dimensional faces is a metric space in its own right, with its own material geometry within itself. But once within a metric space the observer can intend changes within that metric space's material geometric structure. In fact, a measuring process is one of those intended actions (within a system) of which the observer can be a part.

The metric space that we experience is a 3-face on a 4-dimensional fundamental domain which is part of the sun's space-form or simplex structure. There is a set of 3-faces that are associated to the sun which correspond to the planetary orbits, and these 3-faces, which can be thought of equivalently in terms of their moded out metric-space-tubes (or spectral flows on a 4-dimensional space-form), account for the stability of the solar system.

The particular set of 3-faces which compose the solar system may be equivalent metric spaces, eg conformally related. However, other 4-faces associated to the sun's simplex structure may have associated with themselves 3-faces whose metric space structure is not equivalent to the metric space we experience, eg the speed of light in these other metric spaces may not have the same value as in our more familiar metric space structure.

The 3-faces within the observer's simplex can be associated to metric spaces and these metric spaces may be quite different from many of the 3-faces that the sun has associated to itself. That is, some of the observer's 3-faces may be related to the sun's simplex structure and some may be more related to a distant star's 3-face simplex structure than to the sun's 3-face simplex structure. Hence it is the observer's simplex structure and the 3-faces which he contains that are of the most importance when considering perception.

Either within a given metric space or simply within the observer's simplicial structure, the observer can intend connections between systems. The intended systems could be different metric space geometric-spectral structures because such possibilities are within the observer-intender, ie within the observer himself.

Where are the other metric spaces? Is it within the spectral structures of a metric space or does filtration of separate metric spaces occur on the same region? Does each spectral metric structure exist within a common region of every other spectral metric structure? That is, do the different metric spaces share a common region of some other higher dimensional space, or does each metric space that is accessible to the observer occupy a different region in that other higher dimensional space? If the latter is true, then the distinct regions, though accessed by spectral filtration, may very well be best understood as distinct widely separated stars along with the planets that surround them.

The star and its planets form a stable spectral space-form structure in its own right. In fact, the stable 3-dimensional spectral flow of a star's space-form structure identify its planets' stable orbit structures and it defines a (collective) metric space.

If life is based in higher dimensions than three, and they have a space-form structure, then other metric space spectral structures must in fact be part of the life form's space-form or spectral simplex composition.

Rule 2: Highly controlled and/or highly ordered systems imply either linearity or flatness.

Motion along with the magnetic field come from a spin-½ property of oscillation between metric states.

E&M is based within a flat metric structure, namely, within hyperbolic space. Conjecture: The magnetic field, B, comes from (-t)-metric states because that is where magnetic monopole sources exist. Motion also comes from a metric space-state's interaction with the (-t)-metric states. Both motion and magnetic fields depend on the spin-½ rotation of positive and negative (time) metric states. The idea of a metric space state is due to the infinite extent interaction space-forms in the one metric space state within which one is experiencing their perception. This is because the infinite extent space-forms unify the particular metric space state that they are contained within.

By an analogous viewpoint, gravity (of Newton) is also a flat structure with action at a distance within Euclidean space. In Euclidean space, it is the translational and rotational frames which define opposite metric states, thus giving the property of existence to matter and anti-matter. Analogously, in hyperbolic space it is charge vs. magnetic monopole material which exists. Euclidean space has no time associated to itself, hence instantaneous "action at a distance" can occur in Euclidean space.

In hyperbolic space it is the magnetic field which enters the positive time metric state from the negative time metric state, whereas, in Euclidean space it is rotation that enters into the fixed stars metric state from a rotating star metric space state.

Change requires time, or a stable periodic system. Time is both periodicity which can be correlated with a process of change within a metric state, eg change of position, and it is a distinct property of the direction of motion of the hyperbolic metric space itself.

General relativity

Can acceleration cause the speed of light to change, 4as the more general diffeomorphism coordinate changes suggest? The space-time metric space's isometry group is contained within the diffeomorphism group. It seems that this constraint (ie staying within the more constrained isometry group) is maintained by physical processes, eg the fact that a gravitational field seems to change the wave length of light and it does not change its speed of propagation, implies that the material geometry of the E&M containing metric space depends on that metric space having a fixed metric. Thus the rest mass keeps the relation, $E = mc^2$, ie the speed of light, c, remains constant, and the energy of the totality of all the systems contained in the metric space remains constant. The hyperbolic metric space defines both time and energy, so that within itself its energy remains constant. This also suggests that there are simultaneous metric spaces involved in physical processes, eg apparently gravity does not belong in a fixed hyperbolic metric space either in terms of Newton or in terms of general metric spaces. Apparently the affects of mass interactions can be instantaneous, while E&M energy's affects within a fixed metric space are time dependent.

A descriptive system dependent on an indefinite (or general) metric, and the pair of statements that, (1) all frames are equivalent, and that (2) the (tensor) form of an equation defines physical law, has only had success with the one body problem, and that one-body problem has a spherically symmetric geometry.

That the shape of space determines a material particle's motion seems like a good assumption but the question of, "How space is shaped," might have many explanations. Furthermore, spherical symmetry rightfully belongs within a Euclidean metric structure constrained to the sphere, not within a hyperbolic metric structure. So general relativity's "successes" are questionable since its language seems to be inadequately developed. In fact the precession of Mercury's perihelion can be accounted for in terms of space-form interactions defined in a 4-dimensional metric space. This is done in the book "The Authority of Material vs. The Spirit." this new derivation is important since its currently accepted derivation, in terms of general relativity, is done within the Schwartzchild solution which is a spherically symmetric coordinate system that is locally hyperbolic, but this is inconsistent with spherical symmetry and with the apparent flatness of E&M phenomenon. Spherical symmetry belongs in Euclidean space restricted to a sphere, while a constant light speed implies containment of E&M systems to be within a flat hyperbolic metric space.

Note: the Robertson-Walker metric for spherical space may be the more correct version of the affects of spherically symmetric metric structure.

The intentions of the observer

Intending motion:

Intention can exist between two systems within an observer, namely, the current one, which is perceived, and the newly intended system which contains the change. For example, moving one's hand. This all exists within the perceiver-intender, so that one system is being perceived while the other system is being intended.

Does the intended system need to shift to the perceiving part of the mind, ie the maximal torus of spectral sets that have been perceived, and then to the metric space, where change can be caused and registered? If one's perception within a metric space state is to be maintained then it seems that the above chain of structures should be followed, so that perception within the given metric space is maintained. This means that two spectral structures enter the mind and form part of a sequence (or together initiate a sequence) of infinitesimal changes in the metric space's geometry. Namely, a sequence of material changes, ie the body movements needed to realize a new geometric system in the metric space. Thus both mind and internal intent identify and move towards the newly intended system or change in system.

Thus two spectral states of: (1) metric space reality and (2) metric space possibility exist on a maximal torus. Furthermore, within the intender there can be an interaction space-form that exists between these two spectral structures. Each of the two spectral space-form structures represent a physical geometry within a metric space so that the physical geometry is made of both hyperbolic and Euclidean space-forms. The metric space distance between the two systems identifies both the size of the interaction space-form and the E&M energy that is required for the intention to be realized. Whether this (energetic) hyperbolic interaction space-form is compact or infinite extent does not matter because in either case the Euclidean connection space-form is compact hence it corresponds to a spectral representation which is either within a metric space or within the mind. The mind of an observer contains models of entire metric spaces within itself. The pull towards the new intended system is determined by either a compact or an infinite extent hyperbolic space-form as well as (or equivalently by) a unitary transformation push on the old geometry towards the new space-form structure that is intended.

Alternatively: "If the hyperbolic interaction space-form is compact then it has charge and E&M forces." These E&M forces would be made of intent and they reside within the observer's system. The observer has within himself the faces of entire (hyperbolic) metric spaces. The energy of intent equals the E&M energy of the interaction space-form of the intent. On the other hand, if the hyperbolic interaction space-form is infinite extent, then in this process the energy would be equivalent to light energy which causes changes in geometric-spectral state, eg small incremental changes in material spectral-geometric relationships in the metric space. In this second case the light energy can also be interpreted to be the energy of intent. For example, the small changes to which the light connects, are the incremental material displacements that can occur in each spin rotation shift of state. In other words, the intender-observer's mind models two systems, the present state and the intended state. The observer-intender also intends an interaction between these two systems so that this intended interaction space-form is hyperbolic with E&M energy so that it changes the old geometric state to the new geometric state in the containing metric space(s) and this interaction space-form is within the observer.

However, E&M energy (or field) can also be modeled as a unitary group conjugation, or infinitesimally as a unitary (differential) connection composed with itself. Intent is about how E&M forces (which also cause a living space-form to oscillate and generate its own energy), can be (infinitesimally) shifted (changed) within the unitary group in order to direct energy away from the observer's oscillating energy and into the intent.

Conjecture: Space-form properties and flow properties on space-forms, eg charge, current, and sectional curvature, can be used to generate Lie algebra elements of local connections (covariant derivatives).

Intent realized in a new larger context

Other types of intent deal with fashioning of material done without moving one's body to realize new material structures. This would be related to either strong intent (or knowledge of how to control one's intent) and/or shifts in the metric space structure within which one perceives. Such a shift can occur within the observer-intender system. The observer intends the material world he is a part of, a world (or a metric space) within which he perceives and within which he has his intentions.

Variation of Rule (2): The linear differential equations and flat geometric-spectral structures provide order and allow interactions or changes within that order. Note: geometric-spectral structure allow (or cause) changes, while solutions to differential equations describe these changes.

> Moving between metric spaces, and/or holding an intention of a new material geometry within the mind of an observer; both are examples of intending new relations between systems.

There are a number of avenues by which one can intend to other stars (besides the sun).

The general diffeomorphism group allows the speed of light, c, to be variable. The issue seems to be that since light is a space-form which is infinite in extent, so it fills a particular metric space and hence the information carried by light to the living entities is continued to a particular spectral flow (or equivalently a particular space-form). Thus, the eyes of living entities are trained to see in one particular metric space. Thus the living entities tend to stay in the particular metric space in which they have been trained to perceive. Thus the question, "Within what set structure of metric spaces can the general nature of the diffeomorphism group be realized?"

The Mathematical Structure of Stable Physical Systems

Note: E&M structures, eg atoms, light waves, etc, are also confined to a particular metric space.

None the less, apparently light can go slower than the speed of light in our metric space. Apparently light is slowed by the finite time processes associated with its material interactions.

Alternatively there could exist direct interaction between two distinct spectral structures so that a direct interaction can be used to compel spectral-geometric changes.

Conjecture: Changes in metric spaces, brought about by composing (or multiplying) Lie group transformations, with diffeomorphism group transformations together as maps (or transformations) in order to realize mechanical systems whose high speeds can exceed the speed of light.

Euclidean space allows any speed for c, since it has no spatial wave equation, hence it has no light waves. However, Euclidean space does have waves whose differential equations must be defined inside material. These are Euclidean waves that exist within a material medium. Could they go faster than light? This depends on the binding forces of the material? But material geometries are bounded so these ideas seem quite limited since light depends on its infinite extent structure. Material geometries are bounded but perhaps the entire metric space itself can be transformed in one big definitive step.

Perhaps the most promising way to visit other stars deals with the unitary Lie group transformations of simplexes. The unitary Lie group contains the transformations between the various metric spaces that are contained within a spectral simplex structure. Each metric space can contain material systems composed of (similar) material geometries made of space-forms that we experience, along with the pairs of metric space states that compose these metric space structures. The pairs of metric space states are needed in order for material geometries to also have dynamics.

By conjugating a spectral set that identifies one metric space, that metric space can be transformed into another different metric space. Remember, a metric space is both a coordinate space and a face on a spectral simplex. Group conjugation can be used to transform between the various faces of the simplex in which the metric space exists as a face. Furthermore, a metric space can be represented as a spectral set on a maximal torus of the fiber group

The mental side of existence along with mental intentions, exist within a large high dimensional simplex, a simplex that contains all this mental activity [that is, mental activity and the metric space geometry upon which the mental activity depends] in turn (such a simplex) is contained within a large spectral set so that this larger spectral set is also represented and contained within a maximal torus of the fiber group. Thus various new metric spaces become available to a mind because both the metric spaces have a spectral representation on a maximal torus and the observer has a spectral representation on a maximal torus.

By conjugating in the group in order to map between higher dimensional faces of this simplex, where higher dimension means dimension three or greater, one ends up within new metric space structures. Surprisingly, this idea would also include being within a new solar system about a new star.

Solar systems can be used to identify metric spaces and their spectral properties that are associated to stars. The spectral structure of the metric space would be related to the orbital structure of the star's planets.

Apparently, conjugation of a maximal torus of a fiber group can take us to a new mental structure so that, that new mental structure is strongly associated with a particular metric space. The issue becomes that of the realization of a new metric space by breaking our attention from the spectral resonance's of material in a metric space identified by its spectral properties that we tend to consider to be "real."

Are electrical circuitry and antennas diverse enough in terms of being able to create and transmit the spectral signals needed to create a spectral representation of another metric space in some particular region of space? Could this cause a material subset, ie a person or a space capsule, to be immersed within a new spectral structure of a new metric space that exists in a particular region of space so that the forces of resonance are strong enough between the object and the transmitted signals to hold onto the spectral structure (or attention of an observer) of the new metric space, that is being emitted buy the antennae?

Since the Euclidean side of our existence allows action at a distance, this means that the massive part of ourselves (or of a system) can be transported, without a problem, to a new metric space structure simply by conjugating to a new spectral simplex structure in a way in which the transported object maintains resonance with the spectrum of the new metric space, so that the metric space spectra is emitted by the antennae.

Could experiencing a metric space's material geometry also depend only on the property of being immersed in a spectral structure? Such an equivalence between resonance and spectral structures could depend on the question of how to understand group conjugation as a process that we are (or can become more) familiar with. In the section above (and below) the process of group conjugation was described as the process of ordinary bodily movement, caused by a group transformation so as to affect either material interaction or intent.

The oscillation between the two states of time, in order to move in a directed manner to both a new spectral and new material (or new geometric) state, may be modeled as a differential of a group transformation (or conjugation transformation). This local transformation can be gauge transformed so that this differential group structure is aimed (or pointed) at a nearby resonating spectral set (or it even could be a spectral (representation of a) system) so that this new spectral set represents a new metric space.

In fact it is a spin rotation which does the action of oscillation between opposite metric states so that these opposite metric states include the system's opposite dynamical properties. Opposite directions of time states on a hyperbolic metric space allow both material dynamics and, equivalently, that E&M material have both a property of charge and a property of a magnetic field so that the magnetic field slips through the two state structure of existence in metric spaces.

Can a process defined by group conjugation be that much different from metric state spin rotations?

Catalog of spectra

There exist a limited number of hyperbolic space-forms for each dimensional level of a dimensional hierarchy of different signature metric spaces.

In fact it is reasonable to catalog and characterize the discrete isometry subgroups that exist in a dimensional hierarchy of different signature metric spaces.

Spin and unitary transformations

For low dimensions, spin rotations are related to unitary transformations, while (as a rule) the spin group is related to the properties of the metric (or quadratic form), hence they are more directly related to real rotations than to unitary rotations. These can be studied by formally limiting what can change (or be transformed) in a more special example of conjugation eg conformal transformations which it seems would affect a metric space's speed of light.

In fact it should be realized that our mental apparatus should be in command of conjugating one's mental structures, in fact this might be related to imagination, or dreaming, and/or to star travel.

There are a number of possible processes by which one can travel to the stars.

The diffeomorphism group allows changes in the speed of light. The diffeomorphism allows all frames of reference for any observer to be equivalent, in the sense that tensor equations can define the laws of physics, yet the nature of the coordinates may be limited, eg the coordinate of time may not be a part of a general metric space.

The laws of physics are invariably tied to differential forms since it is these structures which are most suitable to integration. Hence solutions to differential equations which can be defined in terms of differential forms, lead more directly to the integrable, stable geometric forms (or systems), ie solutions, that naturally exist in a metric space. The spectral simplexes of discrete isometry subgroups have a natural relation to differential forms. Differential forms (or alternating forms) have the structrue of an anti-symmetric algebra.

Many Lie algebras also have the property of being anti-symmetric just as alternating forms and differential forms are both anti-symmetric algebras. Differential forms depend on a metric structure.

There is no reason to be tied to the metric spaces which only have a fixed value, c, for the speed of light, within some solitary metric space within which general systems are contained. However, our E&M theory is based in a hyperbolic metric space, ie space-time, and E&M theory requires some fixed value for the speed of light in a metric space. This metric space contains the material upon which mental models of that material geometry are created.

A material containing metric space, in turn, is fashioned into a space-form when it is properly viewed as being contained in a higher dimensional metric space. It is within this higher dimensional metric space that the metric space's two metric states are spin rotated into one another. Thus to have a diffeomorphism which changes the speed of light would also have to change E&M theory. Thus diffeomorphism transformations must be applied in conjunction with group conjugation in a higher dimensional unitary fiber group setting, ie acting on complex coordinate (base) spaces whose dimension is greater than the metric space's dimension. Thus both hyperbolic (isometry) and unitary group conjugations must occur in such general transformations acting on the lower and higher dimensional faces of a containing spectral simplex, so that the E&M structure of a metric space can be changed.

Intend to the stars

A new E&M structure would require a new metric space, hence the transformation to a different star system with its own planetary orbital system which identifies a new spectral structure for a metric space. Thus a new star carries the new metric space. Perhaps this transformation to a new metric space can simply be extended in a similar manner up to the next higher dimension, eg going from 3 to 4-dimensions in hyperbolic space. On the 4-dimensional space, both a new concept of time and a new concept for E&M, with a new speed of propagation associated to its infinite extent space-form structures. These infinite extent space-forms carry an energetic signal's propagation. An E&M theory on a higher dimension needs to be created mostly by following in an analogous manner the hyperbolic structure of E&M in 3-dimensions.

Euclidean space has no time associated to itself so it has no natural wave equation other than the wave equation that exist in material where forces of stress hold the material together, causing mechanical vibration in a Galilean frame. Euclidean space has no boundaries on the speed of wave propagation.

The Euclidean space allows action at a distance for material that has the property of having mass. The unitary group acts on the complex number system of metric spaces which contain a pairing of metric space states, so that each (metric space) state is either real or pure imaginary.

The mind creates a spectral set which represents existence on a maximal torus of a fiber group and this spectral representation is associated with a definite metric space.

By intending a conjugation process the next step, in an awareness of a capacity to change things by intent, could be to transform to a new metric space structure within which the speed of light is unbounded (or extremely high).

Alternatively a Lie group conjugation operation (or process (or intention)) may be used which can result in moving to a new metric space structure, and that structure must be, or at least appear to be, on a new planet that orbits another star.

Intention (walking)

When one walks, first the person gathers perceptual data about the world. This is done as the lines of the world converge together in order to pass through a distinguished point at one pole of the living person's energy (or spectral simplex) structure. These lines of the world are infinite extent hyperbolic space-forms, either light or interaction space-forms, that pass through a distinguished point of the sub-space-form of the simplex which composes the observer. The colors of the infinite extent lines of the world, passing through the distinguished point, give spectral data about the geometry that is contained in a metric space and perceivable within the metric space. It is a geometry which depends for its composition on compact space-forms. This spectral data of colors, that exist on the lines of the world, are converted to a spectral set on a maximal torus and thus can be stored mentally.

Then at a distinguished point, at the other end of the polar energy structure of the living being, ie living system's simplex, a spectral structure is created about how a living form may want to change their own geometry or their own position (or change the position of external material). This is done with an interaction hyperbolic space-form, determining an interaction between the same type material space-forms which are in different geometrical locations in the containing coordinate frame. This new interaction space-form is associated to the new hyperbolic space-form structure of the living being, in conjunction with the spectral set of the world that the person can access from the spectral set of the world that is on a fiber group's maximal torus. This spectral set of the world has been placed on the maximal torus due to perception, namely, the lines of infinite extent hyperbolic interaction space-forms converging on a distinguished point of the observer (person).

Note: The maximal torus and its spectral sets are accessible to all points in the base space of a principle fiber bundle.

The intended hyperbolic space-form, which contains the space-form which causes the desired changes, namely, connecting a space-form of the present (or immediate past) to a space-form of the immediate future, is transferred to the maximal torus, ie the spectral structure of the new geometry of the metric space to be caused by the person's own will. Thus there would then be two spectral sets on the maximal torus, (1) the way the world is and (2) the way the world is intended to be by the intender.

This may be thought of as two spectral sets on separate maximal tori. In order to achieve one's will, one of the maximal tori and its spectral properties must be conjugated into the other so as to change the spectral set of the world to that of the geometry that has been willed. If the changes are not very great then the group element used in the conjugation may be an infinitesimal change in the group element of one of the maximal torus's distinguished point (or center) in the group, ie the group element which distinguishes the maximal torus from the group's identity element, (or it could be simply an infinitesimal element about the identity).

More than walking

Thus going to metric space geometries further from (or different from) the one we perceive, would require that we engage in conjugating by unitary group elements that are further from the group identity than usual. This description depends on the context provided by the existence of a higher dimensional spectral simplex, ie fundamental domain, which relates to the fiber group and to the existence of other metric spaces that have the same dimension but which occupy a containing space of greater dimension than simply the given dimension of the metric space itself, ie the metric space where we experience the world of material geometry, dynamics and material spectra. Think in terms of faces on a cube so that the faces represent metric spaces, and the faces can have dimension greater than the usual 2-dimensional faces of a cube.

Light, the mind, and intention

When the infinite extent hyperbolic interaction (or light) space-forms are formed, there are two possible ways in which to describe the changes that occur: (1) space-form interactions and (2) fiber group conjugations which change the geometry of material space-forms that exist in the coordinate frame.

(1) An interaction space-form is created in conjunction with a Euclidean space-form so that the Euclidean torus identifies the distance between the interacting (semi-compact) material space-forms. This Euclidean space-form, ie Euclidean torus, has a non-rational slope associated to itself, and this non-rational value is close to the relative distance that the two interacting material space-forms are separated. Note: A non-rational curve on a torus has an infinite length.

That is, in regard to either a hyperbolic interaction space-form or light, [remember that the light space-form has a Euclidean torus associated to its propagation and this Euclidean torus gives the value of the wave-length of the light, ie related to the light's energy] both of these Euclidean spectral values, ie either the relative separation distance or the wave-length, can be associated with a rational valued slope that is close in value to the non-rational sloped value of the Euclidean space-form's unending curve. This unending curve (whose slope is a non-rational value) is in resonance with the infinite extent arms of the infinite extent hyperbolic interaction space-form. That is, a rational sloped approximation, means a closed curve on the Euclidean torus, and, hence, it is associated to a particular color. This color can be attached to the infinite extent hyperbolic space-form's infinite arms. This color of the hyperbolic space-forms can be translated to a particular rational sloped length for a closed spectral line on a maximal torus. Thus, the color of the infinite extent hyperbolic space-form lines which intersect with the distinguished point of the living space-form can be decoded into rational sloped, closed lines or spectra. These closed curves can be placed by resonance on a (the) maximal torus, eg memory and perception. These closed curves are translated into being closed Euclidean spectral elements which can also be associated to a spectral set on a maximal torus. The rational length of the closed curve (of a particular color) on the torus gives the geometric distance between the interaction material space-forms. For a spectra dominated by finite length curves (on a torus), this provides the structure needed for the geometry of vectors to dominate the description, ie integer spin.

There also exist analogous spectral relations in the other pole of the living system's mental (spectral) structure, ie the living system's own spectral properties. Namely, the living system's various spectral structures, where memory, perception, and intent, can all also be translated into closed Euclidean spectral elements which can also be associated to a spectral set on a maximal torus. Those which come from the living system's high dimensional spectral simplex are also spectrally accessible to the maximal torus. This would include many 3-dimensional faces, all of which can represent a metric space (within which one can perceive), along with other higher dimensional faces, ie 4-dimensional faces. The 3-dimensional faces have a sub-facial structure which gives the spectral range of the material space-forms contained in the 3-dimensional metric space.

Thus, hyperbolic space-forms can be identified and formed within the living system's simplex structure, and which resonate with the external metric space's spectra, so that these hyperbolic space-forms can include both those space-forms related to a material space's material geometry and another set of space-forms related to the changes in the given (or particular) metric space's material geometry.

(2) This allows two spectral sets on a maximal torus (or on two separate maximal tori) which can be associated to one another through intent, ie through group conjugation within the fiber group, and hence also through covariant differentiation. The derivative identifies a function's linear, measurable, structures, within the containing coordinate system, where the function (the formula) would represent geometric and/or spectral changes.

The metric space's two metric states are spun together so that infinitesimal conjugations can change the external geometry of the metric space into the intended geometry. That is, the infinitesimal conjugations would be coordinated with the metric state spin rotations. The inertia would be the inertia of a stable geometry associated with a given spectral structure, which the infinitesimal conjugations would be doing work against (or on). That is, the conjugations (or the covariant derivative's commutation products) would be pushing against spectral inertia as well as the geometric inertia of material.

Indeed an external interaction could be described in terms of both a spectral resonance of a future system to which the present system will tend by the force of resonance, along with a push toward the new spectral system (or spectral set) caused by either infinitesimal conjugations which push so as to cause a movement (or change) between a given spectral set on one torus to another spectral set on (say) another torus.

Similarly such a push could be caused by currents, charge, and sectional curvatures related to interaction space-forms and the differential equations that they induce.

Note: The two spectral sets could be on the same torus so the infinitesimal conjugations could be direct pushes on the spectral sets themselves, and such local pushes would be characterized by the transformation of the highest dimensional spectral-faces, or they could be infinitesimal conjugations between the two separate tori.

Upon what geometric relations of a metric space does the speed of light depend? (two conjectures)

The speed of light is related to the space-form structure of the metric space within which the infinite extent light space-forms exist.

Conjecture: The infinite extent space-forms may go out to infinity, and it is the distance that these infinite extent space-forms travel in order to reach the boundaries of the metric space's space-form [the space-form that controls the stability of planets] divided by some time period which determines a fixed value for the speed of light in a hyperbolic metric space.

This process also defines a containing volume or region, that is uniform in its spectral properties, ie in its change of metric state and its size (properties of material), namely, the region of the metric space. However, since each star may have several planets in orbit about itself, the space-form considered would (must) be related to the 4-dimensional fundamental domain whose 3-faces determine the 3-dimensional space-forms of all the planets about the star. In this case the distance traveled would be either the maximum (or minimum) of all the particular planetary distances to the boundary of each of these 3-dimensional space-forms [or the furthest extent that such a space-form structure reaches from the sun (or star)] divided by some space-form's natural time period, whatever that time period might be. The natural time period on a "given" space-form will now be considered.

The square root of the electric permitivity multiplied by the magnetic permeability gives one over the speed of light, ie 1/c. One over the electric permitivity, ε_0, multiplied by the units of coulombs squared and divided by meters squared gives the units of Newtons (or force). The magnetic permeability, μ_0, multiplied by one over Amperes squared also gives Newtons (or force). Thus

$$\mu_0 \varepsilon_0 = \frac{N}{A^2} \cdot \frac{C^2}{N \cdot m^2} \, (units) = \frac{1}{c^2}.$$

In order to interpret the nature of this expression within the framework of space-forms and metric spaces, so as to understand the speed of light's relation to typical systems composed of space-forms within their containing metric space, the electric permitivity's and the magnetic permeability's relations to forces can be examined.

The force of a magnetic field on a current (or moving charge) at the center of a large loop of current is the picture to be used for determining how to interpret the magnetic permeability. One current is a local current (of a moving charge) while the other current is the source of the magnetic field which is far away. Thus one current is the size of some typical system, eg a "typical" space-form, while the other current comes from the boundary of the metric space's space-form.

The force of a static electric field of two charges, one at the center of the metric space while the other charge is out at the boundary of the metric space. This is in 3-dimensions, so the force is determined by spherical symmetry, hence the distance to the metric space's outer boundary is squared when determining such a force. If the geometry were toral instead of spherical then it would be the product of the two toral radii.

For spherical symmetry the distance of separation is normal to the space-form which carries the sectional curvature that is needed to deforms space in order to cause changes in motion. Thus the "surface" of the interaction space-from is perpendicular to direction in which the distance of separation between the interacting space-forms is measured. This surface defines two radial distances. When the system has spherical symmetry then only the maximal distance is important and defines a sphere, hence the force between the separated charges depends on the square of the radial separation distance of the interacting space-forms. Note: the magnetic forces depend two separate directions important in the interaction, namely, the direction of the charges motion and the direction of the magnetic feild.

Note: In higher dimensions the shapes determining forces are tori, hence, there would be a limit on the distance that the outer charge of a metric space's space-form can affect other charges. Hence there would be two factors, namely the two values for the radii of the torus which defines the electromagnetic limits of interaction.

In terms of measuring unit's for the product, $\mu_0 \varepsilon_0$, the coulombs cancel out, leaving distance squared and two values for current lengths, which are multiplied together. Now the current lengths (related to electric currents) on space-forms can be interpreted to be measures of time. This is allowed since the flows on hyperbolic space-forms are time states, so the length of their flows can be interpreted in terms of giving a measure of time. One of the time measures will cancel one of the radii of the static electric field geometry. This is because they are essentially equal measures, with different units, because they are measured on the same space-form whose boundary is at the outer edge of the metric space. Thus these magnitudes cancel but their units do not, this leaves,

$$c = \sqrt{\frac{r}{t}\left(\frac{m^2}{\sec^2}\right)},$$

where r is the distance to the outer edge of the metric space's space-form, and t is the "time" radius of a typical hyperbolic system's space-form. Thus, if the following critical space-form structures are used then,

$$c = \sqrt{\frac{[Oort-cloud-dis\tan ce]}{[typical-system-size]}\left(\frac{m^2}{\sec^2}\right)} = \sqrt{\frac{[10^{16}]}{[.11]}\left(\frac{m^2}{\sec^2}\right)},$$

If .11sec is multiplied by the speed of light then this gives a Euclidean space-form about the size of the earth, eg $7 \times 10^6 m \approx 10^7 m$. However, this is not consistent with the way magnitudes of time and distance of hyperbolic and Euclidean space-forms, respectively, in the formula for c, are used. The .11 seconds must be interpreted in terms of it giving the size of the hyperbolic space-form, assuming quantities of hyperbolic length are equivalent to Euclidean lengths. Then this space-form system would be comparable in (Euclidean-hyperbolic) size, to about 1/3 of a foot, or about the size of a "typical" life form on earth.

This would give a constant value for the speed of light for a large region surrounding the sun (or star), ie from the sun out to the Oort cloud. The Oort cloud is the region where comets form about the sun.

This entire analysis could be altered if there exists a conversion factor, k, between hyperbolic and Euclidean spectral flow measures, ie $k \times t = d$, where t is a measure of a (1-dimensional) spectral flow on a hyperbolic space-form, and d is a measure of a (1-dimensional) spectral flow on a Euclidean space-form, ie $d \cong 2\pi \cdot r$. Thus,

$$c = \sqrt{\frac{r}{t} \cdot k \cdot \left(\frac{m^2}{\sec^2}\right)},$$

would be the new formula for c in this case.

Conjecture two

Another idea one can use to determine the space-form properties which determine the speed of light could be the formula

$$c = \frac{energy \times time}{mass \times dis\tan e} \Rightarrow \frac{\pi \times r \times k''}{T} = \frac{\pi \times k}{2},$$

Where the distance, r, might be the inner (or shorter) radius of the earth's (or some planet's) orbital space-from structure related to its metric space, ie the outer edge of the metric space space-form, and the time (or time period), T, could be the earth's (or some planet's) orbital period, or the space-form of the earth's natural period, and where energy could be the kinetic energy of the earth's orbit, or the energy of the metric space's space-form structure, and where mass is the mass of the earth. And where k'', along with k, are different constants depending on how the central space-form structure of the metric space is organized with respect to this second formula for c.

In a third case, the speed of light could very well be related to the formula

$$c = \frac{energy \times time \times k}{momentum \times dis\tan e} \Rightarrow \frac{v \times T \times k}{d},$$

where v is the speed of a current in a hyperbolic space-form, T its period, and d, the length of the corresponding resonating Euclidean space-form's spectral flow.

In a fourth case, the speed of light could very well be the constant that connects corresponding measurable quantities of the spectral flows on the different type of metric space space-forms, ie the natural measures of quantities of the same type on the corresponding hyperbolic and Euclidean space-form spectral flow's. That is,

$$c = \frac{d}{t} = \frac{2 \cdot \pi \times r}{T} = k$$

Whereas in the first three cases such a constant is only one factor in the formula for the speed of light.

The relativity of the speed of light

The infinite extent space-forms of light can travel past the metric space's natural boundaries just as other infinite extent light space-forms can travel into the natural space-form boundaries of a metric space from far away. The far away light would appear to have the fixed speed of light of the particular space-form metric space within which the light exists.

This would be a new: relativity of metric space property related to the speed of light. Namely, the speed of light depends on what metric space that the light is in.

Special (general) relativity describes changes in properties of measurable properties between inertial (general) frames. These changes in properties are changes in appearances, ie they are mental or sensory properties, due to the properties of light having a constant speed for all frames, that exist within a particular metric space.

Thus when one is in a particular metric space in which light has a fixed speed, one cannot detect the changes in properties of light's speed when (or if) it exists in another metric space (in some other region of space).

Hence, how can one know what the properties of space are in regions of an abstract coordinate space that are far away from the defining boundaries of a metric space that exists in a particular region in space?

How does one know that the speed of light is universal rather than local (though local over a large region)? How can one say that space is the same everywhere, that a universal coordinate system extends out unbounded in all directions, and that the laws of physics are the same everywhere in this coordinate system?

Such infinite extent uniform coordinate spaces are an integral part of the physical model of existence currently used by modern science. Namely, the coordinates of general relativity define a universe, as do the solutions to the wave equations of quantum mechanics. Perhaps this reach obscures the real, hidden structure of existence.

The structure of the universe the observer and the external world

There is a structure of existence in which the set containment is not what one would expect. Within the new order for (of) existence, on higher dimensional levels the structure of set containment can reverse again and again. A living being, eg a human, has a polar spectral structure. On one pole exists an observer's energy generating spectral structure, A, which can accommodate (contain) many systems, yet perhaps it does not have within itself all the detail. None the less the "external" detail can find its place within A. While on the other pole there occurs the gathering of sense data by resonance between the senses and an external structure, B. This gathering of sense data takes place on a distinguished sub-space-form point of A. The external data can be gathered from a distinguished point because of the convergence of the lines of the world on this distinguished point. The lines of the world have distinct intrinsic colors which can get translated into bounded, closed spectral structures on a maximal torus (or Euclidean torus).

These colored lines of the world converge through (or onto) a distinguished space-form point, and thus allow the perception of the external world's geometry, along with all of the world's pushes, and sounds, and colors, of the details of the external world, B. Apparently this perception allows A to move within (or react to) B. The intended motion, ie reaction, of A, is caused by his intention to actively re-structure a slight change in the geometric-spectral structure of his own body within B. This occurs within both the mind of A, and can be caused by the infinitesimal conjugations in the fiber group. This allows the spectral structure of B to evolve into the spectral-geometric structure that A intends for B, ie motion etc, and equivalently this spectral-geometric evolution is guided by using spectral resonance, sectional curvature and currents of hyperbolic space-forms.

Group conjugation

Conjugations can move between a pair of resonating spectral structures, eg on two maximal tori. On the two maximal tori of A and B there exist two spectral sets (or two spectral sets on the same maximal torus). These two spectral sets exist at the two ends of the polar spectral structure of a living-perceiving being. Group conjugations of an infinitesimal nature can gradually "push" the original spectral set, a discrete group, into the new intended spectral set, another new discrete group. The group conjugations are usually infinitesimal in nature, but this does not have to be the case. The conjugations can be more general, and the resonating spectral sets can be substantially different from the original spectral sets emanating from B, which A is perceiving.

The spectral set of A can push the spectral set, B, toward A's new spectral structure so that B is transformed into a new spectral simplex. This means that the resonating spectral set of the maximal torus associated with the B end of the pole, is at first dominated by A's intended new spectral set, but then the external set of B finds a spectral subset in its larger external domain which, in fact, does resonate with A's new spectral structure. It is a general group element which is involved in the conjugation between these two quite different spectral structures, ie A's new set vs. B's old set and the conjugation which pushes the old B spectral set toward the new spectral set intended by A.

Note: In fact this can be viewed in terms of conjugation to a pair of time states (in say 4-dimensions) so that a dynamic (or inertial) flow can be realized in the containing metric space. That is the dynamical changes can be mapped by the conjugation to a containing system that has two time states. The two time states are "invented" by the conjugation so that these two time states carry the dynamical changes of the material geometry of the metric space, at least in regard to the intended geometric changes in B.

The conjugation which pushes the old B spectral set toward the new spectral set intended by A is possible because A's over-all spectral set is much larger than the original spectral set of B, even though B's spectral set was impinging on one of A's distinguished space-form points, and causing A to perceive B. However, A can perceive B only if the essential spectrum of B is already contained in A. It turns out that B is more dependent on A's spectral set than the other way around. This is because A can move to an equivalent but disjoint spectral set, C, which also represents a metric space. However, once inside C then the spectrum of B no longer has any affect on A. Thus the conjugation between two metric spaces really goes between two different spectral subsets of A. One subset can accommodate the geometry of the original spectral-geometric structure of B, and the other spectral subset of A, creates a new domain (or metric space) for perception. The equivalent external part of the spectra, which was B, must move to C in order to resonate with the new spectral structure of A, due to the intended command of A, within the larger spectral set of A itself (that is a group conjugation which transforms within A). In other words A contains the essential structure which can accommodate the details of many metric spaces which are equivalent to B.

None the less, A itself is contained within an even bigger set. This bigger external set is itself based on living intent. Thus, A has an external containing set but that containing set is not at the level (spectral-dimensional) that is expected, that is, it is not at the level of metric spaces which contain the material world we perceive.

For the two poles of A's spectral structure, one is at a distinguished point of A through which the lines of the world converge. The lines of the world carry colors which can be built into a spectral set of the "external world" on a bounded torus of the fiber group. While the other spectral pole is at another distinguished point of A which carries the larger spectral set of all of A. However at a particular dimension, only one sub-face of the over-all simplex of A may be actively used or actively resonating.

The over-all spectral set of A contains many energy generating spectral subsets so that all these oscillating (energy generating) spectral subsets can also be related to bounded spectral sets that exist on a maximal torus in the fiber group.

The forces, a connection, and choosing a gauge (conjugation)

Never before has the base space that contains all possible aspects of physical systems been so big, in terms of both dimension and distinct separate set (or metric space) containment, nor has the corresponding fiber group been so big. For a given dimension, there exist many faces on an over-all containing spectral simplex so that each face represents a different metric space within which the material it contains has its geometric, spectral, and dynamic properties. Above each of these faces (or metric spaces) there exists a corresponding fiber group, eg the metric space's isometry group.

Within each metric space there are forces, and these forces can be modeled in terms of a connection (or covariant derivative) which incorporates a potential, eg A in E&M, as a Lie algebra element of a fiber group. One can view the connection as the differential of a coordinate transformation by the fiber group.

The derivative gives a linear change (or approximation) of a function dependent on the linear coordinate properties, while the differential of the group transformation, ie the Lie algebra element, gives the changes in the coordinates at that point in the coordinate system. The differential of a Lie group element is a Lie algebra element, A. The Lie algebra element, A, gives the local changes in the coordinates caused by a group transformation on a space-form. Thus $f(x)$, eg the path of an object, becomes $f(x) \exp(iAt)$ on the space-form, and $d[f(x) \exp(iAt)] = (d + A)f(x) \exp(iAt)$. Thus, each metric space has a connection related to the base space's (the metric space's) fiber group, ie particular metric spaces are associated with particular fiber groups. The force field obtained from commuting the covariant derivative,

$$d+A, \text{ gives } dA+[A,A]=\text{Force Field.}$$

If A is from the fiber group's Lie algebra then the force field is in the metric space associated to the fiber group.

Note: If one uses [or $f([\exp(iAt)]x) \exp(iAt)$] then this is saying that the coordinate change at x is important, while the expression $f(x) \exp(iAt)$ says that the path away from x is important in terms of coordinate changes in regard to the dynamical system. It is usually the latter that is considered, ie $f(x) \exp(iAt)$. This is assuming that the coordinate x is inert to the infinitesimal changeless of the system, whereas the object's path, $f(x)$, is what will be involved in the coordinate changes of the system.

Considering the entire base space, ie the over-all containing spectral simplex, then its fiber group is the large group within which are contained all the fiber groups of all the faces of the over-all containing simplex, ie all the faces of the same dimension correspond to all the other metric spaces that have the same dimension. Thus within this large group the idea of a gauge transformation, ie g*Ag+g*dg, can actually transform the Lie algebra connection out of the metric space's fiber group (or base space) and into another distinct metric space's fiber group (or base space), ie a connection in another metric space. The connection Lie algebra element of a particular metric space can transform the spectrum and/or material, ie space-forms and motion, within the metric space upon which the group acts. There also exist connection Lie algebra elements within the fiber group which can transform between different faces of the simplex. Thus a gauge transformation within the context of a bigger group can cause the existence of a new force field which can push between the two spectral structures of two metric spaces. This force could be caused by the commutation term of the gauge transformation of the potential of the covariant derivative, eg [A,dg]=(sum) kA.

The question is: Could dg be chosen so that the, (sum) kA, ie the new Lie algebra element, is a force that pushes the spectral structure of one metric space to the spectral structure of another metric space?

If the commutation term is zero, ie [A,A]=0, then the fiber sub-group is effectively Abelian and the differential forms for that metric space determine the force field for that (flat) metric space.

The world with an observer

There is an observer and an external existence, yet the observer has a much more complicated relation to existence than is thought. This is one of the hidden structures of the world. Though the high dimensional geometric-spectral simplex that mathematically represents the observer is contained in an even higher dimensional simplex, the dimensional geometric-spectral simplex of the observer is higher than expected. This means that the lower dimensional metric space, within which we perceive a material existence, and within which we can move, [and change, or use material to create], is but a subset of the observer-creator simplex. Thus, as a metric space within which one may perceive (the one within which we classify our experiences) is not unique. There exist many metric space sets, within which material-spectral-geometry can be perceived and used, all of them existing inside the over-all spectral-simplex of the observers themselves.

The observer has an "empty" copy of the metric space within his own internal structure. This empty metric space has all the spectral properties of resonance that the external metric space also has. These spectral properties of resonance are intrinsic to the metric space's fundamental domain. The material within a metric space must have resonance relations with the metric space's spectral-simplex to exist within the metric space. This is part of the spectral filtration process which distinguishes (or makes distinct) the material-geometric-spectral properties of one spectral simplex from another.

The mind is connected to a barrier created by the metric space, ie the mind depends for its perceiving properties on this metric space and on the infinite extent (light) space-forms. The barrier hides from the observer his capability (by his intent) of being able to shift to a different but equivalent containing metric space structures. The "observer-creator's" simplex is made up of a complex of equivalent metric spaces descending down from its higher dimensional facial (or space-form, or spectral-flow) structure.

The analogous facial structure at lower dimensions affects the stability of interacting (dynamic) systems within metric spaces, so that this facial structure of metric spaces create resonance's which pull the dual metric-space-state "jumps of change" towards the stable system or stable spectral structure. These "jumps of change" are a result of the discrete spin rotations (Lie algebra elements) between metric space time-states. The pull on the changes can be determined by existing resonance's, ie an over-all set of metric-space spectral faces to which a particular metric space (face) is connected.

The infinitesimal conjugations are guided by interaction space-forms that are out in the world, [and have charge, current and sectional curvature properties associated top themselves] and by the resonance's that exist in the complex spectral structure of metric space faces that exist for 3-dimensional spectral simplexes (or faces) of a 4-dimensional metric space. The interaction system approaches a spectral structure of the spectral space-form set that actually exists in the (4-dimensional) metric space itself, so that a resonance can exist (or occur) between a dynamical system and a stable final state of the system of interacting space-form pairs in the metric space. Note: Many body interactions can also be modeled.

The local group conjugation, ie the derivative of the group conjugation operation, is a transformation which can transform by means of a matrix operation (or Lie algebra transformation) the spectral set and the subsequent spectral changes on a maximal torus. A spectral set will move to the closest (highest energy) spectral face that can resonate with the (Euclidean) interaction space-form. This is a statement about the integrability and stability of solutions to differential equations.

The derivative in which coordinates are transformed represents a connection (or covariant derivative) related to the interaction. In turn, the covariant derivative is represented as a Lie algebra transformation which can be exponentiated so as to act on a set in the group space. Thus, it can also be a transformation of the spectral sets that exist on a particular maximal torus in the fiber group.

Can a particular transformation of one discrete group spectral structure also be a transformation that transforms lower dimensional spectral sets in certain ways? Can transformations of different dimensional spectral structures be combined into one transformation so that the different dimensional spectral sub-transforms will be independent of one another? This would mean that a derivative (or a connection) can have different (dimensional) levels to its structure. Thus it would be a part of a local spectral structure, the type of structures upon which a description of embryonic development would depend.

There are complete metric-space spectral sets which are distinct from each other, and these spectral sets can be either intersecting or disjoint, and they can exist within an observer's simplex. A change or transformation between such metric space spectral complexes requires more than a local conjugation. A local conjugation points to the spectral face related to the metric space within which one is perceiving. A general conjugation transforms at a higher dimension within the group [it will do an analogous operation as a local conjugation namely, in the higher dimensions] so that it will (locally) point to a higher dimensional face of the observer's simplex.

Discrete reflection groups of hyperbolic space have a very restrictive structure, thus one is led to ask the question, "Can the fixed sets of metric spaces, within which the observer can be contained, be known if one knows only the one metric space face?"

Though this knowledge may lead to knowledge of a particular simplex none-the-less interactions between space-forms can extend the nature of the simplex.

The colors of the lines of the world that converge on a distinguished point of a living (or oscillating) space-form are decoded into a geometric-spectrally equivalent set of bounded closed spectra on a maximal torus.

The simplex-spectral structure of a living being contains the metric spaces within which perception occurs. The one particular metric space within which one is perceiving is only one of many such metric space subsets within which the observer's perception can take place.

The observer's system can create a spectrum on its maximal torus, which can define a process of transition between metric-space-geometry so that this transition process depends on resonance, unitary transformations (either infinitesimal or general), and oscillations between metric-space-states. This transformation can lead the old external geometric-state to a new (external) geometric-state. Equivalently, this transition process can also be framed so as to depend on current, charge, sectional curvature and oscillations between space-forms contained within metric space states.

Analogously, when materials interact, the full set of surrounding metric space spectral structures allow for the possibility of a stable system with its own distinct spectral structure to form so that resonance, and oscillations between states cause infinitesimal unitary transformations through which transformations of the spectral system can be made towards a final resonating, stable space-form. However if the dynamical transition system has too much energy then the overall system will move away from such a stable resonating space-form. This is because the dynamics of the system is causing the (potentially) stable system (that resonance's are occurring for) to be over-shot.

Spiral galaxies

In metric spaces of dimension greater that three (dimension > 3) spherical symmetry is no longer predominant and interaction curvatures, ie sectional curvatures, cause forces which are essentially tangent to the "connecting circle" of Euclidean interaction space-forms. Sectional curvature is defined by the radii of a space-form's "circular" spectral flows, ie the spectral circle's diameter for a Euclidean space-form. The sectional curvature gives the metric separation of the two interacting systems. Hence interactions have both a 3-dimensional part, which is spherically symmetric, and a higher dimensional tangential part.

This geometry of forces would tend to cause either orbits or spirals. Both inward spirals and the more likely outward spirals [since the tangential part of the force would tend to make the circle's diameter larger]. This would be for dynamical systems that exist in metric spaces of dimension four and five.

This would explain the galaxy types: outward spirals would give spiral galaxies, while orbital types would give the elliptic galaxies. Inward spirals would be related to something that might be equivalent to rotating "black holes," though an alternative to this notion may also be within the new ideas. Furthermore, the idea of a black hole cannot be a uniform idea if the speed of light is different in different metric structures.

Note: if gravity depends on the graviton, which propagates at the speed of light, then how can a black hole be in orbit about another massive star, unless gravity depends on action at a distance, ie not the graviton?

Problems of Higher dimensions

(no intrinsic geometry because hyperbolic space-forms are all infinite extent)

In 6-dimensional (and higher) hyperbolic metric spaces the space-forms are all infinite extent, hence an outward dynamic might re-cycle into being an emerging process from either a space-form's distinguished point or from the coordinates' origin or identity element.

Hardware

By measuring planetary orbits of distant stars spectral classification of the metric space's spectral faces can be obtained, and hence a unitary sequence can be designed which moves from one metric space spectral structure to another metric space spectral structure.

Could this be simulated by large-scale, 3-dimensional, resonating circuits and antennae whose frequency structure is tuned to the spectral structure of the distant star, can be made so as to surround a material object (space capsule)? Can a geometrically constructed electrical circuit for which the dominating concern is that the circuit's resonance's realistically approximate a space-form structure of a (distant) star?

Thus, if a space-form, which has a conformal relation to a distant star's spectral structure, is modeled as a circuit so that the circuit resonates "conformally" with the spectra of the space-form of a distant star, then whatever is place within this circuit can enter into the spectral structure of the (distant) star due to both resonance and action at a distance. The resonance takes place first within the spectral sets of a maximal torus that exists within the fiber group that exists over each point in space.

Note: a conformal relation is a relation between metrics in which (new metric)=(k) x (old metric), where k is a constant.

Thus an object can be sent to a metric space about a distant star, to which circuit, that exists about the object, resonates. The act of sending is due to both action at a distance and resonance.

Conjecture: Somewhat surprisingly (it is) within each person there is a control over unitary transformations between metric space spectral centers that exist within the over-all simplex structure of all observers.

To move to a new 3-dimensional metric space it seems that one needs to be at least within a 4-dimensional space-form that is contained in a 5-dimensional metric space. Thus this 4 and 5-dimensional spectral structure needs to be mathematically modeled and then electronically simulated so that unitary spectral transformations can be spectrally represented so as to point to the correct 3-dimensional spectral face which, in turn, represents a metric space and its spectral structure, to which "travel" is desired.

There also needs to be some considerations about the frequency range about a certain frequency for which resonance can commence.

References

A New Copernican Revolution, B G P H Bash and P Coatimundi, Trafford, 2004.

The Authority of Material vs. The Spirit, Douglas D Hunter, Trafford, 2006.

Chapter 6

NEW MATH

Though the new, alternative, set of math constructions (described below) use mathematical ideas which are simple... (separable geometries in a multi-dimensional context, where different dimensional levels have geometric independence of the higher dimensional levels)..., nonetheless they are much more challenging intellectual ideas than the concepts used by professional communities of (wage-slave) scientists who so weakly (subserviently) serve commercial interests (where the ideas of the professionals are based on materialism, randomness, point particles, and non-linearity), but the true failing of the professional scientists seems to be (personal)..., ie a belief in, and a will to achieve, intellectual dominance within the eyes of other people in society..., so that they collude with the owners of society to construct an authoritative truth about which only experts have a valid voice, so as to take the power of "free inquiry" away from the people of society, an idea of stripping people of their power, which serves the interests of those owners of society who dominate all of society.

A descriptive language whose basis is probability cannot account for the order that is so fundamentally a part of our observed existence. That is, quantum physics (based on an observed stable, definitive elementary event space), particle physics (based on an elementary event space whose events are unstable elementary particles), string theory (based on both materialism and representing the "unseen geometry" of the unstable elementary events of particle physics), etc cannot account for the stable definitive spectral properties of: nuclei, general atoms, molecules and molecular shape, crystal energy levels and crystal properties.

Furthermore, Darwin's evolution in concert with DNA, is another descriptive structure whose basis is probability, and it cannot account for the origins of life, embryonic development, and "in general" cannot account for the intricate chemical description of life for virtually any extensive chemical process in a life form. Along with this long list of substantial failures . . . , attributable to the currently accepted basis for science, and hence what is also important mathematically . . . , there is also the idea of non-linearity (associated to differential equations) which is a math structure . . . , which implies a deeper form of randomness for which probability cannot account . . . , where non-linearity is a descriptive language whose solution functions provide no basis for describing order . . . , and, in turn, this implies that general relativity . . . , and all non-linear descriptions of physical processes . . . , cannot account for the order that is so fundamentally part of our observed existence, such as the apparently stable solar system, and the motions of stars in galaxies, and how galaxies populate "deep" space.

Furthermore, within this same context "of very complicated math structures which have not shown themselves to be of any practical value," the math structures of algebraic geometry are mostly descriptions of non-linear measuring structures for geometry . . . , and much of algebraic topology . . . , also mostly descriptions of non-linear measuring structures for geometry . . . , have little association with a descriptive language which is useful in a practical sense.

The main property of topology is continuity which seems to mostly describe conserved properties, such as conservation of material and conservation of energy, and in geometry it describes the conserved geometric property of the existence of "holes in a geometric shape," . . . . and though homology and cohomology are fundamentally related to "cubical" simplexes and their associated local alternating-form geometric measures (ie simplexes [homology] and differential-forms [cohomology]), so that this dual pair of ideas have an elementary relation to multi-variable calculus. Nonetheless these two (often) abstract measures of algebraic topology (homology and cohomology) most often seem to have very little to do with the simple "cubical" simplex structures associated to the structure of geometric holes and to calculus, but again they are turned into ideas which have more to do with descriptions of non-linear measuring structures for geometry,

> . . . . but stable definitive spectral-orbital-geometric properties which are observed in existence do not emerge from the patterns of non-linearity.

A huge amount of effort by both mathematicians and physicists has gone into trying to understand the properties of solutions to non-linear differential equations, and their subsequent geometric (or dynamic) randomness, when in fact the fundamental randomness (or fundamental lack of great usefulness) of solutions to non-linear differential equations is a result of the fact that the two measuring structures of the function's values and the values on the function's domain space, cannot be matched-up in a consistent way. This is because the linear structure needed for such consistent relations, between two sets of (independent) values, does not exist for the non-linear differential equation, so that the values on the domain have no bearing on the values of the function, ie one gets discontinuous jumps of value for the function when one moves a small distance on the domain space.

Order cannot emerge from such an inconsistent (or non-logical) set of numerical relations.

Need for new math

However, there is a great complexity of both geometric and physical systems . . . , along with their measuring structures (differential equations given in a new context) . . . , which the geometry of discrete isometry (or unitary) subgroups (or space-forms) can describe . . . , when it has been placed in a multi-dimensional setting . . . , so that this description is not given in either a non-linear context nor in a context of probability. Note: Such a multi-dimensional structure for a descriptive language opposes the idea of materialism.

It is a (relatively) new descriptive language, based on various multi-dimensional space-form structures, which can account for the stable definitive spectral properties of: nuclei, general atoms, molecules and molecular shape, crystal energy levels and crystal properties, where the core spectral properties of these systems are given by the space-form geometries themselves.

It can also provide simple "whole-system" models for: life, intent, mind, and a morphogenetic field, where a morphogenetic field is an idea about which A Turing hypothesized for biological systems and embryonic development, (though Turing may not have originated the idea of the morphogenetic field).

Compare the descriptive capacities of this new language with the descriptive capabilities of the currently accepted language of physical description. The new math language is set-up to describe the fundamental stable systems which are observed while the current language is set-up to describe the particle collisions within a bomb explosion. The currently accepted languages of physical description is based on probability and non-linearity while the new language is set-up to describe order and stability.

Historically, in biology the "organic" molecule which is the primary chemical within a cell's nucleus was found, DNA, and then the biological science community jumped to the conclusion that this DNA was the "computer code" for all of life. The experimental paradigm, since then, has been to identify the approximate properties of life statistically, and then to find the causal relation between these properties and DNA . . . ,

>but
>
>> . . . . such causal sequences only rarely exist, the organization and processes of living systems are emanating from some deeper structure.

The new math identifies simple whole-system models for biological systems through which large complicated systems can be organized. These same (or analogous) whole system structure for the containment set of existence are used to describe stable spectral and orbital properties of material at all size scales.

On the other hand, the current descriptive language says that the fundamentally stable spectral-orbital properties of fundamental physical systems of any size scale are simply too complicated to describe in a useful way, based on our current descriptions of physical law.

Thus, the question is: "Upon what problems do those who "believe in" the currently accepted descriptive language work?"

Answer: Fundamentally they work on bomb and military engineering.

Both the geometric and function-space math relations which were initiated by the invention of calculus and the differential equation . . . , ie beginning with F=ma, where F is a force field with a geometric representation, m is the mass and a is the local change of the mass's velocity . . . , have had their greatest impact on practical technical development, in regard to material creativity, in the context of geometry and not so much in the context of function space solution techniques associated to both probability and identifying discrete spectra . . . , though function space techniques (of analysis) work well in the continuous and more definitively described context of the physical systems of classical physics, they have not worked so well for identifying the approximate spectral properties of discrete quantum systems, eg the solution of the radial equation for the H-atom has not been all that successful, eg the solution of the H-atom's radial equation diverges, and the 1/r potential energy term put into a point-particle model for the components of quantum systems leads to singularities, which have never been successfully managed within the function space (or probability) and energy operator based description of quantum physics.

In the geometric context (of classical physics), it is the linear, geometrically separable, metric-invariant context of a system's differential equation which is the only context in which both the descriptive information and the control of the system . . . , due to having control of the system's boundary or initial conditions . . . , which has allowed a great amount of creative development to take place directly from solving such differential equations which model such physical systems whose description is based on geometric properties.

To <u>not consider the stable</u>, definitive spectral-orbital (and-living) properties of observed physical (and living) systems <u>to be emerging</u> out of the same linear, separable, metric-invariant context . . . , which allows for so much creative development at the material level of due to the control the solutions to these types of differential equations (apparently, after a 100 years of looking elsewhere) . . . , <u>is a mistake</u>.

The new math

The geometry of space-forms, or their "cubical" simplexes are the most fundamental part of the new descriptive language

The geometry of space-forms in a multi-dimensional context (or math structure) are important properties (or structures) because this structure can be used to:

(1) Provide the geometry of the stable, definitive spectral-orbital properties of nuclei, general atoms, molecules, molecular shape, and crystalline (quantum) system structures, which are being directly (related) associated to the geometry of such material space-forms within a particular dimensional level.
(2) Describe both the spectral and interaction properties of material which is contained within a particular dimensional level . . . . [a containment space which is itself a "material" space-form system within its containing metric-space] . . . . In such a "dimensional-spectral-conformal-dynamic-space-form" containing space, there can be (there are) several (other) high dimension space-forms within which any particular dimensional level (which identifies the same subspace) can be contained, and the spectral properties of these (high dimensional) space-form containment sets, along with the material space-forms which are contained in the flows (or metric-spaces) of these (simultaneous) containing spaces . . . , [of any given material system which] . . . . determine (or identify) the allowed spectral properties of the material space-forms which are contained in such a particular dimensional level, and this identification is done through resonance between . . . the spectra of the material space-forms in a given dimensional (and subspace) level and . . . the spectra of its containing "dimensional-spectral-conformal-dynamic-space-form" space.

That is, a material system can be contained in several space-form-spectral structures . . . , so that the relation between material spectra of a given dimensional level and the spectra of the containing set of space-forms . . . , can be used to identify the spectral-conformal structure of the full containing space.

Furthermore, it is such a "many space-form structure" which is needed to understand the physical-chemical-geometric properties of living systems.

If one understands the geometry of the system and its containment set then one will have a good context within which to describe the properties of the system and the relation of these properties to the system's containment set.

What math structures are important in a multi-dimensional context modeled on space-forms?

Answer: The moded-out space-form geometries (or geometries of constant curvature) based on variations of "cubical" simplexes . . . , in all dimensions . . . , where such space-form (simplexes) are defined. These "cubical" simplexes fit-in with the linear, metric-invariant, and geometrically separable math structure of differential equations which are used to define measurable properties of a physical system whose solution functions . . . , which have been so useful . . . , are based on (or given in) the variables of the system containing coordinate space (or domain space).

Identify the geometries or identify their linear differential equations

That is, the properties of space-form shapes can be determined geometrically so that this (various dimension space-form) geometry is based on their rather simple relation to "cubical" simplexes, which form the fundamental domains for (base) space-filling "lattices" which define both discrete (isometry or unitary) subgroups, and their (moded-out) space-form geometric structures.

On the other hand space-forms can also be determined from differential-form solutions to linear, separable differential equations (whose solutions define space-forms) defined in a metric-invariant coordinate space (L Eisenhart, Riemannian Geometry).

Discrete isometry (or unitary) subgroups or equivalently space-form geometries

It is simplest to envision the descriptive structure of the discrete isometry (or unitary) subgroup context . . . , for the use of this type of multi-dimensional space-form math structures along with many space-forms of a specific dimension . . . , when one oscillates between (in one's viewpoint of) the "cubical" simplex structure and its geometric moded-out (space-form) structure, for such discrete isometry subgroup (or space-form) geometries.

It is important to see this math structure (or math pattern) as a geometry of space-forms, but it is easiest to identify simple math properties in the "cubical" simplex structure.

For example, a 2-dimensional "cubical" (or analogously, rectangular) simplex, when moded-out, fits as the geometric shape of a torus (or a doughnut) into a 3-dimensional metric-space (of an analogous metric-function signature).

Space-forms which fit into (or are immersed within) metric-spaces represent the material (space-form) systems contained within the metric-space.

However, the metric-space itself may be, considered to be, a (generalized) face of a higher dimensional "cubical" simplex. The metric-space has a simplex structure so it has: vertices, edges, faces and higher dimensional (general) faces up-to the dimension of the simplex, which is representing the metric-space. However, the "cubical" simplex which represents the metric-space can itself be moded-out to form a space-form, which in turn, can be contained in a metric-space (whose dimension is [at least one dimension] greater than the dimension of the original "cubical" simplex), thus a metric-space can be represented either as a material space-form, or it can be a part of the facial structure of an even higher dimensional "cubical" simplex structure . . . , so that the moding-out of this even higher dimensional simplex structure would result in a high dimensional space-form, of which one of the faces (or now, one of the flows) is a metric-space, which in turn, contains other lower dimensional (independent) material space-forms.

The edges of a (3-dimensional) metric-space's simplex structure represent the natural spectra with which the spectra (of 1-spectra, or 1-flows, or 1-faces) of the (2-dimensional) material space-forms . . . , which are contained in the (3-dimensional) metric-space . . . , can resonate.

For an n-dimensional "cubical" simplex there are $2^{(n-1)}(n!)/2(n)$ (?) paired edges with which the 1-flows of a 2-dimensional space-form can resonate. Eg n=3 implies 24/6=4 paired edges (or four 1-flows for each hole in the 3-dimensional space-form).
That is, (counting the edges [or m-faces] of a high dimensional cubical simplex) / (the way in which they get paired-off in the moding-out process), gives the number of spectral flows (or m-flows) on a space-form.

Conformal factors defined between different dimensional levels

Each given dimensional level can have a conformal factor which relates the spectra of the given metric-space dimension with the spectra of the lower dimension space-forms that the given dimension (metric-space) contains.
Thus there are two spectral structures to consider, the spectra of the high dimension "rectangular" (or "cubical") simplex, and the spectra which is changed by the conformal factor that exists between different dimensional levels.
These conformal factors are the physical constants, eg c, G, h etc.

Metric-spaces and the signatures of their metric-functions

The place in mathematics . . . , where the three properties (the structures of: metric-invariance, linear differential equations, and geometrically separable) are prevalent . . . , is in the context of discrete isomtery (or unitary) subgroups; of "any" dimension, and of any metric-function signature, where the signature . . . of a metric-invariant metric-function on $R(s,t)$ . . . is $s-t$ . . . , so that $s+t=n$.
In a coordinate structure modeled as $R(s,t)$, the full dimension of such a (local) coordinate structure, $R(s,t)$, is, $n = s + t$, where R denotes the real numbers, and where s is the dimension of the spatial subspace, and t is the dimension of the time subspace,

Note: R(s,t) could be C(s,t) where C denotes the complex numbers (with a Hermitian structure defined on these complex coordinates). The complex coordinates C(s,t) are related to the shapes of discrete unitary subgroups (or complex space-forms).

Examples

R(s,0) is Euclidean space while R(s,1) is space-time or equivalently (isomorphically) it is hyperbolic s-space, or H^s.

We usually think either of Euclidean 3-space and time, or of space-time, R(3,1), which is of dimension 4, where space-time is equivalent to hyperbolic 3-space, or H^3.

The space-forms of Euclidean spaces are s-tori, (or (s-1)-tori in s-Euclidean space), where a 2-torus fits into 3-Euclidean space.

The space-forms of hyperbolic space have more than one hole, and the number of holes is called the space-form's genus. The hyperbolic space-forms are very ridged and stable in regard to their spectral properties, ie the spectra of a hyperbolic space-forms are the geometric measures of the general faces of the "cubical" simplex of the space-form.

For example the spectra of a 2-dimensional hyperbolic space-form are the measures of length of the edges of its (polygonal) fundamental domain.

The important spectra of a hyperbolic n-space-form are its (n-1)-flows, ie the geometric measures of the n-simplex's (n-1)-faces, are the important spectral measures of a hyperbolic n-space-form, contained in an (n+1)-dimensional hyperbolic metric-space.

Many properties of hyperbolic space-forms were determined by R Coxeter.
Up to hyperbolic 5-space there are both bounded and unbounded hyperbolic space-forms.
Bounded space-forms have definitive spectral properties.
After hyperbolic 6-space all hyperbolic space-forms are infinite extent and thus they do not have well defined spectral properties.
There are no hyperbolic 10-space-forms in hyperbolic 11-space, ie hyperbolic 10-space is the last hyperbolic space which contains hyperbolic space-forms.

(these properties are from a book about Algebra and geometry of discrete groups,
by J Humphries).

*What is the geometry of interaction dynamics in dimensions greater than both the inertial dynamics of R(3,0) and the velocity-space or the electromagnetic structures of R(3,1)?

There are infinite extent space-forms along with bounded space-forms for space-time dimensions of:

4=R(3,1),
5=R(4,1), and
6=R(5,1),

*These subspaces are (can be) defined by their being faces of a high dimensional "cubical" simplex.

The infinite extent (hyperbolic) space-forms define a subspace out to infinity and they are (or represent) models of light's electromagnetic field structures, ie they are models of photons.

Metric-space states and the need for complex coordinates

Each metric-space has a pair of metric-space states associated to itself. These states are related to the "physical" property that is associated to R(s,t), eg R(n,0) is associated to inertia (or spatial displacements, eg translation or angular) and its two states are related to (1) the stillness of the background stars, and (2) the background stars which are rotating. While R(s,1), or H^s, is related to (1) forward moving time, or (2) backward moving time, where time displacement is associated to the physical property of energy, R(s,1) is also associated to the material of charge (or to current). That is hyperbolic spaces are associated to charge, energy, and time.

If such a metric-space, with two metric-space states defined on itself (as is true of all metric-spaces), were moded-out so as to form a space-form, then the two-metric-space states would be spin rotated into one another on the space-form.

Thus, material has spin-½ properties.

This spin rotation of metric-space states is (also) related to the process of dynamics.

During dynamic interactions between material space-forms the dynamic structure depends on the mixtures of the two time-states of hyperbolic (or velocity space) metric-spaces, so that these mixtures of states affect one another during small time intervals, defined by the period of the spin-rotation of the metric-space states.

This mixture of metric-space states is best modeled on C(s,t), where C represents the complex numbers. Thus, this new description has a natural relation with local unitary transformations of complex coordinates, as well as local isometry transformations of real coordinates, where real geometries (of different metric-space states) form in the pure-real and pure-imaginary subsets of the complex coordinates.

What are subspaces of time whose dimensions are greater than 1?

Conjecture: Higher dimensional temporal subspaces of R(s,t), ie t is not equal to only 1 or 0, result when a new material and (subsequently) a new physical property is added to the metric-space.

A hypothesized example of such a new material could be in R(4,2) where a 3-space-form (contained in 4-space) whose genus is an odd-number would have an odd-number of flows associated to such a space-form. If this odd-number of spectral flows were occupied with material, eg charge or current, then such a charged shape would be electrically imbalanced so that such an imbalance would drive (or cause) the other "charged flows" within the space-form to begin to oscillate, thus pushing opposite states into one another so as to generate energy when matter and anti-matter come in contact. This oscillating space-form, which generates its own energy, could be an example of a "new material" with new properties which is associated to time in a new way, thus necessitating a temporal subspace of 2-dimensions in R(4,2).

Summary of geometry and its relation to the spectral properties of material, and energy generating space-forms

The containment space of a material space-form system is a dimensional-spectral-conformal-dynamic-space-form containing space, whereas there are several high dimension space-forms which can contain a particular dimensional level, and the spectral properties of these space-forms as well as the material space-forms which are contained in the flows (or metric-spaces) of these (simultaneous) containing spaces [of any material system] determine the spectral properties with which the material space-forms in the given dimensional level can possess (or can have), through resonance between the material space-forms and the spectra of the higher dimensional containing spaces, ie these dimensional-spectral-conformal-dynamic-space-form containing spaces can determine the allowed spectra of the material space-forms contained in a particular dimensional level.

These containing space-forms can be oscillating, energy generating structures. This gives a "physical driving structure" for resonance, and through resonance are identified (or one can identify) the allowed material space-forms which can be contained in these given dimensional metric-spaces (or given dimensional faces of its higher dimensional space-form containment set).

That is, the material system can be contained in several space-form-spectral structures so that the relation between material spectra and the spectra of the containing set of space-forms can be used to identify the spectral-conformal structure of the full containing space.

It is such a many space-form structure which is needed to understand the physical-chemical-geometric properties of living systems.

Interactions: differential-forms* and space-form-calculus interaction structures.

If one considers the description of a material interaction, one can define a material interaction based on an interaction space-form where the interaction space-form is a space-form whose dimension is the same as the dimension of the material containing metric-space, and the "cubical" simplex of this interaction space-form is defined by (its [new] composing) faces which are both the material simplexes (of the material involved in the interaction) and a new spatial simplex face which completes the "cubical" simplex of the interaction. This interaction space-form, when moded-out, fits into a higher dimensional metric-space than the dimension of the metric-space within which the interacting materials are contained. The edges (or 1-flows) of the spatial face of this space-form need to be related to the dynamics of materials, which this interaction space-form can cause, that is, they need to be tangent to the interacting material space-forms (or material faces), and the geometric properties of the interaction space-form provide the geometric context which is fundamental to the classical definition of a physical system's differential equation. The changes of the material positions due to the interaction (are to be changes as a result of the interaction) are changes in the metric-space within which the material space-forms are contained and (thus, or it is conjectured that) such spatial displacements are to be related to (or caused by) the local coordinate transformations of the isometry group, (this would be consistent with a connection [a generalization of a derivative] being a Lie algebra valued 1-form on a flat metric-space) but in relation to the mixtures of time-states in the interaction space-form (and in the material containing metric-space). It is the local coordinate transformations, due to the geometric relation that the tangent edges of the spatial face have to the geometry of the isometry (or unitary) group, which can determine the local coordinate changes (spatial displacements) that the interaction causes, as well as the alterations of the relative velocity properties of the interacting material space-forms.

For example, in R(3,0) the tangent edges of the spatial face which are also tangent to the material 2-face, of the interaction space-form geometry, identify a normal direction. The isometry group of R(3,0) is SO(3) which has the geometric properties of a 3-sphere, and thus the tangent 2-face can identify "the natural normal of a 2-dimensional surface in (on) the 3-sphere of SO(3)" as the radial direction for the spatial displacement which is caused by the interaction, and subsequent spatial changes of (or due to) the interacting materials.

This identifies a spherically symmetric interaction in which there are no singularities intrinsic to the descriptive structure. That is, the "radial" changes do not emanate from the material, but rather from both the interaction space-form and the geometry of the fiber group of the metric-space.

The spherical symmetry of the interaction is coming from the geometric relations that exist between the geometry of the interaction space-form and the geometry of the local coordinate transformation space, ie the isometry group.

The geometric structure of an interaction space-form can be viewed in many different ways, eg the material and spatial space-forms can remain in the metric-space as 2-dimensional space-forms, or the space-form can form as an approximate, semi-infinite extent, hyperbolic space-form so that one of the holes in the interaction space-form is associated to the spatial separation of the interacting material space-forms, etc.

Relative velocities of material space-forms which are contained within a frame are related to hyperbolic space-forms. These hyperbolic "relative velocity" space-forms identify a dynamical system's initial conditions at the beginning of an interaction, but at each small time interval (determined by spin rotation of metric-space state) a new interaction space-form is formed as well as a new hyperbolic "relative velocity" space-form.

However, during this small time interval, if the interacting system begins to resonate with the (an) appropriate spectra of the interacting system's over-all containing space, then the interaction may stop (depending on energy values) and a new "stable" space-form (of the appropriate dimension) can form out of the interaction.

The interaction space-form structure can be considered to have many different geometric structures and these geometric structures are important in relation to how the local coordinate transformations of the fiber group and its geometric structures can inter-relate to the geometry of the interaction space-form so as to determine the direction and magnitude of the coordinate changes of the material for that time interval of interaction. This local coordinate transformation can be modeled as a Lie algebra valued 1-form, where the 1-form is related to the geometry of the interaction space-form. That is, the geometry and dimension of the interaction space-form also has a relation to the structure of the differential-forms of the interaction space-form which identify the "force field" structure of the interaction which can be relevant to the dynamics of the interaction.

There are both inertial interaction space-form structures and hyperbolic interaction space-form structures, etc and their shapes and their relation to the geometry of the isometry or unitary local transformation group will affect the dynamics of the material caused by the interaction.

That which causes the apparent random properties of quantum systems

The interaction space-forms of very small systems depend on being defined each time after short time intervals (identified by the period of the spin rotation of metric-space states), for small systems the geometry after each such (short) time interval can be quite different (because inertia is small and electromagnetic forces are relatively large), thus small material space-forms will follow Brownian motion.

E Nelson of Princeton (in 1957) described the random wave-like properties of quantum physics based on a quantum system's small components having the dynamic properties of Brownian motion.

That is, the properties of quantum randomness are deducible from the Brownian motion which is implied by a space-form interaction structure for small quantum system components.

The main properties upon which the language of quantum physics is based can all be explained within the context of space-form geometries.

That is, quantum randomness of the quantum descriptive language, unitary invariance (* in quantum physics, unitary invariance comes from Schrodinger's equation being an energy operator which is a Hermitian operator, while the Schrodinger equation is also a heat operator, or parabolic differential equation), and the stable spectral properties can all be deduced from the space-form model of existence.

Furthermore, the property of spherically symmetric interactions in Euclidean 3-space can also be deduced from the principles of existence described by space-forms.

Thus, it only needs to be shown that each dimensional level is geometrically independent of the other higher dimensional levels in regard to both geometry and interactions, for the space-form description to be consistent with the currently accepted descriptive language of physics.

There are two (main) reasons why each dimensional level is interactively and geometrically independent of the other dimensional levels:

(1) The metric-space (within which material space-forms are contained) is a face on a simplex, thus it identifies a definitive subspace of a fixed dimension within which there are material space-forms and interactive space-form structures, where the interactive space-forms need to have a simplex defined by faces of material of the proper dimension and the faces must be complete or continuous, that is non-continuous changes of a material face of an interaction space-form cannot occur due to projections of a dynamically changing material space-form from a higher dimension are not continuous, ie the space-form from the higher dimension is not conserved and complete.
(2) The second reason would be that the conformal factors (or the actual space-form geometry) is such that for the material interactions in R(4,0) the size of the material space-forms which can interact in R(4,0) is far bigger than the 3-space-forms which can exist around the (dimensional) boundary of R(3,0). Furthermore the interaction space-forms, and its possibly stable space-form which results from the interaction, are not contained in the given metric-space, but are rigidly held to the boundary of the metric-space face, within which the (given) material (of that particular subspace) and its

interactions are contained. Thus these "too small" sized interaction space-forms in R(4,0) do not resonate with the spectra of the dimensional level of R(4,0), thus they can only interact if the entire metric-space of R(3,0) interacts as a space-form in R(4,0).

The interaction space-form is a model of a differential equation in both a particular dimension, and in a particular signature metric-space. Its geometry (related to "cubical" simplexes) determines the nature of the differential-forms which are a part of such an interaction, as well as determining how the local coordinate transformation group and its geometry can be related to both the geometry and the interaction's differential-forms.

Space-forms have a relation to the group geometry of a Lie group.

Parallel manifolds (or separable geometries) are easily defined using Lie groups.

The orbits of group elements (in a group or) relative to a subgroup, can be related to the geometry of the group, or to the geometry of a manifold, ie a parallel manifold or a separable geometry, it can be related to dynamics on a stable space-forms or the dynamics of the interactions of free material.

Loop spaces can be important in a Lie group, ie related to its discrete subgroups, and related to its many maximal tori.

The geometry of differential-forms can be related to a Lie group and to an interaction space-form.

*The relation of Lie group orbital properties to the stable space-forms of a given dimensional level (?)

Spectra of physical systems

In R(3,0) stable definitive spectral structures are given for both the nucleus and the general atom so that these systems are results of a spherical symmetric interaction, and thus they are consistent with the spherical harmonics of a system which has conserved angular momentum, but now there is a new mechanism through which the spectral properties of the 2-space-forms, as well as 3-space-forms, can be found. Namely, such space-forms will resonate in a manner consistent with their observed properties, so that they will be resonating with corresponding spectra of the higher dimensional spectral structures within which these (lower dimensional) material space-forms are contained.

Thus there is a dimensional-spectral-space-form containment set upon which the spectral properties of the material space-forms . . . , which are contained in a particular dimension metric-space . . . , depend, through resonance.

Furthermore, material interactions are a result of (actual) interaction space-form structures which naturally form between (interacting) material space-forms within a metric-space, and these interaction space-forms determine the structure of differential equations of physical systems, as well as the structure of its solution differential-forms, as well as determining the relation that . . . . both the interaction space-form geometry and the geometry of the differential-forms . . . . have to the geometry of the local coordinate transformation (Lie) groups so that this relation determines the dynamic properties of material interactions, where these dynamic interactions take place in complex coordinates so that the dynamics result from both local coordinate transformations as well as being determined by the mixtures of (time) states that are a part of the dynamical system's complex coordinate structure, where the mixtures of metric-space states can be identified.

Life mind, intent etc

Life are the odd-dimensional oscillating energy generating space-form systems. These systems are always related to the fiber group and its maximal tori. The maximal tori can carry on themselves a spectral set (or many spectral sets) which (can) resonate with the spectra of any space-form. Memory can be used to direct energy between different faces of an oscillating energy generating space-form.

Internal external structures of living systems; both systems are composed of high dimensional oscillating energy generating space-forms of various dimensions.

Summary

There are "conformal-spectral-dimensional-interaction" containment sets from which the properties of material and living systems and existing systems beyond material and life are contained, and within which one can find a useful description of these well ordered (physical) systems.

Very simple math but very complex possibilities for systems which possess a great deal of order which is imposed by the math structure.

Social structures

How to get people to identify this new math with high social and intellectual value. One cannot unless they change their minds, but what are the social conditions which select a certain type of person to attain a social position of intellectual authority, so that it is their intellectual judgment upon which the people of society believe, and why is this accepted by the intellectual community?

These are the same issues of communication which are also important in selling soap, developing icons of high social value, eg universities, where universities are the vocational schools for rich corporations, and creating a myth about experts, in which no one understands "what the people of high academic rank" are talking about, due to the specialized technical language of experts, but nonetheless the public believing that the experts do understand what they are talking about.

In fact, the experts do not have any understanding any deeper than their assumptions, which anyone can understand.

> Compare what the problems are in regard to each respective descriptive language,
> or compare
> What the two types of math patterns "used for the description of existence" are capable of asking,
> or compare
> What they are capable of describing.

The current descriptive language says that the fundamentally stable spectral-orbital properties of fundamental physical systems are simply too complicated to describe in a useful way based on our descriptions of physical law, so upon what problems do those who "believe in" the currently accepted descriptive language work?

Fundamentally they work on bomb and military engineering.

> If one viewpoint is asking what relation does knowledge have to creativity in general?
> What is the structure of material and its containing coordinate spaces and it relation to fiber groups?
> and
> What is the nature of interaction (or stable spectral structure) and its subsequent relation to differential equations?

While the other viewpoint is how is knowledge related to the interests of the military industry?

On the one side people accept the assumptions and interpretations which the professional community . . . , whose primary goal is to support the military industry . . . , accepts and uses and to compete and win . . . , and to be promoted . . . , leaves such a "winner" to believe that they are winners because "only they" can do the difficult intellectual activities (of math and thinking) . . . , but which in turn (these competitive people) do not have the capacity to describe (or do) anything except leave people wondering "What the math is all about . . . ? Other than confusion?"

Yet the experts use math which has a very complicated relation to the assumptions and interpretations . . . , but nonetheless, this math is not capable of describing anything of value . . . , eg the stable definitive spectral orbital properties of material systems at all size scales cannot be described by the aggressive experts (they are aggressive to win social position but they are not aggressively intellectually) . . . , is enough to leave anyone wondering about what the description actually is.

But no one is rewarded for pointing out the uselessness of the expert's descriptive language except to the most marginal of applications . . . , ie the probability of particle collision model of a bomb explosion . . . , and that there are other math patterns which can be of a lot greater creative use.

Within the narrow confined structures which are offered to those who want "equal" opportunity to succeed in society . . . , one can use technical language in the particular ways in which those with the greatest invested interest in using that particular expert knowledge want it (the technical language) used . . . , allowing those who are consistent with that narrow viewpoint to expand their dreams about material and space within the narrow confines within which they have agreed to believe.

This type of capture of a category of knowledge, eg physics, and then demanding that it be confined in particular ways, is common in a society in which the fundamental structures of society's necessities and its traditions are controlled by very few interests . . . , so that this results in the creation of the society's sense of very high values . . . , to which everyone adheres . . . , but which are (nonetheless) illusions.

What is affluence and what is poverty and what are "good" living conditions?

Consider the expert economist: Is economics a study about "How agreed upon material value can be established in a society? and in turn, How this established value can be used to organize society?"

"How can societies be organized, around what they value materially, for the best affect on its members?"

> ie What do its members most want to do?
> either Be creative? or Be dominant?

Is economics only about capitalism in which markets and traditions are marked-out and briefly contested and then when the tradition is well established that part of the market is dominated and controlled?

Is the study of economics broad in its scope or narrow?

Basically economists are only employed in order for them to contribute to creating an illusion through which domination and self interest is masked. If one does not play (deal with knowledge) by the rules . . . , imposed by those who fund the use of such knowledge in society . . . , are the losers.

P Krugman (found in, Profit Over People) has pointed out that (economic) knowledge within our society is:

(1) First knowledge is narrow and not well developed in its narrow domain
 (in physics: randomness, particle collisions, spherical symmetry, definitely define a very narrow domain of interest)
(2) Within this context, properties of truth are espoused whose truth is not known
 (in physics: particle collisions are the source of all material forces, strings are the source of all existence)
(3) These properties of truth are constantly shifting and changing
 (in physics: string theory, quantum gravity, grand unification, non-linearity; when in fact, the only part of physics which is consistently useful and successful, in its narrow domain of macroscopic physics, is classical physics, which is still driving all technical development within society today (2010))
(4) Policies and projects seldom are successful
 (in physics: nuclear energy failed since it is so dirty, and it is prone to failure [an obvious flaw which was over-looked], fusion energy failed, use of large scale superconductivity has failed, describing the stable, definitive spectral-orbital properties of material systems at all size scales based on currently accepted physical law has failed)
(5) Yet these policies were a great advantage to the powerful who dominate society
 (in physics: these ideas fit well with bomb and military engineering, the big bang and the anthropic principle, string theory, etc are expressions of intellectual domination, and they identify the high value of our culture, so that this high value is not supposed to be an illusion, rather it is supposed to be science, ie an objective truth). [This provides great value to both the business interests and to the dominant intellectuals of the society, where society's dominant intellects are given this position in return for their submissive obedience to business interests, and their inability to think for themselves.]

Apparently these observed structures of knowledge, and the structure of the institutions which use (expert) knowledge within our society, are true for most bodies of knowledge which are studied by our society.

However, the actual value of the knowledge which our society's dominant intellectuals possess is about the same value as the knowledge required (or which is a part of) for the distinction between one soap or another when one shops at a supermarket. That is, the value of the knowledge possessed by our society's dominant intellects has about the same value as an icon (or brand) has value in promoting one product or another in a market dominated by a few. That is, the actual value (as knowledge) which particle physics possesses is its value as an icon of high value within society, a property which it borrowed from the icon of high value which "the physics department" of universities such as of Princeton or Harvard possess, where a "physics department" is an icon of high value in its own right.

However, scientists today do not give to science a realistic structure as to what science does and/or how science operates. This is because they believe that they have won a contest in which they demonstrated their intellectual superiority over others so that their authority has come to define (scientific) truth.

Science is about being disciplined, but it is mainly about questioning authority and proposing new structures for descriptive language when the authoritative structure of the language of science (eg physics) fails, wherein physics is clearly failing, since it cannot explain the quite prevalent stable, definitive properties of material systems at all size scales which have been observed.

Summary of the new physical science

Though the new, alternative, math constructions (described above) use mathematical ideas which are simple . . . (separable geometries in a multi-dimensional context, where different dimensional levels have geometric independence of the higher dimensional levels) . . . , nonetheless they are much more challenging intellectual ideas than the concepts used by professional communities of scientists who so weakly (meekly) serve commercial interests (where their ideas are based on materialism, randomness, point particles, and non-linearity), but the true failing of the professional scientists seems to be (personal) . . . , ie a belief in, and a will to achieve, intellectual dominance within the eyes of other people in society . . . , so as to take the power of "free inquiry" away from the people of society, an idea which serves the interests of those who do dominate all of society.

Summary of physical science's fundamental failures

The solvable systems in classical physics are the geometries which diagonalize the coordinate space's metric-function so that the function's values and the geometric measures, ie the differential forms, in the domain space can be properly (or consistently) compared with one another by means of the derivative's linear relations (linear relations which are maintained by the system's differential equation also being linear), where a classical system's differential equation is defined in relation to the system's geometric properties . . . , so too in quantum physics it is assumed that a quantum system's function space can be diagonalized by a (complete) set of (differential) operators where the energy operator of a quantum system identifies a linear differential equation so that the spectral function's define the particle component's (of a quantum system) random distribution of particle-spectral events in space and time and the energy (or other) eigenvalues define the quantum system's elementary event space of stable definitive spectral properties, ie the quantum system's measurable properties which characterize the system. If the wave-equation is linear then the probabilities of particle-spectral events in space will be (are supposed to be) consistent with the measurements made in space (ie in the domain space).

While the geometrically separable classical systems are causal and provide geometric information of a system's measurable properties (though the properties may be difficult to measure, eg an object's change in velocity), on the other hand the operators and function spaces of quantum systems are plagued with singularities which are not mathematically resolvable and the eigenvalues are not consistent with the system's observed spectral structures.

Though a quantum system's energy operator is (globally) invariant to unitary transformations (where nonetheless the many functions (of the function space) domain spaces are Euclidean [not complex coordinates]) . . . , as a classical system is locally invariant to metric-invariant local coordinate transformations . . . , by using Lie algebra valued 1-forms in the context of a "singularity generating" particle-collision (model for field interaction) so as to be both a unitary invariance and an expression of hidden particle forces, which perturb the system's wave-function, is an unbelievable contrivance and not mathematically resolvable.

In quantum physics why try to make more complicated a descriptive structure which has serious limitations as to what it is capable of describing?

The fundamental failure of quantum physics is that its techniques have not been successful at identifying a quantum system's fundamental spectral properties . . . , where point-like charge singularities are intrinsic to the quantum model, but they are not mathematically resolvable in the math structure which is given to quantum physics . . . , though rotational symmetry has some merit . . . , nonetheless, the stable orbital-spectral properties of (nuclei) or of general atoms have yet to be identified using the techniques of quantum physics (or particle physics) . . . , where it should (might) be noted that, in R(3,0), rotational symmetry (at "all" size scales) is a part of the new description.

In regard to general relativity, though an engaging idea that inertial properties follow geodesics, its non-linear nature has not allowed this idea to be developed, so perhaps there are other geometric constructions in which this idea can be expressed in other ways, eg that space is metric-invariant and stable physical systems are related to separable geometries where geodesics are followed.

Chapter 7

NEW IDEAS

The public relations spin of deception upon which politics depends was perfected by the sciences, physics in particular. Questionable assumptions and particular interpretations of data, without any alternative voices, as well as a focus on specially devised data (or facts) leads to misconceptions of the truth about what physics is trying to describe and the validity of the language which physics uses. Physics should be trying to describe patterns which are useful, instead it is trying to establish (or maintain) the "truth" of the statement that all material systems reduce to elementary particles and must be described in the language of particle event probabilities.

This is the main dogma for bomb engineering institutions (or projects).

If the institutions which are held in highest value by society, ie physics and science, are, in fact, engaged mostly in fraud and serving the product development interests of the powerful few, then one must know that minority rule, the few people and the societal structure which leads our society's people, has corrupted society so that society is no longer able to pursue truth and develop. Society is only for material magnificence and pleasure of the powerful few and those who are useful helpers to the powerful few.

The destruction of physics is the result of allowing the idea of a market, or selfish dominance, to enter into the context of science. Because science is a description of existence made in a limited language so that the description must be verified, and the description needs to be based on widely applicable principles, ie the hidden structure of existence, and the descriptions should result in great usefulness, this means that everyone can (and must) understand the descriptions of science. Hence there is no one person who understands science any better than anyone else, when science is working correctly. Thus it appears that competitive students (eg Oppenheimer) can be substituted for those in control of the creative process, ie those willing to be freely inquisitive (eg Einstein), and this works very well on particular technical projects (eg atomic bombs). Indeed one must be trained in the language, but once the language is mastered it can be used for technical purposes, but the main idea of science is to find a new basis for the language which is used in the description of existence, a new description which fit's the observed patterns better and which is more useful than the current language.

The point of science is that there are no authorities in science, truth can only be found in an egalitarian structure (context), the description is simple and clear and it is useful for everyone. That is, an expert scientist is one who works on special projects, not one who understands science better than others. But society's rulers now want the public to view scientists as elite authorities. However, science is built from free inquiry and equality.

An idea can be sold by creating a vague sense of value for a particular idea. This can be done by creating an exaggerated, overly strained form of competition within a context of authority and domination. Within this context, ideas are sold (to the new scientists) by giving experts too much credence, allowing their expertness (on particular projects) to determine their authority.

The idea of a market is that the best things, or the best value, comes from the most competitive social environments. Thus the ideas of physics are sold based on the authority of experts who have worked on secret projects (experts who have a secret agenda) and these ideas, along with the fixed language and its dogmas and orthodoxies determine the contest.

The winners of the contest within the competitive market (almost) always make proclamations about properties of irrelevant physical systems (guided by the language of the standard model and general relativity, ultimately to be applied to bomb engineering, the secret agenda), whose verification is based solely on measurement and interpretation of mostly spectral data that comes from remote systems, eg particles from particle colliders and light from distant stars and galaxies, in what has become a Ptolemy-like scientific verification process, and the proclamations are authoritative and dominant, as the experts who are determined by specific technical projects, have come to dominate what is called physics, they follow the secret agenda of physics, ie bomb engineering, yet their technical usefulness is nil (apparently even for bombs).

Ultimately the incompetence and failures of science (devastatingly listed below) can all be traced back to both what society and minority rulers, including its dominant scientists, believe is true.

The basic idea underlying the language of physics is materialism. Society believes in materialism and in an idea of high value, as opposed to equality. The separate belief of religion, which claims to be opposing materialism, is a belief in a vague notion of high value, which depends on vague ideas which are supposed to transcend materialism. But religion does not lead people past materialism, so its language is as useless as the currently accepted language of physics. It is this along with a somewhat hidden belief that civilization should make pleasure easily accessible to its elites, ie life should be pleasurable for the people of value.

The basic ideas of civilization are: materialism, a vague idea of high value, competition for having high social value, and rewards of pleasure for those who obey and achieve what the rulers want the society to accomplish for them (the ruling minority).

The antidote to this dogma of: materialism, high value, and pleasure, (science, the church [or ruling minority], and the porn shop [or that people prostitute themselves]) (which is an existence represented as absolute truth (materialism), and hence it opposes the further development of truth), a dogma which is the cause of the sickness which has corrupted society, while the antidote are the principles upon which the US is supposed to be based. Namely, equality and the freedom to speak to an unselfish truth, or equality and the freedom of speech and freedom to seek truth. Namely, finding answers to the fundamental questions of: "What exists?" and "What is the true context of life?" so that the answers extend past the deception of materialism.

The materialism of a society (run by a ruling minority) is placed within a context of high value and it is associated with pleasure that an external material world can provide.

That is, freedom of speech based on selflessness deals with truth so that it is not based on coercion, eg a person must believe in certain things otherwise they cannot be a part of society, or the non-believers are enemies to those who believe in (the correct) truth (the sentence of extermination for non-believers). A selfless truth deals with truth so that truth is based on free inquiry and selflessness, ie free choice about truth.

It is over selfless truth which a society should be willing to protect an individual from other selfish viewpoints of truth. For example, both materialism and many religions imply a truth based on selfishness, eg survival of the fittest or a (jealous) selfish creator (or transcendent "absolute" ruler [who never grants transcendence]).

The notion of freedom of both speech and religion implies speaking about a selfless truth.

Unfortunately, many religions make statements of absolute truths which define an in-group of believers and an external opposition of non-believers whom are to be fought, thus defining a "selfish truth."

Equality means a democratic government for everyone. So that everyone owns society's materials. Material is from the earth and its use should be in harmony with the earth as opposed to being destructive to earth. Everyone decides "What material will be used," and "How that material will be taken from the earth." Everyone will decide "How it is used" and/or "What is made." And everyone decides "Who will use, what is made."

In order to allow life to be accessible to each successive generation, society needs to live in harmony with the earth.

Law will be based on equality and opposing the selfish oppression of others (this is stated in the Declaration of Independence). In order to have equality, law shall <u>not</u> be based on property rights and <u>not</u> based on contracts between unequal parties, eg unfair contracts will not be allowed by the law.

The justice system will not be used as a counterinsurgency force which protects minority rule by terrorizing both those who oppose minority rule and (terrorizing) those who have lost the game for power in a material centered, selfish society.

People act as if materialism is an absolute truth despite the (religious) lip service paid to the contrary. The religious lip service, really only supports the (vague) idea of a hierarchy of value for people or civilization, ie it supports (the harshness of) minority rule by stating that people are to be subservient to a vague notion of a high value. In religion, it is implied that unfair social laws are to be followed if people want to transcend materialism when they die (a little bit late in the game for such a powerful gift).

A change in society can only occur if a simple, clear (in a new scientific language) and useful scientific description of existence transcends materialism. If this is provided then what is beyond materialism can be explored based on both specific descriptions in a limited language, and on practical applications, and life will not be seen to be based on materialism and pleasure.

If life can extend its experience past the material world then life can, to some extent, transcend fear and violence, because people could then make the very real choice between worldly pleasures and the pleasures of the spirit or creativity which extends life's intent past the material world.

Only when people gain knowledge of an existence which transcends materialism can they relinquish their selfish view of the world, and then the world can be ruled by the idea of equality, such as democratic socialism, in which emphasis is placed on using knowledge and material to live in harmony with the earth.

Society

The control of the ruling few over society is mostly based on what people think is true. If people believe what others tell them is true and if that truth is not described in a clearly understood language, ie not described in a causal-geometric based language (the language of classical physics, the most technically useful of the sciences), but rather truth is described in a language of complicated, abstract ideas built from overly general and vague notions in which most often the description is fit into the language of probability [so that the language is based on the idea that a set can contain all possible events, and thus it is believed to be a description which is all encompassing] then truth can be made ever more vague, yet seemingly ever more all-encompassing (suggesting knowledge has reached a steady state), like the language of religion which describes a vague, all-encompassing, constant truth, but a truth which has no practical value.

The practical usefulness derived from a vague language becomes less and less, both for religion and physics.

That is, such a settling into a vague but very authoritative language which characterizes much of religion, also describes the state of affairs within modern physics.

In such a case where science becomes ever more authoritarian and at the same time lacking an ability to provide knowledge which leads to practical usefulness, subsequently this leaves classical physics and the technical (or laboratory) knowledge of solid material properties to be the "work-horse" for practicality, then the society becomes ever more dependent for its truth being based on the social position and authority of the person speaking.

Science seems to reflect the values of the culture, ie the beliefs of the minority rulers, rather than being a free inquiry into the truth about existence. Science emerged in opposition to the authority of society and its rulers. But now evolution supports the natural dominance of the superior in a competition for materials, and physics is about a Judeo-Christian-Islam myth of creation, ie a point at which creation began, so that both physics and biology are built out of the mathematical laws of probability, yet adaptive actions of living systems are faster and more precise than probability based description of existence allow and the set of elementary events is never defined as a fixed set, but probability depends on identifying precisely the (elementary) set of possible outcomes as in the number of sides of a dice.

The foolish ruling people see the split between materialism and spirit, which they exploit, but which they are not intellectually capable of transcending.
Perhaps the word energetically, instead of the word intellectually, should be used, because the ruling minority are energetically absorbed in protecting and using their good fortune.

If a society is unequal then the upper classes define high value and the society becomes a set of realms of dominion (or institutions of dominion). Each realm ruled by its own set of over-lords, whose words are used to define the set of laws (or truths) for the institution.
Thus it is, that truth becomes the high value of social position, as opposed to the equal right to selflessly speak to truth.
The equal right to selflessly speak to truth is the essence of the first amendment to the Bill of Rights and the Declaration of Independence and it is also the essence of science.

Society's highest rank goes to the overlords who have control over material and/or financial control of markets and economies.
Because the society's highest level leaders are those who control material, thus one can say the society holds materialism up as the highest truth. Society's second highest rank goes to the managers who work for the overlords.
Society's third highest rank goes to the overlords who have control over material knowledge, scientists and engineers.
Such expert people need to be controlled, and this is done by selecting the most competitive and (hence also the most) obedient people to achieve high rank in the science institutions, and then to reward these people very well.

Society is designed to grant, to those people who measure up to society's high value, a life of ease and a life of pleasure.

The institutions of society can be divided into:

> Economy and markets, the structure which feeds power to the minority rulers,
> Government, which, in the US, acquiesces to minority rule and provides many services as well as most of the technical development money,
> Security and justice which maintains and protects minority rule (which uses a counterinsurgency model),
> Media and education which is about controlling language and controlling thought, and
> Technology and science.

Technology is divided into:

> Weapons, which develop bombs and weapon delivery systems,
> Information, electrical engineering and the media,
> Energy, which seems to not be able to develop cheap, clean energy sources,
> Material science, which despite popular belief does not understand the properties of crystals as well as it is advertised in the media, and
> Health and biochemistry, which has no idea about how life began, cannot describe embryonic development, and cannot relate the activity of the brain with thought, though it has faith that the mind must be associated to brain nerve activity. [This might be far from being true.]
> Psychology, or the study of controlling human behavior.

No variation in any of the ideas or dogmas which define these authoritarian institutions (from both lists) is allowed, because the power of the minority rulers is derived from the fixed state of these institutions

Note:
A free and equal society does not need to fear ideas, because in a free and equal society law is based on equality and opposing selfish oppression and/or opposing selfish tyranny.
However, minority rule is all about selfish oppression of all of society by the ruling few, thus they fear ideas which could oppose their power.
What we have is a society centrally planned around the selfish interests of the selfish few, and in doing this thought and language must be carefully controlled.

Brief review of currently accepted paradigm of physics

All of science today (2007) is based on materialism, an idea which means that material defines all of existence, . . . including the space and time that contains material, as well as its interaction fields (ie now believed to be colliding particles).

The idea of materialism is maintained in both classical physics, and in quantum mechanics where probability wave functions determine particle events in space and time. However, the unbounded (global) nature of probability wave functions determine the idea of non-localness as well as the notion of wave collapse when an wave-event is observed. In the standard model of elementary particles, the higher dimensional particle state space is described in terms of probability, but the particle states are viewed as a mathematical fiction, while in string theory, where the particle state space as well as particles are given geometry, none the less, these small strings are still governed by the probability of events in space-time. In string theory the higher dimensional space is wound up into microscopic shapes so that these microscopic shapes do not influence the properties of 4-dimensional space-time, but rather influence only the probabilities of small string (or particle) events in space-time. In the space-time of string theory, either microscopic random (apparently) particle events (still) occur or the regular macroscopic material geometry exists, so that material still determines a 4-dimensional space-time of either material events or material geometry.

However, it should be noted that "membrane theory" (apparently a variation of string theory, whose origins are some paper by Dirac) allows greater (or new) variety of dimensional independence, so as to determine a particular dimensional subspace which can be immersed in an independent manner within a yet higher dimensional containing space. Such dimensional independence is also a property of the new theory and this new dimensional independence is related to the nature of material (space-form) interactions for material (space-forms) contained in a particular dimension metric space (which is also a higher dimensional space-form, that is, higher dimension than the dimension of the material space-forms which the metric space contains.)

All previous models for physical systems of various dimensional spaces, assumed the property of continuity would exist between all the dimensional subspaces, ie material in a particular dimensional subspace cannot be independent of other different (higher) dimensional spaces.

None the less both string and membrane theories are probability based descriptive languages upon which geometry is imposed, and due to the uncertainty principle of probability spaces, at best such a language is an inconsistent language.

The lack of any practical uses for string-membrane theory might be a result of this logical problem, and because of this, string-membrane does not qualify as physics. None the less, such "theories" dominate professional physics publications. This is apparently because a particle-collision and a gravitational-singularity are natural models to be put together for bomb engineering, and string theory claims to unify these two theories.

There are four separate domains of physics:

(1) classical physics (Newtonian gravity, electromagnetism, and thermodynamics (or statistical mechanics)),
(2) general relativity (the 1-body problem with spherical symmetry),
(3) quantum mechanics (the probability waves of microscopic particle events in space), and
(4) the standard model of elementary particles (all material as well as interaction fields are particles which interact by means of particle collisions).

Each of these four domains will be reviewed.

Note: String theory does not qualify as a physical theory, it seems to be a description based on a language whose basis is too abstract, and subsequently either too general or too narrow of a descriptive language so it cannot describe physical properties.

(1) Classical physics requires a containing coordinate space which is a flat metric space, where flat means that everywhere within the metric space the metric has the same form. A flat metric space also requires isometry group transformations, ie local coordinate transformations which always leave the metric invariant.

Material defines space-time (or space and time). A position function of some material can be defined on such a space-time domain so that local changes in scale for the function can be associated to geometric relations between the material position and other material in space so that the changes of the position function are applied in the direction determined by the material geometry. Partial derivatives are related to a tangent space defined on functions, ie derivatives of functions, such as position (eg manifold or dynamic systems) and/or force-field functions, which are related to a physical system's (both local and global) geometry, and whose domain's are contained within coordinate sets which are also flat metric spaces.

These coordinates can have certain "flat shapes" such as cylinders and tori (doughnuts) as well as rectangular coordinates (perhaps the flatness of these shapes is best modeled in a complex number system). Such geometric-coordinate shapes can be related to the system's (or its function's) geometric properties, as well as relatable to the "separable" (tangent vectors of the system's coordinates are always locally orthogonal to one another at each point in space) nature of a partial differential equation's solution. This relation to geometry can be identified (or accomplished) by transforming the local tangent vectors of functions (whose domains are the [geometric] coordinate space) by isometry group transformations. These coordinate and/or system (manifold) shapes can be found by means of associating a function's tangent vectors to a system's natural relations to the geometries of the system's material by means of identifying (or defining) differential equations. The solutions to such differential equations are related to the system's geometric shape. The laws of classical physics describe how to define a differential equation on a function (such as a position function) which identifies physical properties of the system. Note: This can be done for an energy function also.

For gravity and electromagnetism, as well as thermodynamics, the system's differential equations are essentially the Poisson (or Laplace or wave) equations. That is, the differential or Laplace operator acting on differential forms (which can identify force fields). Such equations are invariant to isometry transformations of the coordinates.

Compact geometries can be covered with (small) coordinate regions so that coordinates and their local vector space properties can be pieced together so as to form the system's geometry or its manifold structure.

However, electromagnetism cannot account for the energy level properties of charged micro-systems, while in gravity the many body problem of "solar system stability" has never been accounted for, whereas in thermodynamics the entropy function has always been a mystery as to its origins, though quantum mechanics gives some overly simple insight into the origins of this entropy function.

(2) The laws of quantum mechanics can be formulated in terms of the system being a (Hilbert) function space, which contains all the possible events associated to the system and, whose eigenvalue (or physically measurable) properties can be found by the function space's relation to a set of (commuting) (differential) operators. These (differential) operators identify the type of measurable and/or symmetry properties that a micro-system can have so that for each measurable type there exists a spectral set. The set of commuting operators act on the function space so as to determine the microscopic physical system's spectral set (of functions) which exist within the function space. The function space can be transformed by unitary operators which rotate the spectral subsets of functions in the function space in an invariant manner, while the operators which identify measurable properties are Hermitian (or self-adjoint) operators which have real eigenvalues. One can find a system's set of measurable values, ie the system's complete set of commuting operators, and thus the system's eigenvalue structure, and the probability of being in a particular measurable state. In order to do this one must associate a probability to the event of a particle found at a particular point in space so that such a particle has the eigenvalue and probability of a particular measurable state of the system for which the particle is (was) a component, ie the measurable property must occur somewhere.

The function space of quantum mechanics plays the same role as a classical system's geometric function [upon which the classical differential equation is defined] so that in quantum mechanics differential operators act on the function space.

It is not clear that these ideas define a probability space in a well defined way. That is, (1) the set of elementary events for a system is never really pinned down, so that newly found events seem to always be a part of the elementary event space, but the elementary events of a system is supposed to be fixed, and (2) the (quantum) differential operators which determine a system's measurable properties seem to be specific to each different system (ie there are no general principles which have wide applicability in being able to describe a quantum system's spectra). That is the operators which apply to the H-atom do not seem to work in general for atoms with higher atomic numbers. Thus the description of quantum mechanics does not have general principles which are widely applicable, ie quantum mechanics is not widely applicable, so that each system has its own specific operator properties which identify a system's spectral set of functions in the (event) containing function space.

Furthermore, quantum mechanics has not been useful. Only the bomb and the laser come directly from applications of theory of quantum mechanics, and that is because in these systems the desired event properties have probabilities which can either be controlled or used, ie the desired properties (for particle events) are very likely.

Note: The transistor is mostly a technical development. Its development did not depend on theory because the quantum theory for crystals is still inadequate, eg high temperature superconductivity has no explanation. Small micro-chips depend on optics for their small size, not solid state theory. Furthermore, quantum mechanics cannot be applied to the nucleus. For this one needs the standard model of elementary particles, but even the standard model will not be enough to provide such answers about nuclear properties such as the orbital structures of all nuclei.

(3) General relativity is very limited in its range of description. It can be used to describe the one-bodied problem which has spherical symmetry. But it does not answer the most pressing gravitational question, namely, the many-body gravitating systems which have bodies that have the property of orbital stability. Furthermore, for the non-linear gravitating problems of small satellites in the solar system, it is Newton's laws which are used, not the laws of general relativity. Thus general relativity has not been a very useful theory for physical description, thus its validity is still questionable. General relativity's descriptive language seems to be too general, so that it is not a useful description of (a material) existence.

(4) The laws of the standard model of elementary particles depend on describing the changes in a harmonic wave function which represents the wave function of a physical system, which is derived from regular quantum mechanics. All of space is modeled as being full of harmonic oscillators. The system's wave function is made discrete so that it can be related to any of these harmonic oscillators in space. However, a system is modeled in terms of the particle interactions that take place at only one of these harmonic oscillators, independent of all surrounding harmonic oscillators. That is, a system is defined by its particle interactions at a single point.

The modifications of a discrete wave function are determined by measuring the particle occupation (or collision) properties of the single harmonic oscillator and how they affect the wave function. The wave function is also given the further dimensional properties of a fictitious (material) particle-state space (which are representation vectors of the Poincare group) so that these material particle states are all Fermions. The wave function is modified by measuring its particle interaction properties at a point. This measure is related to a Lie group transformation which, in turn, is related to the Lagrangian density of the system, within which exists a particle-state field-particle interaction term which is a unitary matrix acting on a particle state vector. The transformation symmetries of the Fermion-Boson model of interactions introduce probability based particle-interactions (or particle collisions) which define the system's modifications, which act on the original system's discrete wave which is also represented as a "discrete" particle-state vector, which is changed by such particle interactions, resulting in a perturbation representation of the altered (original) discrete wave function. Note: To re-iterate this occurs at a point and is independent of other surrounding harmonic oscillator points in space.

The standard model cannot predict the orbital properties of the nucleus. However, if the standard model cannot predict the notable properties of the nucleus, eg its orbital properties, of what value is it as a theory?

Within the standard model the logical context of both the unitary field-particle symmetries and the higher dimensional space of Fermion or Boson particle-states is not known. As with all probability based descriptions these patterns are simply formal patterns which fit the empirical data.

The system described by the discrete harmonic function is only vaguely defined, or as in quantum electrodynamics, the system of the discrete wave is defined, but without any practical value. The reason one can say this, is because the corrective perturbations to the discrete wave function are too difficult to compute, and a person can only be sure if such calculations are correct if there is data to which a calculation can be compared.

The standard model only seems to be applicable to particle observations within particle accelerators, and this is (also) subject to great skepticism, as neither neutrino oscillations between particle families, nor dark matter, nor dark energy, were predicted by the standard model, not to mention the problem of its inability to predict the spectral masses (and/or arbitrary parameters) of the elementary particles, a fundamental property, where (it might be noted that) by analogy the H-atom model predicted the H-atom's spectral structure.

Clearly the standard model's strongest feature is its ability to follow (or be fitted to new) data, not predict it.

This is also true of regular quantum mechanics, and it was also true of Ptolemy's theory of planetary motions.

None the less, it is the standard model and the gravitational singularities of general relativity which are considered to be the heart of professional physics.

Thus the authority of physics is being used to deceive, because scientific truth is based on wide applicability and usefulness, not the Ptolemaic view that scientific verification depends on fitting a description to the (measurable) data, rather than the physical description being used to generate data, which in turn, leads one to verifiable properties that were previously unknown.

In the two cases of the standard model and gravitational singularities it is simply an occasional match of highly specialized, or selected, experimental data with the description, that is claimed to be verification of both of these highly abstract theories. That is, there is as much, if not much more (or conclusive), evidence that these two theories are wrong as there is supporting data (see below).

Apparently professional physicists focus on particle physics and gravitational singularities because of physic's strong relation to bomb engineering. Particle physics grew out of the Manhattan Project. Though probabilities of particle collisions work well to model an exploding bomb they seem to <u>not</u> be important in applicable or practical physics.

The logic of description

Descriptions based on geometry are useful, eg classical physics, on the other hand descriptions based on the probabilities of particle events in space (where the probabilities depend on solutions to wave [or function space operator] equations) are not useful. For example, only the laser (1958) and the atomic and H-bombs have been developed from (particle) properties of the world described by quantum mechanics or particle physics. The transistor is an experimental development, it did not result from a theoretical calculation. Spin signal imaging came from optics. Though quantum properties are useful, spin properties and the energy levels of (micro-system) systems, their usefulness is developed in the context of classical physics. Very few things (or inventions) of useful value have come directly from the descriptions of quantum mechanics, particle physics, or string theory. Whereas classical physics is still the main description of existence which drives technical development.

Both classical and probability based physical descriptions are based on the simple measurable ideas of: material, space, and time (where material's reducibility to random particle events is the basis of probability based physics). Re-iterating, physics is based on materialism, ie that material determines the patterns of existence, be it material, geometry, or random particle events. However, the descriptions that physics is now providing to the world are no longer useful. Conclusion: The descriptive language of physics must be revolutionized.

The clear reasons as to why one can say modern physics has failed

A description must conform to data, and be based on general principles, so that it has wide applicability and is easy to understand so that it is a useful.

Based on general principles, or physical laws, physics cannot even provide descriptions (let alone useful descriptions) of the properties of the following physical systems:

Orbits of nuclei,
High atomic number atoms,
Shapes of large molecules,
The properties of crystals,
The origins of life,
Embryonic development,
The mind,
The stability of the solar system,
The motions of stars in galaxies (dark matter), and
The accelerating expansion of the galaxies in the universe (dark energy)

amongst many other systems which go without adequate descriptions.

At each level of scale these physical systems are some of the most fundamental physical systems of all the systems which exist.

Instead of concluding that "what now passes for physical law" is wrong, or that the language of physics as it is now used is not adequate for describing the physical world, rather physics claims the laws and language which it uses are correct, insisting that the descriptions of science can be verified on a Ptolemy-like verification process, but none the less most physical systems of the world are simply too complicated to describe, so that physical description can no longer be related to practical uses.

Physics then proceeds to focus on marginal physical systems such as particle physics (note that both the standard model and string theory cannot describe the orbital properties of the nucleus) and gravitational singularities. Note: Gravitational singularities are systems which are defined by general relativity, general relativity can only be solved for the 1-body spherically symmetric system, that is, it has nothing to say about solar system stability nor about non-linear trajectories of satellites which are flung out to greater orbits, this is done using Newtonian gravity.

The two main ideas of physics, particle collisions and gravitational singularities, seem to be mostly about bomb engineering. They are no longer related to descriptions of "What exists?"

Indeed the uselessness of the language of quantum mechanics and the standard model is also evidenced by its failure to describe the microstructure of the crystal (or is so exceedingly difficult). Note: The critical temperature of BSC superconductivity has been exceeded, ie as a physical theory it is wrong.

But materialism, the "objective truth" of a physics, which is verified in a Ptolemy-like manner (which describes the fundamental state of physics today (2007) Ptolemy-like), is wrong. The properties of material systems cannot be described by a language whose foundations rest on materialism, probability, and material reductionism to particle events.

(Note: Particles are not particles, rather [in the new theory] they are the distinguished points of space-forms about which all the [material] space-form interactions focus)

Yet probability based descriptions are claimed to be verified by a Ptolemy-like verification process. Probability based descriptions can be adjusted so as to fit a micro-system's spectral data which has already been found, but many properties of micro-systems cannot be described by a probability based language.

General relativity is supposed to be verified yet it only applies to a 1-body problem and is highly non-linear, hence it has nothing to say about the most important problem of gravitating systems, namely the stability of the solar system, ie general relativity is too complicated of a descriptive language to account for either a system composed of more than one material body or the orbital stability of the solar system.

If a descriptive language organizes data patterns in a useful manner, then this allows science to develop.

That is, one wants in scientific descriptions to move ever more towards truth, where a truthful description of existence is a useful answer to the question of "What exists?".

Physics is lost

If the vanguards of truth for society, ie the physics community, has become authoritarian, deceptive, and fraudulent so that being in the physics community is like having a badge of intellectual virtue assigned to oneself [similar to the badge of virtue that religious followers claim for themselves, and want to force on others] to be used in a similar manipulative manner, forcing a vision of truth on others, dutifully working on highly technical problems which have no relevance to a widely applicable, useful description of physical systems, ie no relevance to truth.

That is physics has become divorced from the process of seeking truth, and has become concerned about the absolute truth of a fixed, but isolated, language which only has vague meaning in regard to the patterns of existence that are being observed. The most definitive statement about physical systems coming from the physics community is that fundamental physical systems are too complicated to describe, by using what is dogmatically put forth as physical law.

This is exactly analogous to the fixed language and authoritative actions of religion, where religion proclaims that materialism is wrong, but offers only a vague language which is to be used to describe what is beyond the material world and then demands that people be obedient to its authoritative truth in order to experience something beyond materialism.

Similarly, physics is now saying to those who want to do physics, "Agree to these useless patterns of language, which are associated to measurement, they are true, but physical description can no longer be related to usefulness." Yet by agreeing to use this language you will be in the social class which over-sees bomb and weapon technology. So that now virtually all the practical stuff which leads to technical development now deals with classical physics.

Limitations of a fixed language

Godel's incompleteness theorem states that any descriptive language based on a set of axioms (about quantities and geometries) is incapable of describing within its structure of language some (or many) of the mathematical patterns that (do in all reality) exist. That is, there always exist patterns that are outside of a fixed limited language's capacity to account for such patterns. Thus even what might seem to be the most general ideas, such as "set containment" and/or "real numbers and their operations", as being the basis for a descriptive (quantitative) language there will still be patterns which such a language cannot describe.

Though, now it seems (G Chaitin, Metamath! [book]) as if the axioms of real numbers apply to a set of elements (the real numbers) in which no single likely point which compose the set of real numbers can be identified.

Thus the stable properties of existence which allow measurement in what we think of as a real number system, ie the identification of a stable uniform measuring unit (identified with a particular type of number) of metric (or length) measurement and the relation that other measurable units might have to one another (either globally or locally), might really be the set of properties which should be identified, so as to form a basis for a quantitative-geometric descriptive language for existence, as opposed to the mathematical axioms of real numbers and general notions of set containment (and a physics of materialism based in probabilities).

The stability of geometric form (or geometric shape) might be (is) best described in terms of the natural geometric and spectral properties of metric spaces, ie their discrete isometry subgroups (or space-forms), so that the stable (material) geometries of a particular dimension metric space depend on a spectral-dimensional hierarchy of space-form geometries as well as their fundamental domains which are composed of "cubical" simplexes. Thus containment is "relatively global," geometric, measurable, quantitative, and spectral.

There is a new basis for a physically descriptive language in which, stable, measurable geometry is related both to (1) space-form models of material which are contained in (somewhat) independent dimensional metric-space levels and to (2) a very stable spectral containment set (analogous to event containment), which has a (mostly) definite spectral-dimensional structure (or hierarchy), in which uncertainties (or set dependencies) of set containment are identifiable.

Both material and space have the properties of being space-form structures with different dimensional relationships to one another, where space-forms carry on themselves properties of having both geometry and of having spectra, so that spectral properties are associated to space-form (or simplex) geometry.

Truth is about a language and its descriptions (or describable patterns) of what exists which are consistent with the patterns that are observed to exist, so that the description organizes these patterns (both the observed patterns as well as the patterns [or axioms] upon which a language is based) within the descriptive language so that its descriptions of the observed properties of existence, are both widely applicable and useful.

Thus both mathematics and science are about adjusting one's descriptive language so that the language is capable of describing the patterns that are observed to exist, so that the descriptions are useful. Otherwise, why bother describing patterns which are not useful? That is, one can abandon patterns of language which have not been as useful as other patterns.

But then again, who knows from where the useful patterns of a new language will come?

The process of improving the language never ends.

The most pressing issue for society is truth. Descriptions of what exists is the context of determining the truth. First ask, "What exists?" then provide descriptions of existence, and then determine the truth of these different descriptions by both observation and their usefulness. That is, to determine what is true about the various descriptions of existence, requires that the descriptions be both geometric-spectral and measurable. This means that the language is to be quantitative (measurable), and the language must be widely applicable (underlying stable geometry which is related to spectral properties), and it must have many practical uses (the intuitive usefulness of geometric descriptions). That is, to do this requires that the language describe something which is specific and stable with well identified properties as opposed to describing something which is quite vague, such as a set of (properties such as set containment properties or) events within which it is claimed are contained all possible events (for all existence).

Yet this set of all possible events (events which are determined by randomness) keeps getting changed by new observations, eg oscillations between particle families by neutrinos (for particle event families which are supposed to be separated by energy hierarchies, furthermore such particle oscillations between families imply particle events related to the higher energy particle families should be a more commonly occurring events, ie if a particle from a particular particle family is seen then the field particles for that particle should be seen which implies that many such field Bosons should be seen), as well as the (relatively) newly identified properties which require dark matter and dark energy (which are supposed to be related to new particle events) in order for such properties to be described within the currently accepted context of physical description. Yet such particle events were not described, or considered, until the new properties were observed.

If one has the set of all possible events for a system then one can determine the probabilities of those events, but if a system does not have a set which contains all possible events then such event probabilities cannot be defined. Exactly how is the event containment of physical systems defined in the so called probability based descriptions of existence?

That is, the idea of set containment, and/or the axioms of the real number operations, and material particle events and their probabilities, as the fundamental patterns which form the basis for one's descriptive language, might be too general, or these fundamental patterns simply do not take into account a set's type, and/or a set's spectral properties, or a way of describing changes in a set's spectral properties, or the ability to define a measurable unit (stable and uniform) on a geometric shape (or within a metric space) which is used to compose or build a coordinate space, within which a function can have locally measurable properties, in a consistent or uniform system of measurement.

The containing set may need the more specific properties (as opposed to set containment properties) of having a:

> Metric,
> A consistent unit of measurement of length,
> A non-local geometric structure,
> A relationship between geometry and simplexes, eg a cube is a form of a simplex,
> Having a spectral structure based on geometry,
> A spectral structure of containing sets which can change
> A hidden, or unperceived, geometry which also has the property of having a distinguished point, as well as
> Distinguishing types of quantities and/or types of coordinates,
> eg (different types of quantities)
> Material versus space,
> Rotational versus translational displacements,
> Inertia versus charge,
> Momentum versus energy,
> Spatial displacement versus time displacement,
> Euclidean space versus hyperbolic space (hyperbolic space is equivalent to space-time),
> Vaguely defined form (or hidden form, eg on the boundary of a higher dimensional space so that the higher dimensional geometry can be macroscopic) versus stable form, or measurable form, (eg entropy versus solvable [or integrable]).

It might be noted that the distinguish-ability of different types of quantities can be used to associate different types of quantities to metric spaces and then to define different "states" within a metric space, if both types of quantities associated to each state are containable within the (same) metric space, eg translation and rotation displacements both fit within a Euclidean metric space, eg the sphere has a metric which is the Euclidean metric which is restricted to a sphere.

This new set of patterns about which a new descriptive language can be built requires that such a quantitative language be organized around geometry and/or around different types of metric spaces, each associated to different physical properties and each having its own range of spectral properties, as opposed to probability and probability's subsequent elementary event, and event or subset structure of event containment and the relationship between set containment (or containment within a function space) and analysis.

The containing sets, about which a new descriptive language can be (has been) built, need to have within themselves the structure of metrics, that is, a stable unit of measurement, for quantities of particular types of numbers, need to be definable, and the quantities determined by formulas need to be (locally) measurable (smooth), and the object of the description (the physical system) needs to have geometric properties which also determine spectral properties, and furthermore a physical system's spectral properties must be related to (stable) geometric properties which might extend to higher dimensions so that the geometries can be either macroscopic or microscopic (thus allowing a large range of spectral changes), in order for the predictions of the description to be both widely applicable and useful.

In other words it is not enough to have a containing set and its subsets of points (or functions), one also must be very concerned about the types of quantities that the points (or functions, or space-forms) represent, and the relation of those points (or space-forms) have to both geometric and spectral properties (dependent on geometry and higher dimensional geometric [or set] containment) that are intrinsic to the containing set. Furthermore, a stable unit of measurement needs to be definable by means of the set of geometric-spectral properties which fit within the containing (metric) space.

A descriptive language can be based on a containing set having the properties of:

(1) a metric,
(2) so that the signature of the metric (and dimension of the metric space) can determine the type of quantities that such a coordinate (metric) space can contain and subsequently determining the "metric-space states" within which those quantities can exist,
(3) a local smooth tangent space structure of a geometry resulting from (in) a global shape,
(4) the coordinate base space having either a differential structure or a relation to a Lie algebra which determines a coordinate metric space's global geometric properties, eg spherical symmetry, as well as
(5) the global properties of both a geometric and a spectral structure?

In other words relatively stable geometric and spectral structures must emerge directly from the quantitative and geometric-spectral (or set) containment structure within which the physical patterns are observed, and about which the language is determined.

A function space may be extended by extending the domain of the functions in the function space. This is done by both the standard model as well as string theory, but the standard model views the extension as a fiction and string-theory microscopically winds up these higher dimensions so that this effectively limit's the spectral extension of the extended function space whose functions are defined on an extended higher dimensional domain to the spectra of particle events, whereas a macroscopic extension into higher dimension allows macroscopic as well as microscopic systems new ways in which to organize spectral sets. The question is, "Does one want to extend spectral description in terms of the difficult to use language of probability or in the more (intuitively) useful geometric language?"

The process of science is to continually adjust one's descriptive language so that such a language can describe the patterns that are observed, ie describe the patterns that do exist.

Science and free inquiry

Science is not authoritative rather, it is a description which is: (1) verifiable by measurement (Ptolemy's science satisfied this property), and (2) (it is) based on widely applicable (general) principles which describe a wide range of phenomenon, ie measurable properties associated to causal relationships (or relatively stable geometric relationships) between, or about, physical systems (or about a physical system and its components, eg an over-all simplex which defines the system and which organizes the relatively independent components which it contains), and (3) (such a description must also be) useful.

Science emerges from a process which is about changing the descriptive language (of science) so that descriptions of "What exists" becomes ever more useful.

Such a process cannot be contained by an authoritative body of knowledge nor by an authoritative body of experts. There are no absolutes in such a process.

This requires that the descriptions of a limited language, whose words have very specific meanings and so that the underlying assumptions (which determine the context in which the language is used) are easily determined, provide predictions which can be verified by measurement, though such a verification process was also satisfied by Ptolemy's descriptions of the planetary motions.

Thus the core of what determines a scientific description is both the wide applicability of the description and the usefulness of the description and not its verification by measurement, though measured properties which are predicted either into the future or predicted as geometric structures whose properties can be used so as to gain causal control over a physical system's future properties. This last property, of predictability of geometric properties in the future, is why classical physics has been so useful

Science does not emerge from personality, despite the fact that today it is being organized around personality and authority.

Science is a culture of free inquiry which tries to provide clear, useful descriptions of existence for everyone so that everyone can use the properties of existence which have been described. A community can use the descriptions of existence to decide what to do (eg find a cheap clean energy source) as well as how to go about doing it, and if actions of the community (are seen to) might be causing harm to the earth, then the community can use the observed conditions of harmfulness as the basis for stopping such actions on the part of the community.

The use of science is for the benefit of one's understanding, eg providing a context in existence for one's life, and to better live in harmony with the earth, and within this context, provide beneficial conditions for people and life. The new ideas about existence (new scientific ideas) proposed by this paper are not to be used for destructive-ness, nor for singular selfish gain. Useful knowledge is for everyone, as well as the earth, so that all will gain or be improved.

Science is a democracy and it should not allow itself to be used for selfish purposes. But it is democracy which depends on protecting the right to selflessly speak about the truth about existence, and the democracy should not protect the selfish interests of people. Science is about a selfless viewpoint about existence. Unfortunately, it has become a selfish pursuit. This is because of the selfish structure of society.

Namely, we are governed by minority rule.

New ideas

The new descriptions identify systems in terms of "cubical" simplexes, which form the fundamental domains of lattices of discrete isometry subgroups. Discrete isometry subgroups are determined in many dimensions and for each different metric signature. The "cubical" simplexes of discrete isometry subgroups (or space-forms) identify both geometric and spectral properties of material systems. The faces of such a simplex identify both metric-space states (on the "cubical" simplexes) and spectral currents or eigen-flows. The "cubical" simplexes can have their faces made topologically equivalent, that is they may be "moded-out," to form space-forms.

Both material and space are made of the same space-form structures, but always separated by one dimension. Within such a set of spectral-geometric (and separable) structures there also exist the space-form forms for life. Note: Some odd dimensional hyperbolic space-forms are not geometrically symmetric, but hyperbolic space-forms are ridged so when they are occupied with charge they naturally oscillate and generate energy. This is a simple model of life. Space-forms and space are organized around a principle fiber bundle whose base space is composed of various dimensional space-forms and the isometry transformation groups form the fiber groups. The mind comes from combinations of properties of space-forms (both stable and oscillating) and a natural containing set for spectral sets, namely the maximal tori of fiber groups. A maximal torus can store spectral sets so as to become a structure for memory. Note: Space-forms emerge from the topological properties of the isometry fiber groups.

The new description gives the logical basis for the unitary fiber groups and the high dimensional particle states which are to be associated to higher dimensional spectral-geometries (or space-forms) which can have both macroscopic and microscopic properties. Furthermore, the new description naturally accounts for the orbital properties of all systems (including the solar system).

However, this new description requires that one give up the idea of materialism. Both material and space are defined, within the new descriptive language, by discrete isometry subgroups, or space-forms, on a dimensional hierarchy of metric spaces, where the dimensional hierarchy determines a spectral dependence (or spectral control) of the lower dimensional metric spaces by the higher dimensional metric spaces, where the higher dimensional spectral metric spaces contain the lower dimensional spectral spaces. Note: What was a metric space becomes a space-form in the next higher dimensional metric space etc.

When considering a physical system being composed of space-forms, it is the hyperbolic space-forms which are the most stable and which have very stable spectral properties and hence it is them which have the biggest affect on the properties of relatively stable physical systems. However, the inertial properties of a material space-form system are governed by Euclidean space-forms.

In higher dimensions new materials become part of the description, due to new properties of the time subspace which is identified by a metric's signature, ie related to $R(n,m)$, perhaps slightly analogous to the new families of particles in high dimensional particle state space of particle physics, but material never really has particle properties.

The most stable properties of a system as well as its energetic properties are determined by the hyperbolic space-forms.

When considering a physical system contained in a hyperbolic metric space with three spatial dimensions there are three different dimensional levels to consider.

1. The 2-dimensional material space-forms which are in the containing metric space.
2. The 3-dimensional metric space and its spectral properties, as well as the 3-dimensional interaction space-forms which models the interactions between the various (usually a pair of) 2-dimensional material space-forms, and

3. The 4-dimensional metric space which contains both the 3-dimensional interaction space-forms and the 3-dimensional metric space which is but one of the 3-faces of the fundamental domain of the 4-dimensional metric space's space-form structure.

The spectra of the 2-dimensional material space-forms depend on the spectra of their 3-dimensional metric space, and in turn, the spectra of the 3-dimensional metric space, or of 3-dimensional space-forms, depends on the spectra of their containing 4-dimensional metric space.

Each dimensional level can have a (real) conformal factor associated to itself. These conformal factors affect the relative size of the space-forms and/or the range of the spectra for a physical system in its containing metric space, and this spectral range is for both space-forms and metric spaces, from one level to the next. That is, each of the conformal factors determines the size and spectra of both the 2-dimensional space-forms in the 3-dimensional metric space, and the 3-dimensional space-forms in the 4-dimensional metric space, as well as the spectra of the 4-dimensional metric space.

The first conformal factor (multiplying the metric of a 2-dimensional metric space) determines the size of nuclei' and atoms' 2-dimensional space-form structures. The 2-dimensional space-forms have the spectral range of the 3-dimensional metric space within which it is contained. This means that the 3-dimensional metric space has within its spectral structures the spectral range of the 2-dimensional metric space

The second conformal factor (multiplying the metric of a 3-dimensional metric space) determines whether a crystal, which is essentially a 3-dimensional space-form system, has its own (small) nuclear structure or if the "nuclear structure" of the 3-dimensional space-forms are essentially the (large) boundary of the 3-dimensional metric space's space-form structure.

The third conformal factor (multiplying the metric of a 4-dimensional metric space)

The 3-dimensional space-forms have the spectral range of the 4-dimensional metric space within which it is contained. This means that the 4-dimensional metric space has within its spectral structures the spectral range of the 3-dimensional metric space, as well as the similar spectral ranges of the other three faces (along with the given 3-dimensional metric space's facial structure) which together compose the 3-faces of the fundamental domain of the 4-dimensional metric space's space-from structure and hence determine the 4-dimensional metric space's spectral range.

Note: Each of the different 3-faces of the 4-metric space's fundamental domain can have a different value for the speed of light. The speed of light depends on both the size of the metric space (as a bounded space-form) and on the rate of spin rotation between the different metric space states which are related to the natural identification of physical properties with particular metric spaces, eg specific properties of dimension and metric signature are each associated to physical properties.

The spectral ranges of the set of 3-faces of the fundamental domain of the 4-dimensional metric space, determine the spectral range within which the interaction 3-space-form can experience spectral and/or geometric space-form change.

If one wants to use and control the spectral structures of 2-dimensional material-spectral space-forms, ie atoms and molecules, then one must know the space-form and conformal structures of these three different dimensional levels, which together determine the system's physical properties and the range and type of spectral changes (or spectral properties) that the system can experience. The spectral properties of the 2-space-forms are related to the different spectral properties of all the 3-faces which compose the 4-metric-space.

It is certainly a description which is consistent with the orbital and energy level properties that have been observed. However, the system's lower energy levels might well be related to the space-form's lower values of sectional curvature.

The 2-dimensional hyperbolic space-forms have already been listed (there are many), Thurston has shown that 3-dimensional hyperbolic space-forms are more common than was originally thought and he has used them to determine the spatial properties of (flat) 3-dimensional metric spaces.

The four and five dimensional hyperbolic fundamental domains might be less complicated than the lower dimensional fundamental domains, as suggested by some geometric theorem by Coxeter. Whether there is more richness to the number and types of hyperbolic space-forms in dimensions four and five etc is not now known.

To model a system in this language uses the same language as classical physics, namely $d^*dF=j$ or Dirac's equivalent first order equation, where F is the differential form of the system or a (hyperbolic) space-form, but now both the geometry and the boundary conditions as well as the Dirac rotation of (metric space) states exists within a multidimensional context, so that each dimensional level has its own set of independent properties, and where j is a current flow (or spectral current) on the space-form (or on separate interaction space-forms). The description is essentially for physical systems in 3-space so the current flows are mostly 1-flows, although the current flows of 3-space-forms are 2-flows. Dirac's equation requires that the (metric space) states of both metric spaces and space-forms be given a complex number structure. This is a natural pattern for this new descriptive language.

Atoms and nuclei deal with the properties of 1- and 2-dimensional space-forms, while the space-form structure of a thermal system is a 3-dimensional space-form, and crystals fall in between these two different dimensional structures.

In fact, space-forms and metric spaces are all systems which are mixtures of different dimensional level's geometric and spectral space-form structures, just as 2-dimensional space-forms are related to both one- and two-dimensional space-forms that are contained within a 3-dimensional metric space, where the variety of 2-dimensional (hyperbolic) space-forms is determined by the spectral and geometric properties of the containing 3-dimensional metric space (which happens to also has the structure of a 3-dimensional space-form).

The atom has its orbital or energy level structure determined by hyperbolic space-forms and because its inertial properties are contained in a 3-dimensional Euclidean space, which has properties of spherical symmetry, the atom's inertial structure will be affected by spherical harmonics thus its properties will be related to an energy level number and (2) angular momentum numbers (where these identifying number are related to the separability of the solution to the system's differential equation), while a crystal's inertia will be influenced by three translational values (if there is no small nuclear structure associated with a crystal?) of the bounded crystal, and the numbers associated to eigenvalues which identify the differential form (space-form) solution for the Cartesian product of two 3-spheres related to the containing 4-dimensional Euclidean metric space's natural geometric (or "differential structure"). To understand the affect this product of two 3-spheres (which is the differential structure of a 4-dimensional Euclidean space) has on the solution (or differential form or space-form) structure depends on whether the spectral properties of the containing 4-dimensional metric space allows the crystal to have a small nuclear structure or not. This depends on the conformal factor of the 4-dimensional metric space level.

Because the interaction structure of 2-dimensional space-forms is contained in a 4-dimensional metric space this means that many of the spectra of the 2-dimensional space-forms has a greater range than the 2-space-form spectra that their three spatial dimension containing metric space allows. That is, the various 3-faces of the 4-metric-space allow a greater range for the 2-space-form spectra, which can exist in a particular three spatial dimension metric-space. The various 2-space-forms can depend for their spectra on the spectra of the various 3-faces of the containing 4-metric-space. A 2-space-form's spectra which is not directly related to the spectra of its containing metric-space will have different spectral properties from that of a typical 2-space-form spectra which is in resonance with the 2-space-form's containing metric space. For example, different atoms can have spectral properties related more to one or another of the various 3-faces of the containing 4-metric-space's fundamental domain. This can affect the accessibility of these various spectral levels in an untypical atom. Thus when particular atoms with various spectral relations, with the (spectral) 3-faces that can compose the system, are near to one another this could result in different spectral patterns and spectral accessibilities (which are related to the spectra of the various 3-faces of the containing 4-metric-space's fundamental domains) which can be a part of such a system.

To pull a signal from noise, the signal must exist within the noise to begin with. One can use what exists if one knows its relation to the existing spectral structure, ie resonances, transitions or changes in spectral properties, eg harmonic deformations of space-form structures. Resonances can exist between space-forms. The space-forms may be either material space-forms or metric spaces. Spectral levels can be near or far from each other. What is the range of transitions, that is, what are all the different types of space-form or energy level transitions?

Clearly the thermal and crystal and atomic space-form structures will all interact. These interactions will depend on the spectral structures of the space-forms which compose the system. Light might not always be a part of the energy level transitions of the space-forms which compose the system, because the thermal space-form model of the system could be the energy structure related to different energy level (or spectral level) transitions within the system's space-form (or spectral and geometric) structures.

It should be noted that in this new language (or this new description) the spectral set is given a geometric and conformal structure and within this structure both spectral existences and spectral transitions, as well as space-form deformations, can be described once the space-form structure of the three and four dimensional hyperbolic spaces as well as their conformal factors are described. Whereas in quantum mechanics, one can either alter the wave equation until its solutions match the data or the spectral structure of a system can be determined from experiment. That is, what spectra exist and what transitions and deformations can occur is not part of the descriptive language of quantum mechanics. One can only passively find spectral properties and the spectral context is not known. Furthermore, if the context for the quantum description of a system is changed, eg the atomic number of the potential of the Hamiltonian is changed, then the process of finding the new spectra within the new context must be re-done. It should also be noted that transitions between energy levels (or between orbits) cannot be found if the operator structure of the system cannot be established, eg if a system's Hamiltonian cannot be determined because the spectral structure is found experimentally. The old description is passive and chases data, while the new description is active and seeks to describe the complete structure of a physical system's spectral properties so that these properties can be predicted and/or the system's spectral properties controlled.

Note: We essentially see primarily the Euclidean or inertial aspects of a system but the stability of a form (or of a system) depends on its (hidden) hyperbolic space-form structure.

Particle-like appearances result from the distinguished points (where all the vertices of a space-form's fundamental domain comes together) of hyperbolic space-forms, about which interactions focus.

Probability is based on being able to identify the full set elementary events which identify the system, ie all possible events for a system, about which the probabilities describe the system's event's random possibilities.

In this new set of ideas a metric space affects the spectrum of the physical systems which it (the metric space) can contain. However, because a physical system depends on being in both the containing metric space as well as being in relation with the next higher dimension metric space, due to the dimensional structure of interaction, this means that there are several faces in the higher dimensional metric space which are equivalent to the face which represents the actual metric space of containment. None the less, these extra faces allow the relevant set of spectra, ie the relevant set of elementary (spectral) events, for a system to be a larger spectral set, and thus the system's spectral set can change depending on which face of the (highest dimensional containing) metric space's fundamental domain is allowed to affect the system's spectra. Thus a system's spectral character can change if the spectra of the different faces are allowed to influence the system's spectral set.

Or

The spectral structure of one system can be radically different from the spectral character of another system depending on which of the spectral faces of the containing metric space influence the system's spectral properties of composition. Thus, because spectra have both a dimensional (or conformal) relation to space and an (independent) relation to a containing set's spectral faces, not only can the spectral set change its eigenvalue structure, but the entire spectral set of the lower dimensional structure can be reorganized. This is a set which can be recognized if the description is geometric but it is a difficult reach for a probability based description to identify such a structure for elementary events.

Chapter 8

NEW CONTEXT FOR LIFE

A new, many-dimensional, context within which to describe the properties of a living-system

There is a spiritual message that exists within some new ideas about physics, ideas which provide a new language, based in the geometry of discrete isometry subgroups (or space-forms), which can be used to describe the physical world, yet the description goes on to describe a further notion of existence, which points to a new context within which to understand life, namely, life's creative and perceptive relation to existence.

Entering the abstract, ie transcending material as well as life and death

This section gives a "poetic" introduction to some new ideas about how to describe existence. For many more details about the new language see other references (given at the end) as well as the section in this article titled "Opposition to materialism, and The true context of existence, and its description," and the three paragraphs below this paragraph. Otherwise, poetically . . . .

. . . Existence can be modeled in terms of a nested Russian doll set, except that after each successive doll is opened this also results in a successive change in one spatial dimension for the set which each new doll represents. However, these dimensional spatial sets (nested dolls) also have both properties of being semi-unbounded (some of the infinite extent space-forms get through the large space-form shape, which allows us to see the stars) as well as mostly being bounded (as the dolls themselves are bounded) and each of these sets (dolls) is associated with a definite spectral set. Each newly uncovered doll represents the material components of the previous higher dimensional set (the previous doll within which the newly uncovered doll was contained) which represents the metric space for the material sets that it contains, this pattern continues, beginning at dimension eleven (the largest doll) and ending at dimension one (the smallest doll).

This pattern can be more accurately represented (in the new theory) as the various faces of a "cubical" simplex structure which in turn represents one fundamental domain (or checkerboard square) of a metric-space lattice (or checkerboard) representation of a discrete isometry subgroup. Perhaps the nested Russian doll set is better referred to as a high dimensional crystal with a multifaceted structure of faces, ie the "cubical" simplexes of the various different signature metric spaces and their subsequent isometry groups. Note: A dice is a cubical simplex which is 3-dimensional, and has six 2-dimensional faces, which in turn, have four 1-dimensional faces called edges etc. If one of the nested Russian dolls represented a 3-dimensional space the immediately following (smaller) doll which it contains would be an example of one of the many 2-dimensional "space-forms of material (or molecules) contained in the 3-dimensional space. However, such a molecule would be a combination of both a hyperbolic space-form, which is characterized by properties of electromagnetism and time (along with properties of great spectral stability), and by a Euclidean space-form which is in resonance with the hyperbolic space-form's spectra and is characterized by properties of inertia and spatial displacements. We seem to perceive only of the inertial structures which have a certain large sectional curvature, that is, we do not see light (an infinite extent space-form) until it interacts with (inertial) material. This "poetic" description has omitted the properties of interactions.

Life cannot be described simply in terms of material systems in a 3-dimensional space (or 4-dimensional space-time). Rather life depends on being a non-reducible system which is properly and mostly described in higher dimensional space so that it has a fundamentally non-reducible "skeletal" subset structure which define both geometric and spectral sets and which is (also) filled with "molecular" components.

How can properties of such a set (eg nested Russian doll set) be changed or altered so as to extend the properties and possibilities of existence? This is the central question (or quest) of life's creative intent.

The descriptions of such possible changes are not about a personality claiming to be a messenger, rather it is the relatively precise language of mathematics, ie the language of quantity and shape, which is also associated to physical attributes of the world, as well as the various relations between these patterns, so that life is described in a precise manner within this new language so that life is a part of these mathematical-physical patterns [Note: It is a language which can be used to do a better (or more useful) job of describing the patterns and properties of material systems than the current paradigm of physics is capable of doing], so that a good description of the precise control that life has over its own material composition. Furthermore, it allows the description of life to be in a creative relation with existence, and the existence described by these patterns has a definite relation with life's own intent, ie patterns which can be used to describe: material, space, life, and mind and which can also describe a pattern which leaves behind the material world's singular notion of time. This can be called entering into the abstract, that is, entering the higher dimensional existence which transcends the 3-dimensional material existence [which few people believe is the complete story about existence, but never really take their own doubts seriously], in order to bring knowledge to bear on a creative process intended to extend the structure of all existence. This is the essential creative property of life, and living. (1) Intending to create at such a higher level of existence, (2) having a scientific model described in a language based in geometry (not probability) that such a higher dimensional structure of existence does in fact exist, and (3) comprehending such patterns, and then (4) creating one's own patterns, within such a mathematical structure, which transcend a material existence in a 3-dimensional space, are steps which intend one towards the abstract.

The new paradigm

The new language for physics is based on discrete isometry subgroups (or space-forms) whose metric space lattices are defined on a dimensional hierarchy of different signature metric spaces, eg Euclidean and hyperbolic spaces, so that both material and space have the structure of being space-forms (but one dimension removed from one another), so that both material and space are defined on the base space of a principle fiber bundle, where the fiber groups act on the various dimensional space-forms. Furthermore, each distinct metric space has physical properties and subsequently metric-space states associated to itself, an idea which provides the basis for understanding the spin rotations of metric-space states as well as understanding the discrete properties of inertial dynamics, as spin rotations also identify discrete time intervals.

It should be noted that in the new paradigm the discrete isometry subgroups, or space-forms, have both geometric and spectral properties associated to themselves, and both material and space are composed of space-form structures so that the material space-forms have a distinguished point about which space-form collisions center, and thus they appear to be particles when they collide. Furthermore, a metric space has a spectral structure associated to itself. This new language is based in the geometry of discrete isometry subgroups, as opposed to probability.

. . . . As opposed to . . . .

The currently accepted paradigm of physics

. . . . As opposed to the current paradigm of physics, where probabilities of particle events are defined on a base space of a principle fiber bundle so that the base space is composed of both space-time (where the events occur) and an independent set of particle-state spaces which are associated to both independent harmonic oscillators and unitary fiber group transformations of the particle states, where the independent harmonic oscillators fill all of space (or the space part of space-time), so that these harmonic oscillators are separate and independent of one another, ie such a system has no geometry, and where the unitary fiber group transformations of the particle states are related to the probabilities of the type of particle collisions which occur within (or occupy) the harmonic oscillator, so that this mathematical structure adjusts the properties of a wave function, defined on all of space, obtained from a regular quantum mechanical system, ie a wave equation which has a geometric potential function, so that the regular wave equation solution function is represented as a series of discrete annihilation and creation operators. Note, this process of adjusting a regular wave function by probabilistic particle state collisions with the discrete representation of the wave function, always results in a diverging series, which represent the solution functions, unless it is renormalized, ie infinity is subtracted form it . . . Note, this is very complicated "mathematical structure," and its focus is on fitting wave functions properties with measured data, similar to Ptolemy's epicycle structures. One should be skeptical that this is even science, much of it is not mathematics, though it masquerades as if it were really highly rigorous mathematics.

Introduction

The new alternative ideas are not allowed in the literature of physics because they are far from the orthodoxy of physics . . . they are revolutionary new ideas . . . as opposed to variations on the theme of materialism and probability of material events, which along with gravitational singularities (due to a believed spherical symmetry of space), essentially comprise what the current paradigm of physics allows.

String theory is simply the geometrization of the (probability based) particle state space so as to gain convergences of the above mentioned series solutions. Note: If words mean anything, then introducing a small scale geometry into a probability based language should not be attempted, it is an exercise in inconsistency.

The new alternative ideas are revolutionary and subsequently they also have quite surprising implications which provide life a new context for existence, a context which transcends both materialism and three spatial dimensions and therefore it implies that within such a viewpoint, life is very much equal, whereas materialism implies elite, fit survivors in the competition for survival.

Revolutions in science exist outside the social structures of the times, ie revolutions examine the hidden structures of existence which civilization is not examining. Because the current paradigm of physics is well within the present social structure, it is not capable of considering (or adjusting to) the hidden structures which it does not recognize.

In fact because the hidden structure for the description of physical existence is discrete isometry subgroups this results in opening up the realm of the spirit to a real verifiable description, ie a realm beyond materialism, into a higher dimensional realm (or into the realm of the larger nested Russian dolls).

The practical side

If the energy crisis can be solved by new technology, it is more likely that such new technology will come from truly new ideas, described within a language based on geometry, as opposed to technology which can emerge from today's paradigm of physics whose main focus is bomb engineering, ie the probabilities of high energy particle collisions at spherically symmetric singularity points of space.

In fact these new ideas describe, within a geometric based language, the source of energy for both radioactivity and the sun, the relation between a metric space and its opposite state, as well as describing the relation between life's intent and its motion. Can things be moved using higher dimensional structures?

These new descriptions are geometric, and the spectral context of such systems also have a geometric nature (or structure).

Geometric descriptions are more conducive to control and invention than are probabilistic descriptions. Furthermore, the focus of these new ideas about how to describe physics is different from the bomb engineering focus upon which the current paradigm of physics so heavily focuses.

Authority versus science

Imagine Copernicus being "peer reviewed" by the intellectual authorities of his age. These authorities would be the priests and authorities of the church. Were such church authorities ever going to stick their necks out and support such ideas as Copernicus was putting forth? (No)

The intellectual authoritative community of this age are the technical experts in the military-industrial complex, "a community which has one fundamental interest, ie nuclear and bomb engineering, or instruments of war." Such a community of technical experts need the most exacting knowledge in order to build the "most exacting instruments." These experts are the high priests of materialism who possess the hidden knowledge of our society which others, it is generally believed, are not capable of comprehending. No one can question the authority of these experts who dispense "scientific truth" to the society, as they work on the projects and in the institutions controlled by the ruling few.

Will an authority from such a community put their reputation on the line by supporting alternative ideas about physics which are outside of community's orthodoxy? (No)

Consider how the authority of high level social positions, eg the executive branch of the government (which represents the interests of the very rich), can throw a monkey-wrench into the proceedings and considerations of the "authoritative truth" of the scientific community. This can be done for the purpose of big (energy) companies' selfish interests as well as for the "morality" issues (the relation of stem cells to abortion) in regard to corralling voting blocks of people, who insist on their own interpretations of events as absolute truths (who most often call themselves Christians, yet make deals with Caesar, and they do not love their enemies). However, when "social scientists" transcend "moral" issues, namely, when they consider improving the methods of torture, this seems to be interpreted to be within the (moral) law, ie it is not the righteous Christian base which loudly protests.

Note: The main issue of morality is distinguishing between selfish and non-selfish acts.

Extending knowledge is selfless, but if it is done it in the context so that such knowledge is used mostly for the selfish interests of the few, then it becomes selfish. Science has allowed itself to become the basis for the power of very selfish interests. Whereas science could be about living with many inexpensive conveniences but also living more in harmony with the earth.

In regard to weather and the energy companies' interests, we see how the "ideals of scientific truth" come to the forefront so as to manipulate information to support oil company selfish interests.

The earth is too complicated of a system to have any "absolute" knowledge of "the cause" of global warming, yet CO_2 is hypothesized to cause global warming and as the atmospheric CO_2 percentage increases so too does the global temperature. This is certainly compelling enough evidence to take measures to protect society from what could possibly be very dangerous outcomes, in regard to global warming. Yet the selfish interests of the energy companies demand "absolute scientific truth," that in the future there will exist dangerous outcomes will result from increased CO_2 in the atmosphere, before they will cut their oil and gas use, and this debate about the "ideals of scientific truth" is allowed to proceed. The problem is that such arguments for scientific certainty mix classical causality with the limited information of probabilistic models of material systems. Yet its real argument is that within an, at best, probability based model of a complex physical system, CO_2 as the exact cause of warming can be questioned. The capacity of probability models to describe things in a useful (causal) manner is very limited, and perhaps a probabilistic description is outside the descriptive parameters (or boundaries) which should define science.

However, it might be noted that the orbital properties of a nucleus as well as the orbital stability of the planets have no description based on physical law, at least as physical law is believed to be true within the current paradigm and the current language which is used to describe the properties of material systems. Yet the "scientific truth" regarding such an inadequately demonstrated physical law as represented in the standard model of elementary particles is unquestionable. The only "predictions" that the standard model does are about using its epicycle-like mathematical structure to correct wave functions of the H-atom to obtain highly precise fits with data, but when the standard model is applied to the properties of the nucleus (the one system it must be applicable to within its own focus of bomb engineering) it fails to give any results. The authoritative scientists insist on their own interpretations of observations as absolute truths. This is not unlike the "religious" zealots who focus only on certain selfish acts of others (certain selected aspects of the data) to identify their morality, whereas they themselves act mostly in a selfish manner.

Are the abilities to exactly (or approximately) describe weather, and exactly (or approximately) describe nuclear orbital properties the same? No, though both are probability based theories, global warming is more believable (has more truth associated to itself) as there exist theories about global-warming's relation to the percentage of CO_2 in the atmosphere and there is supportive evidence backing up this idea. Though the global warming theory is weakened by the fact that it is a theory based in probability describing both a chaotic and complex system, thus its causality can be questioned. However, there is no such theory for the orbital properties of the nucleus based on the standard model.

The scientific truth of the standard model is more questionable than the scientific validity of global warming.

Even if all the predicted particles are found in very expensive experiments, the standard model is not widely applicable and useful, it applies only to particles, and mostly to particles in particle accelerators. A scientific truth is not demonstrated by correlation with measured properties, otherwise Ptolemy's model of the heavens works fine, rather it is determined by the wide applicability of its scientific principles, reducing all material systems to particles so that the system's properties are determined from particle collisions that occur at a solitary point in all space, and its associated usefulness, which has been quite limited. By these standards of scientific truth, the standard model has failed to have been demonstrated as being a scientific truth.

The ideal of "scientific truth" is important in the "global warming," media debate, yet it is not even considered, in the much more relevant debate (in physics) about alternative descriptions of physical existence in its relation to fundamental unanswered mysteries of physical systems, ie the orbital structure of the nucleus, and the usefulness of a theory. None the less, the authority of the standard model of elementary particles is unquestionable within the professional physics community even though it fails to describe the orbital properties of the nucleus, the very type of system it should be able to describe. Whereas the ideal of "scientific truth," confused by using a classical standard for truth, ie predicting the future state with a high level of certainty, within a probability based description, but rightfully questioning the causality of CO_2, is the center piece of the debate about global warming and CO_2. Though the standard model can fit data, in a very limited class of systems, it cannot describe nuclear properties, hence it should be questioned that particle collisions are the valid "causal source" (affecting the probability waves which are supposed to be used to describe the nucleus) for describing nuclear properties.

Alternative ideas about physics, though greatly needed, are ignored in the more fundamental case which deals with the true ideals of "fundamental scientific truth." The scientific community allows authority, and hence social class, to substitute for "scientific truth." When a way to obtain a description of a system (the orbital properties of the nucleus) has been found, which is consistent with known physics, take a uniform hyperbolic space-form and conformally transform it so as to correspond to the orbital properties of the nucleus, which describes the properties of a system which has eluded description within the accepted paradigm, then this demands a serious review of the alternative description. Yet the alternative description is outside the orthodoxy of physics, which demands particle collisions at singularity points, so the alternative ideas are not allowed in peer reviewed (authoritative) journals.

Unfortunately for the easily misled people of society this means that the alternative ideas are not scientifically true. This is the same social condition of an "authority controlled truth" which Copernicus faced.

The conclusion is that institutional science (institutional authority) is not about "scientific truth" (truth) rather it is about selfish dominance and minority rule.

That is, the property of minority rule for society results in selfish dominance by the high social classes and it results in science which becomes dysfunctional, and in a "justice system" which is a law without basis in principle, (it is based on the property rights of stolen property) except that it defends the rights of the selfish ruling few or those with high social class (the knowledgeable, where their knowledge is based on their proclamations and opinions).

Free speech about physics is not allowed due to the authority of minority rule . . . .

Materialism as well as monotheistic religions, which support materialism together support minority rule

Materialism is socially related to selfish dominance, as opposed a non-materialist viewpoint usually associated with the idea of being nice to one another. The notion of "being nice to one another" implies equality. Minority rule as well as a society based on monetary value are ideas which have worked together to cause the economic and social basis for activity to reinforce the hierarchy of value held in place by the ruling minority, ie money within an economy naturally flows to the ruling minority.

Religion steps into the social structure so as to say that the oppressed minority should be obedient, ie or be nice, to others who are your superiors in human value.

Material control is the basis for the control of money, money flows in the economy to the ruling minority because of their vested interest, and knowledge is controlled through the projects that require technical knowledge so that the projects are being determined by vested interests. The ruling minority has control over both material and knowledge and how that knowledge is to be used in society, and that knowledge is used to develop the interests of the ruling minority.

By believing (so as to base one's actions upon) the idea that material things should be made and distributed by means of a money based "economy" means that such a person's actions are all channeled to support the hierarchical truth which has a selfish intent of the ruling minority, unless one tries to speak to the truth of one's real beliefs. However, though free speech is the basis for law, none the less free speech is not allowed. Instead the "justice system" aligns itself with an army which keeps the ruling few in power based on property rights, and an "our army (the ruling minority) versus their army (everyone else)" mentality.

Those who learn the truths of the hierarchy, eg its high science, in turn, must work for the hierarchy on the projects the elites want built. For "technical matters" within society, it is only the voice of the experts, who can properly give voice to the truths of the elite. Sociologically it is only the experts who have sufficient knowledge to have a voice, and the experts use this voice to channel their knowledge towards supporting the selfish interests of the few.

The experts on physics are supposedly those people who have examined "everything" in excruciating detail, there is not a thing they have overlooked, and they have elucidated the subject of materialism (the physical world) in the only proper and rigorous language which allows insight into "truth."

Thus one must ask, "What is this language which carries the truth of the world as well as the context of life?"

Is it the language of rigorous mathematics?

No there are too many flaws in the mathematics of physics which require the use of physical thought so as to make descriptions of physical systems clear.

"Then the language must be carefully considered physical thought as Einstein used."

No, again, this too would not be the true language, as most of the descriptions use both formal mathematical patterns for material properties which seem to transcend the physical world as well as an inconsistent language of both probability (the standard model) and the geometric potential term of the quantum mechanical wave equation.

This mix of formal mathematics interspersed with key physical interpretations, is really the language of nuclear bomb engineering which came out of the Manhatten project. This language was continued and hence emphasized by E Teller and his H-bomb project. It is a language which describes the probabilistic properties of particle collisions within a small explosive "singular" region, and the physics community, under the guidance of the ruling class, has been trying to knit the ideas of geometric singularities with particle physics. Where particle physics describes the adjustments of quantum wave equations in terms of the properties of particles at one singular point.

Physical description is really in need of a mathematical pattern which is both geometric and spectral, thus avoiding the function space relation to spectra along with the probabilistic interpretation of that function space.

The (hyperbolic) space-forms of discrete isometry subgroups provide both the natural geometry and the natural spectra of "flat" metric spaces.

This is revolutionary science. It is not an idea which the ruling few can corral for their selfish purposes because it leads people into the true context of their lives, ie it leads life into the creative side of the spirit which transcends a 3-dimensional material existence. They are ideas which move life away from the confines of materialism.

Unfortunately, everyone in society has only been taught one idea, namely to be selfishly dominant. Selfless creativity is defined in narrow terms and is always about being a highly paid servant to the rich, which is used to actually define selfishly motivated competition. Thus with freedom there will be many selfish acts, but now freedom extends beyond materialism. The problem with selfishness is that it is based on a belief in an absolute truth, but creative power is about extending beyond what is known.

In our society, all abstract discussions about life's true context, or determining the properties of existence, ie the main subject of both science and religion, are all mediated by the overwhelming belief and backed up by everyone's actions which are consistent with 3-dimensional material existence (or space-time-material existence).

In our society, the discussion of life's true context is carried on within a pair of different and quite separate languages which have very few points in common. The language of science is based on materialism, while the language of non-materialism (or the language of going beyond materialism is usually considered to be religion) is based on a world that we apparently have not been taught to perceive. Though people may want to believe in something beyond the material world. Within the context of society, they are not dealing with such non-materialistic ideas in any practical manner. Minority rule keeps focus squarely on material concerns, and religion seems to mostly provide social services for the ruling minority by keeping people obedient to the ruling minority, so that all such religious leaders actions are consistent with a material world. Thus one can conclude that religion is not about transcending materialism, rather it is about being materialists and serving the ruling minority. Using material in a harmonious manner is more conducive to the non-material viewpoint, than exploiting material to create a ruling hierarchy based on the control of material.

Monotheism is all about attaching a hierarchy of value to a language which is supposed to transcend materialism, but does not, and as such it diminishes the structural value of what life is, within the real context of life's existence, [the proper context of all life is at least that of life being equal].

Monotheism is a bit of a misnomer as there are an entire set of angels and immortal souls inhabiting the same realm within which the said, one-god exists, thus monotheism is the attribute of highest value within such a population of beings which transcend materialism. Where exactly is this realm of angels etc? This is answered in terms of the high dimensional discrete isometry subgroups, or space-forms.

Note: The new ideas about physics describe the realm where beings which transcend materialism can exist. It might be further noted that life itself inhabits that realm.

In civilization "high value" is determined within the context of a material centered culture, and because of this, the notion of a hierarchy of value focuses on the superiority of a few people who have widest range of control over material, and subsequently knowledge and its uses, within the culture, so that such knowledge focuses on material control and material uses.

Perhaps if high value were dispersed throughout nature, instead of focusing on a notion of one-god, then people would be less likely to accept a ruling minority. Having a ruling minority is also an idea which is supported by the material centered notion of "survival of the fittest" in which the competition is defined in terms of getting the scarce materials. Thus minority rule is also an attribute which logically descends from materialism. Both monotheism and material based science support the idea of a natural dominant class of people who should rule the masses. Yet supporting the ruling minority seems to be the main function that monotheistic religion has, though it claims to be describing an existence which transcends materialism, however the description it provides is wholly inadequate.

It should be noted that the most powerful entity of what a hierarchical God might be, that is, an entity which is aware and who is intent on controlling material events related to the lives of other beings, can be modeled as higher dimensional energy generating spectral-geometric (space-form) system along with its attendant containing sets (or metric spaces), however this entire system could, in fact, be lower dimensional than the true dimensional structure of life itself. However, one cannot be selfish, or believe in absolute truths, and attain practical use of a high dimensional existence.

We are born with the capacities to use our true high-dimensional living system's properties within their proper context, but no one in civilization is doing this. Instead we are all taught to only pay attention to the relation our life has to the 3-dimensional material world, within which we subsequently become expert perceivers, and our will focuses on our intending within this lower dimensional context.

Opposition to the idea of a ruling minority, and The true context of life in the US

The ideas which form the basis for our society are that:

We are unequal and our life and freedom is directly related to the property we own and the contracts that we can (or do) make.

There is no law in the US, this is because law is based on judicial review, as well as using or enforcing the notion of "the letter of the law," which has its basis in property rights, which subsequently implies minority rule. Hence powerful judges are picked who serve the interests of the ruling minority. Legal precedents are established which protect the oppressive rights of the ruling minority, ie those who own the country, or who have positions of societal power, are allowed to oppress the people of the country in an unequal manner. However, US law is really based in equality freedom of speech, freedom to not have fixed standing armies which maintain a social hierarchy, and the right to have criminal behavior well defined not by the letter of the law but by its spirit, ie defined in terms of selfishness, and a formal method of judgment provided.

Law does not apply to the ruling minority. This is because law and enforcement has subsequently been based on protecting property rights. Yet the Declaration of Independence and the Bill of Rights make it clear that the spirit of the law of the new country, born of revolution, is to be based on equality and the law is to be in opposition to the oppressive forces, or oppressive capabilities of institutions or individuals. Yet the police-justice system is a counterinsurgency operation which protects the property rights and overwhelming power of the rich few, and thus it protects the oppressive powers of the ruling minority. That is the intent of the second amendment, is that there is to be no standing army, because armies are designed to protect and follow order that the high ranks give. As it is now, community police first and foremost protect the property rights of the high ranking members of society. This is not equality. The spirit of the law clearly resides with the need for law to oppose those who have, and subsequently use, oppressive force within society. Yet it is the letter of the law related to property rights which is enforced, and the processes of the "justice" system as well as institutions' right to essentially clobber those who are considered to be either "out of line with the interests of the ruling minority," and/or are those people who are without power.

It is the freedom to express selfless ideas about knowledge, and to not have to (or not try to) make deals with Caesar which makes life worth living. Being in harmony with the earth from which we came.

Control of society based on the control of material, so that material considerations are nearly always at the core of one's actions, is a behavior characterized by selfishness.

The power of minority rule rests with science, physics in particular, as this is the basis by which people can control material processes and use materials to build things, hence an important motivation of physics to adhere to the idea of materialism.

We can aspire to be either "top managers" or "top scientists," ie the masses can compete over a narrow vision of knowledge and one's ability to aggressively reach the top of that institutional discipline.

There are at least two weaknesses in this scheme, (1) the society can be based on harmony with nature (as opposed to master of nature) and on equality, and (2) science has failed as a material and probability based language to describe many of the fundamental properties of material systems, eg the orbits of the nucleus, the stability of the solar system etc, hence science itself should be striving for ideas which transcend materialism.

The new ideas presented in this article provide an example of science based on ideas which transcend materialism, namely both material and space can be modeled in terms of discrete isometry subgroups (or space-forms).

One can be sure that if people were encouraged to exercise their freedom (without repressive retaliation, eg loss of career) to seek and express new physical ideas then the new physical ideas expressed in this article would have already been expressed. However, society allows only a narrow range of physical ideas (as well as economic ideas) to be expressed, otherwise one is not allowed to have a career as a scientist, ie repression. This narrow idea of what free speech means is a result of the use of the oppressive power which is possessed by the ruling minority, such power can also be called totalitarianism, ie having the oppressive power to control what people are allowed to think, so that this control is gained through economic processes, ie derived from property rights.

The idea of only having limited property ownership rights within society, might be imposed, especially on the very selfish, so that the society is first and foremost about living in harmony with a world which belongs to everyone, allows for the notion of equality to be believed.

None the less:

With minority rule, no one will believe in the idea of equality. People's worth can be easily measured by practical considerations of their accomplishments, their knowledge, their actions, their material wealth and their social position, all within a hierarchy of value controlled by the minority rulers. This hierarchy focuses on [the fear that those without the seal of approval of the ruling minority, concerning] the truth of information which an "unauthorized" person provides, that is, those outside the hierarchy might be presenting "lies as being true" so as to gain some selfish advantage. This concern is ironic (or perhaps natural) because that is what the ruling minority is doing all the time, ie presenting their version of truth, within education and media systems, for their selfish advantage, eg physics does bomb engineering, economics does free markets, etc. Note: The technical disciplines are more constrained than the humanities, such constriction is considered a natural aspect of mathematically based descriptions, none the less art is also very much tied to technicalities.

If one, who believes in (or is acting within) the social hierarchy, has a new idea then they always want the approval of the ruling minority, about the truth of the idea, so as to verify the truth value of the idea, and to gain a monetary compensation for such new ideas. Thus these new ideas must give the ruling minority some advantage. If a new idea is detrimental to the interests of the ruling minority, eg that materialism is wrong and an alternative to materialism, ie non-materialism, has a scientific basis, and this idea in turn supports the notion that people are equal within the greater context of life's existence, then the ruling minority will never allow such ideas to be expressed.

A society based on equality

Distinctions need to be made between creating at the level of existence, and creating at lower levels.

If equality is believed by society, then the society must make a distinction between (1) those who want to reproduce and those who want to gain creative knowledge, and (2) those who are motivated by selfish interest and those motivated to seek knowledge in a selfless manner. Note: (1) and (2) are separate and distinct ideas.

For one who wants to gain creative knowledge in a selfless manner, at what level does such creativity take place? It takes place at the level of determining how "material" geometry can exist and how fundamental geometric structures related to the fundamental properties of existence can be changed. Is the spherical symmetry of 3-dimensional Euclidean space a result of its differential structure being SO(3) which, in turn, is isomorphic to the 3-sphere, or is it because the simplest three dimensional group is SU(2) which is also isomorphic to the 3-sphere?

On the other hand the drive to reproduce moves people to distinguish differences and create hierarchies of value. That is, their reproductive partners need to be both strong and valuable.

Some hierarchical structure is alright, in fact quite useful, but unfortunately, it leads to dogma, orthodoxy, traditions, and illusion which can be exploited especially if the culture floats (or rests) upon illusions, such as the illusion of materialism, as our civilization now does.

We now have minority rule which rules over both material and knowledge. A dogmatic illusion, which is a good description for civilization, is held together by the dark sensual under-belly of sexuality. It is within the context of breeding that much of the illusionary values of a society, as well as the illusionary value of reproduction partners, are reinforced. The need for reproductive contact can hold one at both a material and selfish level. Transcending these constraints is a challenge of knowledge.

Opposition to materialism, and
The true context of existence, and its description

The new scientific ideas are based on the properties of discrete isometry subgroups of the Euclidean, hyperbolic, and spherical coordinate spaces (where spherical coordinates have a metric which is the Euclidean metric that is restricted to the sphere) which are expressed as being parts of the base space within a principle fiber bundle so that the lattices of the discrete isometry groups (or space-forms) are carried on the metric-space base space . . . (so that some of these space-forms are oscillating and energy generating space-forms and represent the creative force, or the intent, of life) . . . and the spectral properties of a maximal torus of the fiber group which can carry all the spectral information of a galaxy within itself (the maximal torus) so that this represents the set of all possibilities, and such a set can be used to hold together energy generating spectral sets of the base space by means of resonance, through their (spectral) connection in the maximal torus.

The two major ancient forces of existence, the male (creative or intent) and the female (the nurturing and supporting, or all possibilities), are represented by these above two aspects of the principle fiber bundle, ie the energy generating lattices of the base space and all possible spectral relations of the maximal torus that is inside the fiber group, within which all existence can be described (contained).

The base space is a multidimensional space composed of space-forms and metric spaces, so that most of the metric spaces also have the structure of a space-form. The hyperbolic spaces are very spectrally stable. The energy generating space-forms can activate, through resonance, other spectral subsets (or spectral faces of the "cubical" simplexes) which compose the metric space lattices of discrete isometry subgroups. This allows changes to occur (changes of spectral structure as well as changes of material structure) which can be guided by the intent of the energy generating oscillating space-forms which is connected to the mental structure of the spectral sets of the maximal tori, which exist in the fiber bundles, ie are associated to every point in space.

Change occurs in discrete actions (of interaction space-forms acting on material space-forms within the context of two metric-space states, as well as the actions) of the fiber group acting on the material interaction space-forms' spectral-vector properties, so that these matrix transformations exist within time periods of spin rotations of states (metric-space states) which are defined on space-forms so that these spin time periods are crucial to material dynamics. That is, changes in space-form structure occur in two time states, which mirror one another, ie time and opposite-time metric-space states, and the discrete interchange of properties, in these states, caused (or allowed) by spin rotations.

The types of space-form changes which can occur are related to the spectral and geometric properties of the energy generating space-forms and the spectral (and/or material) sets which they contain. This over-all spectral set (or spectral structure) is placed in a stable resonance relation that exists, between the contained material components and metric spaces, by means of the resonating spectral sets of the maximal torus, which "by the way," also provides the structure for memory.

Material is modeled as a space-form whose distinguished point has a spin-½ rotation of (metric-space) state properties, so that such a material space-form is contained within a metric space which is defined as an independent dimensional set. The property of dimensional independence is related to the infinite extent (interaction) space-forms, contained in the metric space, which define an independent dimensional coordinate subset. These infinite extent space-forms are defined out to infinity and such a coordinate subset represents the geometric metric-space states whose rotations are regular rotations. Hence the distinguished point of an infinite extent (interaction) space-form would be associated to an integer factor related to its rotations of states because they are regular geometric rotations in a metric space, ie not rotations between the metric states that exist on lower dimensional space-forms than the dimension of the containing metric space. These infinite extent space-forms also define material interactions within the dimensional subset which they also define. These metric spaces become the independent metric spaces of a particular dimension within which material space-forms of lower dimension can interact.

On the other hand, bounded (or compact) space-forms allow positions of material space-forms to be determined, and thus allow for the existence of material geometry.

Resonance defines the stability of spectral sets, eg material space-forms as well as metric-space space-forms, which gain their energy through the energy generating spectral space-forms within which they are contained, and with whom they resonate. The spectral set of the maximal torus also provides the memory of spectral structure of "other" space-form structures, so that such a memory of particular spectral sets can be used to intend space-form changes.

There can exist many simultaneous spectral structures which do not influence one another due to the filtering structures of space-forms (or discrete isometry subgroups) which determine which spectral sets are consistent with one another. This process is done by both the energy generating spectral space-forms and the sets of resonance determined by the spectral sets which are activated (or connected by an energy relation) in a maximal torus of the fiber group.

Intention is a flow of energy between space-forms (or faces of space-forms), force is a geometric space-form relation that material space-forms have to one another (this can be used to describe dark matter whose interaction structure depends on scale (or size)), and existence is the set of spectra (and associated space-from geometry) which depend on both the energy generating space-form, as well as the spectrally resonating space-forms which it contains, as well as the spectral set in the maximal torus which identifies the over-all resonance of such a spectral set, ie it allows the spectral connection through which resonance can occur, ie it gives the spectral set coherence.

Within this context, life is a high dimensional, energy generating spectral structure, yet we have convinced ourselves that we exist in a 3-dimensional material-spatial existence within which material defines the dimensional confines of our perception. This low dimensional viewpoint of existence is an illusion.

However, the will to access and use higher dimensions is also about the will to create existence itself, as our higher dimensional structure is related to our higher dimensional capacity to create, and this requires knowledge.

How do space-forms explain:

Classical physics?

Both gravity and electromagnetism can be modeled in terms of Euclidean and hyperbolic space-forms which are the material sources (for fields), and a space-form interaction structure has been devised which models, in 3-space, the same inverse square force fields. But the inverse square force field depends on the property of spherical symmetry in 3-dimensional Euclidean space. Such spherical symmetry is not a property of higher dimensions, thus resulting in tangential, as opposed to radial, forces in higher dimensions.

Because space-forms come from a "cubical" simplex structure this means that they naturally define the same differential form structures upon which both gravity and electromagnetism can be based, eg $ddF=0$ abd $d*dF=j$, where F describes the properties of an interaction space-form.

Note: We see and touch only the heavier Euclidean space-forms.

Material space-forms in metric spaces have the size of atoms, thus it models material according to the same atomic hypothesis upon which most classical physical science is based. Thus the new ideas are consistent with many of the statistical methods which apply to systems made up of atomic components. However, the new theory also provides a macroscopic space-form structure (for macroscopic systems) whose sectional curvature is small hence its mass is small, thus not perceived, but which is non-reducible. Thus it naturally explains the crystal energy levels as well as the density of states within a crystal, or for a confined (say thermal) system.

Determining the wave equation for crystals has not been easy and apparently still deficient as superconductivity is without explanation, ie the critical temperature above which superconductivity is not supposed to exist, has been exceeded.

However, higher dimensional space-forms, such as those associated to the sun, will have a larger size, including the 3-dimensional flows, which exist as spectral flows on the relatively large 4-dimensional space-forms of the sun. Such 3-dimensional spectral flows define the metric space within which a planet is contained, and it is this flow structure which also determines the orbital stability of the planet.

Quantum mechanics?

The (wave)-(particle) duality of quantum mechanics has a similar relation of (lattice)-(space-form) duality in the new language. The ideas of E Nelson can be used to derive quantum mechanical properties of probability from the stochastic interaction properties that exist between atomic systems.

The diffraction patterns of material and waves can be understood in terms of the stochastic interaction processes of both space-forms and the diffraction grating as well as the diffraction grating and the screen, and these interactions along with non-local space-form properties of interactions within such a system.

The H-atom's energy levels can be understood using similar rules as Bohr (and Sommerfeld) applied to the H-atom where the orbits are the closed spectral flows on the atomic space-forms.

It should also be noted that the new ideas allow the shape of relatively large molecules to be changed, when the spectral structure within which they are contained is also changed, where the new ideas allow such a spectral change based on local spectral structures, ie the local influence of energy generating, non-local, 3-dimensional space-forms which can be used to model living systems.

The standard model of elementary particles?

In the new set of ideas, a particle is a distinguished point on a material space-form. Space-forms exist in a dimensional hierarchy within which existence is defined. Thus the higher dimensional particle states of the standard model are in fact the higher dimensions of existence which space-forms and "cubical" simplexes define. The new theory uses isometry (real), spin, and unitary (complex) fiber groups with which to transform interaction space-forms. This has a mathematical structure which more properly uses the mathematical structure of the fiber groups of a principle fiber bundle in a more proper geometric context, as opposed to particle state probabilities at a (harmonic oscillator) point in space so that there exist no other geometric relations with other points in space. Yet the particle state transformations model the very geometric "particle collisions" at the site of the harmonic oscillator. A principle fiber bundle should be describing a non-local geometric context, not a hidden higher dimensional particle states which only apply to the probabilities of particle collision data so that its geometric context only applies to a single point. Particle-state space is hidden and associated to particle-state probabilities so that the idea of "material" particles can be maintained. Material particles are the obvious interpretation of the data, while space-forms with distinguished points determine the hidden structure of micro-existence. Copernicus found the hidden structure and Ptolemy used the obvious interpretation of the data.

Interaction space-forms, which are usually infinite extent, can be curled up within a bounded (or compact) space-form, so that within such a context, these interaction space-forms would appear to have the property of mass, as mass depends on a space-form's sectional curvature. Furthermore, if they determined interacting charges within the flows, then such charges cannot radiate because they are confined to a metric space.

Many interacting 2-dimensional space-forms contained in a 3-dimensional metric space, when interacting, form into 3-dimensional space-forms which exist within a bounded 4-flow, so that there is uniform motion, eg motion or acceleration, within this bounded 4-flow. This 4-flow defines a flat space, ie the metric of such a flat space is invariant within the space-form. Uniform motion for an entire region is not detectable by the principle of both general and special relativity.

The general relativity?

In the new theory a 3-dimensional Euclidean space has the property of spherical symmetry, yet the spherical symmetry can be dominated by space-form interaction geometric structures when the size of the space-forms are sufficiently small. This is because all material is defined by its space-form structure, hence a spherically symmetric singular point must be adjusted to allow for the true shape of material which in general relativity is the source of metric properties. The shape of space is determined mainly by the geometric shape of space-forms, in particular when their sectional curvature is greater than the sectional curvature of the spherical metric space's spherical sectional curvature. The space-form shape of space also applies to "singularities" of gravitationally spherically symmetric systems, thus gravitational singularities are put into question.

The precession of the perihelion of mercury's elliptic orbit is caused by a 4-dimensional tangent force due to a space-form interaction force, which projects into the 3-dimensional space, and which is not mediated by the property of spherical symmetry as happens in 3-dimensional Euclidean space.

String theory?

There are a number of properties that come from ideas about strings which are consistent with the new theory.

A compact 1-dimensional space-forms would have a similar geometry as the geometry of a closed string.

String theory might be best fit into an 11-dimensional space-time, whereas a geometric theorem of Coxeter identifies 11-dimensional space-time (or 10-dimensional hyperbolic space) as the last dimensional space within which hyperbolic space-forms exist.

It has been hypothesized that membranes can exist as dimensionally independent subsets contained within higher dimensional spaces due to the string relations that exist between the membrane and the strings. Similarly, in the new theory, dimensionally independent subsets can be contained within higher dimensional spaces in an independent manner. This is due to the fact that material interactions, between the material that is contained within such a dimensionally independent subset, are mediated by infinite extent space-forms so that the interactions and the infinite extent space-forms together define an independent coordinate subset of a fixed dimension. Furthermore, the subspace has measure zero in its higher dimensional containing spaces, hence higher dimensional interactions would appear as gamma particles in the lower dimensional space.

Creativity related to the new higher dimensional context of life

The real context for life is that of "life creating existence." This is an idea which material based science attributes to stars with special properties which allow the formation of new atoms. However, life may have the same dimensional structure as the stars, and it also can create new aspects of the fundamental geometry of existence.

By intention life can change space-form structure so as to create something new.

Within a dimensional hierarchy of existence, the property of a hyperbolic space-form existing, stops at hyperbolic dimension ten, ie space-time dimension eleven.

What properties of existence can be changed?

What are changes which would affect what can be created within the confines of existence?

For example, creating the property of spherical symmetry in higher dimensions than three. Though this may be detrimental.

Extend the dimension within which of hyperbolic space-forms exist past hyperbolic dimension ten.

The hyperbolic space-forms in spaces above dimension six are all infinite extent. Thus the geometry of material in higher than 6-dimensional hyperbolic space does not exist because there is no well defined position for material space-forms as all material space-forms (If material space-forms do exist in this dimension metric space?) are infinite extent.

Thus, can compact 6-dimensional space-forms be invented (or intended)?

These are some ideas about the challenges of the creative process which has its context within the context that life's will (or intent) naturally belongs.

Chapter 9

MODEL OF INTENT

Model of intent and freedom

The United States of America began with a promise of equality, ie the Declaration of Independence. Compelling demands can be made (and obeyed) between individuals who are not equal. Freedom is about making decisions (or judgments) and creating things within a context in which one is independent of the influence of others, ie determining for oneself what is true and then expressing or using this truth. Freedom can only be obtained after one is in a society of equals. However, once in a society, all cultural actions will be judged and inequality will be determined by judging the performance of these actions (or processes). Though this might be true, it should not affect either an individual's decisions or their creations or actions, if they are free, ie influences of such judgments should not be dominant. Truth can be determined by people who are free because the external world does not (other people do not) compel them in any way, so that truth depends on an individual's integrity.

Can the ideal of freedom based on equality, and its subsequent drive towards truth, ever be realized?

This depends on how independent one is when one judges truth, and then is capable of acting on this judgment.

Existence and truth

The main issue concerning truth and knowledge is about, "What exists?" In turn, existence can depend on either material or intent one being a subset of the other (where intent is, what could be the beginning of a creative act). Materialism says that it is material which determines existence, while intent is another way in which to understand what determines existence. Intent is related to a knowing or living existence within which material is to be a proper subset.

Knowledge is about the true statements that can be made about existence, ie either about what will exist, or what exists, or what did exist. Knowledge which provides information about the present and in particular about the future is the more valuable knowledge. Science is a method for gaining knowledge. The scientific method is about creating a quantitatively based description of existence which can be verified by observation, or by measurement, where again, the knowledge has a stronger form if it describes what is, or even more importantly, what will be, rather than what was found to be the case in the past. Geometric language can be used to describe the future, while probabilistic based language (always) describes what was found in the past, or (only in a weak sense) what is present, ie the "'is' part" (or the new case for which data has just been collected) may find new events present (just as Ptolemy found new planetary motions for which he added new epicycles).

Material is related to science but intent is not considered to be a part of science. However, patterns of a quantitative language can be used to model intent, so that material is a proper subset of this model of intent (see below for the basis of that language).

Intent is available to everyone (each person has their own intent) and this implies equality. When one has equality one finds the freedom needed to make decisions about both what knowledge is important and about what is true, and thus one becomes independent. With truth, one can successfully create, not only for oneself, but for others, or for one's appreciation for existence. Furthermore, what one knows influences what it is that one creates.

Materialism implies scarcity and hence competition, a competition in which the winners become the controlling elites and hence powerful in a society, with concerns mostly about material things, and/or (subsequently) selfish things. The powerful elites naturally subjugate those who are lowly (or not elite). Truth is equivalent to authority, and fundamental absolutes which are used in particular ways, and other truths (or other authorities) cause factions to form. Factions of people squabble and fight amongst themselves. Knowledge is to be used for personal gain, to create in a manner which makes one ever more powerful, within a context of materialism.

There are two natural divisions of people in America, the egalitarians versus the materialists. Many would question this separation or distinction, claiming many egalitarians are also materialists. None the less materialism implies a state of existence in which there exist those who grab and dominate material and then dominate the ability to have a voice. This is contrary to American law because it is contrary to equal-ness and it is contrary to the first amendment right to have a free voice. Placing voice with the rich and powerful creates a totalitarian state. Surprisingly, this practice of allowing only an authoritarian voice seems to have begun with physical science, and it was ushered in with the regression of physical science back to a Ptolemaic verification process of a probability based description of physical existence. A language based in probability has caused science to simply validate data. This is allowed because probability is based on a set of elementary events, so that if the elementary events change then the probability must also change. It is not a causal description, and it does not predict the future, where the future might be thought of as a place of a very large elementary event space, ie a space with many probability distributions associated to itself, in other words a place where systems cannot be distinguished from each other by a complete set of commuting operators. That is not only does one system exist in the elementary event space but also a different system of elementary events also exists which is the future, which is not accessible by means of the (original) system's original set of elementary events. A search for a language which describes the future state of a physical system, ie a geometric basis for the language of physical description should have been protected by both physic's containment within philosophy and by an education system based on finding an aspect of knowledge one finds interesting and pursuing that knowledge with all integrity.

In other words an education system which is "hands-on" and whose subsequent "experts," are those who sincerely and with all integrity follow a mystery, and on "graduation" (from an education institution) can be (subsistent) funded by society. So that "graduation" is based on mystery and integrity and not authority and using language inconsistently. In other words, because we do not know the origins of life, or really what life is, "creationists" qualify as academics. However, their language needs to become specific, and a measurable description put forth, if they are to be funded by society. Though it will be argued that politicians and business men are supported by society for being authoritative, dishonest, and supporting those already with social power. However, more to the point, those deeply interested in dinosaurs need to be allowed by society to subsist because of their honest search to find and understand dinosaurs, as opposed to having a handful of dinosaur experts in society. Education needs to be based on honesty and interest in mystery, as opposed to authority, or equivalently dogma, and competition. People engage seriously in cultural activities, ie agreed upon activities with agreed upon processes, thus social judgment enters into well established social activities, so that within this context competition is always about social class, truth is implied by the agreements of the competition. Competitive education system is about seeking a piece of the societal pie, not about finding truth. Within a competitive context, truth is what the authorities, and those who control material, say it is. Copernicus, Kepler, Galileo, Descartes, Faraday, and Einstein were all marginal people within the authoritative educational structures of their times. It might be noted that Einstein allowed the good ideas of many marginal people, by academic standards, de Broglie, Kaluza, etc to have a voice.

Those believing in equality tend to be the ones who appreciate how the Declaration of Independence, a proclamation of both independence and equality, created America. The egalitarians versus their natural opponents, the materialists, is the natural division of people in America, and the law, ie the Declaration of Independence, is on the side of the egalitarians, but alas, power resides with hypocrisy and the materialists.

Perhaps people are not strong enough to realize such a noble state of existence as the state of equality.

They cannot control their selfishness, which is expressed in many ways, most notably by believing that they are better than others. It is believed that within the material context that their value of being better than others can be "proved." However, this is possible only if people value material, or equivalently (due to the propaganda of media) money, above all else in their lives. What does propaganda have to do with people worshiping money? The question forces one to distinguish between the materialists and the idealists.

How is value determined? By scarce material? By product? By knowledge of spirit or by knowledge of controlling material? By controlling property? It is not clear that Moses, Jesus, or Mohammed sided with material power, as opposed to the spirit, but the religious institutions which claim to carry on the teaching traditions of these prophets all have supported the idea of materialism. Though Moses and Jesus did not believe in staying within civilization, preferring the "wilderness" or its margins, none the less the institutions which represent these prophets wholeheartedly support the materialism of civilization, as they go on in an inarticulate manner, though with great faith and reverence, for something beyond the material basis for existence. There are some religions which come straight out and say, that the material world that we have learned about through civilization, is not the real world. We have been taught to believe in an illusion and the basis for that illusion is materialism.

In our material centered society, if one controls material and/or property then one gains enormous power. However, this leaves the sincere idealists without any power. This destroys equality and undermines the freedom of the idealists. Idealists can seek to articulate their beliefs and in so doing create knowledge with practical value, but to do this they need to be given a modest existence for their (sincere and verifiable) efforts.

If faith in, say, a creator (or idealism over materialism) does not lead one to as powerful a description or useful set of ideas or processes or behaviors as materialism does, then what is the value of a belief in an inarticulate idea as a vague notion of a creator?

Language,

Language can lock thought into confined regions of thinking and attention that the words allow and suggest (or demand). A language, and its associated judgments and actions (or behaviors by people), constructs a world which is compelling and keeps people constrained to cultural norms. For example all aspects of acquiring material in a culture, may be related to an exchange of money. Food is a form of material, so to eat, one must devise ways to get money. Having control over either language or material (or over both) can result in having control over other people. Ultimately control over language also means control over law. Organizing material (and subsequently law) for the selfish advantage of a few results in the apparent ease of life style for others. For example, farmers exploit land and water supplies to produce and sell food so that others no longer are required to gather food for themselves, which they cannot do as easily anyway because they are restricted from the farmer's lands. The argument is always made that those who produce what others buy have a knowledge which is superior to the knowledge of those who buy the goods, thus the producers deserve their higher social standing. Thus it is claimed that an economic distribution of material (or moneyed) wealth is based on the acquisition and development of knowledge. However, in fact, it might be the opposite. It might be that the earth can sustain a greater population based on harmonious biological systems as opposed to farming. It is the consumer's expediency, related to ease of product acquisition and use, which in turn is most exploited by the rich against the low class consumer. It might be that a planned economy for life's necessities might itself be a necessity to bring the earth back to balance, which the exploitative materialists are so selfishly willing to ignore and subsequently destroy (the earth).

A possibility about knowledge that those with elite knowledge of material will claim is absolutely false

The expression of an idea, which elite materialists are so adamantly in opposition they might quit reading at this point. People are afraid of certain ideas because of that to which they think such ideas are connected.

It might be that people have within themselves powers of communication and travel (and control over material) from which they actively separate themselves. It might be that the knowledge from which they separate themselves is the proper knowledge about their own internal existing structure which they may use to develop further powers of communication and travel. Such seemingly impossible power might very well be allowed by an understanding (or description) our true existence which is within a higher dimensional structure, but in order to use such power, a knowledge of this structure must exist. It might be that the barrier to this knowledge is mostly due to a belief in, and obsession with, materialism and subsequent class struggles which ensue, where it is mostly the upper class which unfairly uses their advantages to dominate and use the others, who comprise the lower classes. It might be noted that many from the lower classes are complicit with the upper class in this struggle.

If those absolute materialists can actually describe the origins of life and how it is that we have a mental experience, then perhaps their claims (which oppose the above ideas), that people do not have (and existence will not allow) such extraordinary properties, can actually be taken seriously. However, as it now stands materialism cannot even describe the patterns within which material organizes itself. Eg nuclear orbits, subsystem interactions within crystals, high temperature superconductivity, the apparent stability of the solar system, the motions of stars in galaxies, etc. As usual the authority of one's language, and proclamation of truth, mostly lies with the authority of one's social class, ie The will to power.

Freedom and truth

The freedom to judge truth is not given freely, though the Declaration of Independence and the Bill of Rights make these claims. Truth is usually presented in terms of absolute truths, so that these absolute truths depend on how language is used. By means of language, used in absolute ways, factions of fundamentalist groups, or factions based on absolute truths, are formed. There are various factions including factions of intellectual disciplines. It is believed that in these intellectual factions all ideas are examined and all possible ways of getting at the truth are considered. However, these intellectual factions do themselves only consider a constrained form of truth due to the fixedness of the language with which they form their ideas. In fact physical science is a form of fundamentalism. Namely, as it is now organized physical science is based on the absolute belief that the statement, "Material determines existence," is true, ie physical science represents the absolute truth of materialism. On the other hand there are those factions which use language to describe their belief in a creator which is not well defined, but none the less is also an absolute.

If one compares the intellectual faction which says that, life originated so that life is based on laws of materialism, with the faction who believe in an origin that is based on a "creator," one finds that though one side articulates its ideas precisely (the scientific material centered side), none the less this (relatively articulate) side, in fact, has no ideas about how life originated, while the opposing side does not articulate anything, yet claims there is an absolute creator. Based on science, neither side can be said to posses more truth than the other. However, the inarticulate side might embrace the idea that it base its descriptions on a language, which in turn is based on mathematics, ie the language of quantity and shape.

It is within this context that the statement, that "there exists a mathematical model of intent," will be disregarded, ie opposed, by both factions, either both fearing that their absolute world will be destroyed, or both absolute in their resolve that they possess the one absolute truth, hence both sides tend to ignore such claims.

It is believed (by many people) that the idea, to explain intent with the patterns of mathematics, is already being done by the intellectuals within a very rigorous framework of materialism. However, such endeavors are not successful because they begin with the wrong assumptions and the wrong language. It would be like Ptolemy trying to show the earth goes around the sun based on the assumption that the sun goes around the earth.

Free voice

Simply the fact that, "a new mathematical structure can be related to physical description," makes such ideas about intent very interesting. Unfortunately they are (apparently) new ideas, hence no peer would be willing to stake their career on the "wild ideas" of an unknown person, and apparently no celebrity scientist or mathematician is willing to consider these ideas either. This is in keeping with science being more about making information remote from people and hence validating a society of elites. In turn, this has caused science to become more about personality cult, than true science.

One must remember that it was Einstein who allowed the "wild ideas" of the unknowns deBroglie and Kaluza, as well as others to be published. It was deBroglie's ideas which opened the doors for quantum mechanics, and the wave-particle viewpoint. It also should be noted that Einstein opposed physical description being based on probability, because he knew that probability causes the language to follow data just as Ptolemy followed the unexpected motions of the planets with his epicycles, so to, now mathematical operators are adjusted by the scientist to conform to the physical data. The technology uncovers new physical patterns and the scientist rushes to adjust his mathematics to adjust to that new data. However, there are many known physical patterns which the mathematical structures cannot be adjusted so as to be put into agreement. In these cases it is simply stated that such systems are simply too complicated to describe. However, the exact precision, within such a small period of time, with which life can control itself does not suggest a very complicated of a system, but rather it suggests a very simple linear or perhaps flat basis for description.

Einstein also said there were three types of physicists 1st those who see physics as a contest, 2nd those who want to obtain personal (selfish) gain from physics, and 3rd those who love to know. It is clear that the 1st type of physicist dominates physics today, and it is they who are most likely to win the contests they need to, in order to become a physicist today.

Though Einstein was the main architect of quantum mechanics, he also warned against using probability as a basis for physical description. Perhaps he was aware that probability would cause science to degenerate back into a Ptolemy-like structure for science.

It is certainly not the usefulness of new scientific ideas that is giving the current physical scientist their high position in society. Note: See the "End of Science" by J Horgan. They are not dealing adequately with material systems, and they are not providing truth in any convincing manner concerning living systems, or neural (or mental) systems. All their investigations are based on authoritative materialism within a probabilistic language with statements which are being supported in a Ptolemaic manner, yet the scientists are allowed to take the role of being the beacon's for an "open minded" truth. Von Neumann proved that "There could not be hidden variables." However, today there is data to support a hidden variable theory, Lord Kelvin proved that "Man was too heavy to ever be able to be lifted off the ground by a mechanical means of flight," then of course the Wright brothers flew, and E Kant proved (and P Erhenfest enhanced) that, "If there were higher dimensions then interactions would deviate from an inverse square law." This is an idea he proved by using Stokes theorem in higher dimensions and assuming that macroscopic material interactions in higher dimensions, so that material is modeled as a scalar quantity, would also satisfy the property of spherical symmetry, as they do in three dimensions, ie higher dimensional spheres cause the inverse square law for forces to change. It is this idea which is being used to interpret experimental evidence to show that higher dimensions do not exist.

Yet with a (new) space-form structure, for both material and its interactions, there is no reason to believe that the higher dimensional material interactions will have the property of spherical symmetry. As is often the case, authoritative assumptions are shown to be wrong, eg the sun does not go around the earth, though each day we witness the sun going around us (over our heads). The better question might be, "Where does 3-dimensional spherical symmetry come from?"

The three examples of the previous paragraph are all examples of how languages and words, whose assumptions are not questioned, can be used in an authoritative manner to lead into the unknown in definitive manner which could none the less be quite wrong.

It might be noted that back in about 1630 Descartes said that it was impossible to learn anything from the experts of his day because they made everything too complicated. He said that it was better to learn things on one's own. This was 40 years before Newton's Principia. Today this is even more true and the experts want everyone to believe that only they, the experts, can even have any right to the knowledge in their field. Their fields are narrow, overly precise, use old languages which have lost their usefulness, ie often the old language cannot describe the patterns which their field is trying to describe thus leading to great complications, and thus mostly their fields are not related to useful applications, or only a very narrow range of applications. Learning is a difficult issue today, especially when scientific truth is based on questionable criterion, namely, a Ptolemaic version of verification of a descriptive language based in probabilities of events coupled with a strong sense that authority and authoritative assumptions determine truth. The following general interpretation of Godel's incompleteness theorem will be dismissed by people who believe truth and language can only be overly limited with detail and very narrow in its scope, ie it can only be dealt with by authorities, none the less the interpretation does capture the theorem's essence in a general sense.

Language and authority mislead people at all levels and it is the principle of equality which must be used to offset dogma and destructive authority.

That words can have a logical structure to themselves which can be used to prove true statements, within the word usage of the language, but which may not be true within the context of external verification. That is, other sets of patterns may be associated, not with some well known language, but rather with different languages and different ways in which to use words. This is a technical issue related to Godel's incompleteness theorem, whose resolution is that truth is about usefulness, not about authoritative use of language. Today authoritative use of language is being used to define academic contests, and scientists are competing for the social label of "Most intelligent." Though Godel's incompleteness theorem is a statement about language systems which are related to axioms and definitions, none the less, it applies to any language used in specific ways, and the conclusion of Godel's incompleteness theorem is that there are always patterns that exist outside the descriptive realm of languages which use words in specific ways, ie certain patterns cannot be described or proved by some given (or fixed) language structure.

The language of physical science, in particular the pattern of materialism, needs to be changed. When such a change is made then much of the formal attributes of the old language are lost. One cannot then use the old language and prove the attributes of the new language. Similarly the attributes of the old system are not necessarily a part of the new patterns which the new language can describe.

The underlying assumptions of so much of what current science so dogmatically proclaims are never questioned. If physical science was pushing the envelope of knowledge back in a convincing manner, this would be easily over-looked. But the truth of the matter is that physical science is failing at describing the fundamental properties of physical systems. At best, the current paradigm is claiming to be able to describe the properties of physical systems after all the data has been accumulated for those physical systems, though this is not even the case. At worst they are overly touted bomb engineers who have lost sight of truth (they are inventing a language which no one can understand, and applying it to model marginal physical systems which are still considered to be relevant to bomb engineering, it is a language which has Ptolemy-like evidence supporting itself) because they are so eager to impress their funding masters.

If one wants to try to examine new aspects of patterns, that do exist, in such a way so as to be related to physical description, ie the description of the structure (or the further possibilities), of what exists, then one needs to find hidden structures of existence which can be used to devise new ways to use language.

Though engineers and experimentalists keep finding new properties of material, eg high temperature superconductivity, the theorists can not now, even fit the data to their descriptive structures. The theorists pontificate about marginal physical systems within a language which has a very narrow range of application (if any range at all, eg string theory) which apparently has outlived its usefulness (except for academic contests), and now within a standard for scientific verification, in which Ptolemy would have felt right at home. For example, scientists claiming their ideas are scientific when, in fact, their ideas are supported by a Ptolemaic process of verification. The inability to describe fundamental physical systems becomes attributed to such systems "Being too complicated to describe."

Equality and truth

Therefore, if one believes in equality and freedom then one is certainly allowed to deal with this problem of describing intent in a revolutionary manner, eg by finding a hidden structure which is used to form a basis for a new language which can be used to describe intent. Indeed the origins of life may be so difficult to understand because life itself is not being described in its proper language, a proper language which allows for a mathematical pattern of intent to enter into the discussion.

None the less it is believed that one does not have the freedom to consider such ideas, rather that the scientists are the guards of truth and no one else has a right to take their authority about truth from them. However, the vanguard of truth has adopted the criterion for verifiability from Ptolemy, ie as long as there is a quantitative language which is correlated with the data, then such a weak method of verification is supposed to validate the descriptions scientifically. If a description is not extending knowledge in a useful manner, rather the description (or method) only consolidates what is already known, then that knowledge is not dependable and of very little use. Einstein fought against quantum mechanics because it was based in probability. It is a probability based language for physical description which is what has caused this weak form of science to be re-instituted.

Einstein thought quantum mechanics would eventually be transcended, and now it has, but the authorities are fighting this transcending language, and it is they (the academics), who the poorly educated public looks to for truth.

Education should teach skepticism and to believe in one's own abilities to decide issues of truth, rather it has become a contest based on absolute truths, mostly "truths" that are related to things that the rich few want examined and developed.

But the new ideas are not allowed, only scientists who use the language of science (based in materialism) are allowed to do science, which includes describing intent, and this description must be based on materialism. The sciences are dogmatic and totalitarian. They exist in as totalitarian a state (of authority) as does fundamental religion. Just as religious fundamentalism exists, so too scientific fundamentalism also exists, especially since the criterion for "scientific" verification has again become Ptolemaic, as opposed to Galilean. Note: Faraday's (or Galileo's) notion of science, in which the future properties of a system's geometry are predicted, is the type of scientific description which is most conducive to useful applications.

On the one hand there is this (a) fundamental belief that science must be based on materialism, whereas on the other hand there is also a belief that the intellectual world of science and mathematics know essentially everything there is to know. Between the scientists and the mathematicians all the interesting and important questions have been asked, and together they have considered all possible ways in which things can be described. This is an illusion, and far from being true. It is the arrogance of the academic contest winners.

Namely, there are examples which show that this cannot be true, eg (three of many such examples) the nucleus with its orbital properties has __not__ been described, current within a crystal interacting with charges at lattice sites in the crystal has __not__ been described, the stability of the solar system has __not__ been described.

Apparently these systems lie outside the range of patterns that the current use of language is capable of describing, ie perhaps the intellectual endeavors are not as thorough as both the academic intellectuals and "the people in general" believe.

The question about how mathematical patterns can be applied to physical systems is still as central a question today as it was back in the days of Copernicus. And the solutions will be of the same type, an obvious interpretation of data will be found to have a hidden structure with which a new way to use words can be determined, so that descriptions of physical systems can be improved, improved in the sense of the new description becoming very useful. The newly found hidden structure is that, particles are not particles, rather they are space-forms which have distinguished points on their geometries about which they appear to be particles (see below).

This new viewpoint opens up questions about creation and intent to a new philosophical level. It asks that people make judgments about truth for themselves, and it opposes both religious and scientific fundamentalism, ie belief in absolute truths.

Equality

If equality is to be assured, then materialist centered individuals cannot be given (complete) control over either the material of society or over the means by which people can express themselves in society.

In order to assure equality: (1) people need equal access to material, eg for necessities and creative purposes, (2) language cannot be controlled by a few with special (selfish) interests, and (3) equal accessibility to knowledge, as it is known, ie experts cannot turn knowledge into fundamentalism, or equivalently (in our culture) into personality cult. That is, treating knowledge as a basis for a fundamentalist truth about which contests can be formed to identify the winning scientists, identified by other dogmatic individuals (the professors) within a fixed framework of what they think the world is.

Knowledge is to be communicated as simply and directly as possible. If the discipline is based on a fixed language then the communication of this fact must be made clear.

The language used today in science is not a language which has a logical structure which one can follow, rather it only makes sense in the context of the right authorities using the language, eg in the language of the standard model the measure for particle occupancy in a harmonic oscillator is an undefined idea, it is sustained by authority, not by what the words, that are used to define it as a concept, actually mean.

Only if a new language is being formed, ie new axioms etc (not simply new definitions), should a survey be allowed which cannot be backed up by clearly communicated papers which give much detail, ie the new ideas suggest a range of applicability. That is, the rigor of Copernicus cannot be that which is demanded by Newton or Maxwell. Note: Faraday provided the central idea, while Maxwell provided the "academic" rigor. Rigor does not denote powerful thinking, rather it represents the icing on the cake for authority. The ideas come first, then as they gain in authority also comes the rigorous language by which they are taught so that, as Descartes stated, they will be more difficult to understand. Though rigor always claims clarity as its basis, sometimes this is true, but a probability based language is opposed to lucid descriptions which enjoy a wide range of applications.

Once people feel the excitement of knowing, and accept the responsibility for realizing truth, then knowledge can be pushed to its limits. Materialistic concerns should not dominate how knowledge is used. This is the principle of conservation (using knowledge in a limited sense, ie carefulness) and harmony (of knowledge with the world within which the knowledge is being used). Though (creative) description depends for its truth value on its usefulness, but how is one to know all the different contexts in which an idea can be used, especially if higher dimensions and new material types in those higher dimensions are possible?

Advertising needs to be about product specifications not about a dominating voice, while some markets need to be controlled to preserve harmony and conservation, eg food.

Materialism is advertised in terms of a dominating voice, but this is a voice it does not deserve.

Science for everyone

Science is a public endeavor, with descriptions that should be simple, direct, and clear with evidence presented in a public manner. That is, the properties determined in the physical descriptions should be made available for any one to use in order to create things.

The ultimate validity of scientific descriptions are the variety and/or range of their useful applications.

Science is about finding verifiable descriptions of the world so that the language is (and remains) useful. The hidden structures of existence need to be found and used as a basis to improve the language of a verifiable physical description, ie improve by describing a wide range of physical patterns such that those descriptions are very useful in creating other things. The descriptions should point out new properties as opposed to substantiating the existence of things that are already known to exist.

There is a scientific theory which describes, in a quantitative language, intent, such that material is a proper subset of intent. Philosophically, or logically, this is an idea which favors egalitarianism and opposes elitism and tyranny. This is because intent is available to everyone.

Problems in physical science

The current paradigm of physical science is suffering from many problems. Its descriptive language is based in probability and this has resulted in the re-institution of a Ptolemaic type science, in which the descriptive process tries to fit the data. This is the same way that Ptolemy tried to describe unusual motions of the planets by adding epicycles to his description.

Furthermore, physical science [based in materialism, its reducibility to elementary particles, and a probabilistic language] cannot adequately describe the fundamental orbital properties that characterize material's organization at all levels of scale (size), eg from the orbits of nuclei, to molecular properties, to properties of crystals, to the stability of the solar system, to motions of stars. At each (size) level physical science now has major failures in its attempt to describe the (orbital) properties of fundamental material systems.

Despite this, the major focus of physical science (physics) is on the standard model of elementary particles, gravitational singularities (eg the big bang), and string theory, though all of these theories have very narrow ranges of applicability.

The promise of non-linearity (ie non-linear differential equations)

There is another side to the materialist viewpoint, namely the side that deals with non-linear equations and non-linear techniques which might be used to model physical systems.

The standard model of elementary particles is a non-linear model

For the standard model this (non-linearity) is a fundamental part of its deceptive nature. Particles in particle-accelerators define non-linear systems, this can be deduced from the fact that the particle patterns are not reproducible even if the same energy of collision is re-created, yet the mathematics that is being used is linear. The assumption of quantum mechanical systems is that on must use linear equations (or operators), otherwise a complete set of commuting operators will not exist to diagonalize the function space (if non-linear then they will not commute). The Lagrangian density, ie L of $\exp(itL)$, is invariant to unitary transformations, yet the system acted on by such a (linear) transformation models an interacting force between the colliding particles, ie curvature and non-linearity (which gets modeled as a discontinuity). Furthermore, the perturbation expansion of $\exp(itL)$ is a linear expansion which is modeling a non-linear structure. In other words the structure is described by an inconsistent language.

While classical systems can be modeled as linear differential equations so that the "solution function" can describe the properties of forces, ie curvature and non-linearity, because the solution models a geometry in the containing coordinate space. However, the difference between classical linear equations and quantum is that in classical differential equation "group transformation invariance" allows curvilinear coordinates to be found in a (local) linear context, ie locally orthogonal vectors, and subsequently the system's (non-linear) geometry can be found from the solution to the differential equation, while in the standard model differential equation, invariance causes forces to emerge directly from (local) linear (collision) relations.

The difference between classical linear equations and quantum is that a classical solution is geometric, hence its geometry, if governed by a force, will display the system's non-linearity or curvature. However, in the standard model no such geometric solution function exists, rather solution wave functions can only determine (energy) states and/or probability of events of the wave function based on the spectra of particles as determined by the patterns of a Lie (fiber) group. The quantum solution relates (probability of events) to integral (quantum) operators and at that level, the mathematics is still linear, ie integrals are linear operators. The solution or wave function does not determine a geometry which is not linear, ie one can say that the geometry of wave functions is flat. The description is confined to both the properties of local particle collisions as well as to a system (or global) wave function. In the standard model, the local collision properties determine a discontinuity, but the discontinuity emerges from a transformation of a linear (differential) operation which leaves the differential operator invariant. Thus, one finds a mathematical inconsistency, if the function is (globally) differentiable then the function should be continuous. However, the standard model is designed for this (locally) discontinuous event structure, because the wave function, though defined on a sea of non-interacting harmonic oscillators (where it has the property of being smooth), is composed of a series of annihilation and creation operators which act discontinuously at the site of the one harmonic oscillator (where the system's interactions all take place), the same harmonic oscillator whose zero occupation defines the vacuum for the system, but on the sea of harmonic oscillators the wave function determined by a series of annihilation and creation operators, in turn, determines a smooth wave function for a regular quantum system. In regular quantum mechanics a smooth wave function is found from a solution to the differential operator (or [complete] set of commuting differential operators), however, in the standard model the wave function (the series of annihilation and creation operators) represents both the data of the particle events, that the fiber group models, which is to be used to adjust to the correct wave function, as well as the quantum system's wave function which needs to be adjusted so as to better fit the known data. What can a wave function (of a system) be, when the system (it models) only interacts with continuous fiber groups in a locally discontinuous manner? (Answer: Apparently a wave function which can be adjusted to the data) The fiber groups applied to the wave function are chosen to be consistent with the interaction data that is known about system (or the wave function). Does such a system, represented as a wave function composed of a series of annihilation and creation operators which is smooth in space and discontinuous at a limit point where the interaction takes place, make any sense? (Answer: As a language which describes something new it appears to be meaningless. It is about adjusting wave functions, whose energy levels are slightly off, by introducing (local) particle interactions.)

The standard model might match the event data for a few systems, but what about the data of the systems which cannot be matched such as the orbital structure of the nucleus? None the less such matching of data, by small adjustments, does not lead to any control over such a system except perhaps allowing the determination of particle collisions rates, but particle collision rates exist within a statistical context usually related to the transition between two quite different stable systems, as in chemical or nuclear reactions. Ordinary quantum mechanics (supposedly) gives a systems energy structure for light (or Boson) capture (or occupation), however, the standard model does not do give the orbital structures for the new Bosons, which are different from light, that are a part of its Lie algebra event structures. In the standard model the Boson energy structures deal with different families of particles and a Lie algebra pattern which is associated to (particle) collision energy levels. Usual (or regular) quantum mechanics gives both an energy and an orbital structure for the system. Such usual quantum energy structures can be the energy structures that one wants corrected with the standard model. The statistical context of particles is that of fermions and Bosons, and it is a particular wave function which is selected because of the data that certain particles are involved in interactions related to that orbital structure (is known) so as to then adjust the wave function to match the (again known) fine differences in the energy structures of (what should be) the system's orbitals. However, the solution of the standard model does not give new information about the Boson (orbital) energy structures, for Bosons which are not models of light.

With data from particle accelerators the physicists are only asking if the Lie algebra (and scalar potential of the Higgs particle) is the correct pattern, but the language of the standard model does not give (have) a description of this pattern which is useful, it only follows data to try to adjust the event probabilities of systems in order to fit the data.

The standard model is a language of inconsistency, which is what one should expect from a probability based language describing the geometry of colliding particles. A very wide variety of mathematical ideas are used in the standard model, ultimately because certain unitary fiber groups reflect the patterns of particle events in particle accelerators. (The real question is, why does a Lie algebra relate to patterns of particle data?) However, it is not clear that the standard model is describing anything definite. Is the wave function, composed of a series of annihilation and creation operators, a valid representation of a system, or is such a mathematical model, simply inarticulate, allowing the system to be whatever the authority chooses it to be, usually for which data about particle events is discovered in the lab before choosing what system to describe using the standard model. For example, the fiber group transforms smoothly in the classical case, but not in the case of the standard model, whose fiber transformations have the properties of causing particle discontinuities and forces. However, the fiber group chosen for the transformations depends on there being a known set of particle data for the system being described. It is a way of using language much like a person claiming to read someone's mind, if the language used to describe the mental state is general enough sometimes the subject himself divulges the information, but not all subjects divulge the information for the mind reading trick to be convincing.

Classically: Given a material geometry the classical language of physics can then describe its future state. However, in the language of the standard model if one can say that certain data is present for a system then, for a few systems, then these certain types of fiber transformations can be used to fit the data, but the data set of the description essentially remains the same, before and after the standard model is used to adjust the wave function to the data. The standard model is about the spectra of particle events, for the particles that are claimed to exist as a part of our experience, eg for electrons and photons and phonons, the standard model's ability to better fit data is limited to a few of the electron energy levels of the H-atom and perhaps (optimistically) a range of superconductivity, while for particles that are not as much a part of our experience, protons and photons and other so called (nuclear) particle spectra it is not capable of fitting the data to the nucleus's most fundamental "orbital" properties. To go one step further it is claimed that the standard model describes the nuclear energy generating property of radioactivity. The original description used the neutrino particle as a part of this model, however, when the detection of this particle was attempted, for radioactive decay associated with the sun, it was not detected with its proper frequency. The other families of neutrinos were considered and, surprise, they were detected. However, this implies that the other families of particles are more prevalent than previously thought, but what is their orbital structure? For this question the inability of the standard model to articulate anything is clear. The meaning of the words of the standard model is to fit data, nothing else. Should the orbital properties of these higher families of particles be related to the nucleus, but then they would not be out in the world as the detection of the different neutrino types suggest. So is this orbital structure for the other families of particles part of the sun? This would imply a macroscopic structure for what appear to be high energy particle families and so apparently this is forbidden. [The new theory, based on discrete isometry subgroups, does supply the orbital properties of higher dimensional energy generating material structures that are related to Lie algebra transformations, and these structures are macroscopically prevalent, namely they can be used to model both radioactivity and life.] Thus the question about (for) the standard model is, even if the Lie algebra patterns for so called particles are still identified at higher energy levels by larger particle accelerators which achieve even greater energy particle collisions, of what value is this information if it cannot even be incorporated into a data fitting scheme? That is, the oscillations between the different families of neutrinos conveniently fit's the solar neutrino data, so this means that for the language of the standard model this is the end of the story, mission accomplished. However, the existence of such data implies a mystery which demonstrates the uselessness of the language of the standard model. How do these other families of particles fit into our world, the only data about this possibility is the solar neutrino data itself so the data fitting language of the standard model has no structure to deal with this problem. At an even deeper level, of what good (or use) is a data fitting scheme? At best, the whole structure of language for the standard model is focusing on collision probabilities for transitioning systems, thus determining reaction rates for such system transitions, ie reaction, or system transition, rates in bombs. Furthermore, as a data fitting scheme, the language has no meaning, that is, "Why particles?" they are simply an obvious interpretation, "Why are there orbital structures?" the answer given by the current paradigm is very weak and is only provided in the framework of regular quantum mechanics, ie because potentials exist, yet the solutions to the most realistic, $1/r$, potentials are divergent (the series solution to the radial equation of the H-atom diverges), so this answer holds no water, and "Why is the micro-structure of energies based on unitary Lie groups?" again it is a magical potential, furthermore, the series of annihilation and creation operators, which represents a wave equation in the standard model, causes vacuum points to be independent of one another, a problem of an absurd (or meaningless) language.

Perhaps one should look for hidden structures of what has been interpreted to be a particle spectra, perhaps the Lie fiber groups should be modeled in terms of a more geometric context, especially one with higher dimensional structures. This is provided by the discrete isometry subgroup viewpoint, which needs a forum for expression, which is the center of the idea about equality and freedom of expression for ideas, so that new ideas are allowed to be expressed so as to be considered by others, even if they challenge the fundamental basis for civilization, namely, materialism, and hence they represent the most dangerous idea to ever confront the ruling class. The people who compose civilization unfortunately adopt the same elitist viewpoint about the knowledge that they think civilization (supposedly) possesses, just as the dominant elites of civilization believe. This causes the people to believe the so called "truths" that the authorities espouse, as opposed to questioning and challenging these "truths" . . . as they should. Unfortunately the education system is organized as an elitist hierarchy, as opposed to an institution which challenges authority and seeks to engage in mystery and places a higher value on truth and honesty.

The new theory is providing geometric descriptions where descriptions did not exist before (eg relations between dimensional levels). In the new theory a 2-dimensional (Euclidean) space-form can be flat in a 2-dimensional containing space, while the same 2-dimensional (Euclidean) space-form can have curvature in a 3-dimensional containing coordinate space, eg it certainly would have a sectional curvature property.

Non-linearity again

For other physical systems, some non-linear systems can have a relative amount of stability, namely those that have singularity (or critical) points, and thus such systems define limit cycles, though the convergence process to those limit cycles can vary. Whereas other non-linear systems, depending on (stable) boundary conditions, can show properties of self organization.

These are all interesting mathematical properties that are relevant to the description of physical systems, but they cannot be used as a catch-all, by saying that eventually all (un-understood) physical systems will be understood in terms of non-linearity.

The new ideas, that are being presented which have a relation to flatness and to curvature depending on the dimensional viewpoint, and higher dimensions are allowed in the new mathematical structure. Furthermore this new description can be interpreted so as to provide a description now, not a promise of a future description. That is, if non-linearity can describe the many mysteries of the world then they must provide the descriptions of those systems today, for the new descriptive structure is providing those much wanted descriptions now, today.

To put stress on the current paradigm based on its own societal value structure

That is, the voice which supports the material basis for science (and existence), whether linear or non-linear, needs to describe the orbital structure of material, if it wants to truly "compete" in the endeavor to describe the properties of existence, ie both material existence and beyond, because the space-form basis for both language and existence, is consistent with the valid aspects of both classical and quantum physics, and furthermore, it has described material's orbital properties at all (size) levels, as well as much more.

Authority vs. freedom

It is difficult to see where the authority is based, upon which people should take the speculative <u>descriptions of material based language</u>, of either fundamental systems, or speculative descriptions which mostly deal with marginal physical systems, <u>so very seriously</u>. It seems to be more an issue of social class than an issue of science. Clearly they are drawing their authority from being called scientists, and from the fact that no one knows what they are talking about, including themselves. This, language based, class distinction is expressed in terms of the scientist being able to understand the world while the non-expert is unqualified to determine the truth. As is usual, in the will to power, the hidden truth (upon which the powerful; base their power) is a very weak truth, in fact it is a lie, just as Ptolemy's truth was a very weak truth.

It is also true that the upper intellectual classes must be materialists, though as is usual with the people of civilization they often do not distinguish themselves as egalitarians or elitists, and they also do not distinguish themselves as materialists or those who believe in intent, however, these ideas actually require that people distinguish themselves one way or another, because their acts have consequences. None the less all people who throw themselves at the mercy of civilization are by that action materialists, unless a new language can be devised which allows for an alternative description of both intent and material so that material is a subset of intent, as has already happened, ie a new language has been devised.

Religion as an abstract idea

"Treat all people as equals," and "Do not make agreements with civilization for the most important decisions in one's life," are sayings which might be considered the equivalent of Jesus saying "love one's enemy" and "Render Caesar's things unto Caesar," ie do not enter into agreements with civilization in order to establish one's (absolute) truths.

The creed of the Christian is love one's enemy and do not bring one's beliefs into the political arena, a creed of equality, freedom and early (real) town hall democracy. Judaism as well wants freedom, breaking away from a pharaoh in order to maintain their beliefs. The creed of the Moslem includes the exploration of the beauty of mathematical expression and interpretation, and to do so in reverence. A creed for freeing oneself by loving the earth.

The existence of a healthy useful scientific descriptive language is an indicator of a level of "freedom to find truth" that exists in a culture.

Conclusions about America

How is America doing at fulfilling its promise of equality and freedom, such that freedom is based on equality?

Answer: A few individuals have been allowed to control all the access to material. By means of this control over material, these same individuals, plus perhaps a few more individuals, essentially control how language is used over the entire media (ie the education system, the press [including professional journals], the legal system, and entertainment), so that those allowed to express themselves within the media, do so in terms of a very constrained language of elitism, even alternative outlets use a language which portrays and emphasizes social hierarchies and social norms.

The education institutions have represented science in terms of absolute truths (which happen to be very weak). These absolute truths are used to define an academic contest, and in doing this, knowledge is kept confined to the few who win the contest.

On the other hand, the "non-scientists" view scientific knowledge as being too difficult to be of any value to themselves, thus they depend on experts to decide what is true about the world for themselves and thus for all society. The few who control material and how language is used in society, also control the knowledge that society possesses, in particular they also control scientific knowledge.

It is clear that social Darwinism, the survival of the fittest, and the virtues of competition in society is still one of the strongest voices in society. Any theory which proves this idea (social Darwinism) to be irrelevant will be opposed with great force. Though most of the opposition will be of an unconscious type, simply because the way in which knowledge (and language) has been structured socially, new ideas will be opposed on authoritative grounds because of an illusion that exists about a rigorous useful mathematically centered language for physical science.

That is, America has essentially become a totalitarian state. It is currently undermining its own power, and its science has become ineffective.

Outline of the currently accepted paradigm of physics

Classical physics is about using derivatives, vector fields, and boundary conditions to find unique solution functions to (linear) differential equations defined on coordinate frames, so that the solution functions, defined on (a containing coordinate) space, represent the unique geometric properties of a physical system. These ideas can be adjusted so that classical physics becomes about finding differential forms, or metric invariant differential equations, whose solutions represent the geometric properties of physical systems.

The atomic hypothesis and subsequent statistical techniques can also be used to describe thermal physical systems.

Such a classical description provides a geometric description which allows a great deal of control over physical systems, and the development of these ideas is upon which most of our technology is based. However, it fails to describe the microscopic spectral properties of systems, and it also fails to explain solar system stability.

General relativity has only been applied, in a limited extent, to the spherically symmetric 1-body properties of physical existence, as well as the bending of light.

Micro-systems relate material particle-events, which identify discrete values of physical properties, to both probabilities and spectral properties. A (complete) set of (commuting) differential operators, which represent physically measurable properties, are applied to a function space (which represent s a probability space of functions) in order to find the spectral set (or spectral properties) of a micro-system. The only general principle that quantum mechanics can give, is that (micro) systems have discrete (energy) state structures. This descriptive basis is too broad and the needed details are hard to obtain. It is a mathematical structure which chases data.

The use of the energy level properties of systems has allowed some new ways to control physical systems, which is determined by the energy level at which the system operates, eg lasers and semi-conductors. However, it is classical (geometric) physics upon which our technical control of physical systems mostly depends.

In the standard model of elementary particles, the locality of the descriptive language centered at a single point, which characterized much of classical physics, was re-instituted by applying an occupancy (integral) measure, $\exp(itL)$, to model particle collisions (or interactions) within a harmonic oscillator. The harmonic oscillator is in a sea of (other independent) harmonic oscillators which, together they carry a global harmonic function of a further containing system, eg electron interactions within the energy levels of an atom. The vacuum is the zero occupancy of a harmonic oscillator.

The occupancy measure identifies higher dimensional particle states which make the local event structure somewhat complex. It is (1) no longer actually mathematics (not a measure), and (2) it has arbitrary parameters, ie it is no longer really science, (3) it has extra unwanted infinities, as well as (4) an unverified notion of mass, (as of yet no Higg's particle) and (5) seems to have an overly narrow range of application, dealing only with particles in particle accelerators, electron energy levels in a H-atom, and (so it is claimed) with particles in the mythological "big bang."

It might be noted that the standard model has not been able to describe (1) the orbits of charge in the nucleus, (2) current within a crystal lattice or current interacting with a crystal's charged lattice site so that a useful description has been obtained. The micro-chip industry could use such a description. And (3) its model for superconductivity in crystals is in turmoil.

This is after over 50 years, with armies of "standard model scientists" (or particle physicists) working on these apparently very limited ideas of the standard model of elementary particles.

Note: What these particle physicists really are is bomb engineers.

The physical description of the standard model allows some control over particles in particle accelerators, and can be used to fit the data of electron energy levels in H-atoms, and it can determine the critical temperatures for some superconducting systems, ie its usefulness is very limited. Whereas the usefulness of classical physics is vast. This is because the descriptive language of classical physics is geometry, it can predict future states, and is useful.

Truth

There exists a new mathematical pattern upon which physical description can be based. It is implicitly already a part of the existing mathematical structure. Namely a principle fiber bundle, whose fiber groups which are mostly isometry groups, which is (or can be) a part of classical as well as a part of the standard model, and can be easily related to the mathematical structure of general relativity. In fact the shapes of (n-1)-dimensional geometries (mainly space-forms) in an n-dimensional metric space is the natural context for general relativity. Note: Describing 3-dimensional spherical symmetry in a 4-dimensional space-time metric space, as is now done in general relativity, has not been that useful.

Furthermore, quantum mechanics can be derived from this new structure, in terms of the Brownian motion which results from the microscopic interactions between material space-forms. E Nelson used microscopic Brownian motion to re-derive quantum mechanics.

It is the geometric-spectral structure of discrete isometry subgroups, which identify the geometric-spectral geometries of space-forms and "cubical" simplexes. Note: This language naturally describes the orbital properties of physical systems. Its mathematical structures are applicable to: (1) classical physics, (2) it pushes general relativity beyond the limits of spherical symmetry for one body, (3) it is applicable to quantum mechanics, as well as (4) giving the proper logical foundation for the standard model's mathematical structure in a new geometric setting. Note: The vacuum is replaced by space-form structures.

Classical systems as well as the micro-systems of the standard model (within the new mathematical structure) are both modeled in the same way. There is nothing in the new mathematical structure which requires that many classical systems be modeled in terms of space-forms, one can choose between the old classical structure and space-form structure, though as one approaches the microscopic size, or if one describes thermal phase transitions, then space-forms will be helpful, thus crystal current and lattice site interactions will be related to a set of space-forms.

Furthermore, the general relativistic description of how an n-dimensional material system curves space is rightfully modeled in an (n+1)-dimensional metric space. It might be noted that metric curvature and curvature due to a Lie algebra valued connection 1-form within the geometric setting of this new description will be equivalent.

On the other hand quantum mechanics based on the Brownian motion of many micro-interactions is similar to statistical mechanics being based on the many atoms of the atomic hypothesis. None the less the spectral structures of micro-systems are accounted for with space-forms without being dependent on function spaces, although function space methods can be applied, but the set of functions will be defined on space-form domains.

Strings

Surprisingly (but unfortunately) these new ideas share many mathematical properties with string theory, eg dimensional independence of the different dimensional levels, and an 11-dimensional hyperbolic-space (or 12-dimensional) cut-off for generalized space-time, amongst other things. However, string theory is based on both geometry and probability, hence at best it is an inconsistent language, where geometry is used for the technical purpose of mathematically controlling convergence problems. However, probability and geometry are incompatible mathematical ideas, (see Powell and Crasemann, Quantum Mechanics).

Strings, like the idea of non-linearity, cannot simply promise great knowledge in the future, if there already is an existing language (namely, the new language provided in this book) which is describing the systems that those string and non-linear beliefs are promising. A new descriptive language has already surpassed what the language of strings can even hope to accomplish, so the string people need to either provide everyone with the great wisdom of the string-theory language (and belief), or acknowledge the new language which is different from the language of strings.

Another language which is descriptively functional, and provides useful geometric descriptions of physical systems (and beyond) has emerged.

It did not emerge from the class of people who believe they are better than everyone else, subsequently it destroys any belief that any one should have about the power of social class. Social class is determined by materialism and a fixed way of using language. A social class simply provides dreams and promises, none of which it (as administered by the ruling-class) can keep, just as the politicians play-off class conflicts to gain their power, but it is a power which is defined in terms of their service to the rich.

Higher dimensions

These new spaces and shapes form into a dimensional hierarchy of metric spaces and space-forms in which the dimensional levels are geometrically independent, yet spectrally dependent, ie spectra is given a dimensional structure. Metric spaces are given spectral structures. Regions in space can also be given spectral structures, an idea that can be used to describe the changeable geometric properties of molecules in living systems. In particular at the point of embryonic development. There exists a dimensional hierarchy of metric spaces, each with its own signature, and each metric space is associated with different specific physical properties. Lower dimensions are either contained in, or excluded from (by resonance), higher dimensional spaces, while some aspects of higher dimensions can project down or occasionally interfere with lower dimensional spaces, eg high energy particles, gamma rays etc, but lower dimensional spaces have measure zero in higher dimensional spaces, thus largely invisible.

If (solar) neutrinos do in fact oscillate between particle families more often than previously described (in the standard model), then this implies that "those other families" of material particles are also more common that previously thought. That is, if more material from other families exists, then this implies more field interactions, which implies the existence of more Bosons. However, Bosons attract other Bosons to the same energy level. So where would those extra Bosons (where would that energy level) be located? Perhaps these "rare particles" of the other families can be found if one has a geometric picture of their space-form structure in higher dimensions. Thus one could then know where and how to look for their properties. In fact, this is what the new language does, it provides, the high dimensional geometry of the "perhaps sparse" existence, of this other material in the low dimensions. This would mean that there are further rules or geometries involved in the relations that the different particle families have to one another. Therefore it cannot be claimed that the description of nuclear properties of nuclei in stars has a definitive description. In fact it cannot be correct, because now one must know what non-reducible geometry in higher dimensions is associated with the different particle families, and how it naturally projects down into the lower dimensions.

Space-forms are essentially the solutions to differential equations and they represent the unique structural properties of physical systems, both material as well as the shape of space.

The new issues in the new structure for physical description are the relations of spectra to other dimensions, as well as questions about how the properties of different differential structures, eg $SU(2)$ for $E(3)$ and $SU(2)xSU(2)$ for $E(4)$, can enter into physical description. (eg Can the differential structure $SU(2)$ determine the property of spherical symmetry for a particular dimensional level of description?) [Can the $SU(2)xSU(2)$ differential structure of 4-dimensional Euclidean space, $E(4)$, which is related to two spheres, be used to model how life manifests its will, down into 3-dimensional Euclidean space? Note: $SU(2)$ is isomorphic to a 3-sphere.]

It might be noted that, the spectra of systems that exist within a particular dimensional level, are associated to definite geometries, ie the geometry of hyperbolic space-forms. Furthermore, the hyperbolic metric spaces, ie those metric spaces related to the physical property of charge, have both compact and infinite extent space-forms. The infinite extent space-forms can be used to model light, as well as other interactions. Hyperbolic space-forms also have very limited, yet very stable, set of spectral properties associated to themselves.

These new ideas use geometry for precise descriptions of the geometric properties of a system so that these descriptions can predict what will exist in the future, ie it pulls science away from the Ptolemaic nature of science that the currently accepted paradigm of physical science accepts.

Perhaps language has limits as to what it can describe, but the precise control that living systems have over themselves, suggests a linear controllable system which is describable with a predictive language. Furthermore nuclear systems are demonstrating stable orbital geometry, which suggests that reducibility is not correct, and the new theory suggests that its new language provides (or requires) a natural (geometric) orbital structure for physical systems.

These new ideas allow for a non-reducible mathematical pattern which can be used to model both life, and intent, as well as mental activity (see other papers and other books). Something of which the current paradigm of science only dreams. None the less there is an army of authoritative (well paid) people who are trying to describe these systems using the assumption that they can only be described in terms of material reductionism.

The currently accepted paradigm of (physical) science fails at describing material properties, fails at describing living systems, fails at describing properties of material in living systems, as well as failing at describing mental activities within living systems. None the less all the enlightened scientists are dogmatically sure that only their way of using language, only their absolute view-point, will unlock the absolute truth about existence as they have been taught and (handsomely) rewarded for envisioning. This currently accepted view of science is a fundamentalist vision of science, equivalent to a fundamentalist religion.

Illusions and deceptions

Though it was believed that the industrialists were increasing wealth, thus permitting America to revert into being a tyranny in the 1800's, now the industrialists have become militarists and they are turning the world into one battlefield. It should be clear to many that selfish exploitation, control of both material and intellectual capital by the few, was never creating wealth (as a sustainable state) for any but the few. Just as water supplies are being decimated so that the rich can get richer at the expense of everyone else and the ruination of many water supplies. Other countries, as well as America, are being looted, and now shredded, so that it is all now done in the name of absolute good, ie free markets creating wealth for the few so it can trickle down to the many (or Is it supposed to be knowledge which is being developed by the few? So as to be generously given to those below.), all still being supported by a feudal hierarchy based on materialism, still the few lying in order to dominate the many, still in place is a feudal hierarchy of the type the American revolution fought against. Fundamental to this structure of absolutes is the absolute value of materialism (in biology this idea subsequently implies the survival of the fittest) which happens to also be the absolute doctrine of the sciences which are all based on materialism. Thus the scientist is being used as a pawn in a game of feudalism, allowing many to believe in the high cultural value of their community's scientific understanding of the physical world, but in fact this is more of a deception, gladly supported by the scientist who has been given a high social status for his adherence to the orthodoxy of materialism, though it is exactly materialism of which science should be most skeptical, as physical science is now dominated by its failures to describe system's properties.

The culture that the Europeans encountered in America in the 16th century was the superior culture, based in harmony and equality, it was a sustainable life and it was full of wealth or abundance and beauty for everyone.

The power of free speech

In a totalitarian society where someone has (enough) voice to change an election (2000), ie stop the "announcement of exit polls," and someone has enough voice to change science reports, with collusion of other scientists, as well as the fact that the citizens are poorly educated in a non-egalitarian manner about science.

There needs to be a voice which would allow the exploration of patterns which transcend the basic current paradigm, "Materialism as well as a 3-dimensional spatial structure which is consistent with a scalar model of macro-material interacting in a single vision of spherically symmetric interactions." There needs to be a voice for a language which has a dimensional structure as well as patterns which move outside of both the classical (material as a scalar) and probability based (material as a set of events) dogmas or assumptions. As it is now, assumptions in science are based on authority, but that authority is <u>not</u> being supported by accomplishments, rather it is being supported by a Ptolemaic vision of scientific verification.

A language needs to be created which opens up to new patterns so that the issues of materialism versus intent can be explored on a rational basis. Otherwise these two views will for ever be separated and will end up being antagonistic to one another. Ultimately, it will be an argument between those who control material and who justify their positions based on competition (and the best raising to the top (survival of the fittest) as well as the pronouncements of materialism being an absolute truth which science supports) versus those who believe in an intent which they cannot articulate, except in vague statements of a belief in oneness or a belief in an initial intent, in a language in which words are not used in specific ways. Such believers in intent, possess a belief which alienates them from society's basis in power, thus creating factions within society which struggle for power.

Sociology

Thus there are those (rich few) who posses control over material, who owns it and determine how it is used, and hence, also have control over human value (these are the "brave and the free," or more aptly described as tyrannical rulers) and the only issue for them is maintaining the control and the power that they possess by opposing both the masses and the (other) few who are true enemies.

The material master's true enemies are those people who also accept materialism as an absolute truth and are driven by a need to demonstrate their worth and prove their sense of importance to both the evolutionary order and to all of society, that they too can ruthlessly climb to the position of being in control of material and human value. Those who accept the "law" of absolute materialism so it compels them to have a competitive manner, are the true enemies of society's de facto rulers, while the masses are easy to befuddle. The masses are easy to befuddle, because over half of them believe in a truth that is not a functioning part of the social structure of power, ie a belief in intent.

None the less a clean efficient, abundant, energy source has now become a necessity.

The industrial militarists are able to rape the planet because so many blindly follow the monetary value structure that the rulers control.

Globalization is an idea which destroys all sense of national community, and simply allows the rich to get richer.

The issue about power in society will need to be about ideas. But can a voice exist with the clarity needed to identify the ideas before the political experts come in and divide everyone based on irrelevant ideas usually about self centeredness, sweeping away both the masses lost in irrelevant thoughts, as well as those true real threats (the ruthless materialists seeking to also win).

None the less there are those in society who see science and knowledge as a chink in the armor of the elite rulers, who are de facto rulers by means of their power in society over human value. Much as Copernicus changed the basis for authority within the church. If that view of science and knowledge is upset there is a brief window open for change, perhaps it will remain permanently open.

The rich few can control their power bases of people whom seek to prove their worth to both the rulers and the external world.

None the less, the absolute belief in materialism, if abandoned would destroy the basis in power of the rich few. Materialism is a truth at least as deep as a belief in an inarticulate vision of intent.

Divide and conquer

The world is about the two factions those who believe in equality and intent, and those who are materialists and subsequently elitists (and some who are confused about what these words and ideas mean).

The rich few, or the highest elites, get their power from controlling society's material, and subsequently also gain control over what people value in society and subsequently also gain control over the intellectual capabilities of society. If one believes that people are not equal, or if one believes in materialism, which through evolution's notion of the survival of the fittest, implies inequality, then one belongs on the side which supports the elites.

Elites and other elites

Perhaps those in opposition to the rich few, are themselves elitists. Believing that the smarter and/or more moral people can run things better. However elitism implies selfishness, hence they could not be more moral. Smartness depends on winning a contest based on fixed rules upon which a contest can form. However, the smartest thing that the opposition can do is find an alternative to materialism as the basis for science.

The tyrannical elites are selfish, immature, dependent, and they are certain of an external truth. It is petty, immature, highly dependent behavior in society which is being rewarded. It is this model of behavior which is reflected in the people who listen to the "main voices" which are allowed to be expressed in society. This voice reflects the interests of the few, and it encourages the people of society to faithfully help the few realize what they want.

Only by deposing the external truth of materialism can the reign of the few controlling so much, be disrupted, and egalitarian rule be instituted, ie egalitarian rule, not rule by the masses, nor rule over the masses.

Scientists as an elitist group

Scientists are materialists hence they are allied to the few elite who control the wealth and material of society. Scientists provide society with a measured truth, but now the same standard of truth and measurement is that of the old standard of Ptolemy. That is, today probability operators are adjusted to fit data just as Ptolemy adjusted his epicycles to fit the data of planetary motions. With such a weak descriptive structure for physical science, now, as in the time of Ptolemy, the truth has become dogmatic and authoritative.

To insist on maintaining such a language based on materialism, probability, and reducibility means that science wants to: (1) maintain its elite status, (2) be dogmatic so that it cannot be truthful within its own language, (3) be an ally to the rich, (4) oppose a language which can be used to make a quantitative description of intent, (5) represent class interests.

New voice

What is needed is an intermediary language between classical geometric language and function space spectral language of probability. Such a language has been found and it is consistent with classical physics and quantum physics, it clarifies the standard model of elementary particles as well as general relativity, furthermore it can be used to describe intent.

PART III

Old Book III

A Book of Essays II: Science history, and the shapes which are stable, and the subspaces, and finite spectra, of a high-dimension containment space

PREFACE

Note (book III) (A Book of Essays II)

These essays span 2004 to 2012 and some old ideas (along these same lines) expressed around 2004 may not be ideas considered correct by myself today (2012), but I have not re-edited them.

Ideas are worth expressing, and the development of ideas can have interesting histories, and old ideas can be re-considered.

Today there exist experts of "dogmatic authority" who are represented in the propaganda system (as well as in educational institutions which also serve the interests of the owners of society, ie the modern day Roman Emperors) as being "always correct," yet they fail to be able to describe the stability of fundamental physical systems, and their descriptions have no relation to being related to practical creative development, since they are descriptive constructs based on probability and non-linearity, and deal with systems made-up of only a few components which possess unstable properties.

"Peer review" checks for the "dogmatic purity" of its contributors. However, such a situation in science and mathematics does not express the (true) spirit of knowledge. Knowledge is related to "practical" creativity and knowledge is about equal free-inquiry with an "eye on" "what one wants to create." In this context knowledge should be as much about re-formulating, and re-organizing technical (precise) language as about learning from the current expressions of knowledge [with its narrow range of creativity associated to itself].

A community of "dogmatically pure" scientists and mathematicians is not about knowledge, but rather about the power structure of society (a society with a power structure which is essentially the same as that of the Holy-Roman-Empire, ie fundamentally based on extreme violence) and the scientists and mathematicians are serving the owners of society (the new Emperors) by competing in a narrow dogmatic structure of authority, so as to form a hierarchical array of talent to be selected from by the Emperor, and then used within the narrowly defined ranges of creativity, about which the owners of society want attended.

That is, scientists function in society as elite wage-slaves for the owners of society, and they are trained experts, similar to trained lap-dogs.

This structure of knowledge is the opposite of "valid" knowledge, which should be related to a wide range of creative efforts, by many people, expressing many diverse interests.

It is good to express a range of ideas.

Whereas, "being correct" is associated to a "false," or at best, limited "knowledge," which serves "the owners of society" and is mostly used to express in the propaganda system, the (false) idea, that people are not equal. It is this type of idea which the Committee on Un-American Activities should investigate, since the US Declaration of Independence states that all people are equal, and this should be the basis for US law, and not: property rights, and minority rule, which is the same basis as for Roman-Empire Law.

Copyrights

These new ideas put existence into a new context, a context for both manipulating and adjusting material properties in new ways, but also a context in which life and creativity (practical creativity, ie intentionally adjusting the properties of existence) are not confined to the traditional context of "material existence," and material manipulations, where materialism has traditionally defined the containment of material-existence in either 3-space or within space-time.

Thus, since copyrights are supposed to give the author of the ideas the rights over the relation of the new ideas to creativity [whereas copyrights have traditionally been about the relation that the owners of society have to the new ideas of others, and the culture itself, namely, the right of the owners to steal these ideas for themselves, often by payment to the "wage-slave authors," so as to gain selfish advantages from the new ideas, for they themselves, the owners, in a society where the economics (flow of money, and the definition of social value) serves the power which the owners of society, unjustly, possess within society].

Thus the relation of these new ideas to creativity is (are) as follows:

These ideas cannot be used to make things (material or otherwise) which destroy or harm the earth or other lives.

These new ideas cannot be used to make things for a person's selfish advantage, ie only a 1% or 2% profit in relation to costs and sales (revenues).

These new ideas can only be used to create helpful, non-destructive things, for both the earth and society, eg resources cannot be exploited to make material things whose creation depends on the use of these new ideas, and the things which are made, based on these new ideas, must be done in a social context of selflessness, wherein people are equal creators, and the condition of either wage-slavery, or oppressive intellectual authority, does not exist, but their creations cannot be used in destructive, or selfish, ways.

PREFACE

The math descriptions, about which what this book is about, are about using math patterns within measurable descriptions of the properties of existence which are: stable, quantitatively consistent, geometrically based, and many-dimensional, which are used to model of existence, within which materialism is a proper subset.

In regard to the partial differential equations which are used to describe stable "material" systems they are: linear, metric-invariant [ie isometry (SO, as well as spin) and unitary (SU) fiber groups], separable, commutative (the coordinates remain globally, continuously independent), and solvable.

The metric-spaces, of various dimensions and various metric-function signatures [eg where a signature is related to R(s,t) metric-spaces] have the properties of being of non-positive constant curvature, where the coefficients of the metric-functions (symmetric 2-tensors) are constants.

That is, the containment sets and "material systems" are based on (or modeled by) the simplest of the stable geometries, namely, the discrete Euclidean shapes (tori) and the discrete hyperbolic shapes (tori fitted together), where the discrete hyperbolic shapes are very geometrically stable and they possess very stable spectral properties.

One can say that these shapes are built from "cubical" simplexes (or rectangular simplexes).

Both the (system containing) metric-spaces and the "material" systems have stable shapes of various dimensions and various metric-function signatures, where material interactions are built around the structures of discrete Euclidean shapes (sort of as an extra toral component of the interacting stable discrete hyperbolic shapes), within a new dimensional-context for such material-interaction descriptions, and there are similar interaction constructs in the different dimensional levels. The size of the interacting material "from one dimensional level to the next" is determined by constant multiplicative factors (defined between dimensional levels) which are (now) called physical constants.

Furthermore, the basic quantitative basis for this description, ie the stable spectra of the discrete hyperbolic shapes, forms a finite set. The quantitative structure is, essentially: stable, quantitatively consistent, and finite.

This descriptive construct can accurately, and to sufficient precision, and with wide ranging generality, describe the stable spectral-orbital properties of material systems of all size scales, and in all dimensional levels. It is a (linear, solvable) geometric and controllable description so it is useful in regard to practical creativity.

The many-dimensions allow for new high-dimension, well organized, controllable models of complicated systems, such as life-forms. These ideas provide a "map" to help envision these geometric structures.

These new ideas are an alternative to the authoritative (and overly-domineering) math patterns used by professional math and physics people which are based on non-linearity, non-commutativity, and indefinable randomness (the elementary event spaces do not have a valid definition), where these are math-patterns, which at best, can only describe unstable, fleeting patterns, which are unrelated to practical creative development, and whose measured properties can only be related to feedback systems (whose stability depends on the range of validity of such a system's differential equation).

That is, it is a math construct which is not capable of describing the stable properties of so many fundamental (relatively) stable physical systems, eg nuclei, where within this authoritative descriptive context it is claimed that these stable fundamental physical systems are "too complicated" to describe.

There are many social commentaries, in this book, this is because such a "new context of containment, in regard to measurable descriptions," which possesses so many desirable properties, one would think that such a descriptive language should be of interest to society. But inequality, and its basis in arbitrary (and failing) authority, and the relation which this authority has to extreme violence (in maintaining its arbitrary authority, and in maintaining a social structure (as Mark Twain pointed-out) which is based on: lying, stealing, and murdering) have excluded these new ideas from being expressed within society.

People have been herded, and tricked, into wage-slavery, where deceiving people is easy with a propaganda system which allows only one authoritative voice, and that one-voice is the voice of the property owners (with the controlling stake), and the people are paralyzed by the extreme violence which upholds this social structure, where this extreme violence emanates from the justice system, and whereas the political system has been defined as "politicians being propagandists" within the propaganda system (politicians sell laws to the owners of society, for the selfish gain of the politicians, and then the politicians promote those laws on the media).

Chapter 10

SCIENCE HISTORY

There are three main ways in which to describe existence within a quantitative based descriptive language: (1) the measurable properties of geometry (2) the measurable properties of averages determined on a constrained system composed of many particles (or many components) (3) the measurable properties of a system's spectra which determine an "elementary event space," upon which probabilities can be determined.

Probabilities defined on a continuum can be represented by harmonic functions in a function space, and the functions can be related to the spectral set through eigenvalue equations, and/or through a "complete sets of commuting Hermitian operators" which act on the function space so as to determine the eigenvalues, which correspond to both the set of operators, and the eigenvalues correspond to the same set of values which compose the "elementary event space." Can the eigenvalues of the operators correspond (or correlate) with the observed eigenvalues of the elementary event space?

Quantum physics is based on probability and a probability based descriptive language is designed (is set-up) to follow data, thus it is a language which effectively has an epicycle structure built into its fundamental descriptive structure. Thus to distinguish quantum physics as being an ineffective descriptive language . . . ,

> when the act of "'following data' is considered to be a verification of a scientific description,"

. . . . has already (essentially) been ruled out by the assumptions of the scientific community. But such an assumption clearly identifies the science community as a dysfunctional (scientific) community, since Ptolemy's epicycle structures also followed data quite well, so by today's standards Ptolemy's idea would be a proven scientific fact (but we know this cannot be true).

Data is interpreted through the theory, whereas a scientific fact is a widely applicable and very useful descriptive language based on both measured values and principles, but ultimately a descriptive language must be based on geometry if it is to be a useful language (or a useful set of principles). Only classical physics is scientifically true (since it is the only widely applicable and very useful descriptive language in science), but it cannot describe the spectral properties of quantum systems. Nonetheless, classical physics is based on geometry, ie conjecture: only a geometric based description can be "very useful."

History of science (briefly)

Ptolemy placed material in a spatial geometric context, wherein material moved, and Copernicus provided new interpretations for the geometry of planetary material and its motions.

Galileo identified the idea of inertia and changes in motion related to forces and mass, Newton explored the idea of local measurement of an object's geometric properties within space (where changes of position are considered geometric properties), and related these changes of material motions to the distant geometric properties of material in space, ie the inverse square law of a radial attraction, and an assumption of spherical symmetry, in regard to the material geometry of material interactions (these are differential-forms in 3-space).

Newton explored the math structure of local measuring, namely, the derivative and its inverse operation of integration which (both the derivative and integral) act on a "domain space of functions," ie not a domain space of space, while the domain space for a function is space.

The math of calculus of Newton in regard to solutions of separable differential equations fit with (into) the answers provided by Copernicus and Kepler, and Newton placed the local measures of differential equations into the context of local measuring of a mass's motion which was devised by Galileo.

Forces, translations, rotations and levers were placed into a language which described the "mechanics" of material wherein forces were applied to material in geometric contexts.

Faraday, 200 years later (after Newton), determined the correct descriptive structure for electromagnetism, also within the language of calculus, where relatively still charges also have spherically symmetric and associated radial force fields, but now there is also a field which depends on the motions of interacting charges, ie the magnetic field, but now the space is not Euclidean space but rather the containing metric-space is space-time, wherein the linear, separable, solvable differential equations (whose solutions are 2-forms, where 2-forms are consistent with the metric-invariant second-order differential operators defined on metric-spaces) of electromagnetic systems exist in a space-time metric-space, where the fundamental equation to be solved (for electromagnetism) is the wave-equation [the metric-invariant differential operator of space-time] . . . , (where a wave-operator is a hyperbolic differential equation, as opposed to elliptic [gravitational] equations and parabolic equations [or heat equation, or thermal differential forms, or quantum physic's heat equations, {though the electromagnetic 2-form can also be modeled as being a 2-form in a Euclidean 4-space where the extra spatial dimension is multiplied by either, i, or (-1), so as to organize the magnetic field in the higher, 4th, dimension}])

..., and quickly technical developments, in regard to electronics, started happening for electromagnetic systems (ie as soon as Faraday developed the "correct" language of electromagnetism technical development began), eg based on electromagnetic-waves and circuits and circuit-oscillations.

In fact, modern technology is (mostly) all about continuing to develop the ideas about electromagnetism which were begun by Faraday, and it might be that Tesla guided it into its current form of technical development.

Indeed the rich people of the world should be the families of the above mentioned key personalities in the inventive exercise of science (or understanding existence in relation to practical creativity) within a culture which belongs to everyone.

However, the world is governed by the selfish, murderous, coercive, thieves of society and the arrogant (over-valued) institutions which support these rogues. (I suppose this could be a quote from Mark Twain)

Thermal physics has also been developed, based on the measurements of physical properties which are average values defined on reservoirs of many material components confined within some closed geometric structure and at equilibrium. The laws of thermal physics are represented as differential forms, where the function values, such as energy or entropy are fixed (average) values (as are also) while [most of] the domain values, except the number of components and the volume, are average values. This descriptive language of thermal physics (mostly about the properties of heat, eg heat flows from hot to cold) is very much related to technical development.

That is classical physics is best characterized in regard to its close relation to linear differential forms which can be used to identify the laws of the force fields, ie 2-forms which affect inertia, and the laws of thermal physics in a geometrically closed system (in equilibrium), where energy (or entropy) is a zero-form.

Then it was found that the laws of classical physics could not describe the observed properties of the small stable discrete definitive material systems called quantum systems. Should the working mathematical aspects of classical physics be considered and then extended to form a more fundamental stable geometric model of existence, a model which embraces a many-dimensional structure of existence or should a descriptive structure based on randomness be adopted so as to account for the random point-particle events in (classical) space and time (where the particle-events are associated to definite spectral values) that are associated to quantum systems? But the assumption of randomness as a basis for description has not been able to describe the stable discrete properties of general quantum systems in a convincing manner.

In order to be consistent with accepted physical law it is assumed that particular types of differential operators should be used and placed in certain descriptive contexts, eg classical physics is related to geometry while quantum physics is related to function spaces and randomness, and general relativity is related to non-linearity, but this needs to be reflected upon and analyzed in new ways.

When satellites are guided through the solar system it is Newton's laws and action-at-a-distance, which are still used. That is, when there exist great distances between interacting material satellites in the solar system then the local causal space-time model of gravity, as given by general relativity, is not used to model "how to control a satellite's motion in the solar system," rather action-at-a-distance is used.

While A Aspect's experiments have confirmed that quantum systems have (possess) the property of action-at-a-distance (in Euclidean space), ie non-locality.

One might ask, are certain types of operator properties related to the space within which the physical property (being modeled by the operation) is contained?

Do some physical operations (or physical properties) belong to one type of metric-space and other physical properties belonging to other types of metric-spaces?

For example, the property of position belonging to Euclidean space, while the property of (stable) energy belonging to space-time space (or hyperbolic space).

On the other hand, descriptions based on randomness either particle-waves or the particle-collisions of particle-physics have the limitations of not being accurate in regard to describing the properties of general quantum systems, as well as not providing a practically useful context for technical development (or for technical creativity).

The development of quantitative descriptions based on indefinable randomness and non-linearity (where both general relativity and particle-physics are non-linear models of material-energy systems), have not been capable of describing the observed stable discrete definitive spectral-orbital properties of quantum systems (or for all [any] material systems at all size scales, ie "why does a general nucleus have its stable spectral structures?" "why is the solar system stable?"), and the meager sets of descriptions which exist for quantum systems have not been related to technical development in any significant way.

Induction, supposedly the logical basis for science, requires either over-whelming accuracy for many systems or over-whelming relation to useful development, but neither of these conditions is met by modern physics.

However, when the descriptive language is correct, then one sees wide-spread technical development occurring almost immediately. If technical development is not significant, in regard to a descriptive language, then the descriptive structures being used to describe the properties of material systems must be wrong (at least this is the best conclusion which one can make as a scientist).

The claim quantum physicists might make is that the observed phenomenon of quantum physics does not allow a useful descriptive language.

However, the quantum systems that the laws of quantum physics are supposed to be able to describe are stable definitive and discrete, and they formed from material interactions. But the properties of these general quantum systems have not been described by using the laws of quantum physics.

If these interactions were so dependent on the inaccurate descriptive language and associated physical structure described by the laws of quantum physics, ie if the descriptive language of quantum physics is both accurate and unique, then there is no reason to believe that some general atom would possess a spectral structure which could be used to identify it from some other type of atom, yet in the interactions associated to the particular (but general) quantum system always lead to the same type of quantum system with the same type of identifiable spectra, a property which the currently accepted descriptive language of quantum physics cannot describe.

This means

There is a descriptive context in which material order emerges in an essentially unbounded and "open" spatial-material structure (or context).

What is that context?

But social forces are getting in the way of seeking a new context for physical description, ie finding new sets of assumptions on which to base physical description. That is, social forces do not allow certain types of questions to be asked.

This leads back to "what the state of society (or human behavior)" which science requires so that science will most readily be able to get-at "what the context of an emerging stable material system" at all size scales "actually is."

One wants a society which is equal, and a society in which many ways of structuring the assumptions of a precise descriptive language are being put forth, so that this is done by many different (equal) people. Such a society is a society which depends on equality so that free-inquiry has the acceptance needed for many new ideas to be considered by the many people of society, where the people are interested in using their knowledge to create in many new (selfless) ways.

That is, science is really about equality, so as to allow people within society to develop new ways of developing language (in an equal manner), to get-at a descriptive structure which is useful in practical creative ways, while science should oppose a society based on: hierarchy, fixed authority, and a narrow view of creativity. Yet the professional scientist is defined as a "supremely intelligent" wage-slave in a hierarchical society within an authoritarian institution of professional science, and this social set-up is designed to allow the monopolistic economic interests to: control knowledge, to control creativity, and to support the idea that people are not equal.

There are sections of:

> General discussion of language and science
> Histories of classical and quantum physics
> Societal influence on free expression and free inquiry
> (the main section) New ways in which to use language to describe physical systems, and
> (an additional [short] section) An alternative way in which to organize society (at the end)

The characterizing property of an overly authoritative science community is its rigidness in using its descriptive language in conjunction with a fixed way of data interpretation and particular assumptions or contexts, eg the context of materialism, and its over emphasis on rigor.

However, in the development of science new descriptive languages and their (new) contexts are more important than proofs of properties, which at first might be (slightly) helpful, but as time goes on they (usually) have ever more narrow ranges of application.

Copernicus did not prove his ideas by starting with the Ptolemaic axioms (or assumptions, or data interpretations) he invented a new way in which to use language.

These are the conclusions of Gödel's incompleteness theorem*, and they are (also) the nature of scientific revolutions:

> Copernicus (1543),
> R Boyle (thermodynamics, 1670) and S Carnot (1824),
> M Faraday (electromagnetism, 1820-1850), and
> quantum physics (Planck, Einstein (De Broglie), Bohr 1900-1930),
> but quantum physics has not worked, it has not led to greater development. Though Einstein built it, in the end he opposed the probability structure of quantum physics. A 100 years of quantum physics has led to a very short list of technical developments.

* Note: Godel's incompleteness theorem essentially says that there will always exist true patterns .... (found from observing the properties of material) .... which lie outside the capacity of a particular precise language to describe these true patterns.

Quantum physics seems to not be able to describe the definitive spectral properties of general quantum systems. Particle physics cannot describe the orbital properties of a general nucleus. These are very troubling flaws.

The list of achievements for quantum physics is (embarrassingly) very short compared to the achievements of classical physics, and that short list is:

The Hall effect (1879, 1839 (?)), quantum Hall effect (1975)
The development of the photoelectric cell (1883, before quantum physics),
Electron microscopes (1931)
Fission and fusion "reactor" and bomb (1940, 1950)
Lasers (1917, 1928; 1950, 1960), and
The tunneling microscope (1981).

Transistors were developed (1947) through lab work, similar to how the photovoltaic cell was developed, not so much from technical quantum descriptions.

By 1985 . . . , with the help of the micro-chip (1959, J Kilby, R Noyce, K Lehovec) which was developed through innovations on the semiconductor transistor (and whose development almost entirely depended on classical optical, thermal, physics as well as chemistry) . . . , there were (and still are) very powerful and cheap personal computers.

Most technical development of quantum properties is about coupling to quantum (or discrete) properties of quantum systems with highly controllable classical systems, such as circuits, magnetic fields (spins), eg magnetic (spin) resonance imaging, and light.

Use of particle physics in bomb technology

It is incorrectly assumed that if general relativity is true and quantum physics is true then they should be unified in order to have a single descriptive language to describe all of the material world.

This idea of unification is actually being motivated by the interests of bomb engineering so as to unite a geometric singularity with the properties of particle collisions.

The two places in physics where the particle collision model of material interactions make sense are in statistical physics and subsequently in bomb explosion rates, which are related to particle collision rates.

Note: A chemical or nuclear explosion is an unstable, high energy system in which the independent material components (which compose the explosion) are furiously colliding with one another so that the collision rate depends on the width (cross section) of the colliding components, and the density of components. The collision rate is related to the energy given off in an explosion. A system of furiously colliding components is not a stable system, rather it is a system in transition from an unstable (exploding) system which gives off energy, to a system which is cooling and finding new stable structures. What is observed in particle accelerator colliding systems is also that of an unstable system in transition to something stable, so the data from such colliding systems is not relevant to stable quantum systems.

That is, (it may well be that) collisions in explosions are not fundamental to the physical description of stable quantum systems, rather they are only fundamental to describing bomb explosions.

This shows how much control the rich oligarchs have over knowledge within our society. The elite scientists are not evaluating science in a valid manner.

Who creates new ideas, and who benefits from new ideas in society?

For example, ideas and interpretations of experimental data about electromagnetism were developed by Orsted, Faraday, and Hertz, and their descriptions and lab inventions, in turn, were taken (stolen) and developed by others. Marconi made a lot of money from his radio business but he did not invent the radio, it was Tesla who improved Hertz's lab-radio, but Marconi got rich from it. Tesla got rich from Faraday's electric motor, etc.

If intellectual property rights were properly implemented and enforced, then the families of Copernicus, Newton, and Faraday would be some of the richest families of the world, rather than the families of the Medici, the Rothschild's, the Rockefeller's, and the Walton's, the king of Saudi Arabia, or China. However, the ideas of Copernicus, Newton, and Faraday themselves emerge from a culture which has done as much to frame their ideas as they themselves have framed their own ideas. That is, it is the culture itself which develops ideas. One wants a culture where ideas are always being developed. This occurs best when the people "who compose the culture" are equal.

The Medici, the Rothschild's, the Rockefeller's, and the Walton's represent the high priesthood of materialism (and its "legal" theft) and they (symbolically, if not in reality) have great control over markets and economies, (economies which are overly controlled). Their (symbolic) dominion over society . . . .

> (though oligarchs do dominate society it is just not known exactly who the individuals are along with the corporations)

. . . . depends on property rights, and intellectual property rights which are both sets of laws which are essentially about legalized stealing.

The inventions of a society belong to the culture, not to its thieves (and/or land owners). The material resources of a society belong to the society not to its thieves.

Oligarchies (oligarchical societies) do not reward those aspects of the culture which allow a culture to expand and develop, rather they reward those who lie and deceive and steal, in a world dominated by the selfish material concerns of the oligarchs.

Invention history of classical physics and quantum physics

Consider the following seven components in a sequence of technical development, they are (today, 2009) central to technical development and the control of the market, the media, and the university.

(1) M Faraday did much of his work on electromagnetism from 1820 to 1850, in which time he mastered the electric motor and his other ideas led to the idea of electromagnetic waves and then to the radio. The telephone was invented by A Bell in 1875 (or there about), H Hertz invented the lab radio in about 1890 (N Tesla's radio in 1891 and 1894).

(2) The dye and chemical industries (1856).

(3) The steam engine (around 1700), the internal combustion engine (which became common in 1860) and oil dependence, and subsequently the automobile and energy generating turbines. Carnot's work on thermodynamics(1830).

(4) The airplane (of the Wright brothers 1903) and the air-foil of an airplane wing, and the jet engine (1930-1945).

(5) The TV was invented in 1907, ten years after the cathode ray tube was invented, and the computer was developed out of the TV technology (fast switches) by 1936.

> [It should be noted that when a description is geometric and clear then development is rapid. In 20 years the theory of electromagnetism led to great technical development.]

(6) Particle collisions are relevant to bomb explosions, both chemical and nuclear. The property of nuclear fission, accompanied by the emission of multiple neutrons, along with the chemical model of a chain reaction based on particle collisions together led to atomic bombs (1940). Thus, physics has become bomb and military engineering. This has meant that the ideas of physics are related to: particle physics, general relativity, non-linear physics, and string-theory, all related to nuclear explosions, so that these ideas have come to be the ideas which define physics (engineering) departments at universities. This is about "national security," or equivalently military business interests, and it is a condition which stifles the development of new scientific knowledge.

(7) The transistor was invented in the lab experimentally (1948), as a way in which to couple a quantum system to a classical system, ie semi-conductors to an electric circuit. The transistor was not developed as a result of a quantum description, rather it had to do with the chemical (or atomic) composition of the semi-conductor material, so that this atomic composition of the semi-conductor is controlled by thermal processes. Solar cells were also developed in the lab, not as a result of the quantum description of material but in processes similar to how the transistor was developed. Micro-chips were developed due to classical techniques of optics and etching, again not so much due to quantum description. Thus, through circuitry and miniaturization of circuits, along with computer programming, came about the development of our communication and information systems.

Knowledge of quantum physics (1925) has led to nuclear bombs, lasers, and micro-processors and these developments seem to have been developed mostly within the lab, not as a result of an opening-up "of insight" due to physical theory.

Whereas electronics (1890) moved quickly from radios, to TV's, to computers (1936). The description of electromagnetism provides a way in which to control the properties of the systems which it describes so circuit structure is directly related to the control of electric signals within the circuits.

Energy and transportation systems are not being dealt with at all (fossil fuel burning generators) and renewable energy sources are being stifled.

Communication and information systems are well developed businesses . . . .

> (this allows the ranks of the minions of the oligarchs to be shrank, and this might explain the collapse in terms of gathering ever larger pools of capital to deal with the world market as defined by the 7% around the world who compose the minions of the oligarchs),

. . . . and armaments and spy equipment are well developed businesses.

This short list of seven fields of technical development pretty much identifies the main aspects of the technical development of our society. Only a few of the inventors actually make money, and "the idea for an invention" usually has a long history of development. Although one could say that, H C Orsted, and M Faraday were the fathers of the modern electrical technology.

These are very relevant and very deep considerations and statements about math and science.

Does one want to arrogantly consider if these "are true" (where truth is defined by authority) or of "high enough value" to talk about, so as to promote a myth of absolute high-truths which only all the intelligent, properly qualified and properly educated people must be able to recite to the authorities in order to get their certificate of being knowledge-worthy (yet, it seems to be the case that the experts function better when they have a check-list. That is, it is a myth that the "better people" can recite the proper truths, so as to identify their knowledgeable value and hence their proper position in society), or does one want education to be about talking freely about many ideas.

That is does one want education to be about the dysfunctional abstractions about quantum physics, particle physics, general relativity, and their derivative theories such as string theory, whose (almost totally non-existent) usefulness has shown this type of descriptive knowledge to be irrelevant, and most-likely a description of an illusionary world.

Where it should be noted that this irrelevant knowledge has come to our society through the funding and interests of the absolute high value of the NSS.

Knowledge is mostly about being wrong, when descriptive knowledge, concerning a particular context, is put into the correct language then development happens fairly rapidly, eg Faraday's electromagnetism (1870) (within 40 years there was the TV (1910) and within 70 years the programmable computer (1936), the electric motor was invented in about (1835) (by Faraday)).

In comparison, quantum physics has been around for over 100 years (except for Bose-Fermi statistical properties, relevant to the laser, quantum properties are only used when they can be coupled to a highly controllable classical system and in these cases the quantum description is unnecessary), particle physics has been around for about 80 years (it is only related technically to the explosions of bombs, which is actually the model of chemical reactions which link two relatively stable systems [before and after the reaction]), general relativity about 90 years (may be it is applied to bending light, but it has nothing to say about the stability of the solar system, the most important problem in all of astronomy), and string theory about 40 years (seems to predict something that is not true the instability of the proton), yet very little technical development has resulted from any of these theories, which are considered to be absolute truths.

It should be noted that the new descriptive structure directly and very simply provides the structure through which the stability of the solar system and the spectral stability of both the general atom and the general nucleus can all be described. This is a fundamental problem which modern physics and math (2011) have not been able to solve.

It is of great negligence that this fundamental dysfunctional-ness, that modern science and math are not capable of solving fundamental problems in science, is not widely known. Instead, one only gets hype and misleading statements of "public-relations science," and not true information about the true failing-state of science and math. This is mostly a smoke-screen, used to create a sense of "high value" for NSS science and math. If "new science" enters the scene, then there could be some "true competition" in the "free market" of the "highly controlled" business world, and the owners of the US's "highly controlled" economy oppose such a scenario, they could become destabilized.

This is a fundamental error in analysis of the NSS (national security state) by those on the side of reason, and a tribute to how well the NSS does control language and information and how they use the fake idea that these science and math people possess some type of "high valued esoteric knowledge," but knowledge is directly related to its wide applicability and its great usefulness, and modern physics and math are obviously failing on these attributes, see J Horgan "The End of Science," (1995) where Horgan (a former writer for Scientific American) though Horgan does not feel qualified to identify science as failing (which it is), yet he does identify that science and higher math is completely and totally irrelevant to the inventive or technical development in our society.

Descriptive structures need to be of a valid (or of a manageable) size. In a large-size set, "Are the properties which are related to the large-set actually relevant to descriptions of stable systems?"
How do stable systems form? Do they form by means of random wave-structures which "follow" energy properties (where the quantum-energy properties are associated to classical ideas about measurable properties, such as energy)? However, the quantum description is outside the context of geometry (where in the quantum description all aspects of existence and interactions are based on randomness).

or

Are stable properties of material systems all about discrete (separable) geometries, which interact in a discrete geometric context, and which are also separable geometries?

These are fundamental questions.

Classical physics is about locally measurable properties of (relative) motion being related (by some unknown means) to geometric properties, which were the causes of changes in motion, but the actual cause still is not known.
General relativity tries to answer this question, where motion is based on inertia and the (non-linear) shape of space, but no one knows how to put two (or more) bodies in such a descriptive structure.

Quantum physics claims that systems are (random) oscillations defined around (vague models of) a system's geometric and energetic structures, where the actual interaction is described in a context of elementary-particle-collisions whose random properties are guided (or caused) by non-linear, random particle-collisions. But, this descriptive structure has not been able to describe any of the observed forms of stable material systems.

In both of these alternatives (general relativity and quantum physics), to the descriptions of classical material interactions (and subsequent system formations), the descriptive structures are very large sets (highly detailed, and non-linear [and thus unstable] geometric properties, or huge sets of localized interactions whose relation to a system as a whole is not defined [in any way which is useful]).

In fact, they seem to be sets which are "too large," so that it can not be understood (or described) within these large sets, "how physical systems acquire their stable properties."

If one claims to be able to describe the properties of the world, consider that both Ptolemy and Copernicus made these claims and both were measurably verified descriptions, but if the descriptive language does not describe any "real physical systems" in any believable way, or in any useful, way, then the descriptive language does not describe anything "at all."

Thus the description is either closer to being an illusion or closer to being a religion, as is (or as in) the case of current physical description.

That is, the principles of physic cannot be used to describe any "real systems," in any believable or useful manner (either descriptively or in relation to usefulness), and current physical description is based on randomness and non-linear connections, but the measured (or observed) properties of physical systems are stable, definitive, and discrete. These stable properties imply the existence of a linear, metric-invariant and geometrically separable context of system containment, as opposed to random and non-linear.

That is, the useless descriptive structure of modern physics is treated as a religion, and it provides a clear example of how "state sponsored religion" (the science-religion interests of society's ruling class and its relation to their business interests and their interest in domination) is dangerous, and in opposition, to the idea of developing knowledge, where a state religion (based on math) is about developing knowledge along a narrow and dogmatic context (of a specific type of usefulness, but which, nonetheless, is a very narrow usefulness).

Instead one wants all assumptions to be challenged and new sets of assumptions and new contexts (continually) developed (unless a descriptive structure relatively quickly leads to new technical (or new practical) innovation and invention), so that knowledge can lead to technical usefulness . . . , in a context in which people live in harmony with the earth.

Instead of destroying the earth in order to exploit "partial knowledge" for "personal gain" at the expense of both destroying the people in society and the destroying the earth for the generations to come.

Though the set which contains the transitions between stable systems seems to be big, and it is a set which is full of both non-linear and random properties (whose elementary set space is: unstable, undefined, and is bigger than can be identified in a finite time), nonetheless (or despite) this focus on very large descriptive sets (whose properties defy description)..., the best descriptive structure for describing "the transitions between relatively stable physical systems (which seem to define very large sets)" is a descriptive structure which is based on: linearity, and separable geometries, within a consistent structure of metric-invariance and geometric measures (on the domain space of functions, where functions represent physical properties and) which identify both the real (descriptive structure) and the consistent (or continuous) descriptive structure of measurable sets, which compose both the description (or the descriptive language), as well as the properties of the mathematical (or measurable) entities which compose the system.

Though many math patterns may be interesting (or complicated), and there is a hint that some of these descriptive structures (ie a descriptive structure is a set of assumptions and a context for interpretation) may be applicable to physical description (hopefully a useful description), nonetheless truth must be related to widely applicable usefulness.

Thus the lure of both complicated descriptive structures and very large containment sets must be analyzed in a skeptical manner, demanding a capacity for the description to be related to great usefulness.

It is not enough to hint at (what might be) a valid description.

Ptolemy's ideas did not lead to usefulness, while the ideas of Copernicus were related to useful ideas.

What does work?

That is, the useful aspects of physical description are all related to linear, separable geometries, and contained within a metric-invariant context. These simple properties may be used to identify a new math category which can be studied in the hopes of determining a useful description of the physical world. This new category both allows for useful descriptions, and it also allows for measurements which can have the property of being stable, definitive, and discrete and/or continuous.

A description needs a property of continuity, the system or the new system to which a description transforms (or transitions) need to have a property of relative stability so that the description possesses a property of continuity (or stability). That is, the essential property of conservation of energy and mass and charge are implicitly a part of a descriptive structure which possesses continuity within itself. When these properties are conserved then the description has a continuous context. But for the description to be continuous it also needs to be both quantitatively consistent, and set-wise consistent where the geometric measures of the domain space need to be consistent with, and directly related to, the values of the functions, which are modeling a system's measurable properties, where local measures, eg derivatives, are applied to functions (or equivalently to the system's measurable properties).

In the context of metric-invariance and geometric separability and metric-invariance, what is possible, what is being described?

In this context, it is only the two-body problem which has been solved for both gravity and in quantum wave-function physics, where in both cases there is spherical symmetry (a separable geometry). But the radial equation diverges for the H-atom, and only three or four atomic quantum systems can actually be solved, but only after one assumes discreteness for the radial equation.

The critical temperature of superconductivity has not been accounted for (a critical temperature was predicted by the BCS model of superconductivity, but this predicted value was exceeded by high temperature superconductivity), and the descriptions of crystal structures is not adequate for guiding development in the semi-conductor systems in regard to micro-chip electronics, eg having the current couple to spin-states of atoms in the crystal.

The non-linear dynamics of satellites in the solar system (defined using Newton's equations) can be controlled using feedback, but such a many body system cannot be considered (solved) in the non-linear context of general relativity, and this is because only one element in system (whose description is based on general relativity) can have mass.

The spectral properties of the nucleus cannot be described using the principles of either quantum physics or particle physics. That is, not only are the theories of modern physics not useful at describing observed physical properties, such as the stability of the solar system, but it is also a description which is not useful in regard to guiding technical development.

That is, Quantum physics, particle physics, general relativity, and all other derived theories, such as string theory, are "a bunch of baloney," which cannot be used to describe systems' properties, nor do they guide technical development.

Rather development is being led by finding, and describing, material properties which are observed in the lab.

Whereas, within 40 years of the development of the (presently understood) laws of electromagnetism (about 1870) the TV was invented (1910). Whereas quantum physics is over 100 years old (and general relativity is almost 100 years old) and the technical development which has resulted from these descriptions is very limited (essentially, only the laser).

When a descriptive language is accurate then very interesting technical development occurs rapidly.

Measuring, and the context of stability in regard to measuring

Yet the stable definitive discrete structures which are observed in physical systems fit into the context of measurability which is associated to metric-invariant containing spaces within which measurement is modeled. If the containment sets were fundamentally dependent on non-linear properties it is not clear that the discrete properties would be so definitive or so stable. Consistency between quantitative sets, eg domains and function values, occurs when there are the properties of linearity, metric-invariance, and separable geometries. That is, measuring scales, eg energy, exist in a context of metric-invariance and measurable properties (or function values) and domain measures. Thus, the structures from which the measured values emerge (or originate) as a description, must exist in the same context of metric-invariance and separable geometries in which the systems are linear, ie the context in which the measuring scales are defined.

Chapter 11

SCIENCE

There is not a valid understanding of the relationship between language and logic, nor between quantitative data (or measured properties) and relationships that such data supports, ie the descriptions and interpretations of the data. There is the wish (or purpose), then knowledge is considered, then material is obtained, and then material is fashioned and creativity can take place. If descriptions are not related to practical, useful creativity then any other (descriptive-verified) truth should be challenged, even if the ideas one is challenging are both logical rigorous (Platonic truths, see below) and measurably verified (remember Ptolemy's ideas were measurably verified).

Science and inequality

Science and the aristocracy of intellect, the difficult abstractness of science is used as proof for the inevitability of inequality, in fact, the inequality of (subjective) value can only be defined in a narrow fixed context, and a narrow context leads not to a useful truth, but rather it leads to a non-functional set of authoritative illusions.

The main issue facing society is the question of equality or of inequality. Is freedom to be defined in a context of equality or in the (dark) context of inequality, and elitism, eg the aristocracy of intellect?

Is freedom about the freedom to know and use that knowledge in order to build, or is it about the freedom to steal from others and (then) coerce others so as to dominate them. Is law to be based on equality and creativity, or is it to be based on inequality and selfish domination, which is implied by the property rights of those who are thieves.

Scientific knowledge (or its criterion for its truth) should revolve around the pragmatic uses of the mathematical descriptions of existence. However, science is now used in society to identify an elite and authoritative "truth" which has very limited uses, and because this authoritative truth is not useful this means that it has greater relationship to illusion than to a widely applicable and pragmatic truth. Note: The science of Ptolemy was measurably verifiable, yet it was still an illusion.

Science is about determining knowledge within a context of equality, and scientific equality is best described as finding the best descriptive language in which to describe a measurable system's properties so the description can be, at least, measurably verified, but such a criterion of measurable verification is not demanding enough, because descriptions based on probability can always be modified to follow data (this is the true history of quantum physics) and Ptolemy's descriptions (of planetary motions) were measurably verified and they were (at first) more precise than the descriptions of Copernicus, thus the best descriptions, the descriptions which should be considered true, are those which are (1) measurably verifiable and (2) widely applicable and (3) very useful, (useful for the purpose of creating new systems or new instruments).

It should be noted that quantum physics, general relativity, string theory, and non-linear systems are not widely applicable to general systems, as classical physics has been widely applicable to general classical systems, nor have the theories of: quantum physics, general relativity, string theory, and non-linear systems, been all that useful (electron microscope, lasers, tunneling devices, ie a rather short list compared to the technical development related to classical physics).

If one analyzes (or compares) which scientific descriptions have been (and continue to be) the most useful descriptions for the purpose of developing technology, one sees that it is classical physics, which is the most dominant physical theory in terms of its usefulness. That is, classical physics is still the main driving force behind technical development. Thus the theories of quantum physics, general relativity, string theory, and non-linear systems should not be considered to be scientifically true rather they are speculations which have not shown themselves to be either widely applicable or all that useful.

If one accepts that quantum physics, general relativity, string theory, and non-linear systems are based on the true laws of physics, then the main statement that this theoretical structure makes is that "nearly all fundamental systems are too complicated to describe in a useful manner."

That is, the currently accepted theories of physics lead to the laboratories . . . to find the properties of the system in question because its descriptions, based on the currently accepted laws of physics, do not provide useful information about the system's properties, and instead this information needs to be found through measurement in the laboratory . . . and they do not lead to new inventions, as the descriptions of classical systems have led to rapid multi-various inventions.

Note: Most technologies which use quantum properties are about coupling highly controllable classical systems to quantum systems so that the quantum properties can be used in the system, eg the transistor, magnetic spin imaging, furthermore, nuclear bombs are based on 19th century models of chemical reactions (related to collision probabilities of atoms (or particles) etc.

Nonetheless these theories of: quantum physics, (particle physics), general relativity, string theory, and non-linear systems have been given inordinate amount of authority by society given their limited usefulness. This is because society is based on inequality, and authoritative knowledge (of absolutes) is one measure of social position, and the verification criterion of authoritative scientific knowledge is the same criterion as the authoritative age of Ptolemy, namely, it only requires measurable verification. Science has been made authoritative and elite in nature because the owners of society want to control scientific knowledge.

This is a property of a highly controlled society run for the selfish interests of the few owners of society, and it opposes free inquiry and it opposes the development of useful knowledge.

What is science?

Science is related to descriptive knowledge, where the descriptive knowledge is described in a precise measurable language.

Science is (perhaps) best described as being about the newness of descriptive knowledge and developing the relations that these descriptions have to creativity.

Physics is about:

> Containing spaces of material (positions and motions, where calculus operators inter-relate positions and motions of a material object) where materialism implies a particular dimensional relation for the domain space of a function,

> Measuring distinguishable types of properties,
> either
> local measures of a physical property represented by a (solution) function,
> or
> an oscillating (random event) property of a particle which possesses a particular discrete spectral value of a system being described, where the system is being described with a function space and a set of eigenvalues of the stable discrete definitive spectral-orbital system,

> Geometry (today, a general manifold geometry of an unstable non-linear shape, ie general relativity)

Discreteness (atomic hypothesis, as well as the observed stable discrete properties of a quantum system's energy structure), and

Randomness (it is assumed that quantum systems are fundamentally random in regard to both (1) finding particles, which have particular spectral properties, as random events in space and time, and (2) material interacts are modeled by means of random elementary particle-collisions)

Force and energy are related by calculus, by integrating force acting on a test particle along a path in the test particle's containing space,
Force is easier to think about for motions of a material object
Energy is easier to consider for thermal systems or in regard to continuous (or conserved) systems.

The main problem of physics today, but which is ignored, is about "how to relate a discrete system's observed properties to the assumed set of descriptive structures?" That is, the observed stable spectra-orbital properties cannot be derived from the descriptive laws for general systems, and this is true for all size scales. There is only one idea which is applied, wherein $1/r$ is used in an energy operator (potential energy operator) along with some estimated spectral cut-offs, are used to obtain all the possible spectra, but the cut-offs seem to "not do the job."

However, in some cases $1/r$ can be related to the value of the kinetic energy, but this would give a continuum of spectral energy values, and explicit relations of a spectral continuum to radial spectral values seems to evade these types of descriptive methods.

The angular momentum values for a spherically symmetric interaction give some crude estimates of a quantum system's spectra, but the many-particle quantum system usually does not have spherical symmetry.

The main idea in physics today (2012) is about how to determine stable properties from an assumed random context. This is logically backwards.
Because, hidden in the idea of quantity is the idea of stability, a hidden axiom.

Randomness is mathematically valid only when the random set of events is finite and each event is stable. Randomness is not used in this context in modern mathematics, and this can be the reason that math has become so ineffective at solving for fundamental general physical systems.

Thus, the logically consistent structure of physical-measurable description is to begin with stability and derive the properties of randomness from an assumed stable structure.
 This has already been done, but it is ignored.

What most opposes science is:

> social hierarchies,
> learning hierarchies,
> fixed authority, and
> a narrow view of creativity
> (ie the view of monopolistic economic control of knowledge and creativity, and
> society, ie central planning based on narrow selfish interests,

subsequently, knowledge becomes an institution of domination and authority so as to become rigidly fixed with its own ponderous authority, (yet) and servile to the creative interests of the rulers of society who want knowledge to serve only their own narrow interests, thus an aristocracy of intellect emerges which is:

> arrogant, and
> has become essentially useless in regard to creative development, and
> useless in regard to valid descriptions, and
> this aristocracy of intellect is in opposition to free and equal inquiry).

Science is about precise, or measurable, descriptions about the shapes and structures, or properties, of material systems, or descriptions of some more encompassing model of material existence.

The ability to measure (so as to have a measurable descriptive language) depends on defining numbers and defining their properties of order and their operations which seem to ultimately depend on identifying a uniform unit which measures a particular type of measuring quality, and then using this uniform unit to count, and thus to identify an ordered number system which can be used to measure and compare values of measurements of particular types of properties, but both the number-type and the size-scale, of the uniform unit can change based on "grouping" and/or multiplying (which is grouping) so that numerical comparisons can be consistent (with the order defined for numbers in the process of counting). In the development of the idea of number (or value) and its properties of order (based on counting) the process of counting also determines the math operation of addition (as well as subtraction), while the grouping of numbers (or identifying a unit's size, such as a fraction of a unit) determine the math operation of multiplying (and dividing).

The implied inverse operations of adding and multiplying are given (ie subtraction and division respectively).

In the terse algebraic axioms of algebra the notions of zero and one, ie the uniform unit of measuring, are identified as the "identity operators" of the "two" math operations (adding and multiplying). This does not identify the significance of these numbers (ie 1 and 0) in the definition of numbers, though zero seems to be more significant in regard to set theory, ie in regard to the empty set, but in numbers zero is a marker of where number-value begins and can define the two opposite (positive and negative) directions of a line (or of numbers).

It is in the elementary context of numbers and arithmetic that the only valid way in which to consistently relate different number systems with different measuring scales and this is by means of the linear property of multiplying either of the two of these different number scales by constants.

In science-math there is materialism related to either causal geometry (upon which essentially all technical development depends) or probability (related to quantum physics, with a very limited relation to technical development) wherein random particle-events (in space) are local but the eigenfunctions . . . , to which these random particle-events are associated as eigenvalues . . . , are global (or non-local). That is, a non-local relation exists between a random, local, spatial-material-particle-eigenvalue-event and the global system-based eigenfunction associated to its related random eigenvalue-particle-event.

Non-localness is observed for quantum systems, a property which can be interpreted as the existence of action-at-a-distance.

But the random eigenfunction structures of quantum physics only have validity if they have been defined by geometric relations which exist in the quantum system's defining operators, but these geometric structures within defining quantum operators seldom exist, eg the H-atom has a geometric (charge) structure in quantum physics but "in general" (many-electron) atoms do not have a well defined geometric (charge) structure.

Particle-physics is an inconsistent model of geometry (the particle-collision) and randomness, while general relativity is non-linear, and thus it is also inconsistent. These inconsistent descriptions have neither descriptive accuracy nor practical value associated to their descriptive properties, if, in this descriptive structure, any properties are actually described (at all), or (as often as not) if the system simply cannot be solved.

Therefore, one must question, how do stable structures for material systems emerge?

The idea that there is a mind and the idea that "the mind has a structure" which affects one's perception is a vague idea, (this is often) usually related to some vague notion about the structure of language, ie the way words are put-together or organized in phrases and sentences. But more specifically, in the context of precise descriptive languages one can focus on "descriptive knowledge" and consider the direct relation that assumptions have to descriptive knowledge, in relation to subjects or categories of thought, and how this knowledge relates to practical creativity, where things must be measured and fit-together so as to function in a particular way (the meaning of language is about a (precise) description's relation to practical creativity).

In the context of descriptive knowledge the basis for descriptions are . . . :

> assumptions (axioms) and
> definitions of words or concepts, as well as
> contexts (or containing sets) of a descriptive subject, as well as
> interpretations of observed patterns
> . . . , in regard to a subject of interest.

Thus it might be better to consider how assumptions affect descriptive languages, which in turn guide one's attention and perception, rather than to attribute knowledge to the structure of a mind, since a model of a mind has never been all that definitive and precise.

That is, it is clear that language confines the range of descriptive possibility.

The incompleteness theorem of Godel says that descriptions with a precise language (1) depend on assumptions, and (2) have a limit as to the types of patterns which the language can describe (before new assumptions are needed in order to be able to describe the [new] patterns which are being observed).

If one wants to claim that "perception is related to the properties of a mind," then one needs an idea of "what a mind is." Furthermore, one may not be able to describe the range of "the possibility for perception."

In fact, this is the point [what perceptions are possible?] of the more articulate religions (or more objective), such as Buddhism, where the point of one's struggle in life is to be able to perceive "the world as it really is," so that perception becomes more related to direct knowledge of the world (without a need for an interpretive language), where knowledge is about possessing a useful truth in regard to both intent and to creativity (in relation to one's actions in the context of their perceptions, "What is the range of one's actions, or the 'true range' of one's perceptions?"). or the true range of what one can measure.

That is, as good a model as any of a mind is that "both the external world and the internal (mental) world of a living person (or living system) can be in correspondence," so that perception is both being aware of this correspondence and intending (or using knowledge) is also about being and acting within the (that) same context within which one perceives.

Such a pattern of existence, where both the external world and the internal world of a living system can both be modeled as high-dimension discrete hyperbolic shapes (perhaps of adjacent dimensions), and where the lower dimensional discrete hyperbolic shape (the living system) would be of an odd-dimension and have an odd-number of holes in its shape (so that it would naturally oscillate and generate its own energy) where both shapes (the living system and its containing metric-space, which are stable discrete hyperbolic shapes) are (can be) represented as being related to spectral values (or spectral flows) upon the maximal tori which is contained in a high-dimension unitary fiber group (of (or over) the high-dimension complex-coordinates within which both the world and the living (perceiving) system are contained). This (very big and/ or highly ordered) spectral set (of complex systems) which exists on a maximal torus is central to the construct of perception, so that resonances can exist between the internal structure of a living system and the external structure of the existing world (where both spectral sets exist on the maximal torus) would certainly allow for resonances to exist between the lower-dimensional material and lower-dimensional metric-space structures which are (can be) contained within and comprise the two higher-dimensional systems (both the [energy generating] living system and its containing metric-space).

That is, western science is about descriptions of existence and these descriptions do depend on assumptions, but (unfortunately) western science is exclusively about an assumption of materialism, which would exclude, due to materialism's dimensional constraints, the type of model of a mind given in the above paragraph. But if pressed its ability to justify its belief in materialism has no foundation.

There is no reason to maintain the assumption of materialism, unless one is very attached to one's own material advantage within a model of the world which is based on materialism, ie existence is defined by material existence, as would be the economically monopolistic interests of western society. Materialism is a model of existence which supports a model of human society which is stuck in its traditions and it possesses social inertia which is of most use to the rulers of society. In fact, descriptive knowledge, in regard to precise measurable descriptions of patterns, would be both better able to describe the properties of the "so called" material world, as well as being much more interesting (humans creating outside the context of the material world, ie man's relation to the spirit could be developed) if "physical" science would break away from the basic assumption of materialism, ie "Is 'what we talk about in a precise manner' always dependent on the shapes of material systems and the properties of physical systems which are conveyed by material?" No. Material could be stable (discrete) shapes which are closed metric-spaces, in turn, trapped in a dimensional level due to its own dimensional structure.
This opens-up a new realm of measuring (new possibilities for measuring and creating).

In order to be consistent with accepted physical law it is assumed that particular types of differential operators should be used and placed in certain descriptive contexts, eg classical physics is related to geometry while quantum physics is related to function spaces and randomness, and general relativity is related to non-linearity, but this needs to be reflected upon and analyzed in new ways.

When satellites are guided through the solar system it is Newton's laws and action-at-a-distance, which are still used. That is, when there exist great distances between interacting material satellites in the solar system then the local causal space-time model of gravity, as given by general relativity, is not used to model "how to control a satellite's motion in the solar system," rather action-at-a-distance is used.

While A Aspect's experiments have confirmed that quantum systems have (possess) the property of action-at-a-distance (in Euclidean space), ie non-locality.

One might ask, are certain types of operator properties related to the space within which the physical property (being modeled by the operation) is contained?

Do some physical operations (or physical properties) belong to one type of metric-space and other physical properties belonging to other types of metric-spaces?

For example, the property of position belonging to Euclidean space, while the property of (stable) energy belonging to space-time space (or hyperbolic space).

On the other hand, descriptions based on randomness either particle-waves or the particle-collisions of particle-physics have the limitations of not being accurate in regard to describing the properties of general quantum systems, as well as not providing a practically useful context for technical development (or for technical creativity).

The development of quantitative descriptions based on indefinable randomness and non-linearity (where both general relativity and particle-physics are non-linear models of material-energy systems), have not been capable of describing the observed stable discrete definitive spectral-orbital properties of quantum systems (or for all [any] material systems at all size scales, ie "why does a general nucleus have its stable spectral structures?" "why is the solar system stable?"), and the meager sets of descriptions which exist for quantum systems have not been related to technical development in any significant way.

Induction, supposedly the logical basis for science, requires either over-whelming accuracy for many systems or over-whelming relation to useful development, but neither of these conditions is met by modern physics.

However, when the descriptive language is correct, then one sees wide-spread technical development occurring almost immediately. If technical development is not significant, in regard to a descriptive language, then the descriptive structures being used to describe the properties of material systems must be wrong (at least this is the best conclusion which one can make as a scientist).

The claim quantum physicists might make is that the observed phenomenon of quantum physics does not allow a useful descriptive language.

However, the quantum systems that the laws of quantum physics are supposed to be able to describe are

stable definitive and discrete, and they formed from material interactions. But the properties of these general quantum systems have not been described by using the laws of quantum physics.

If these interactions were so dependent on the inaccurate descriptive language and associated physical structure described by the laws of quantum physics, ie if the descriptive language of quantum physics is both accurate and unique, then there is no reason to believe that some general atom would possess a spectral structure which could be used to identify it from some other type of atom, yet in the interactions associated to the particular (but general) quantum system always lead to the same type of quantum system with the same type of identifiable spectra, a property which the currently accepted descriptive language of quantum physics cannot describe.

This means

There is a descriptive context in which material order emerges in an essentially unbounded and "open" spatial-material structure (or context).

What is that context?

Math and physical law

Math is about both (1) "identifying the patterns of shapes (of physical systems) and numbers (or measurable properties)" as well as (2) using numbers and math operations to identify (both functions and) equations, and then solving these equations.

If the equation is a (linear, solvable) differential equation then the solution is a function. However, in algebra, finding a formula of a function is about identifying the math operations which are used to form a function's formula (and relating them to measured values of physical properties). Either finding a formula algebraically by measuring average physical measured values or solving a physical system's differential equation distinguishes the "scientific methods" of Boyle vs. Newton (where Boyle experimentally discovered the thermal formula of $PV=kNT$).

In physical theory, math is used to determine an equation based on the properties of a measured property's relation to the system's containing coordinates and the geometric measures on these coordinates, where one wants to determine a function's values in relation to the geometric measures of the (containing) coordinates by means of "local measurements," in turn, related to some measurable property, eg geometric properties or spectral-energy properties as related to functions (or to measurable values of the containing coordinate space). If there is a differential equation being used to model a physical system, then the equation depends on a local measuring process, for which one wants to find an inverse operation "so that the equation can be solved," so as to find a solution function, where the solution function provides a global set of measured properties for the function's values (or the measured property) defined on the entire set of coordinates (see next paragraph).

It has been thought that non-linear equations and their solution functions can be defined in a local manner, and then pieced together, but this would depend on quantitative inconsistencies between the functions-values and the geometric measures on the domain space (due to non-linearity), and (but) these inconsistencies cannot be resolved.

Subsequently, the information (built upon these non-linear inconsistencies) is not useful, ie the global solution is necessarily to be given in a linear, geometrically separable, and metric-invariant context wherein the metric-functions can only have constant coefficients (where the solution function is defined globally, if the function is defined in an (apparently unbounded and) open context of a containing metric-space of coordinates).

This is the ultimate reason that the very stable discrete hyperbolic shapes [which exist over a hyperbolic dimensional range from 1-dimension to 10-dimensions] hold the key to the properties of the physical world (in a many-dimensional context, as a containing space) which are both mathematically represent-able, solvable, and physically stable, and the information (of the global solution) useful for both "system control" and for further creative relations of the system to other aspects of the world context.

Calculus and the classical physics of differential-forms where the relatively simple relations that measurable properties (or a function's values) of a physical system have to geometric measures of the function's domain space, based on the relation that local linear measurements' of a function's values have to local (linear) geometric measures on the domain space. The math structure of differential forms (tend to) tie-together the solutions differential equations (defined on differential forms) to the boundary properties of the region upon which the differential equation is defined. The discrete shapes of both Euclidean space and hyperbolic space are differential forms.

Discrete hyperbolic shapes, and discrete Euclidean shapes (ie space-forms) are linear, separable geometries of differential forms, which have holes in their shapes, thus they are simple stable models of the (but unstable) math ideals of great generality.

Perhaps the question should not be about considering holes in space (or holes in shapes) in some general context, rather the question should be "does the idea of a hole in a shape (or in a space) have any validity in an unstable and inconsistent descriptive context of the containing structures of general math considerations?" That is, holes in space require a model of space which is not simply "local," rather it must be about functions which possess more than a local definition. Does this discussion about "holes in space" make any sense if the functions involved in identifying the property of being a "hole in a space" are not in fact globally defined functions, ie defined on the entire containing space of the space (or shape), as opposed to only "partially non-local functions" (or functions defined on small regions of the function's domain space and/or pieced-together)?

An important question in math is about "the value of generality vs. specific structures."

When are specific math structures dominant to the (a) description . . . , eg discrete hyperbolic shapes are central to stable descriptions and stable physical systems, and . . . , when is "math generality" of value? One context in which specific structures are more important than general structures seems to be when one is concerned about the property of stability of both descriptive math structures as well as the stability of physical structures.

Simply identifying the useful context of classical physics because it is both accurate (within the range of macroscopic phenomenon) and very useful . . . ,

Since essentially all of technical development of our modern age is (all) about developing the ideas of classical physics, though sometimes classical systems can be made to couple to the properties of quantum systems, but this has essentially nothing to do with quantum descriptions and depends only on the qualitatively observed properties of quantum systems, where the precision of the coupling (between classical and quantum systems) may depend on experimentally determining the energy range of the quantum properties (since quantum calculations can not describe the energy properties of general quantum systems).

. . . , and thus the math context of these classical descriptions should qualify as valid topics for serious discussion in the sciences (about the best structure of scientific descriptions), since the logical basis for science is induction, thus new interpretations of a traditional language may well lead to both a more accurate and more useful descriptive structures.

That is, science is not (supposed to be) based on formal assumptions (validation of inductive descriptions depend on repetitions, as well as valid descriptions based on (successfully) applying the laws of physics so as to provide the accurate descriptions of many general systems, and the information provided also needs to lead to a great amount of technical development) and thus there are many topics about science which could well be very relevant to discussions about the best structure for science descriptions, yet they are excluded from the (very much overly) authoritative professional sciences, which has taken on a character of dogma, where physical science is now being based on deducing from sets of assumptions or from the laws of physics, which is the new (deductive) position which the "so called" laws of physics have now taken on in professional science.

That is, "peer review" of the professional science and math journals is about an inside (or authoritative) vision, an overly narrow vision, about how the observed patterns of the world should be described, so as to (not serve creativity in general) but rather to serve military industrial interests. Science is now socially structured so as to serve narrow social interests, rather than science being about the mystery of existence, and science having an interest in the many possible ways in which to describe that existence, in an ultimately practically useful manner, in regard to a wide range of creative possibilities.

These are science and math issues which are about both logic and assumptions, ie "where does one begin a descriptive structure?"

That is, making considerations about, "How the (descriptive) foundations of a scientific subject should be determined?"

Should some arbitrary context (or containment set, such as the idea of materialism) and subsequent interpretations of observed patterns be sufficient to determine the foundations of descriptive languages.

A context and an interpretation could be in regard to contrasting the beginning viewpoints of existence being:

> either
> material, space, and time (Euclidean space, ie Newtonian descriptions), or material and space-time, (space-time or hyperbolic space), but still in 3-spatial dimensions,
> or
> it is about a high-dimension, hyperbolic shapes, as well as naturally oscillating hyperbolic shapes, ie an alternative model of existence which might be interpreted as a mind.

Whereas math is more formally about axioms and definitions, and subsequently, the deduced descriptive patterns or theorems.

Furthermore, math is formally (traditionally) about quantities and shapes but some important issues about quantitative (or measurable) descriptions go ignored, namely, those issues about the basis for consistency and stability of description, including issues about "set size," whereas stability of quantitative descriptive structures may very well depend on the stability of shapes.

However, there are other "less important" sides to science such as geology and earth science and biological science, where some of these sciences, such as geology, allow a wide range of ideas to be expressed so as to try to describe the observed patterns of the earth. Whereas biology, though it should have a wide range of ideas to consider, nonetheless biology stays fixed to the idea of mutation of DNA and survival of the fittest naturally selects the strong individuals and this process controls biological development even though the control that DNA has over development cannot be described and supported by observed patterns of biological systems. The idea of evolution, mutation and natural selection (of the strongest) supports the idea that competition and the domination of the many others by the strongest should be the model by which society should be organized.

More explicit descriptions of science and math

The language of probability cannot be used to describe the properties of a system, so that the described properties have any relation to useful creativity, due to either control of the system or due to the correct information which a calculable description (based in the probability of random spectral events) might provide.

The probability based descriptions of quantum physics provide neither of these qualities (neither control nor accurate calculations of a quantum system's spectra).

General relativity seems to be irrelevant since it only deals with a one-body problem, and it is non-linear, thus not useful, yet the many-bodied solar system appears to be stable for billions of years.

Einstein's contribution seems to mainly be the statements that: (1) electromagnetism is contained in a metric-invariant coordinate space of space-time, and (2) due to E Noether's displacement invariances time is associated to energy, and subsequently in relation to space-time (or momentum-energy) coordinates an adjusted function related to mass (where mass is associated to a stable electromagnetic structure) is associated to the time variable and thus it is associated to energy, thus mass equals energy, and (3) when light is placed in a spherically symmetric space-time coordinate system, as identified by (say) Schwarzschild, then light follows slightly spherical curves.

In the precise language of math, it is the containing (identifiable) sets which identify systems, which are to be described based on measurable properties, and thus maps: from these system sets to quantitative sets, give the system quantitative properties, so that the measurable values of the "key properties" of a system's description are defined by further:

> Maps (or functions),
> convergences, and
> limits in regard to function's values (Note: Needed to define derivatives and integrals),
> converging series (eg power series),
> converging sequences of functions, and
> distribution (functions), etc.

This context of quantitative description, related mostly to properties of convergence, exists on very large sets, ie sets which have a large number of elements. In such large sets the type and value of numbers (can) come to have ambiguously identified properties. That is, the context of descriptive stability seems to need a new basis. That is, continually adjusting the context of convergence within very large sets seems to become an elaborate mechanism for data fitting, similar to the structure of Ptolemy's epicycles.

In classical physics it is:

> Materialism,
> The properties concerning a material object's position, velocity (or motion), and
> A local linear relation between measuring motion and geometric properties in relation to the moving object, so that these structures identify a material system's differential equations.

That is, classical physics resulted from the definition of both calculus and the differential equation, where the idea was that local linear (or consistent) models of measurement (which could be represented as vectors) could be equated to linear (or vector) models of geometric properties. This changed, with M Faraday's description of electromagnetism, to the calculus of differential forms, which also apply to thermal physics.

But the only useful part of this structure of classical physics are related to the properties of linear differential equations, defined on separable geometries, in a metric-invariant coordinate space, where the metric-functions have constant coefficients, ie spaces of constant curvature.

Nonetheless, this useful part of classical physics contains all the relationships "which science has" in regard to technical development, in all of western society over the last 200 or 300 years.

The technical development of material systems has been, and still is, almost entirely dependent on the knowledge of classical physics (only discreteness of state, and the Bose-Fermi-ness properties of system components which fill the system's energy states, are new quantum properties in regard to useful material description).

The useful part of classical physics is about defining inertia (or mass) and applying linear differential forms to the description of material (both mass and charged material) defined on metric-invariant coordinate spaces, where the metric-functions have constant coefficients, and the material geometry is separable, and the dimension of the coordinate spaces is determined by materialism.

The use of the idea of materialism to determine the dimension of the containing space "of the system properties which do exist" is unnecessary. Material systems or material components defined as space-forms (of various dimensions) in a many-dimensional context, where material (or space-form) interactions are still defined by differential forms, but now in a many-dimensional context. Bose components fit into metric-space system structures (integral spin) while Fermi components fit into material space-form system structures (½-integral spin) (Note: This general idea about Bosons and Fermions includes the idea of super-symmetry because the metric-space becomes a material space-form in higher dimensions).

In quantum physics the description is based on the (apparent) randomness of material reduced to random particle-spectral events in space and time, where the descriptive structure is linear and based on metric-invariant coordinates whose metric-functions have constant coefficients. A quantum system is a function space, and its spectral representation, where the random events are the random spectral values of observed particle-spectral properties associated to a quantum system, and the measurable properties of the quantum system are related to a set of linear (complete, commuting) operators which act on the quantum system's function space.

The function space of "say" the quantum property of particle-spectral position is related to a dual space determined by a Fourier transform so that the properties of variance defined on both the dual space and the original space define an uncertainty principle, where the consequence of this uncertainty principle is that a piece of material defined at a particular point will have greatly uncertain momentum properties which will destroy the material geometry of which the material (of the quantum system) is a part. That is, though the quantum system's spectra is defined by an energy operator so that the geometric constraints . . . , which the energy operator identifies (by means of the potential energy) . . . , determine the quantum system's spectral values, nonetheless a direct geometric relation between a quantum material system and its geometry is not allowed in the descriptive structure of quantum physics. This is due to the uncertainty principle and the fundamental randomness which quantum physics assumes.

The fundamental discrete spectral properties which quantum physics identifies (or upon which quantum physics depends) are stable, discrete, definitive spectral properties of quantum systems, which are properties which are contained in a linear description, defined on a metric-invariant coordinate space, where the metric-function has constant coefficients, and the two distinct types of material components which fit into the discrete spectral structures of quantum systems are the Bosons (many particles per spectral-level) and the Fermions (only one particle per spectral-level [or only two particles in an approximate spectral-level]). Furthermore, rotation properties of a quantum system are important in spherically symmetric systems, or in systems composed of atoms, including the spin representation properties of rotations. Note: Within the context of spin the Bose and Fermion components, which fill quantum (energy) states, are defined.

Unfortunately, the fundamental principle of randomness for quantum systems seems to require that the spectral structures of quantum systems cannot be determined by mathematical methods. That is, the spectral structure of a quantum system seems to depend only on the geometry of the system's potential energy term, but the geometry of the potential energy term cannot be identified in relation to the geometry of the components which compose the quantum system (both the structure of quantum description and the uncertainty principle do not allow this to be possible), nonetheless the geometry of the potential energy term does depend on the geometry of the components which compose the quantum system (but this is not allowed), thus there is a logical gap in quantum description which has not been (or cannot be) bridged. Using (or finding) sets of linear operators to act on a function space so as to find the set of spectral functions for the function space has not been capable of finding the observed spectral sets of quantum systems. One of the main problems is that in many-component quantum systems (composed of charges) there are many $1/r$ singularities which interfere with the convergence properties of the math models.

Particle physics says that all material interactions reduce to particle-collisions of particular types of elementary particles, and these elementary particles identify families of particles, where particle-families are associated to energy levels, and where the families of particles are modeled as inner-particle-states which define new (mathematical) dimensions (or particle-states define dimensions beyond the dimensions of the material-containing coordinate dimensions). During a particle-collision interaction the particle-states of particles are changed during the collisions, where the changes are governed by non-linear connections, so that these non-linear structures are imposed on linear wave-functions, whose spectral values are being adjusted by the particle-collision quantum interaction (math) process. However, "tying together" the non-linear properties of particle-collisions with the original linear properties of distinct quantum systems seems to not make sense, where particle-states of particle-collisions are only vaguely related to the ranges of energy levels which the (original) quantum system possesses (before the non-linear particle-collisions adjust the original quantum systems linear energy structures), and then these "new," point-defined, "particle-collision properties" (or wave-function adjustments) are to be extended out to a "global wave-function," . . . (where these particle-collision wave-function adjustments are defined at solitary points of the wave-function's domain space so that these single points of the domain are associated to a wave-function's solitary value) . . . , by means of function sequences and subsequent function convergences which make no sense. How does one attach a general, non-linear particle-collision (model of general material interaction) . . . whose properties are random (due to non-linearity) rather than systematic . . . to a specific linear, global, wave-function of the infinite-number of global wave-functions which share the same point of space which is the site of the particle-collisions?* Answer: There is no obvious (causal) inter-relationship between global wave-functions and random elementary-particle-collisions at a specific point. Particle physics is a probability based description and in this context it is really an elaborate epicycle structure whose only purpose is to try to fit data in a complex math process, which is incoherent* and has no relation to useful applications.

The fundamental properties upon which particle physics depends (or which particle physics identifies) are unitary-ness, ie containment of the system's description in complex coordinates, and many-dimensions . . . , [as well as the many layers of different containment spaces which compose (or contain) the descriptive properties of a physical system]* . . . , which particle physics claims are the new dimensions of the internal symmetries of elementary particles . . . , but it is a unnecessary to relegate the higher dimensions, whose attributes seem to suddenly and discontinuously appear in high energy particle-collision (experiments), to the internal properties of elementary particles, and this assumption which attributes the new particle structures (which occur at collision vertices (that is, vertices [of discontinuously appearing new particles] which can be attributed to a set of hidden higher dimensional structures)) is an effort which is motivated by the desire to maintain the ideas of both materialism "along with the need for data fitting."

The derived theories such as string theory, grand unification, etc are equally useless descriptive structures, as are the useless descriptive structures of quantum physics, general relativity and particle physics (which are all useless descriptive structures, in that they do not describe observed properties, eg the spectral properties of general nuclei, and they have no relation to creative development, they simply fit data).

The failing of science

In these descriptive languages, ie quantum physics, general relativity, and particle physics, there is no basis for describing the stable, discrete, definitive spectral-orbital properties of the general nuclei, general atoms (even the radial equation of the H-atom diverges), molecules, crystals . . . , ie the containment in a box made-up of infinite-potential walls is the simple fundamental, but unreal, approximation upon which the discrete crystal spectral properties are based.

That is, these descriptive structures (quantum physics, general relativity, and particle physics, etc), which are the authoritative "absolute truths" of the physical sciences (upon which the contests [whose purpose is to filter-out from the public our society's most capable intellects] of our authoritative educational system are based), do not work at any level of simplicity or complexity.

The arrogant claim of the authorities of science is that, in regard to the "laws of physics," the fundamental physical systems which compose our world are simply too complicated to describe in a useful manner, but nonetheless these experts (who are incapable of describing anything of practical value) are held-up by society as superior people, because they have memorized (but clearly do not comprehend) all the complicated irrelevant descriptive math structures of these useless descriptive structures such as quantum physics, general relativity, particle physics, string theory, grand unification, quantum gravity, etc, ie these descriptive structures do not describe any actual system and they do not lead to technical creative development.

This seems to suggest that the experts of society's educational institutions are pre-dominantly a competitive set of autistic types who love to have authority, but whose understanding of the intent of the descriptions of science over which they are the experts, ie the descriptions over which they are responsible, is only a surface understanding, and they never question assumptions in any meaningful way. That is, these people can only increase the complexity and they do not even consider that the descriptions are supposed to be both useful and widely applicable.

Nonetheless if the fundamental authoritarian assumptions of our society are questioned then the owners of the society loses control of what has high value, ie icons of high value, in the media.

*That is, the same problems in regard to communication which exist for politics and the structure of power in society, also affect (in a very deep way) all aspects of the knowledge which a society formulates and uses. Thus, the institutions which deal with knowledge and creativity and their relation to either equality or selfishness (through the law) are the institutions which most affect the changes of a society. That is a sense of a high value for a science which has limited relationship to creativity is a mis-placed sense of value.

Chapter 12

SCIENCE AND RELIGION

Illusion, religion, and scientific authority

People would rather believe high sounding authority than believe that these authorities could be following illusions. But authority whose criterion for truth is similar to the science of Ptolemy, namely simple measurable verification of a description, is most likely following an illusion. Modern physics determines truth by verifying measurements, but a description should only be considered true if it is widely applicable and very useful, which identifies classical physics as being a very useful but limited truth.

If a precisely determined language is used in a fixed (unchangeable) language structure and its descriptions of existence are neither widely applicable nor (very) useful, then such a fixed description must be considered to be based on illusion and not based on truth.

Such limited truths, eg general relativity, are upheld by authority, and an authoritative truth is not a valid scientific truth.

When descriptions of existence are neither widely applicable nor (very) useful then it is (or should be) within the nature of science to put forth alternative descriptive languages which might lead a descriptive structure to have greater usefulness. It is exactly such an alternative description of the planets orbiting the sun which Copernicus provided as an alternative descriptive language to the "earth centered descriptions of planetary orbits which are associated to Ptolemy," and this new alternative descriptive language of Copernicus, became the beginning of classical physics.

Ptolemy's descriptions of planetary orbits were both authoritative and measurably verified descriptions which were supporting the authoritative illusions of the church.

Today's overly authoritative and measurably verified scientific descriptions of (an assumed material) existence given by science institutions, in particular provided by physics and mathematics, are neither widely applicable nor useful, ie they are illusions and, in this, they are no different from the illusions of religion.

The (non-useful) descriptions of physics and math are supporting the idea of authority and elitism, and have the same qualities of illusion as the descriptions of religion, which religions put forth in "the vague descriptive languages of religions," ie the words used in religious description do not have precise meaning subsequently religious descriptions never have any practical usefulness.

Today the verified illusions of precise language (just as the illusionary descriptions of Ptolemy were measurably verified) are supporting authority and elitism and subsequently the destruction of science, the earth, and earth's creatively intentioned people.

If the determination of scientific truth is based on wide applicability and usefulness of a precise descriptive language then science must also require that the act of finding alternative languages, for non-useful languages, to be a central goal of science institutions.

When descriptions of existence are neither widely applicable nor (very) useful then it is best to think of these descriptions as not being verified (or not being true) even if these descriptions are measurably verified. The illusions of Ptolemy were measurably verified.

Knowledge about existence is the basis for creativity, thus the freedom to describe existence must be based on equality so that wide applicability and great usefulness in a creative context determine truth, not simply measurable verifications, since measurable verifications can lead to elitism and authority, as the example of Ptolemy so clearly shows, and in this context "authority becomes the same as knowledge," and then "'knowledge' becomes illusion."

Using the criterion for the truth of a description about existence to be "wide applicability and great usefulness" then this supports equality, and then freedom can be defined as the requirement that all alternative descriptions of existence presented in the language of mathematics be published and presented in public institutions of science. Equality, and "freedom based on equality" are directly related to a descriptive (or language) based knowledge which is given in a precise (or mathematical) language, so that only widely applicable and (very) useful descriptions identify the truth of a particular description, and thus determine the (truth) value of the language upon which the description of existence is based.

Otherwise, descriptions which are not widely applicable, and which are not useful, are descriptions of illusions. Claiming that such descriptions of an illusionary world be considered to be true is a way of supporting both authority and the hierarchy of religion and elitism, ie selfish domination of the few (who, it is claimed, know the truth) over the many.

Is following the description of an illusion (though related to a form or a described pattern, which may or may not be verified by measurement) any different from following the authoritative illusion of religion? Answer: No.

For example, consider descriptive patterns which have not yet shown themselves to be both widely applicable and have not yet shown themselves to be (very) useful, that is, consider the following basis for (or beliefs in) illusionary descriptions which are (nonetheless) associated with authoritative truth:

(1) the geometry of non-linear dynamics (a description which is dominated by randomness [due to chaos] and which is sensitive to boundary conditions). [Thus, the only definitive geometry of such a non-linear description is the geometry of the critical points of the solution of a non-linear differential equation, and the subsequent effects of how these critical points influence the chaotic dynamics.]
(2) believing in the importance of diffeomorphic local coordinate transformations. (Whereas, metric invariance may be more important than general metrics in describing the properties of existence, in the context of containing coordinates upon which a general metric function has been defined. Discreteness in the context of metric invariance is [may be] more important than the property of local inversion of locally measurable properties (or locally measurable [coordinate] functions.)
(3) believing that probability based on linear operators (differential and a few multiplicative operators) acting on function spaces, which represent quantum systems . . . , (so as to determine a spectral set of functions in the function space) . . . . is the only way in which to describe the elementary event space of discrete random systems assumed to be composed of (or reduced to) elementary particles [but the actual elementary event spaces for quantum systems found by using these techniques are almost non-existent (or difficult to come by, in the sense of finding a Hamiltonian which fit's the data)]. (The language of function spaces fits into the descriptions of analysis, and analysis fits into the context of smoothness in relation to solving differential equations. That is, descriptions based on function spaces are mostly descriptions of smooth systems rather than being descriptions of discrete systems.)

Because these three descriptive languages are not (both) widely applicable and very useful, in the context in which they are being used, this (then) means that these three different types of descriptive languages are describing illusions, because they are not describing existence in a useful way.

Note: Discrete mathematical systems [or structures] can be used to describe smoothness, but not vise versa, yet the "hard to describe" properties of the world are properties of discreteness. Perhaps discrete mathematical patterns should be used to describe discrete [or quantum] physical systems.

That is, the basis for questioning the validity of the authoritative descriptions in relation to . . . . ; non-linear differential equations, coordinate transformations based on diffeomorphisms, and function space descriptions based on finding the spectral function of (observed) existence . . . , is that they are not (both) widely applicable nor are they (very) useful.

Each of these three bases for physical description seem to require unique mathematical techniques for each different system they are used to describe (if a description is actually forthcoming), as opposed to widely applicable principles for a system's description which classical physics provides (namely, the principles used to define a physical [classical] system's differential equations), but the main basis for the criticism, of these three bases for description of observed existence, is their lack of usefulness.

Except for non-linear dynamical systems which have (real time) feedback, these three, difficult to obtain, descriptions of individual systems have not been useful, or their best use is to manipulate their descriptive basis so as to fit data, eg changing the elementary event space or changing boundary conditions.

It should be noted that quantum physics has led to only a short list of technical developments:

(1) the electron microscope,
(2) tunneling devices (the tunneling microscope seems to have eclipsed the electron microscope),
(3) some geometric-chemical understanding of the H-atom, and
(4) the laser,
(5) . . . . otherwise technical uses of quantum systems is mostly related to coupling a highly controllable linear classical system to the discrete properties of quantum systems, eg transistors and magnetic spin resonance imaging etc.

This is a short and limited list of technical achievements for the "scientific" description of quantum physics which has been around for over one-hundred years.

That is, most technical development is still dependent on the descriptions of classical physics as opposed to other physical descriptions.

It is not known how to mathematically describe either the orbits of (say) the oxygen nucleus, or the atomic (electron) orbits of the oxygen atom, with the precision of (say) the Schrodinger description for the H-atom, nor can the spectral properties of crystals be described mathematically with any precision using the techniques of quantum physics, eg the critical temperature for superconductivity predicted by BCS has been exceeded and a valid alternative description has not been found.

A description of "How an electric current running through a crystal can be coupled to the spin properties that exist at the crystal's lattice sites" cannot be determined with the mathematical structures which are now being used, ie particle physics.

The description of some such crystalline-lattice site properties are needed to develop quantum computing.

The stability of the planetary orbits of the solar system have (also) not been described.

And general relativity only describes a 1-body problem with spherical symmetry, with limited uses, ie it has nothing to say about the stability of the solar system.

The stable, definitive orbital properties of matter at all size scales is the fundamental property upon which our experience of the world depends, and none of these structures have been adequately described so that the descriptions provide useful information in either a technical or engineering context.

The existence of stable definitive spectral (or orbital) properties suggests that the description of such properties should be related to linear differential equations defined on "separable geometries" in a metric invariant context, where it should be noted that "separable geometries" are curved coordinates where at each point, in the curved coordinate space, all the tangent directions of the coordinate curves are pair-wise perpendicular to one another, thus allowing the differential equation (used to model the system) to be separated into separate differential equations for each independent (or orthogonal) coordinate, and then solved independently, each in one variable.

The stable and definitive orbital properties, at all size scales, are the central properties of the "observed" properties of the existence we experience (and the properties which we want to use for creative purpose). Finding the general principles through which these stable definitive orbital properties can be described so as to have the widest range of (creative) use, is the intent of knowledge (or truth).

The point of life is to break free of illusion and then realize one's power to create (in a selfless manner).

Illusion is used by selfish people to dominate others whom are confused and whom let others decide things for them, decisions about things which they should decide for themselves, things such as, "What is true about existence?" This is the point of both cultural knowledge and of education based on free inquiry for a free people. Whereas today the education system encourages the belief in the idea that only experts have the necessary knowledge and the necessary mental capacity that is needed to determine what is true.

The function of today's (7-09) education system is to filter out from the public those people with the "necessary mental capacity to determine what is true." The filter system is based on competition which is defined around a fixed and authoritarian descriptive language. This language includes the mathematical patterns of non-linearity, quantum physics, general relativity, and string theory. As well as associated illusionary patterns in mathematics: diffeomorphisms, and improperly defined probability structures etc. Those who are supposed to be capable of determining truth are engaging in a competition (or a filtering process) based on illusions. No wonder they lie and mislead, in turn, for their social rewards.

It is claimed that the European empire was possible because of superior scientific knowledge, and subsequently superior technology, but this was because the science was useful, at least useful in making the tools of war, but today science is no longer useful but now represents a social class which defends illusions and authority, yet its duty is still "to make the tools and communication systems of war." this limited duty results in the limitation of science.

Authority which follows illusions is quite similar to religious authority, and reverence for the high illusions of academic authority is used in very much the same way as religion is used to control people, by getting people to adhere to a language (if they do not then they cannot win the contest) which is used in very limited and controlled ways within society.

This opposes equality, and it opposes the freedom of the individual, and is a form of state sponsored religion. Belief in the undetectable quark and belief in the anthropic principle are not beliefs about science but rather they are religious beliefs, similar to beliefs in an undetectable god and a belief in the book of genesis.

Focusing education on authority rather than widely applicable and (very) useful descriptive languages encourages both the notion of dependence and the need for wage-slavery in society. That is, it limits knowledge so much that creative development is stifled so that only a few (oligarchs) are allowed to direct what things the society creates.

Whereas creativity is about having knowledge which enables one to create this is best done in the context of free inquiry and universal ownership of a society's natural resources.

Classical physics

is defined to be smooth, and function spaces are defined within this context of smoothness.

Local measurability of one property (of material) related, by equality, to a geometric property (of material) so that both properties can be related to some given point in the system's containing coordinates (or containing space), is the basic principle of classical physics. This principle describes how differential equations for a classical physical system can be formulated.

However, the local operators (which determine local measurability), and their inverse operators (ie derivatives and integrals), are (also) defined on function spaces. That is, the function space description of derivative and integral operators is a different context from the derivative and its inverse operator, being described in the geometric context of a domain space and the set which a function's value defines, or equivalently, of a material containing metric-space (or coordinate space) and the properties (as functions) which the material might possess. Where it should be noted that these operators (ie derivatives and integrals), are defined (or their development motivated) by geometric properties within the coordinate spaces on which these operators (of calculus) are defined.

Quantum physics

Function spaces, in particular the $4L^2$ spaces (or Hilbert spaces), may be associated to spectral (or harmonic) properties and probabilities (for continuous random variables). Differential and multiplicative operators (ie smooth and continuous operators) can be used to act on function spaces so as to determine a set of spectral functions for the function space and thus to determine a system's spectra. The spectra also determine the assumed particle-event probabilities of (usually) small material systems (which are assumed to be composed of elementary particles).

But extending differential operators to act on function spaces in order to find a system's spectral-probabilistic properties has not been useful. The probabilistic sample space of spectral properties [or spectral events] cannot, in general, be found by these methods.

Note: The low dimension unitary transformations (beyond the unitary transformation structure of the Schrodinger operator acting on the quantum system's function space) of a quantum system's wave-function (single point function representation, along with internal dimensions) is supposed to correctly identify the complete sample space (of elementary particles) for all quantum systems, where it is assumed that all material systems can be reduced to elementary particles. The unitary invariance of the wave-function for low dimension unitary group transformations is supposed to identify the absolute laws of physics for all material in low energy systems. But these laws are not sufficient to describe the orbital properties of a general nucleus. If these absolute laws cannot describe the properties of the nucleus then they are not useful, and they are simply illusions.

These spectral-probability methods of quantum physics have not been widely applicable, each system has its own function space or spectral structure which are found by unique mathematical techniques (most often experimental), as opposed to being found by general principles such as identifying and solving a differential equation, or systems of differential equations (as is done in classical physics). The Hamiltonians for general quantum systems seem to elude description, ie the laws of quantum physics are not sufficient to describe the properties of quantum systems.

It is claimed that the fundamental methods of quantum physics are always about formulating a quantum system's Schrodinger equation (or Hamiltonian), but apparently there are no general principles which can be used to determine each system's Schrodinger equation, whereas in classical physics finding a system's differential equation has not been the main obstacle in developing the classical description of physical systems, in many very useful cases.

Can the full set of events for a quantum (or random) system be found using function space techniques, and are these events well defined, ie are these events identified by well defined states (like the heads and tails of tossed coins)? That is, are all the particles and their energy states which are supposed to define the system well defined events (or well defined entities)? It is not clear the quantum description can answer these questions, and often the elementary event space is arbitrarily expanded by scientists when new data is found for some particular quantum system. This is similar to the epicycle descriptive structure (or process) of Ptolemy.

Non-linearity

The non-linear aspects of physical description:
In non-linear descriptions of physical systems, local measurability is being related (through equality) to geometry. Non-linear descriptions have not been widely applicable so as to be useful because non-linearity (usually) implies probabilistic or chaotic motions of a system's components, though it is formulated in a more or less geometric context, ie since both linear and non-linear differential equations for physical systems are defined in a geometric context.

Furthermore, in the non-linear context, adjusting the boundary conditions (for such a non-linear system) can be used to change the solution, and thus adjusting boundary conditions can be used to adjust the (non-linear) description so as to fit data.

This adjustment of the/a description is typical for random systems which are not about causes, but rather random systems are about finding probabilities and the statistics (means and variances) for the full set of events of a random system.

Though there are a few (rare) examples of non-linear differential equations having solutions which are well defined manifolds (or geometries).

If non-linear patterns are so fundamental then it is the duty of the non-linear advocates to actually provide the precise models which can be found to be so useful, eg the actual non-linear structure of the stable solar system, etc.

However, providing separate and unique non-linear descriptions for each separate system is no longer science in the sense of classical science. Where classical science requires principles which are widely applicable to (nearly) all systems whose subsequent descriptions (based on these principles) result in great usefulness.

That most physical systems are definitive, discrete, spectral systems suggests that such widely applicable principles exist which can be shown to be very useful, and these principles should be related to linear, separable differential equations, since such equations characterize stable, definitive physical systems.

Does the society want to do classical science and find useful descriptions of existence, or does it prefer authoritarian based truths, truths which are no longer useful but which can be used effectively in the social context to identify social class and essentially religious authority.

Conjecture: Usefulness requires either "cause and effect" or a geometric description.

Probability based quantum physics opposes material geometry, ie the uncertainty principle, but probability based systems can sometimes be a part of a causal context, eg for a large population of random events in a uniform state so that this uniform state can be induced to change its state, eg laser light excited to a higher state, and nuclear chain reaction (of radioactive atoms) caused by inducing random neutron-nucleus collisions by forming a critical mass.

The split as to what types of mathematical patterns should be used to describe most physical systems which are definitive, discrete, spectral systems can be elucidated:

Given that most physical systems are definitive, discrete, spectral systems, and because neither the smooth techniques of classical physics nor the quasi-smooth techniques of quantum physics have been useful in describing these systems, there is, thus, no reason to believe that local measurability (or smoothness) is a fundamental property of existence, and subsequently there is no reason to believe that function spaces (in the context of discreteness) should be fundamental to physical description (though local measurability related to geometry is fundamental to the widely applicable and very useful classical physics).

One can believe instead that smoothness comes from discreteness, ie discrete isometry subgroups (or "cubical" simplexes), where it is conjectured (or axiomatized) that specific signature metric-spaces identify geometric and temporal measures related to specific physical properties, and subsequently to metric-space states distinguished by these properties, eg $+t$ or $-t$ for space-time (where 4-space-time is equivalent to hyperbolic 3-space).

In other words, how one uses a descriptive language and its axiomatic structure are all open to alteration in order to find a (more) widely applicable and (more) useful descriptive truth, a truth upon which new creations can be built, (new creations where one does not seek selfish gain from the creations).

Science versus religion

To dwell on descriptions which are illusions, opposes identifying truth as being characterized as it being a useful truth. That is, all speculations about describing systems, in terms of a precise definitive language (which may or may not have yet to be described in a widely applicable and useful way) must be considered, as opposed to only considering the speculations of authorities, ie only considering the speculations of expert authorities or religious authorities.

The nature of a truth, which is based on a precise or definitive (and hence useful) descriptive language, requires that truth be determined by its wide applicability and its (very) usefulness. Furthermore, it requires that its (or a descriptive language's) axiomatic basis and its precise definitions always be open to alteration, because precise languages have limits as to what patterns any given language can describe (ie the conclusion of Godel's incompleteness theorem).

However, complicated description within a fixed use of language, is being upheld by science institutions because such a fixed language can be used to determine a set of elite people, those who want to selfishly dominate the material world, or who want to ally themselves with the (selfish) dominant people of the world. (They want the context of science to always be materialism, ie the religion of materialism.)

This action is more of a religious action and a religious belief in illusions than an act of science which seeks wide applicability and great usefulness for its measurably verified descriptions. That is, science does not fit into a certain set of assumptions, the assumptions for its descriptive language can always change.

Mathematics and the description of observed patterns of existence

> The three choices for the mathematical structure of physical description
> (there may exist more such choices, but
> either they are not being expressed in a cohesive and comprehendible manner,
> or they are being excluded by the media,
> or they do not address observed patterns of existence)

What type of mathematical structure is most related to the patterns that are observed? Here are three choices.

(1) Locally measurable properties which are linear and separable (or non-linear) and are subsequently related to geometry within a spatial context (or a context of a containing coordinate system) of either isometry or diffeomorphism. That is, containing spaces which are either metric invariant or there is a general metric which is defined differently for each point in space.

or

(2) Function spaces and differential operator actions on the function space related to finding a system's spectral structures, ie spectral structures which are occupied by either anti-symmetric (Fermion) or symmetric (Boson) solution functions, which give the solution-probability properties for the material particles' (which are supposed to compose all matter) spectral properties. Note: The domain spaces for the functions which compose the function space are isometric (and almost always Euclidean).

The main observation about (or one of the main patterns of) existence is that micro-systems (or quantum systems) are organized into definitive stable spectral structures.

(3) Isometries with their symmetric metric structures (related to coordinate geometry and to light) and their discrete structures related to alternating-form geometric measures or "cubical" simplexes (which model material), [ie "cubical" simplexes from which both global smoothness and local (or micro) randomness are derived], so that stable properties of physical systems fit into metric invariant metric-spaces and material interactions and dynamics fit into a unitary mixture of metric-space states in a complex coordinate space.

Religion and science

Religion is a belief in abstract descriptions which cannot be found to be widely applicable and (practically) useful, ie a belief in illusions.
 Science is about mathematical (ie measurably verifiable) descriptions which are widely applicable and (very) useful.
 What now passes for high valued science, ie quantum physics, general relativity, non-linear structures, and mutation and natural selection . . .

> (though [it appears that] now environmental influences are now being acknowledged as being as important as DNA, ie environment (and now the DNA is seen as a memory source for protein creation), in fact the memory might exist elsewhere, namely, in the maximal tori of unitary fiber groups)

. . . . do not supply widely applicable principles which result in great usefulness, as classical physics is useful. But classical physics has its limits so that it cannot be used to describe the stable spectral properties of most material (quantum) systems.

Each physical system which is being described by the accepted (but overly authoritative) scientific methods, eg either "non-linear (1)," or (2) above, seems to be unique and specific in its description, ie the system's description is not being based on general principles.

This could be because the true basis (in math) for a system's organization (and described properties) is not being expressed today (7-09) in the accepted authoritative institutional description of science, eg either non-linear (1), or (2) above.

If what is called math and science is in fact most often belief in illusions, ie no different from religion, then the authority of science distorts social forces and pulls people away from truth, ie widely applicable and useful descriptions of observed (and used) properties of existence.

Alternative answers to questions so that the alternative answers are based on the mathematical patterns of discreteness.

Why is the geometric property of an object's motion (or local changes in geometry) at one particular point in space and time related to the relative geometric properties of the given object to other geometric properties of material distribution which can also be associated with the same particular point (in space and time)?

Possible answer (alternative): The small scale structure of the material which identifies a geometric distribution in one dimensional level, in turn, forms the faces of a discrete geometry (of interaction) in the higher adjacent dimension. This new higher dimensional geometry is related (though energy) to a new stable definitive system which might manifest in lower dimensions if the energy range is correct and there are resonances (in relation to the containing set) which allow stability for the new system.

For a new stable definitive system to manifest from interactions there must exist a resonance between the lower dimension (original material containing) metric-space and the new system.

It should be noted that the different (but equivalent) metric-space faces of the higher dimensional metric-space . . . .

> (within which the interaction geometry manifests [where the material within the lower dimensional metric-space form the faces of an interaction space-form in a higher dimensional containing metric-space])

. . . . have different spectral properties, and these spectral properties determine which resonances of existence can manifest in the various dimensions, including the lower dimensions.

There are always various discrete geometries in several adjacent (or related) dimensions which are ultimately determined by spectral-dimensional hierarchies of containment (or, equivalently, resonances) within a multi-dimensional containment set whose structure is determined by "cubical" simplexes, or equivalently discrete mathematical patterns.

Though abstract speculation about how a description of existence can be organized in a mathematical language is a good exercise, nonetheless placing authority on a particular description that is not widely useful (that nonetheless defines scientific institutions) is not valid science.

Note: Though classical physics is widely applicable and very useful it has limits which require that descriptions of existence look to other mathematical patterns which might be organized into a new "widely applicable and useful" description.

Excluding other math descriptions of existence from the media, or equivalently from the public education system, is the same as the state supporting one illusionary description of existence, to the detriment of other equivalent (or better) speculative descriptions (which may or may not be useful, ie true), and thus it is supporting a non-useful description, as if it were supporting a particular religion.

Religious descriptions of existence are not in precise languages (their descriptive words can have many meanings) though religious ideas can be associated to a mathematical (verifiable) language, as the example of Ptolemy shows, but as a description, religion has not provided any practical uses.

On the other hand looking at a discrete multi-dimensional description of existence allows new ways in which to understand the religious experience, and new ways in which to describe the religious experience, what life is, and what life can do, ie what new contexts in which life can exist. New ideas about creativity, etc.

When people believe in illusions rather than widely applicable and useful, but specific, math principles (or mathematical patterns, eg discrete mathematics) then they become controlled by authority and elitism. Nonetheless illusions can be described by specific mathematical patterns and they can be constructed by making agreements about how to use (descriptive) words. Only the principle of a widely applicable and useful description of truth can distinguish between useful scientific knowledge and illusion.

State sponsored religion (which is what today's science really is) allows an artificial control by the state (or by those who control the state) over the people of society.

Summary:

All material existence depends on the stable definitive orbital structures of material at all scales of size.

If a description of existence is based on discrete isometry subgroups (or equivalently, based on space filling, non-over-lapping "cubical" simplexes in a multi-dimensional structure and in a multiple metric-space setting (ie different dimensions and different signatures) then the existence and stability of orbital structures for all the various size scales of space and material, ie particle, nuclear, atomic, molecular, crystalline, and planetary orbital structures, is described in a direct manner as space-forms (resonating with the rest of existence) except perhaps some of the details of crystal space-form structure. The remaining issues of determining the resonances between the different dimensional levels. It is the resonances between the different dimensional levels which determine the over-all definitive spectral properties of a particular spectral-dimensional-space-form composition of existence. In turn, these resonances between various space-forms in various dimensional levels provide spectral-geometric information about the different dimensional levels of a particular spectral-dimensional-space-form existence (of many such existences).

All orbital properties of existence are accounted for by space-forms which fit (by resonance) into a greater dimensional space-form structure upon which resonances are identified between both different dimensional levels and between simultaneous space-form structures.

Pursuing the belief that . . . . non-linearity, diffeomorphism coordinate transformations, function space transformation techniques (including low dimension unitary matrix transformations) . . . , are fundamental to physical description is like believing in the illusions of religion, eg "existence in god," that "there is life after death" whose truth is based on authority.

(Though intending creative activity beyond a "material level death" may be possible, but this is related to the multi-dimensional structure of our space-form existence).

Platonic truths, ie pattern description based on abstract word use agreement, which are not widely applicable and not very useful, are illusions.

That is, Euclidean space has value because it is useful.

Knowledge and society

Education should not be about authority which serves the dominant selfish interests of society, ie the oligarchs, so that society uses material in particular ways which support the power of the few. Rather education which is involved in seeking knowledge based in descriptive language, is about free inquiry, integrity, realizing the limits of language, knowing the limits of knowledge, and questioning authority.

Knowledge is directly related to creativity. Precisely (or mathematically) describing existence so that such a describe "truth" is useful in a creative context.

Knowledge and creativity (is) are not about authority and individual elite-ness, rather it is about the selfless nature of people to know and create.

Our heritage as people is to create existence (perhaps at a dimensional level beyond what the idea of materialism demands) thus we need to know about existence, thus a need for free inquiry.

The nature (or limitations) of language when determining truth through mathematical descriptions of existence requires that truth be determined by wide applicability of a few principles (if possible) and a good amount of usefulness, in terms of creativity which is directly related to the description.

Measurable verification of a mathematical description is not sufficient for the determination of a description's truth as the example of Ptolemy (as well as quantum physics and general relativity) demonstrates.

Note: If non-linear description can be useful then (the authorities must) demonstrate this usefulness, but (the authorities) cannot claim an authority for such descriptions which does not exist, because their widely applicable usefulness has not been demonstrated. The authorities of non-linear physical description must provide the actual, useful non-linear model for the: solar system, the origins of life, etc.

Indeed claiming authority for institutional proclamations of a verifiable truth without the description also possessing the property of wide applicability and great usefulness is fraud.

The failed probability based math models of the economic system which are represented authoritatively as being true is an example of this fraud, where the fundamental issue of defining a fixed sample space of random events, which (in fact) is supposed to contain the (economic) system being described, are not being satisfied in these many probability descriptions (many of which are descriptions similar to the descriptions of quantum systems). These probability methods are related to the probability methods of the descriptions of quantum systems (both the probability descriptions of the economy and the probability descriptions of particle quantum systems share some similar descriptive failings).

Indeed the development of clean cheap energy through the currently "believed in" authority of institutional physics (which is really formed into bomb engineering by the military-industrial complex) has failed to allow clean cheap energy systems to be developed. That is, the ideas within the current paradigm of physics are not capable of developing a clean cheap energy source, much to the delight of oilmen.

Development of such clean cheap energy systems based on a knowledge of probability of particle collisions which is related to chemical-nuclear explosions is a fraud perpetuated on the public.

This fraud is maintained by public educational institutions supporting an authoritative truth which is really speculation which has not been widely applicable, nor has it been useful . . . ,

> . . . . yet nonetheless an army of "knowledge-able" people accept the illusions presented to them by the education systems so as to compete in a game which, when won, demonstrates their intrinsic intellectual worth to the world and gives them social position (such an authoritative viewpoint about knowledge is part of class warfare).

This limited "knowledge" also determines what types of things can be created with society's institutional knowledge.

This adherence to authority also causes the public to believe in the vague notion of high value placed on a particular speculative form of physics that is quite similar to the science of Ptolemy, just as in the case of religion, where people are encouraged to believe in the authority of some vague notion of high value.

Freedom of religion (or meaningful freedom of speech) must allow other mathematical speculative ideas to be part of the discussion in public institutions.

It should be noted that the math structure of multi-dimensional and material-spectral-spatial model of existence based on discrete isometry groups (or "cubical" simplexes or space-forms) is a description of existence which those interested in the spirit, the mind, origins of life, a morphogenetic field (eg R Sheldrake), creativity, and the (true) nature of existence which mathematically models existence to be beyond materialism, should all be interested, ie religions should appreciate (and be interested in) such new descriptions.

> Illusions, high value, and truth.
> Religion and free speech.
> Who has the right to claim "What (expert) ideas are relevant to any particular context of 'truth'?"
> Much of authority has to do with how people's (creative) actions are organized around language.

> What is true? Is this determined by:

> Authority (and selfish interest),
> Revelation,
> Verifiable description, or
> A verifiable, widely applicable and very useful mathematical description?

Differential equations (or calculus) allows the association of geometry (or a geometric property, eg length) with local measurability of a measurable property (ie or a function, eg coordinate curve [or coordinate function], or a measurable position of a material object in a coordinate space).

The "Science" of the oligarchs

Consider the drivel about science which comes out of our society's premier university education institutions.

Physics assumes that materialism is an absolute truth (absolute truths are not true) and it has two basic ideas about how the measurable and useful description of the physical world should be organized.

The classical idea, which is both widely applicable and useful, and is the idea upon which nearly all technical development is based, namely, that physical description should be based on geometry where differential equations describe the geometric relationship that local measurement of a property (or function) has to geometry of material in space, so that the interpretation of the solution functions (to the differential equation of a physical classical system) are done in regard to geometry, so as to tell the observer-describer what property to expect and in what place to measure.

On the other hand, quantum systems are described in the language of probability where differential wave equations (or sets of operators) determine the spectral and probabilistic properties of a quantum system's function space. This is a descriptive method which is best suited for following data. Furthermore, the practitioners of quantum physics need to be able to clarify exactly what their elementary event space actually is. It should be noted that this is not done. [This is something that has not been done.] For example, though particle physics did not predict dark matter this is the only peer reviewed method of its description which is allowed in the professional literature. That is, all of a sudden the elementary event space of particle physics is expanded to contain "the particle" that causes dark matter to manifest. This is clearly an epicycle technique. The oscillating (and now assumed to be massive) neutrino is a similar method of using epicycles to follow data, within a probabilistic framework in which the elementary event space is never identified, or it can be modified at will.

The main statement of quantum physics is that "quantum systems cannot be described in a useful manner because they are too complicated." Yet quantum systems organize themselves in precise and stable ways. If the laws of quantum physics lead to complication (then if that is correct) then quantum systems should not be so well organized and existing in such a precise and stable spectral (energy) states.

If a quantum system can be coupled to a controllable classical system then the energy level structure of the quantum system can be used (primarily done in regard to semiconductors in electric circuits [transistors, microchips, solar cells]), but other than this type of use, in which cases it is sufficient to experimentally measure the quantum system's energy levels (because quantum physics has not been successful at describing these energy levels based on the laws of quantum physics), the usefulness of quantum physics has been:

> the electron microscope,
> the tunneling microscope,
> the tunneling diode, and
> the laser.

The MRI is again about using quantum spectral properties in a classical context, etc. that is, quantum physics is not that widely applicable, and it has not been all that useful, though quantum energy levels are prevalent and can be used when they can be coupled to classical systems.

Thus an alternative description of quantum physics based on geometry, for example based on discrete isometry subgroups, needs to be considered.

The funding of alternative ideas should be the basis for new funding.

However, the peers of the physics institutions are not going to give a pass to an idea which questions quantum mechanics in a fundamental way, because that would end their careers.

Immediately the claim will be made that this will encourage all types of fraud. Well. Capitalism encourages all types of fraud. Institutions of learning are not institutions of absolute truths, and today the institutions of learning are victims of their own fraud due to their adherence to absolute truths and their arrogance.

Nonetheless:

> Thus the person who is supported as having alternative ideas needs a history of expressing alternative ideas.
> Do the new alternative ideas make any sense are they really different from what is already being expressed are natural questions which could be easily distinguished by (library or Google-type based) search systems as well as analysis based on the philosophy (or logic) of language, eg Godel's incompleteness theorem, ie the development of knowledge depends on the development of a new way of using language.

On the other hand the collapse of the economy and society is all about fraud, so what qualifies this criticism of fraud directed to alternative ideas to have any validity? The critics of alternative ideas are people who support the established structures. Why should the supporters of a failed system be taken too seriously. Though if they want to offer alternative ideas perhaps they will have some validity, but dogmatists usually can only offer permutations and innovations on the same dogmatic ideas that they hold.

Clearly the best road to development is to seek new voices, to honor free speech, and to try to seek and develop new ideas.

Those who support the economic system and justice system that we have lived under are the architects of the ruin, the track record of their judgment should exclude them from these type decisions.

Both the constitution and the justice system and the inability of the justice system to both stop the economic collapse (mostly caused by fraud) and to be unable to maintain the Bill of Rights, deserves very severe criticism, and a subsequent need to seek new ways to organize the laws of society.

Seeking the experts of our failed system (of the government, law, and the economy as well as science) to re-establish the system is like seeking the expertise of the utility companies for alternative energy systems, they best understand the system which collapsed and want to save that system.

The myth of objectivity

There is a myth that the press is supposed to "objectively" cover the facts, ie the material events which took place between people or institutions, though for some meetings whispers might be the most important material event which took place. Nonetheless this claim to objectiveness is a fallacy. The corporate media has all sorts of assumptions built into its stories. Assumptions about the economy about markets about the good winners as opposed to the seldom talked about lowly losers. Furthermore, the media lies all the time, journalism should be related to ideas, however, journals about particular ideas (as they should be) nonetheless they should seek evaluation by those of opposing ideas, yet facts are a wild card in the press. The only person who can really determine truth is oneself.

That is new voices in the press (or the media) should be supported (with money) so that these new voices clearly describe the assumptions and viewpoints they hold and describe the types of programs and institutions which they want to be realized in society.

For example, the curriculum of the third grade should be about how to determine the differential equations of physical systems because functions (solutions to differential equations) allow predictions which can be measured, so learning becomes part of learning about the world.

The corporate media needs to join the game of clarity of expression, that is they need to state that they are the voice of the winners of capitalism and a crooked justice system. The government should make the corporate media be at a disadvantage because they claim to be news but they really are propaganda industries, which are not interested in the public interest.

That is the whole notion of spending to redevelop society should be all about funding the disciplined and knowledgeable yet marginal people of society who can build upon new notions of how to put together technology. (Apple began in a garage, did Ford begin in Ford's barn?)

The reason alternative energy has never got into the market is that big business sabotaged it, along with the undeserved faith in and reliance of the public upon the knowledge which the public has been led to believe that big business possesses, so the public has relied on the knowledge of corporation to determine what society should be doing. This has been a big mistake.

Tax cuts will simply put money back into the experts hands who brought about this collapse.

When corporate becomes government

The Vannaver Bush NSF model of using science for industrial development has failed. It has resulted in the vision of the institution which represents science to no longer have the vision of (real) science, ie questioning authority and trying to find a better language in which to describe the world.

There exist disciplined and educated people around the margins of the academic world whose ideas should be funded.

The system, as it now is (under W Bush [but really since Truman]), is not going to be able to function properly. This is because it is too dogmatic and too domineering, and it is arrogant in its view of its own importance and intellectual power. It has failed both in science, and now it is clear that our economic system cannot sustain a war of extermination and destabilization in which the US now engages around the world.

Indeed the ideas to be considered in science are not those which support a focus on particle properties or string properties or general relativity properties these have had an army of developers and they have not supplied the world with anything which is widely applicable or useful.

That is, ideas which reduce to quantum physics or general relativity are ideas which lead nowhere.

Grants or investments

Note: There already exist hun4dreds of papers about using discrete isometry subgroups to describe physical systems which the professional peer reviewed journals will not publish because these professional journals are dogmatic and represent the interests in technology of the oligarchs ie the peers all want to compete to work for on the important projects of the oligarchs. To have a good reputation as a good scientist is all about the existence of a name of a competitive "scientist" whose name can be given to the oligarchs (or their minions) so that the scientist can work on an important project of the oligarchs.

Many of the best scientists worked as unknowns because they love science. An anonymous institution of science might best respond directly to ideas as opposed to science responding mostly to social forces.

Competition

The idea of competition was an invention of the oligarchs of the 1800's and was made valid by Darwinian evolution. However, competition depends on a fixed or absolute idea about truth . . . (the "laws of particle physics" are not the laws of physics because physical laws must be widely applicable and useful and particle physics is not true by these measures of validity (or measures of truth)) . . . but science has no absolute truths, though that is what our world of reality "based on propaganda" now claims.

Justice's relation to business

Though the justice system4 is good at policing the mandated auto insurance requirement amongst the public, so as to help the profits of auto insurance companies, on the other hand the justice system seems to reward medical insurance companies which disqualify insurance holders who have been paying their premiums (undoubtedly based on the letter of the law of some deceitful contract) and let them die so as to improve the medical insurance company's profits.

The real value of new ideas in a society strangled and dissolving by its own lack of ideas

However, the real worth of new (or not allowed) ideas should be estimated in the billions, because this is how much has been, and is now being, pumped into quantum physics and string theory, and these probability-geometry (necessarily an inconsistent language) based theories have a very slim (if any) chance of resulting in the development of technology. Furthermore, there is also a large amount of money being pumped into general relativity, but general relativity can only describe a system which contains one-body, and thus it is limited in what technologies to which its descriptions would be relevant.

Whereas a description of spectral systems which is based on analogous geometric techniques as classical physics is an idea which has a better chance in resulting in great technical development, because it is geometric.

Yet the new theory also makes statements about the spirit, again something in which people say they are interested.

New ideas in science and technical development

Again this pres4entation of new ideas will be ridiculed by those of the high social classes, those who are arrogant with a belief in their intellectual superiority, yet what is their belief in their great value of their own vision of themselves based? Answer: Failures and destruction.

What revolutionary development has taken place because of their ideas?

Note: The atomic bomb can exist because there is a critical mass for U235.

Whereas the development of electromagnetism in the 1800's brought about a revolution in technical development which is still being realized.

Those in the high social classes, including the society's overly touted scientists, represent the same privilege that the ruling class represents and that privilege represents an amoral vision of a world of people who maintain their privilege.

Chapter 13

QUANTUM PHYSICS

Start with a review of:

(1) Did a 1-Dimensional Magnet Detect a 248-Dimensional Lie Algebra?, D Brothwick, S Garibaldi
and
(2) String Theory and the Geometry of the Universe's Hidden Dimensions, ST Yau, S Nadis,

Both articles appeared in September 2011 Notices (of AMS),

> ..., demonstrate the weaknesses of modern math and physics, especially since (1) is better written and better organized than are most articles by experts concerning particle physics, thus the short-comings of math and physics can be clearly pointed-out, while (2) has an honesty in regard to how social forces in math-science are used (manipulated), and how they affect (professional) social communities, where personality cult and its association to authority is a main component in professional communities of people, rather than an honest attempt to discern the truth. In "professional math-science communities" truth is about extending the discussion of authority (the narrow discussion allowed by peer review) and within this authoritative context identifying and solving prize problems, where only the "top authorities" of a tested community of proven elite experts can identify the prize problems for the community to solve.

Math-scientists today are pleased to expand on, and compete within a context of a: dysfunctional, irrelevant, failing, and useless model of existence. That is, math-science is best described as meaningless, but highly competitive, babble (where the babble is quite analogous to the church authorities in the age of Copernicus).

This dishonesty is required within math-science social community due to (or caused by) the way in which math-science is managed.

Note: The only aspect of quantum theory . . . , (that material systems are organized around stable discrete definitive energy levels, and it has been assumed that this discrete structure is a result of quantum system descriptions being based on function space descriptions, and subsequently being based on randomness)

> . . . , which has led to technical development (ie invention), is that there are two types of components which fill the discrete quantum levels so the levels are filled based on different statistical rules, namely, the Boson (metrically consistent, integral-spin, no limit on occupation-number for any particular energy level) and the Fermion (spin-½, at most two-components can occupy any particular [distinguishable] energy level), and knowing these two different properties of energy-level-occupation has led to the development of a few types of various systems, such as the laser. Technical development has not been a hallmark of quantum physics (the transistor and micro-chips have been developed by classical thermal and optical processes, eg annealing), most of the development related to quantum physics is similar to the development of fusion energy, such as quantum computing, where quantum computing has been touted for over 20-years now (2011) but has not come to fruition.

Nonetheless the authorities always claim that "it is only by means of developing their (absolutely true) ideas from which greater creativity will be achieved," but the cheap, clean energy (through nuclear fusion) promised by the authorities back in the 1950's has yet to materialize (ie clean cheap energy, as yet, has not been brought into existence for the society,

That is, just as in the case of capitalism, where a few people (maybe 10 people) have enough money to control and manipulate all markets, so as to have central planning based on selfish interests, an organization of society which has consistently failed and led to ever greater destruction due to its reliance on narrowly defined product-markets where one of the main products is military equipment to enforce ever greater violent control over the narrowly defined society whose energy is based on oil.

though there are claims that such useful systems already exist (these types of claims have a relationship to the continual publishing of descriptions of UFO's and "ancient aliens,") where one would guess that in this scenario (where cheap energy are actually technically available [such as thermal solar energy]) the oil companies either buy-up the patents or gain control of the research, and subsequently, trash the systems, so that oil profits will stay high).

Article (1) Describes the complicated context of a particle-field description where the data is difficult to interpret. The data is provided in the article, and thus its tentative claim to verification can be made (or judged) by the reader.

In this set of data, for an essentially particle-physics model of "interactions of spin-properties at lattice sites of a 1-dimensional magnet in an external magnetic field" (a context only slightly associated to a robust context of physically useable properties), the data only approximately suggests that the description is describing a pattern which might exist. That is, the second mass (or second peak in the data) seems to have similar properties as other data (other similar peaks) which are supposed to [which may or may-not] exist within the "uniformly random," [or in an "incoherent continuum (whatever that might mean)] part of the data. That is, the signal for the second mass is very weak and not be much different from the signals which appear in the "incoherent continuum."

This would not be the first time where the ideas of quantum physics were given respectability by "dry-labbing" the results which "scientific consensus" demands (or the community wants, or expects, since it is an intrinsic part of the (their) assumptions [of the authoritative community]), consider the "dry-labbed" data which R Millikan used for proving electrons have a fixed quantum of charge, a property which P Ehrenfest could not replicate (apparently, because Ehrenfest was honest).

> the "verifying objective" of this "particle-physics description" requires the experimentalist to identify the "predicted" ratios of the masses, so that the so called "prediction" is calculated after an experiment is made. That is, the prediction is that "an order" exists which is based on algebraic patterns (which string theory claims are associated to a hidden geometry), which in turn, can be imposed on a random structure, but "Does it matter if the randomness is fundamentally random or if an algebraic structure can be discerned within the randomness?" This type of description is about collecting information which cannot be used in a practical creative manner within the material world (if it could then it would be used to describe the relatively stable spectra of general nuclei), it is only placing a pattern on a random set, yet the random set remains both random and unrelated to creative usefulness and uncontrollable.

Furthermore, it is still unknown in regard to both:

(1) the full set of algebraic patterns which can be imposed on all the possible ways in which physical systems can be put together and
(2) how these patterns of particle collisions are related to the order which is observed in general quantum systems (eg describing the stable spectra of stable nuclei).

That is, this is similar to reducing science to that of studying the patterns of spots on individual cows within a herd of cows and neither identifying the herd nor associating either the individuals or the herd with creative usefulness.

(though patterns of particle collisions can be used in a destructive process in regard to the transformation of one relatively stable system into new, less ordered, but relatively stable system)

Perhaps spots on cows can be associated to vague social properties within the herd or in regard to other herds, but such an analogous level of usefulness is not realized in particle physics, when the algebraic patterns of either E(8) or SU(3) or SU(2) are being identified by the theoretical physics community (particle-physics properties are only used in the context of destruction, in regard to nuclear bomb engineering, they are not even useful in regard to describing the ordered properties of relatively stable systems such as general nuclei etc).

Note: The BCS theory of superconductivity (based on phonon-electron interactions) has been falsified since the critical temperature it predicts has been exceeded.

Note: However, even if algebraic patterns exist at the level of elementary particle collisions . . . ,

(since at best it only predicts ratios of masses which exist between the expected particles involved in the quantum interaction, where it is assumed that there are particular particles, associated to an E(8) algebraic pattern, which are to be a part of the interaction)

. . . , it is a pattern which has no uses, since it is a "probabilistic-experimental" based descriptive model, developed for systems with relatively few components (thus there is no relation of this description to controlling the system, nor does it give useful information about the stable order which quantum systems possess), and all the data needs to be completely extracted from the system by experiment before the data can be compared with "predictions," ie the predictions cannot be verified unless all the information is already known about the system, ie the (any) descriptive structure based on randomness can only follow data as Ptolemy's epicycle structures were used to follow data.

In other words in the context of particle physics (as well as quantum physics) the whole point of looking at a quantum system is to check if the (pathetically insignificant, in regard to practical uses of these descriptions) "predictions" of this theory are correct, ie the only use of the description (with one [noted] exception) is to find quantitative correlations with the description which sheds no light onto the reasons and/or causes of the observed (very stable) order which exists in quantum systems [otherwise this data can be used in models of chaotic transitions between relatively stable systems so as to possibly change the rate of the transition process due to the properties of particle-collisions, and this is how this data on particle collision properties is used in bomb engineering]. Thus, if a particular algebraic pattern, ie Lie group E(8), is related to the masses within a particle-collision model of interaction then (or which are involved in an interaction then) this type of algebraic order is imposed upon an (un-useable) random pattern, which is un-useable except (perhaps) in the context of controlling [through the statistics of particle collisions] the rate of bomb explosions.

On the other hand, the standard which particle physics has established for its verification process is that of "obtaining a precision to 16 (or more) significant digits" in regard to a pair of particle-physics "calculations" and subsequent experiments, ie where one of these "experiments" is about examining the stable, definitive spectral properties of an H-atom (perhaps the other is about the free-energy of an individual electron (?)). That is, since no information of any practical usefulness can be derived from particle physics and only two or three particle physics calculations have been accomplished (thus) the standard for measurable verification has been "claimed to be" the precision of the experimental agreement with the particle physics calculations.

This is not science.

Yet it is this type of thinking (which is destructive of science), which has led to an expensive development of communities of people who are dedicated to pursuing these algebraic patterns of order, to be placed on random systems where the algebraic order cannot be used to determine the order which quantum systems do possess, ie the description is left with its random (useless) basis, (nothing about quantum systems are clarified by this discovery of an existing algebraic order to a quantum interaction process which is remote from the material world and remote to a relation to practical creative development. Rather the algebraic patterns of particle collisions are only related to bomb explosions, yet it is claimed to be an authoritative truth about existence, but it is only examining the spots on individual cows, eg the cow's spots which represent the H-bomb, without even identifying the herd of cows.

That is, now for E(8), the new particle-physics criterion for verification is establishing only an approximate relation between the data of this experiment, and the set of "predicted" particle families which are assumed to be a part of a random particle-collision model of a particular type of magnetic interactions, while on the other hand . . . , since one cannot calculate (or predict) anything which is of any significant "informational value" with particle-physics, such as determining the stable spectral levels of general nuclei . . . , the "standard of verification," which is used in particle physics [based on finding unitary algebraic patterns for particle-collisions], is the two or three calculations verified (correlated) experimentally to sixteen decimal points (or to sixteen significant digits).

The authors of (1), along with writing in a very clear manner, (to their credit) try to critically consider the interpretation of the data, and to determine "what" both the data and the description, itself, is really describing (these field-particle descriptions are so distantly removed from a useful geometric context that a lot of effort is necessary to vaguely relate the model to a physical context, let alone to a practically useful physical context). Note: It is claimed by the math-science communities, that these ideas are useful because they become the basis for other articles about particle physics, or string theory etc. that is, they are patterns which only have literary usefulness, but they are not useful in regard to practical useful creativity.

The authors of (1) are clearly cheer-leaders for this type of (pretty-much useless) type of abstraction, and they do not question the legitimacy of the over-all model of quantum interactions described in the context of both randomness, ie a quantum energy operator, which is associated to a system composed of a relatively small numbers of components and which is related (relatable) to a second layer of randomness, namely, a particle-collision interaction between (in the case of the article) the energy operator and a magnetic field (where such a model can only [at best] follow data). Yet these randomly based descriptions are used to describe the properties of systems composed of a small number of components, such as the H-atom (and in the case of article (1), a 1-dimensional spin-lattice) and thus can only identify probabilities which should be related to the well defined order of the quantum system, but these probability techniques cannot be used to find the order which quantum systems possess.

Thus, the only type of information which can be used from this description is the type of information which can be collected by experiment before the precise description is even developed (it can only follow data).

That is, the descriptions based on randomness can only follow data.

It should be noted that probability techniques, really only work for probability systems which possess a finite "elementary-event space," composed of stable, easily identifiable events, and it is not clear that any quantum systems possess such an elementary-event space structure. Furthermore, the elementary-events which compose a particle-physics system are definitely unstable and undefined, eg dark-matter was not predicted by particle-physics, yet it is now provided with an elementary-particle structure as a quantum interaction, ie another epicycle was introduced by means of an unknown and undefined elementary event space where randomness is defined by any algebraic order imposed on the "internal structure" placed onto particle-collision data, which in turn, are modeled to perturb a quantum energy operator.

This "fitting of data" is similar to how Ptolemy's epicycle model followed data.

The probability structure gives no control over its descriptions (of a system composed of relatively few components), it can only follow data (which has already been observed). That is, the "predictions" of what is to be expected in an experiment are virtually always wrong (eg the atomic energy structure of a general atom such as Fe cannot be obtained from calculations based on the laws of quantum physics [or particle physics]), since one is trying to determine probabilities of unknown events (the operator-function-space reduction to the precisely observed spectra of the Fe-atom cannot be established), ie it is not a well defined mathematical context.

The entire exercise of quantum physics and particle physics, in which all aspects of descriptions which are based on randomness and which are not associated to a "correctly (or explicitly) defined elementary-event-space," is at best a "shot in the dark."

It is this type of math structures, used in this same random but undefined context, which have recently been used to calculate risks and which subsequently such calculations have failed so miserably with catastrophic results, eg the recent (2008) economic collapse.

The idea that there is two layers of randomness with one (lower) layer of randomness with its math structures to be perturbed by a second layer of randomness, where there is a set of algebraic patterns which "might be associated to this second layer of randomness," and this vague pattern of (algebraic) order, in turn, (it is believed) can be used to determine order within the other lower (or first) layer of randomness, so as to find the probabilities and spectral-values of known events (energy levels in the H-atom), has not shown itself to have any practical (creative) value. This effort centers around the idea that the correct number of physical symmetries can be found . . . , eg rotational symmetry (where the rotational symmetry is represented in an algebraic manner) . . . , so that the function space structure will fit together with the observed data.

Yet these techniques, of finding algebraic order to be imposed on a random particle-collision quantum interaction model, cannot be used to describe the stable spectral structures of general nuclei, which is the type of system to which particle-physics is supposed to be most applicable.

Instead of identifying "patterns within randomness" by means of algebraic patterns for the absurd particle-collision model of quantum interactions (the new epicycle structures), which seem to have no (robust) relation (if any relation 'what so ever") to physical description, why not (instead), begin with a geometric order, from which randomness can be derived, so as to find the stable patterns of existence (which are observed in the world) through inter-related stable geometric properties which can also exist between different dimensional levels?

The particle-collision data suggest higher dimensions, and unitary symmetries (invariances) in the interaction process. String theory is one way to consider a geometric model for quantum systems, but string theory maintains the idea of materialism.

Why not consider other geometric models of existence?

Perhaps one should consider some geometric models of physical existence which do not assume materialism, nor do some new math models assume the geometry of spherical symmetry, ie spherical symmetry is a result of the pattern of the interaction process it is not a property of space itself (within the new model of existence).

The probability models of both quantum physics and particle physics, where string theory is consistent with particle physics, is an elaborate epicycle construction similar to the epicycles that Ptolemy used to fit the data he collected (before the epicycles needed to be developed).

Einstein, though central to the development of quantum physics, rejected quantum physics as a valid science description.

Peer review had rejected DeBroglie's matter-waves, it was only Einstein, who was interested in ideas . . . ,

> [(not so much interested in performing for his paymasters and participating in full agreement with the obedient and highly-tested and very competitive "professional scientists," eager to expand on what professional authorities are expected to believe and discuss, and where these professional communities are formed by the paymasters so as to develop the business-investment interests of these paymasters [where as often as not these interests are about stifling intellectual thought])]

. . . , and who allowed matter-waves to be considered by the professional science-class. Subsequently quantum physics relation to function space structures related to solutions of wave-equations was developed.

However, he (Einstein) rejected quantum physics as a valid science description exactly because probability is a data-following construct,

> (an epicycle structure, but now the epicycle structures are algebraic patterns imposed on function spaces (which have been made discrete), which in turn, identify the randomness associated to (essentially) unknown in regard to "what function-space operators can be used to identify the fixed properties of the spectra of quantum systems?", ie the observed, stable, spectra of quantum systems).

Science is about using math to follow measurable data, but the question is, "Are the data-following structures useful in regard to practical (useful) creativity?"

Practical creativity is only relatable to relatively stable geometric shapes, ie related to discrete hyperbolic shapes (which are shapes which are: linear, separable, and metric-invariant so that the metric-functions have constant coefficients).

Article (2)

Unfortunately, Einstein's attempt to develop physical description "based on geometry" led him to consider non-linear shapes (or spaces, manifolds) which are associated to curvature (his idea was that acceleration was related to curved geometry to be defined by inertia [on geodesics]). Apparently, he did this without he himself understanding that descriptions of non-linear systems are "all about" descriptions of the types of systems which have the property of being fundamentally random (or chaotic), in their quantitative, or measurable, structure. This is because local non-linear properties do not allow for (or are inconsistent with) the definition of a well defined stable "uniform unit" of measuring within the quantitative descriptive structures.

That is, expending a great amount of effort to develop descriptions of a wide array of possible (non-linear) geometric properties associated to the observed world is lost effort, because these non-linear shapes are not stable and cannot be described in a quantitatively consistent manner, the result of not being quantitatively consistent is that such (non-linear) shapes (or such non-linear systems) become chaotic, and thus unstable (where these non-linear shapes either dissipate or they deform into stable discrete hyperbolic shapes, ie this is part of the Thurston-Perelman geometrization theorem).

Einstein, non-linearity, and the equivalence of frames (a description which has no practical creative uses)

Einstein spent a great deal of effort in justifying his acceleration-geometry model of inertia, but which by the spatial displacement symmetries of E Noether (in the 19th century) inertia should be represented in Euclidean space (not local space-time space).

Einstein was interested in the idea of equivalent frames for (of) description, he wanted to determine the most general properties of coordinates within which a physical description can be related to general physical laws (so as to be applicable to a wide range of conditions).

That is, because there is a "local space-time metric-space structure" associated to Einstein's general manifold metric-spaces, it was believed that equivalent frames cannot have instantaneous "action-at-a-distance" field structure associated to themselves,

(even though the descriptive structure used in controlling "satellite orbits," in both a feedback and non-linear context, depends on an "action-at-a-distance" gravitational field structure, furthermore the dynamics of stars and galaxies are best calculated (easiest to calculate) in Galilean-frames and assuming "action-at-a-distance" property for the gravitational field which is contained in a Euclidean space).

However, since Einstein's gravitational theory has not been associated to predictions which identify the properties which are a "substantial part" of our experience, such as the stability of the solar system . . . , one should consider how Einstein's general attributes of equivalent coordinate systems might be wrong.

Einstein was concerned with how to describe inertial frames, where inertia was placed into (locally) space-time coordinates, and material was a scalar property (subsequently formulated as an energy-density property) associated to its geometric distribution structures in space.

The wave-equation of quantum physics is based within Euclidean space as the domain space for the functions which compose the quantum system's function space.

On the other hand, probability based descriptions have resulted in a short list of developments, (short) in relation to the long list of development which has been fostered by classical physics: the electron microscope, tunneling electronic devices, and the laser, otherwise the discrete properties of quantum systems can only be used if they can be coupled to a highly controllable classical system, eg transistors. That is, probability cannot describe, in general, the stable definiteness of quantum orbits and the quantum description has not been useful in fostering development.

When a description is based on probability, as is quantum physics (and string theory), and that description cannot calculate the spectra of a quantum system with some general widely applicable set of principles, as classical physics is capable of describing the macroscopic world with a small set of widely applicable principles (which describe how to determine a system's differential equation), as opposed to methods of calculation which are unique to each system, then such a description's basis in probability does not seem to be valid, because the elementary event space is not determinable other than by experiment.

Perhaps the idea that has been overlooked is the idea that stable, definitive, separable geometries are fundamental to stable definitive observable properties at all size scales, however, for small systems such geometries (as well as their discrete interaction geometries acting on adjacent geometries) leads to Brownian motion, and these random Brownian motions are (related to) the source of quantum wave-particle properties. However, the particle is actually a distinguished point of a small space-form geometry used to model atoms and nuclei as well as stable electrons and protons, though geonium shows that size and charge may vary for leptons. That is, space-forms and the differential structures of isometry groups determine material, space, and the interaction process between (space-form) material, where in 3-space the higher dimensional interaction space-forms are unseen (the interaction space-forms [which mediate material interactions] do not exist in 3-space).

These same interaction space-forms, when acting on an atomic scale, have many types of local (or adjacent) interactions which change from instant to instant, and thus result in Brownian motion for small (or quantum) systems. However, when an interaction results in a stable new quantum system that system is best modeled as a space-form.

The problem with quantum physics, is that the spectral properties of quantum systems are not related to the geometry of the system, nor are they related to placing constraints on the domain space of either the geometric (domain of the solution function), or constraints on the eigenvalues of the eigenfunctions of the function space (the domain space for the set of operators which are supposed to identify a quantum system's spectra), and, furthermore, the mass spectra of elementary-particle collisions, being imposed to adjust the quantum eigenvalues does not provide enough (spectral) structure for the particle-spectra to be of any value.

Rather the spectral properties of a quantum system are (can be) related to (1) a discrete hyperbolic shape for a quantum system, and (2) the spectral properties of an over-all high-dimension containing space, where a high-dimensional over-all containing space can have very significant geometric-spectral properties associated to itself.

The problem with function space techniques as a basis for describing locally determined measurable properties, ie either values of functions or spectral values associated to a function space's basis set of spectral functions (both of which are) defined on domain metric-spaces, is:

(1) their focus is on global functions, which collapse when local discrete measures are "taken," so Is it a local or a global model?
(2) the focus is also on the phase properties of a wave, which are supposed to be related to the globally measurable properties of the physical system, eg energy of the system, thus
(2a) it should be noted that different geometric properties in different spatial directions can affect the local properties of waves, and their relation to wave-phases, in different ways (especially for mechanical and electromagnetic waves), yet the phase is the means through which the global system properties, eg energy, must depend (or must be identified), but this energy would be an average property, yet the individual particle-spectral events associated to the system are particular eigenvalues, which can be a part of the average energy value of the system. How can a wave-phase be associated to two different values for the system's energy (an average energy, and a particular energy level)? or How can a spectral (quantum) system have an infinite number of spectral values associated to it simultaneously?
(2b) if wave-phase properties are independent and random (and perhaps unrelated to local geometry) then they are properties which cannot be linked to a system's global properties, and thus the question, "Where, then, is the basis for a (quantum) system's ordered spectral structure to be found?"
(3) thinking inductively; the only descriptions of physical systems which are based on randomness . . . , and are composed of relatively few components, and which have a clear relation to a set of observed stable, discrete properties of spectral systems . . . , are those systems which have definitive geometric constraints associated to their operator structure, (eg 1-body boxes, and the 2-body H-atom) ie it is the geometric boundary conditions which force the system to have certain types of stable, discrete spectral properties.

Despite what certain authoritative figures "might have believed," and so "stated those beliefs," about the observed properties of (quantum) system being dependent on "the phases of waves," such as H Weyl or S Weinberg, or S Hawking, this simply means that they believe that sets of operators can be found which can be used to manipulate wave-phase so as to fit data, but it is a model which has not worked after over 100 years of effort . . . .

> [based on the laws of quantum physics, where are the sufficiently precise descriptions of the electron orbits of (say) Pb-atoms, ... based on the laws of particle-physics, where are the descriptions of the stable orbits of the components which compose (say) the Pb-nucleus? Etc.]

(the physics-math professionals have reduced global (function) structures to local structures (ie observed particle-spectra data) by means of the basis structure of function spaces (wherein material geometry cannot exist), and then the physics-math professionals have defined material interactions as random, non-linear, point-particle collisions (an extreme form of a definitive geometry which, nonetheless, is non-linear, and thus [still] indefinably random) which in turn adjusts the (phases of) global functions by means of "sets of particle-collision possibilities (identified in the non-linear context of differentiation)" defined at a point (the point of the particle-collision), ie the local phase changes [or local adjustments to the wave-phase], in turn, are to be used to determine the properties of a global function),

> [perhaps there needs to be considered a mechanism through which some action-at-a-distance "changes in phase" might exist, but this would imply the existence of stable geometry],

. . . . , and it is a model which has no relation to technical development; despite the observed spectral properties of stable, discrete, definitiveness, which these ("so called, randomly based") quantum systems are observed to possess (that is, a stable descriptive context has not been devised so as to describe, and subsequently use these stable properties) ie but they are definitive systems whose descriptive context is assumed to exist in a context of indefinable randomness (and thus the properties of these few highly controlled components of the spectral system are forever inaccessible to practical creative use).

New ideas need to be considered about the following systems;
the point of science description (the practical use of information about physical systems), and new descriptive contexts need to be considered in regard to the stable discrete properties observed for the spectra of general and fundamental systems as diverse as:

> general nuclei,
> general atoms,

> general molecules,
> general crystals,

Then there are the:

> linear, solvable, macroscopic classical systems (the part of the description which seems to be correct), and
>
> the stable properties of the planetary motions in the solar system.

and

> (Perhaps) Other math structures, which are different from "a continuum defined by the idea of materialism" (ie new models of material which fit into higher-dimensional containment structures) . . . ,

need to be considered:

(1) in regard to the "ordered" structures of material motions within galaxies, and
(2) the relative (expansive) motions of galaxies,
 . . . , such as stars having a higher-dimensional relation to the materials,
 eg each of the stars having a higher-dimensional relation to the material interactions within their associated solar systems.

The currently accepted paradigm of physics

If it is material which determines existence, ie materialism, and this has been the dominant viewpoint of science, then in this case materialism implies a containing space for the descriptions of material properties which are either the metric invariant spaces of space-time or space and time, or in the context of a manifold then containment is local so that one considers local space-time or local space and time in a context in which the manifold (or its space) is curved.

Where curved space is the main idea of general relativity, and general relativity has been neither widely applicable nor all that useful, thus curved space can be discarded as a scientific truth.

Remember that the precise ideas of Ptolemy were measurably verified, thus measurable verification of a precise description does not determine a scientific truth.

Thus within the context of a metric-invariant 4-space-time [or 3-space (and time)] it is assumed that material properties are either

determined by classical physics, so that differential equations determine the physical properties of a material system

or

.... that material (spectral properties) reduces to elementary particles, which identify random material-spectral events in space, so that the elementary particles have interaction fields which are spherically symmetric, where interactions depend on:

either

(1) elementary particle collisions at one particular point in space (the point of the collision [or observed] event in space) where internal particle states are transformed by actions of unitary matrices so that subsequently the different types of particle collisions (identified by the unitary actions on the internal particle states) perturb a wave function's spectral properties, due to both the different identities of the particles during the collision and the dynamic effects (of the identified particles) during and after the collision (this defines the structure of a derivative, or a connection, acting on one of the two types of manifolds (locally metric-invariant 4-space-time [or 3-space (and time)]) which contain the particular particle [or material] types which exist),

or

(2) interactions are determined by . . . :

> Adjoint,
> Number,
> Parity,
> Charge conjugation,
> Time,
> Hermitian,
> Unitary,
> Differential and
> Position

.... (wave function) operators which act on a quantum system's function space so as to determine the quantum system's spectral set (of functions which span the function space), where the spectral functions determine the random particle-spectral events of the particles which compose the system, so that these spectral functions can be perturbed by the connection when placed in the above spatial-material manifold.

Unfortunately, despite all the complexity and careful construction of this (above) abstract scenario for describing material, which reduces to random spectral-particle events, the spectral sets of general quantum systems, that is, the elementary event spaces of quantum systems, cannot be found by means of using this (above) method.

That is, the foundation of the probability based description of quantum physics cannot be established within the material-particle-probability based descriptive language.

Thus, the spectral properties of quantum systems are (can only be) found experimentally, or with the functional density method, which is an iteration process, where if a solution function to the system's Schrodinger equation is guessed, so as to be close enough to the ground state, then iterations converge (if the Schrodinger equation is bounded), ie but this is simply another form of experimentally determined spectral quantities (or properties) for quantum systems.

That is quantum physics is not widely applicable nor is it very useful (the only system to which the ideas of particle physics can be coupled in a useful manner is the nuclear bomb, and it is not clear if particle physics has even been useful in bomb engineering) and thus quantum physics can be discarded as a scientific truth. That is, the general principles of quantum physics cannot be used to find descriptions of spectral systems which are either accurate or (subsequently) useful (with one main exception, the H-atom).

Note: For 30 years after WW II, from 1945-1980, the physics departments of universities were dominated by the ideas about particle collisions of physicists who worked on the Manhatten project (development of the nuclear bomb) thus causing quantum and particle physics to dominate US physics, even though Einstein opposed quantum physics, and he was the main author (and facilitator) of quantum physics.

Note: Space-forms are consistent with Bohr's orbital model of the atom, yet space-forms allow both spin and spherical harmonics (while Bohr's model did not), but space-forms have more variety as to what the (radial, or principle spectral value [or principle energy level] of a) spectral-orbital structure of an atom might be, Sub-note: the spectral properties of a general quantum system cannot be described using the principles of quantum physics.

The spectra of quantum systems which are related to Plank's constant, h, are describable by the Bohr method, but only the spectra of the H-atom have been associated to h.

Note: Within the dysfunctional context of quantum physics, it is claimed "with the wave of a hand" that classical physics can be recovered from quantum physics if the quantum numbers are high enough (or equivalently, the energy of the physical system is very high), but classical fields are determined by the differential equations of differential forms, and this requires a local geometric structure which the uncertainty principle of quantum physics does not allow, where the uncertainty principle is a natural construct of both a quantum system's wave function defined on a function space and its definition on a Fourier transformed dual space, where under the Fourier transformation, x (position) goes to k (momentum [or p]), so that the uncertainty principle is defined in relation to the statistical construct of variance of a system's wave function defined on these two dual spaces, and where because of the uncertainty principle the physical property of momentum can locally tear the geometric properties of position apart at a particular point in space, (or vise versa) so that the classical differential equation at a particular point in space cannot be defined, ie the classical differential equation assumes exact values and/or extremely precise (small) values in its definition, and this is not allowed by the uncertainty principle of quantum physics.

That is, classical physics is a geometric context not a probabilistic context . . . , except for statistical physics and in some statistical contexts also thermal physics, though thermal physics is mostly described in the language of differential forms, as are both gravity and electromagnetism.

That is, the only widely applicable and very useful description which can be associated to physics is classical physics, the rest of modern physics is speculation driven mostly by the narrow selfish military-industrial interests, and most of these interests are about controlling scientific knowledge not developing scientific knowledge.

The authoritarian dogmatic language of modern quantum physics (2009) is used to create contests within educational institutions.

These bomb engineer quantum physicists should not be in education, rather they should be in some professional-industrial guild, because they are dogmatic, and because of this they oppose the education of free inquiry that Socrates rightfully identified as the road to the development of (new) knowledge.

This populating (of) physics departments with bomb engineers is an example of the destruction of knowledge which is associated to oligarchies, where the power of the oligarchs is derived from highly controlled markets, and in turn (based on a similar organization of power), highly controlled intellectual oligarchies.

The only widely applicable and very useful description which can be associated to physics is classical physics, the rest of modern physics is speculation which is all based on an assumption of materialism and quantum physics, the dominant descriptive language of modern physics, is based on materialism and probability.

Quantum theory can only provide a few qualitative properties which are reliable in regard to describing the physical world:

1. discreteness of energy structure, and
2. spin-related statistical rules for the occupancy of energy levels by the spin-related components,

> which can only be useful (in today's technology) if classical systems can be coupled to these discrete energy structures, then these quantum properties can be used in the classical system, most often in either electromagnetic systems or statistical systems

\
3. The only descriptive validity of the quantum energy operator of quantum systems and group representations of the 2-sphere . . . ,

> [which is a feature of the nuclear model of an atom and the spherical symmetry of the interaction]

. . . , is related to the spherical symmetry of material interactions in 3-space, and the nuclear structure of the atoms, and molecules, as well as the periodic "atomic" structure of crystals.

Furthermore, the math methods (developed in quantum theory and particle-physics), when applied to financial models of risk, have failed quite miserably, furthermore, these failures (of identifying risk accurately) have resulted (or aided in) in the destruction of the US economy.

If one bases a math model on instabilities and inconsistencies then the description is not measurable. But for a person to build something based on precise descriptive knowledge the person must have measurements which are reliable in the world in which the person exists, not in some abstract world which results form the assumptions of math wherein instabilities and inconsistencies are allowed in the precise descriptive language. If the logical foundations of a mathematical descriptive language are unstable and inconsistent then the patterns which are being described by such a language are not valid (the observed patterns are not sufficiently described) and they have no meaning (the patterns cannot be used in a practical manner).

Science is about measured properties (or data) concerning either "material and geometry" or "randomness in relation to material being reduced to (random) particle-events" interpreted within a logical structure which is consistent with the measured properties.

Logical structures are based on sets of assumptions, and through these assumptions are developed Platonic truths, ie truths based on word agreement (but Platonic truths may not be related to practical useful creativity, and thus such Platonic truths might have very little practical value, but nonetheless mathematicians adhere to such practically useless math structures, such as both the ideas of non-linearity and fundamentally random sets (which are never properly (or adequately) defined). The point of science and math is to always provide alternative sets of assumptions upon which precise, and thus useable, descriptions can be based. The logical structure within which a description is interpreted or based is to always be challenged especially when the current language lacks any relation to practical usefulness.

> (Such fundamentally random math structures have been used to model risk, and were subsequently instrumental in the economic collapse. This is because such math structures are invalid [they appear to be rigorous descriptions of an illusionary (or undefined) world], nonetheless, (apparently) the managers blindly followed their math authorities).

These invalid math structures are the basis for quantum physics and particle physics as well as non-linear models of physical systems.

Measured properties are either about elementary event spaces (which need to be well defined sets of identifiable events, if one wants to determine probabilities which are valid) or are measurements of properties of material systems identified in a geometric context.

It is only where measurements are geometric or the measurements are averages..., such as temperature, of well defined statistical averages, ie the numbers of particles is fixed within a confined system structure, within a system structure which has geometric properties..., that the measure information of descriptions of physical systems can be used for practical creativity.

Can the random sets be related to other properties (or other quantitative sets)? Yes, if averages are the measured properties and the system is confined or sufficiently restricted. Otherwise it is difficult to (causally) relate a set of random events to other sets. Nonetheless, it has been believed that this is what a "complete set of commuting Hermitian operators" acting on a function space..., where the function space represents the measurable random properties of the fundamentally random system..., can do.

Apparently this is not true.

Apparently this use of operators acting on (harmonic) function spaces can be used to fit data to some systems which are sufficiently restricted by a set of operators and a set of physical conditions, and there are no singularities in the equations (a rare property for small physical systems composed of charges) which are defined by the operators. But the reliability of the correlation between the calculated properties and the observed properties is always open to question, and trying to interact with the system, other than passively through observations, changes the system so that the original description is no longer valid and thus the interaction has no valid description.

These are not surprising results, it is more or less putting enough constraints on a set of harmonic functions (oscillating functions) can cause the set of harmonic functions to take on a structure of averages for the interior of the (mathematically) highly constrained system.

The problem with these harmonic average descriptions is that the physical systems which are most common are built out of charged-particle, or charged-components, where it assumed that the force field interaction geometry of these components is spherically symmetric and thus the operators and subsequent equations have singularities which cannot be resolved.

Thus, quantum physics cannot describe any of the fundamental physical systems in the world.

Before describing the quantitative descriptive structure of quantum physics and particle physics consider the:

Problems with quantum description

1. Quantum physics cannot find the wave-functions (or spectral sets) for fundamental quantum systems. It cannot find the spectra of: general atoms, nuclei, and the fundamental quantum properties of crystals goes without valid descriptions the critical temperature for superconductivity has been exceeded (1986) and as yet (2010) there are no corrections to this problem. There are many wave-equations which are found and it is claimed that they fit data, this seems to be for marginal systems composed of a few particles (one or two) or related to a statistically similar set of components in a simple geometry.
2. The probability based descriptive language of quantum physics follows data, its math structure is too complicated and too abstract to be able to create (a new abstract) wave-equations (or determine complete sets of commuting Hermitian operators) and then identify the quantum system, ie the limiting geometry and energy range, to which, such an abstract, wave-equation applies.

Before Dehmelt, isolated a "still" electron, no one predicted the electron would form into an orbital geometry on its own, but once observed, it is claimed that the wave-equation can be determined which fit's the data.

Probability follows data, it does not provide the structure from which ordered systems have their basis (in order).

Conjecture: To be able to predict the order of a system then the functions which represent the system's properties must be given in the variables of the function's domain space, ie the space within which the properties are actually observed.

It is not all that surprising that one can follow data with a probability structure built around an energy operator, especially since Hermitian operators of quantum physics are analogs of classical properties, but what is surprising is how inadequate the spectral representations of the solutions of these quantum wave-equations actually are (see 1).

3. The probabilities given by the quantum description are not useful. The model of quantum interaction does not describe, directly, the properties of any quantum system, eg the orbital stability of a general nuclei. Both the solutions to the wave-equation and the adjustments to the wave-function due to the particle-collision model of quantum interactions are descriptions which are remote from any capability, identified within the description, to control the geometry, the potential energy function, or the components, though it is claimed that the components in the particle-collision model of interaction are selected by nature based on the energy levels of the quantum system (or the quantum wave-function). The context of energy levels in a random structure (ultimately) means that it is unknown what components are a part of a quantum interaction (based on random particle collisions) at a single point, which carries with itself (the single point in space and time) a set of energy levels specific to the quantum system, but of the many "quantum systems" which can be identified at that same point, "To which 'quantum system' do the set of particle-collisions apply?" the property of randomness implies that particle collisions of any of the particle families whose participation in the particle-collision structure is to be determined by energy levels of a global wave-function collapsed to a point.
4. In the physical world probability is associated to order only in the context of large reservoirs composed of "many" components in relation to properties which represent average values. This is the context of thermal physics as well as classical statistical physics (where in classical statistical physics the existence of a structure for quantum states is related to the idea of entropy [though this structure of quantum energy states is the rather tenuous structure of a "subjective" box into which the constrained thermal system is placed] (space-form structures provide a more substantial model for a structure for energy states for a confined set of components)). The highly ordered quantum system is composed of relatively few components, from 1 up to about 500, so that these components identify (or are a part of) a geometric and bounded structure.
5. The highly ordered quantum system implies the existence of a descriptive language which also provides a new context for the control of these highly ordered quantum systems, such as a geometric description.

The idea of separable geometries within a many dimensional containment set within a metric-invariant context, where different signatures of the metric function at different (spatial) dimensional levels plays an important role in understanding the properties and structure of material systems and for descriptions which extend beyond material systems.

What particle collisions identify is a vertex of a stable hyperbolic space-form (occupied by charges), the unitary transformations, which it is assumed identify random particle-state fluctuations (or oscillations), are not related to a fundamental random and harmonic structure which an over-all unitary invariance engulfs (or bounds), rather they are local transformations related to the two (or more) fundamental "metric-space-states" associated to a spin rotation of metric-space states on both metric-spaces and on space-forms (thus it can be a many dimensional structure to the spin rotation of (metric-space) states which is given the geometric context of material space-forms and interaction space-forms and metric-space space-forms. This spin-rotation of metric-space states has a relation to dynamics which in turn, is related to unitary Lie algebra valued 2-differential-forms of local groups transformations (of spatial displacements) associated to the 2-form force fields.

The hidden higher dimensional particle-states is evidence for the existence of higher dimensional macroscopic structures, so that the structure of interactions within the context of space-form geometries which exist at all dimensional levels (perhaps stopping at hyperbolic dimension eleven, due to a theorem of D Coxeter, about hyperbolic discrete groups) defines a property of geometric independence between each dimensional level. That is the dimensional structure which particle from particle collisions seem to follow is not an internal particle-sate structure but rather a (real) macroscopic dimensional structure of existence, where geometry is the basis for describing the properties of quantum systems.

The interaction space-form structure also identifies the source of an apparent randomness associated to small material systems, where this randomness is the main structure of quantum description based on probability, but instead geometry is the best language with which to describe the properties of quantum systems.

II. What is quantum description, and what are some of its (other) problems and limitations?

In quantum physics measured properties are the random spectral-particle events in space and time. A quantum system is a function space with a Hermitian-form and which is L^2, ie the square of a function integrated over all space converges (or it is a Hilbert function space). A quantum system's measurable properties are associated to a complete set of commuting Hermitian operators, ie operators with real eigenvalues, associated to the observed random spectral-particle events in space and time.

That is, the observed spectral values form a quantum system's elementary event space.

Unfortunately, a meaningful correlation between a general quantum system's observed spectral values and its complete set of commuting Hermitian operators does not exist.

After (not) finding the quantum system's elementary event space . . . , or the function space's set of spectral functions which satisfy the wave-equation (or which are the eigenfunctions of the complete set of commuting operators) . . . , then one must adjust the measured (or observed) spectral values . . . , [or adjust the harmonic eigenfunctions (or spectral functions) of a wave-function, which "diagonalizes" the quantum system's complete set of commuting operators, which act on a quantum system's function space, (or adjust the measured values which define a quantum system's elementary event space)], . . . , so that this (spectral) adjustment is based on a further (or second) set of oscillatory properties of energy-dependent local unitary transformations of particle-state (transformations which define the second harmonic structure) during (an assumed) process of particle-collisions, where elementary-particle-collisions identify the quantum model of material interaction.

This model of second harmonicity is modeled as a connection (derivative), which is usually associated to a "derivative" on a non-linear geometry, yet the harmonic functions of quantum description do not allow the existence of a local geometry, nonetheless, in the particle-collision model of quantum (field) interactions, if the collision is that of charged particles, then this (collision) defines the point of singularity for the $1/r$ energy term of the energy operator related to the first harmonic approximation of a quantum system, which contains two charged particles.

What is the basis for the second harmonic structure (or hidden particle-state transformations), used as a small adjustment to the energy levels associated to the Hermitian operators (or to the energy operator) of a "regular" quantum system? When elementary charged particles are collided with one another in particle accelerators a wide variety, but a finite number, of elementary particles can be identified in regard to, mass, charge, and spin, but which are not usually seen without both high energy and collisions. These sets of elementary particles fit into a particle-state vector structure (of a higher dimensional space) associated to local vector transformations of the particle-states by low dimensional unitary Lie groups. This is the basis for a "reduction to elementary particles" model of quantum systems, which is forced into quantum description based on a collision model of quantum interactions.

Harmonic functions and Hermtian operators, ie operators with real eigenvalues and which act on the harmonic-function's function-space, and a quantum system's defining space of "stable spectral properties" associated to a complete set of commuting Hermtian operators, along with a secondary harmonic structure of elementary particles and a unitary transformation of a "hidden space" of elementary-particle-states, assumed to define quantum interactions based on particle collisions, so that both harmonic structures are unitary invariant and thus preserve the real eigenvalues of Hermitian operators so that this first and second sets of harmonic structures form the basis for a probability description of quantum systems.

This cannot be taken seriously (1) first quantization cannot find valid models of wave-functions for general quantum systems (2) that quantum interactions are to be based on the collisions of elementary particles is simply wishful thinking, where unitary particle-state transformations simply follow particle-collision data "from particle accelerators," and has no obvious relation to the ordered structures of quantum systems, otherwise it would be used to identify the stable orbital properties of (general) nuclei, but this has not been done.

Thus, none of quantum physics and none of particle physics should be taken seriously since containment in a probability based description can be accomplished by "proclamation," if there exists a stable elementary event space (and it is done in practice without a stable, well-defined elementary event space, where such probability based descriptions make statements which have no reliable content, and this type of "probability" language still seems to be the basis through risks are assessed), but unless the observed stable elementary event spaces of general quantum systems are either explained or "identified" by a descriptive language then the description is irrelevant.

A connection operator and second harmonicity might be interesting in regard to the claimed harmonic covers of all (separable [?]) geometric shapes if it was a structure which had widely applicable and valid applications, apparently there are a few such applications but one wonders if its uses are the result of assumptions which result in errors which cancel each other out. That is, consider the question, "What is the probability for this to be the case?" Perhaps this question is "the main question" which quantum physics should be asking itself.

Other descriptive contexts for (apparent) particle collisions

However, space-forms have a distinguished point about which interactions center . . . , as is also the case for a center of mass . . . , and about the distinguished point (on the space-form) there are separable hyperbolic coordinates. Furthermore, the space-form can be occupied in one of its spectral flows. Thus, if a high energy collision breaks such a space-form apart, then the occupying material would naturally shift over to hyperbolic coordinates (which are properties of the hyperbolic space-form) and the occupying component (or occupying energy) would have a definite energy or it would break apart into definite energies (or masses).

Thus, there is a different description of collisions of small material systems which can account for the same set of data, but it does not need an assumption of a hidden particle-state space and the property of unitary patterns could be related to the unitary invariance of the space-form, thus the dimensional structure of the break-up due to the high energy collision could be influenced by the space-form's relation to 8unitary groups.

Note: The geometric potential energy terms are the only aspects of a quantum system's wave-equation which allow a spectral structure to come into existence.

Yet, the harmonic (or random) structure of quantum description (along with its dual function space) does not allow geometry to have any meaning, or any local structure (due to the uncertainty principle defined on a function space in relation to its dual function space).

The second harmonic adjusting process does not depend on the adjustment having a relation to the domain space of the "first quantized" quantum-spectral functions, and only adds function (or spectral) values to the "first quantized" wave-function (or to its spectral set). This is not a valid math process for locally adjusting a function's values (or adjusting eigenvalues), that is, local adjustments (usually) require a relation to the function's domain values.

This structure of quantum description, based on a second harmonic structure, is an elaborate and very complicated data fitting process, which is irrelevant since there do not exist any (or only a very few) valid first harmonic approximations of quantum systems upon which one can apply corrections (or to adjust the original wave-function), where energy bounds on the second harmonic structure are (or can be) imposed so as to fit data.

Thus, one sees an idea in quantum physics that spectra are about energy bounds and energy values not represented in the domain variables, rather than geometric structures which are associated to the domain space. Nonetheless it is only energy structures which are related to the geometry of the potential-energy term in the first quantization wave-equation which actually affect a spectral structure of a quantum system.

Clearly this is a descriptive language which has gone delusional, or hey-wire, or gone batty.

There is a set of first harmonic operators, the wave-equation, along with what is considered a more fundamental . . . (because these second set of operators define the structure of all material interactions in terms of the collisions of the elementary particles, into which it is assumed all material systems reduce) . . . second set of harmonic operators, ie particle-state transformations, so that all these harmonic structures are unitary invariant.

This descriptive structure requires that such a description can have no practical applications and it applies in its actual calculations to a very few quantum systems.

That is, the basic structure of the quantum language is that:

The properties of a quantum system are contained in an elementary event space,
Material-geometry of a quantum system is contained in the harmonic structure of a quantum system's function space, and subsequently,
The harmonic structure of (hidden) particle-state oscillations are introduced to be consistent with the reduction of (quantum) material systems to elementary particles, and their (assumed) collision interaction structure, ie that of low dimension unitary Lie groups,

where

The common property of the descriptive structure is that it is unitary invariant, which implies random harmonic functions, of quantum systems which are composed of a relatively small number of components, and which are energy dependent, but quantum systems have no distinguishable geometric properties except for the potential energy term in the first quantized wave-equation (or Schrodinger equation). Note: There are two basic geometries for the potential function in quantum physics: (1) the box, and (2) the, $1/r$, potential with spherical symmetry.

The only aspect of this descriptive structure which could have any relation to control is the potential energy term in the wave-equation, and the geometry of this term must be geometrically separable or the wave-equation will not be solvable.

That is, the first and second harmonic structures, ie function space randomness and energy eigenvalues and particle-state transformations for particle-collision interactions (which are energy dependent), are both unitary invariant math (harmonic function and/or particle-state) containment structures, thus they preserve the real eigenvalues of Hermitian operators, thus the elementary event space of spectral values is preserved, yet the harmonic structure and its particle-state structure and its associated operator-eigenvalue structure is not capable of identifying the observed spectral properties of general quantum systems.

That is the actual elementary events of general quantum systems are outside what the descriptive structures, ie the Hermitian operators, can identify, nonetheless, a second adjustment to the function (or spectral) values (already [hypothetically] filtered by the system's regular quantum operators) so that the second adjusting harmonic structure is based on an elementary event space of elementary-particle-states, a space which is randomly fluctuating within itself, so that these fluctuations affect quantum interactions modeled as particle-collisions, because these fluctuations are dependent on the energy levels of the general quantum system, within which the adjustments, due to elementary-particle-collisions, are to be made. These adjustments are to be assessed at a single point [if a system whose components are charge] the single point is the site of an operator singularity, the particle-collision interactions change all the values of the quantum system's spectral functions globally so that (supposedly) a "better" representation of a hypothetical eigenvalue or eigen-function can be found.

If it were true, that a random oscillation on a hidden elementary event space (or particle-state space) occurs globally and it is sensitive to a quantum system's energy (levels), then because in a quantum system there is (can be) a distribution of energy, either due to the motion of its components or due to the distribution of the system's material (or mass, ie mass=energy) then, How can a calculation at a single point be valid in relation to global function values?

This is a description which does not provide a context for (about) how physical systems acquire their observed ordered properties. They are models which try to follow data, but it cannot do this

In classical description one has properties, eg unknown functions of values of a particular type, then containment, ie the set of measurable values, eg material, geometric measures, motion, and finally (unknown) measurable properties can be found by means of a local linear measurement of properties placed in a context of separable geometries and geometric causes.

In quantum description one has properties which are either Hermitian operators or sets of spectral values, an elementary event space, obtained from the observed spectral values of random particle-spectral events related to the quantum system, so that these properties are to be described in a random harmonic structure of functions, ie in a function space whose "spectral functions" are to be found by applying to the function space a complete set of commuting Hermitian operators so as to diagonalize the function space, ie finding the (a) set of spectral functions which cover the function space, and are consistent with the quantum system's spectral properties (apparently this cannot be done).

Thus, Hermitian operators equal [or are associated to] values of the elementary events to which the probabilities of harmonic functions are related. The spectral properties of quantum systems are the eigenvalues (or set of operators) and the concept of containment is (believed to be) fulfilled by the quantum system's function space.

Thus material, reduced to the particle components of a quantum system, is contained in a harmonic structure so that material equals particles which are equal to random particle-spectral events (in space and time, the same structure which material defines in classical description, but the spectral values are separated from the domain space).

The function space is associated to Hermitan operators which act on the function space so as to determine a quantum system's real eigenvalues but the function space also has a Hermitian structure (function norm) defined on itself, and subsequently the function space is unitary invariant.

The model for material interactions in quantum physics are elementary particle collisions which are modeled in a context of non-linear local connection operators associated to local (low dimension) unitary particle-state transformations. This is (the source of) a second level of harmonicity associated to the description, and all aspects of the description are unitary invariant, and thus the descriptive structure preserves the "spectral elementary event space" which represents the probability interpretation of which the harmonic functions are the basis (in the descriptive language).

In the second harmonic structure, energy is no longer an energy operator but rather it is related to motion and mass of the colliding particles, where motion is, in turn, related to, an assumed, particle confinement, or orbital containment.

There are two types of particles to consider in the collisions, (1) the charged Fermions, which because of the known structure of quantum systems are usually of the same charge-sign, thus causing instability, while (2) the Bosons, or field particles, eg photons. The Fermions and the Bosons are of two types of particles which collide with one another. The Fermions are those particles associated to either the same charges or those associated to opposite charges. Bosons tend to be common in the state which suits them, ie their natural confined state, while Fermions are rare in the same state, because of the spin-½ properties, unless the Fermions are modeled not as "being in a specific state" but rather modeled as being in a universal (rare) gas-state in a box (as free electrons are modeled in crystals, without a need to consider spin-½ states).

However, Fermion collisions would involve singularities and divergence, but motion of confinement suggest that some Fermion collisions (of the same sign) do (or could) occur. (Can such a model be consistent?)

The Boson particle collisions, for the Bosons from the same charge, will be the most prevalent (since the same charges are the closest together in quantum systems, such as atoms and nuclei) and the confinement in motion that these "same charge" field particles (Bosons) imply is not confinement but rather divergence, so that this implies instability, and thus (in a stable, confined quantum system) this set of collision events, which imply divergence, must adjusted with collisions of Bosons from something else, for example, anti-matter, but if anti-matter particles collide they give-off energy which will cause divergence. Thus, the epicycle structure becomes ever more intricate and complicated and subjective, where claims in which some energies are favored (confining Bosons) while other energies are excluded (matter-anti-matter collisions) becomes the rules by which quantum interactions are interpreted and modeled.

It should be clear that the occupation of (same) charged components in stable flows of "hyperbolic space-forms" . . . , where the spectral properties of the hyperbolic space-form are determined by its resonance relations with a greater containing space . . . , leads to a description of a stable quantum system in which geometry and spectra are consistent with one another (and they are properties which are contained in one containing space whose variables are inter-related).

Whereas the current quantum description is an elaborate math structure whose only purpose is to follow data, but which fails at following data for general quantum systems, and subsequently results in the statement that "fundamental quantum systems are too complicated to describe" in the current orthodoxy of physical description. Furthermore, the description does not lead to control or development of quantum systems, and it cannot describe the orbital-spectral properties of fundamental quantum systems, eg nuclei and general atoms.

Some social comment about science

The main function of quantum physics and particle physics (and all their derivative theories, along with the theories driven by non-linearity) is to provide a "difficult to memorize" set of irrelevant math structures, ie math structures related to illusions (but which are claimed to be measurably verified), which are of great interest for aggressive, competitive "autistic types of people" who faithfully . . . (within the authority which they keep alive through their memories and their competitiveness) . . . . compete for good salaries (wage-slaves), concerning a descriptive language which is given great authority, since the ideas of quantum physics are used to improve the ideas of bomb engineering (where fundamental to understanding an exploding bomb, are particle collisions). Quantum physics has an authority which is similar to the authority of the church in the age of Copernicus, and this authority destroys science (physics).

These difficult to memorize but irrelevant math structures for the physical description of quantum physics and quantum interactions, ie particle-collisions and a second harmonic structure related to particle-states, isolates a few people and it channels them into the business interests of the military-industrial complex.

Science and knowledge in society

The great authority and the huge irrelevant edifice held in place by (overly) obsessive people who do not (or cannot) critically assess "what it is that they are describing," nor do they create new contexts for new descriptive languages . . . , which when based on the conclusion of Godel's incompleteness theorem . . . , is an imperative of both valid science and a valid education system.

New knowledge and creative possibilities go "hand-in-hand" with one another. Knowledge is not developed through authority and illusion, nor is it developed through competition, where competition defines narrow authority. Knowledge is developed through free inquiry, freedom to believe, free expression, and freedom to use knowledge to be creative (in a selfless manner). All of these freedoms require (or are based on) equality.

Back to descriptive languages

A math model which implies stable structures is needed, eg space-forms or separable geometries of discrete isometry or discrete unitary subgroups. This quantum description states that one cannot influence, or one cannot have any control over, quantum interactions, ie particle collisions, in quantum systems which have particular energy ranges.

The geometry of the potential energy term . . . of first quantization (or at the wave-equation) level of description . . . . can have some affect on a quantum system's spectra, but this seems to be restricted to potentials which are containment boxes, in which context, it is believed that crystal lattice interactions are important, and to be modeled as a second quantum harmonic affect, but how to model these interactions so as to be consistent with observed data is not known, eg the critical temperature of superconductivity cannot be predicted.

On the other hand, the rotational symmetry of $1/r$ potential energies seems to be a prevalent property for quantum systems composed of many particles in 3-space (3-Euclidean space), but the geometry of the components is not known, and thus the equations for these types of quantum systems are also unknown.

That is, geometry seems to be important in understanding the spectra of quantum systems but how to model this geometry is unknown, however, the energy levels within which particle-collision interactions take place is also important for this description, but this information of quantum energy levels of general quantum systems is beyond what this type of quantum descriptive language can provide.

Perhaps collisions of elementary particles are not common in quantum systems.
The truth or falseness of is idea cannot be determined.
That is, both the wave-equation and the particle-collision model of quantum interaction are both (all) speculative models of physical descriptions which have not proven themselves to be either widely applicable nor are they "at all" useful, though a probability based descriptive language can always claims to be relevant to a system's eigenvalue description, but the actual usefulness . . . , "that this 'claim for relevance' has," . . . . is so limited (so limited a level of usefulness) that only classical physics should be identified as being science, in regard to the currently accepted quantitative descriptions of physical systems. Whereas, the properties related to the descriptions of quantum randomness as well as non-linear descriptions should both only be labeled as speculations, and poor speculations "at that."

Perhaps another (new) descriptive language can be used to describe the order that is observed for quantum systems, and perhaps the new alternative language can actually identify the spectral values that quantum systems possess, and be related to greater control of quantum systems.

Basically quantum physics is based on describing both the observed randomness of particle-spectral events in space and time (or in space-time), and in preserving the idea of materialism, where material defines the space where observed material events exist (or observed material properties are measured). That is, it is material-particle-spectral values which identify a quantum system's defining elementary events spaces. The observed spectral properties are separated from the variables of the space in which they are observed. Function space of harmonic functions identify probability distributions of random material events in space while Hermitian operators act on the function space so as to identify the observed spectral values. The elementary event space upon which a quantum system is defined are the set of observed stable spectral values which Hermitian operators identify. Hermitian operators are invariant to unitary transformations and the function space is defined to have a Hermitian invariant structure, where the function space "must be" L^2 (or they must be Hilbert spaces) to allow this.

Within this context it is claimed that there is no local geometry, and where a harmonic function space can cover (or can converge to) any geometric structure. Thus, the constraining potential energy term related to a box makes sense, but the singularity, at $r=0$, of a potential energy term of k/r makes no sense near $r=0$. Thus, the harmonic descriptions of the charged components of a small bounded charged systems cannot be handled in a consistent manner.

This is the vast majority of quantum systems.

Unfortunately, the only quantum systems which have valid approximations to the observed spectral structures are those qualitative systems with potential walls and spherically symmetric systems with a, 1/r, potential energy term in the defining (energy) wave-equation, and even then, the formal math procedures related to solving the radial equation for the separable wave equation for the H-atom diverges, so that the solutions to the radial equation are truncations of the diverging radial series. These truncations allow the principle quantum values to be given a formula in quantum numbers, ie these quantum numbers are not given in the variables of the function's domain space. Thus one sees a descriptive structure for a quantum system's spectra which is detached from the space where the spectral values are observed.

The only quantum systems which have anything close to a valid wave-function with spectral formulas, represented in quantum numbers, are:

the harmonic oscillator,
the box,
potential walls, and
the H-atom, and
maybe three other atoms: H, He, He^+, and Li (?).

That is, the general quantum system cannot have its elementary event space, ie its spectral values, identified by finding the "complete set of commuting Hermitian operators" which represent the complete set of measurable properties which identify the quantum system's elementary event space, ie the eigenvalues, so that when applied to the quantum system's function space, these properties can be found from the implied calculations (ie of solving the equations determined by the complete set of commuting operators acting on the quantum system's function space, so as to identify the system's eiegnfunctions and eigenvalues).

Thus only a few wave-functions for quantum systems are known, but in relation to this limited set of wave-functions for quantum systems, there is also defined a second harmonic structure which is used to adjust the wave-function and its eigenvalues so as to better fit the calculated spectral values with the observed spectral values of these few quantum systems. This second harmonic structure is based on unitary invariance, and the unitary changes in particle-state which occur during the particle collisions of the particle-collision model of quantum interactions, ie the interaction of a quantum system's components with the field structure of field particles involved in quantum interactions. This is a hopelessly complicated scheme for the calculation of adjustments to only a very few actual wave-functions and it is claimed that this epicycle data fitting structure is true because it gets two calculations to within 15-decimal places of the observed values. This is a joke. This descriptive structure is not widely applicable (in applications) and it has only the most limited uses, namely, as betting odds for games of chance one might want to design around the descriptions of the random quantum events of these few calculable quantum systems.

The only possible value of this descriptive structure for quantum systems is if it could identify good approximations to the actual observed quantum structures of general quantum systems, as this would be a useable form of information, especially if it were difficult to determine, through observations, a quantum system's spectra.

But quantum physics cannot do this.

Remember, if precision were the basis for determining the validity of a (quantitative) descriptive truth then Copernicus's ideas would have been discarded over the more precise Ptolemaic system, when Copernicus first proposed his ideas.

The higher dimensional containment of physical description placed in at context of separable geometries, ie discrete isometry subgroups (or space-forms), can be used to describe the properties of quantum systems.

The relation that the "set containment of descriptions of measurable properties" has to separable geometries contained in higher dimensions, is a speculation about the nature of physical description that implies a more definite descriptive structure than does quantum description and is equivalent to classical description so as to extend classical description to higher dimensions, and suggests a geometric structure which can more easily be used (than quantum description can be used) in relation to identifying separable geometries between adjacent dimensional levels which can be patterns which can be put to use either as information about the patterns that exist (in relation to the description of the properties of physical systems) or in terms of control and geometric causes of the physical (or existing) systems one is describing with the new language.

Chapter 14

PARTICLE-PHYSICS

If the (particle) events are not stable (just short lived resonances) .... by which it is determined, in a probability based description, that particle states (or a quantum system) transform (upon collision) within the pattern of unitary transformations.... then again it is not clear that an elementary event space (of elementary particles) is truly being distinguished..., an elementary event space through which physical systems of colliding particles can be described in terms of probability..., so that within this probability based description (it is not clear that) the events which are being described (detected particles with short lives, ie unstable spectra), in fact, actually exist.

Perhaps the existence of such particle resonances in high energy particle collisions are simply evidence that the containing space (when placed within the correct context) is a complex Hermitian (or some variation) space.

Eg Bizarre function properties, which relate spectra to (geometric) infinities, in the dual pairs of a function-space and its dual-function-space, such as the Dirac delta functions and Green operators

In the context of differential equations, Dirac delta functions are (most) often associated to singularities of a geometric system (also to the singular point of point-particle collisions), either of a spherically symmetric, $1/r$, point-charge (when a "solution function" is found), or dropping a pebble into a pond to initiate a wave, or the (untenable) elementary point-particle collisions (of particle-physics), where in both cases this is really about insisting that, there exists a very stable system structure, so that geometric infinities can be ignored, or the geometric infinities are placed in a new context, which is related to spectral properties, but without really knowing (in regard to being able to discount infinities) what aspect of the system is being related to spectral properties, which would imply that certain aspects of the (now assumed) spectral-system are stable, eg (hopefully) the spectra, yet the set of spectra which are assumed to be present are really not known. Why would particular aspects of a (spectral) system be stable? or Why would such an approximation (of infinity) work?

According to Manin, particle physics began with Feynman path integrals which were applied to Feynman graphs of particle interactions defined in regard to the Yang-Mills connection and unitary invariance of the wave-function, where the graphs identify the particle interaction pathways (as graphs), and where some of these Feynman graphs are loops, thus identifying holes in space (related to homotopies) within the hidden particle-state space, and this "connection" structure was applied in the form of, Fe^{iL} =(so as to get)= (sum) $[a(n)(A)^n]F$ (see above), which represents a set of spectral perturbations applied to a wave-function, F, of a quantum system, where the perturbations are due to particle interactions, ie particle collisions.

This is the only model of quantum interactions other than interactions identified by potential energy terms in the Schrodinger (unitary) energy operator.

Manin points out that particle physics has no mathematical foundation. A statement of fundamental educational value, and though there are a few hints of this in the physics literature (Dirac is also honest), the physicists seem to be unaware of this problem, but then again they seem to not be allowed to be physicists if they point out the utter uselessness of particle physics, as a description of observed patterns, in a practical context.

String theory grew out of the (Platonic) mathematical determination of Feynman graphs, eg homotopies (ie continuous deformations of "surfaces" in space where the deformations get caught on the holes which exist in the space which contains the "surface") and the subsequent geometry of the hidden particle-state space, as well as the energy level of the collision which is related to determining which group transformations are relevant to the wave-function which is (supposedly) being perturbed, ie mathematical machinery associated to particular interpretations of the data in attempts to match the mathematical process (associated to particle-collisions and wave-functions) to the observed data.

This is no different from Ptolemy building his mathematical model based on the authority that the earth was in the center of the solar system.

Particle physics is simply an extension of the idea of fundamental randomness, and the randomness is even more substantially related to the reduction of all material systems to elementary particle components whose only attributes are

"to randomly collide with one another (based on cross sectional collision probabilities) and which have unitary patterns of particle transformations depending on both the energy-range of the particle-collision (where a particle-collision is an extremely definitive geometry, in a theory which cannot (is not supposed to) define geometry on a small scale) (transformations) to the different particle-types."

Though this particle-collision model is supposed to describe a small corrective value to the energy structure of a quantum system's wave-function, it has as yet to be related to the stable discrete definitive spectral-orbital properties that are observed for a nucleus, (exactly) the one type of quantum system where the strong force is supposed to be a fundamental determining property of the nuclear system's properties.

On the other hand these particle-collision properties do fit into the engineering attributes which might be important in an atomic bomb. Furthermore, the assumed properties of an atomic blast is a highly dense, and high energy, center which might (also) be modeled as having spherical symmetry (in the core of an atomic or thermonuclear bomb blast), thus identifying a motive to consider the gravitational singularities of general relativity to be so important so as to be continually funded as a research interest, yet this is also all about secrets of political power within the world by oligarchs who interfere so much with the lives of people everywhere for their selfish interests, ie many seek the great power of nearly absolute domination (of the many by the few).

That is, particle-physics is far from being proved, yet it is the dominant idea of interest to those who own the society. As a matter of fact, with the geometrization ideas of Thurston-Perelman (where the main stable geometry to be described in a bounded context is not the geometry of the sphere but rather the geometry of the discrete hyperbolic shape) it is clear that geometry should not be introduced at the level of strings [modeling exceedingly small particles of quantum physics (whose force fields are supposed to be spherically symmetric so as to define singularities whose presence makes solving differential equations which model quantum systems impossible)], rather the geometry of all quantum systems needs to be understood as being based on stable, geometrically separable, and (the description) contained in either the hyperbolic or the Euclidean metric-invariant metric-space context, and in a many-dimensional context (which also happens to be a unitary [or complex-coordinate] context).

It should be noted that quantum physics (including particle-physics) and all the theories derived from quantum physics, eg string theory, adhere diligently to the idea of materialism, where the many dimensions of string theory require that the structure of the higher dimensions be (so that) the 3-spatial, or 4-space-time, dimensions so as to be consistent with the idea of materialism.

On the other hand, the many-dimensional structure, which one can use to describe existence based on the stable discrete hyperbolic geometries, has other means of identifying what the property of material is in higher-dimensions so that there is little geometric interference . . . , (though a high-dimensional over-all containing space is a well defined finite spectral set [where, because of the properties of discrete hyperbolic geometries identified by Coxeter, the highest dimension hyperbolic containing metric-space is of dimension-ten, ie 11-dimensional (general) space-time])

. . . , between dimensional levels at the lower dimensions, where this non-interference depends on both the structure of material and its interactions as well as the values of the physical constants.

CHAPTER 15

BASIS FOR PHYSICS

The math structure of physics:

How many ways can one organize a many-dimensional containment set, along with the set-constructs of both system interactions, and the systems (themselves) . . . ,
  eg operators defined on functions, where functions can model:

> either
> measured properties, ie solution function to differential equations (classical physics),
> or
> probabilities for random local-spectral events (quantum physics),

. . . , so that continuity for physical processes (eg conservation laws (for quantum systems), and changes of motion remain consistent with small changes in the domain space (for non-linear geometries)) can be defined differently for different containment constructs?

That is, one already has within the dogmas of physics

either

> I. Geometry of a fixed dimension (caused by materialism),
> eg the laws of Newton, Maxwell, and thermal physics:

Defined

either

over space, along with an assumption of spherically symmetric material interactions and current dependent force-fields, ie gravity and electromagnetism,

or

over a containment set of measurable physical properties which are averages over many components, which fit into a large reservoir contained in a Euclidean space, ie thermal and statistical physics.

or

II. Randomness of local-spectral-events in space (these events are modeled as particle-collisions)

1. Function spaces (the probability structure of the random events of measured values which define a quantum system, described by averages but composed of relatively few numbers of components, a context for averages quite different from the structure of averages in classical thermal physics)

2. Sets of operators, the set of measured properties which identify the system's spectral set (which cannot be

   [in general, have not been]
   found for general quantum systems, which, in turn, implies non-commutativity of operators, or
   equivalently, non-linearity on the function space, from which the solution function is to be constructed, where the solution function is to be represented as a series of spectral functions)

3. Rules for formulation of the relation between a function space and the measured properties (defined on the function space, ie the eigenvalues)

   or, equivalently,
   The (algebraic and set structure) rules for solvability, ie finding the spectral set, or diagonalizing the function space with a set of operators. {A complete set of commuting, Hermtian operators}

4. The spectral function as representing a local-global-random set-of-spectral-events in space (where such a detection of the system's spectra, results in the collapses the quantum system's set-and-function structure)

5. Material interactions which possess the properties of both of a, 1/r, spherically-symmetric singularity points, as well as non-linear particle-collisions.

Note: This set structure is "doubly" non-linear in both (2) and (5).

(1) In the linear, metric-invariant, geometrically separable context, differential equations based on geometry are solvable, and this is the context of classical physics and it is the basis for virtually all of our society's technical development (TV's, computers, Micro-chips, aero-space, including classical thermal and statistical physics, where classical statistical physics is based on a finite elementary event space, based on a fixed number of N particles contained in a closed system so that many of the measured values are averages over N, [see (3)] etc).

(2) In the linear, metric-invariant, geometrically separable context, for mechanical or electromagnetic waves, or for a few quantum systems, the equations for these systems are also solvable, and are either very useful in classical descriptions, or slightly useful in a limited (probabilistic) context of quantum descriptions. But, this context is not common for the set of general, many-component quantum systems, and determining the spectra of quantum systems based on a more abstract math context, eg many components placed within a context of 1/r potential energy (but where spherical symmetry can no longer be assumed to exist [since there are many components] and geometry is no longer valid [and then there is the quantum-foam which is a part of particle-physics renormalization techniques, etc]) has not been useful. Yet this method, along with the "even worse methods of particle-physics" are, essentially, the only math methods allowed, in regard to determining the spectra of general quantum systems.

(3) In professional math communities there is the belief that indefinably random events, which are (nonetheless) distinguishable "elementary" events, can be consistently described in a quantitatively consistent context, but this belief seems to only lead to failure. [This is because such improperly defined events are believed to be consistently related to a stable counting process (because they are distinguishable events) which (it is believed) allows probabilities of these unstable events to be determined, but this is not true (counting is not well defined for events which can change as time changes).]

Science is about measured properties (or data) concerning either "material and geometry" or "randomness in relation to material being reduced to (random) particle-events" interpreted within a logical structure which is consistent with the measured properties.

Logical structures are based on sets of assumptions, and through these assumptions are developed Platonic truths, ie truths based on word agreement (but Platonic truths may not be related to practical useful creativity, and thus such Platonic truths might have very little practical value, but nonetheless mathematicians adhere to such practically useless math structures, such as both the ideas of non-linearity and fundamentally random sets (which are never properly (or adequately) defined). The point of science and math is to always provide alternative sets of assumptions upon which precise, and thus useable, descriptions can be based. The logical structure within which a description is interpreted or based is to always be challenged especially when the current language lacks any relation to practical usefulness.

(Such fundamentally random math structures have been used to model risk, and were subsequently instrumental in the economic collapse. This is because such math structures are invalid [they appear to be rigorous descriptions of an illusionary (or undefined) world], nonetheless, (apparently) the managers blindly followed their math authorities).

These invalid math structures are the basis for quantum physics and particle physics as well as non-linear models of physical systems.

Measured properties are either about elementary event spaces (which need to be well defined sets of identifiable events, if one wants to determine probabilities which are valid) or are measurements of properties of material systems identified in a geometric context.

It is only where measurements are geometric or the measurements are averages . . . , such as temperature, of well defined statistical averages, ie the numbers of particles is fixed within a confined system structure, within a system structure which has geometric properties . . . , that the measure information of descriptions of physical systems can be used for practical creativity.

Can the random sets be related to other properties (or other quantitative sets)? Yes, if averages are the measured properties and the system is confined or sufficiently restricted. Otherwise it is difficult to (causally) relate a set of random events to other sets. Nonetheless, it has been believed that this is what a "complete set of commuting Hermitian operators" acting on a function space . . . , where the function space represents the measurable random properties of the fundamentally random system . . . , can do.

Apparently this is not true.

What is the mathematical basis for the measurability of relatively stable structures?

> Is it the new ideas presented in this short paper*
> or
> Is measurement about defining measurable values by defining convergences, (eg sequences, bounds, least upper bounds, limits, derivatives, integrals, function spaces, distribution functions), which can only be defined on sets which are too-big, so that these "too-big" sets, can no longer be logically consistent sets.

Measurable consistency requires metric-invariant coordinate metric-spaces, linear differential equations for systems, the system being defined on separable geometries, and descriptive continuity, eg conservation of energy and material.

The issue is to describe in a useful (or geometric) way..., so that description leads to a lot of creative development..., the general, stable, discrete, definitive orbital-spectral structures of material systems at all size scales.

Despite all the hype of a complete and dominant knowledge possessed by the physics community: quantum physics, particle physics and general relativity and all their derivative theories (such as string theory), cannot do this, ie they cannot describe the general stable orbital-spectral structures of the fundamental systems which compose the vast amount of our experience. It should be pointed-out that classical physics is a very useful description in the limited range of its solvability ie the linear, metric-invariant, separable geometric contexts of systems differential equations on a large size-scale, though nano-physics is approximately the lower size-scale limit where classical physics is supposed to be valid.

That is, the general nuclei, general atoms, eg Fe-atoms, molecules, and the properties of crystals, and the stability of the solar system..., ie where classical physics fails in the large size-scale (is) in relation to non-linear differential equations..., are all system whose properties cannot be described by using the laws of physics as they are currently expressed. Though the stable spectral-orbital properties of these systems can be determined experimentally, they cannot be described using the laws of physics as they are currently formulated, eg using gravity, electromagnetism, weak, and strong forces and in a containing set which is consistent with the idea of materialism, ie material determines all of existence.

The point of a measurable precise description is not about mapping a system's set into quantitative sets and then identifying sequences and bounds on sequences and other processes which depend on convergences which are defined in very big sets (ie sets which are too big so that they cause logical problems). This process seems to not result in useful quantitative information, especially when the quantities are related to either function spaces whose eigenvalue structure is not known (or is unstable), or when it is related to non-linear models.

Rather, the point of a measurable precise description, is in identifying an underlying relatively stable geometric structure which is fundamental to existence and geometric measurements, where all quantum values are related to classical measuring processes, and subsystem coupling can be defined and subsequently used due to the useful information, in creative processes. The geometric structure of stability allows a context for both measurable information and geometric coupling to be useful. This new geometric structure can be considered to be a way in which to extend the ideas of classical physics where the idea of materialism is re-defined.

In order to be able to describe these above mentioned properties; describing in a useful (or geometric) way..., so that description leads to a lot of creative development..., the general, stable, discrete, definitive orbital-spectral structures of material systems at all size scales, then consider the following new axiomatic basis for physics and math.

The properties of materials and the properties of coordinate spaces are the same, but one-dimension removed from one another, and that material properties are identified by space-forms, these space-forms are placed in a spectral context of an over-all high-dimensional containing set, where each dimensional level is associated to a stable space-form shape and thus associated to a spectral set.

However, space-forms also have the same: linear, metric-invariant, separable geometric descriptive context as does the useful (or solvable) part of classical physics, along with the classical idea of descriptive continuity, ie conservation of energy and material.

The stable boundaries between the different dimensional levels of a many dimensional containment set (containment set for all existence) are discrete separable geometries related to the discrete subgroups of the sets of semi-simple Lie groups (or space-forms), which are in turn associated to metric-invariant coordinate spaces on a base space, whose metric-functions (of the different semi-simple [fiber] Lie-groups) have constant coefficients.

Due to the interaction structure as well as the conformal factors between dimensional levels . . . , eg the physical constants, ie c, G, h, etc . . . , each dimensional level, and distinguishing spatial and temporal subspace structures (of the coordinates) . . . . [Note: This subspace structure is related to a metric-function's signature.] . . . , have descriptive continuity, eg conservation of energy and conservation of material.

This is a statement which is equivalent to the assumption that existence is determined by material.

That is, now material is determined by space-forms, but space-forms also determine coordinate spaces, of certain dimensions and certain metric-function signatures, and these material space-forms are contained in a certain spectral set, defined by the entire (over-all) many-dimensional containing space, where each dimensional level is associated to a space-form geometric structure, or spectral context.

In this new context, measurement is defined as being local, linear and based on (or consistent with) the metric-function and the differential-forms . . . , which measure local geometric properties on the coordinates of the base space . . . , as well as the measures of material, (eg amounts and types of material associated to space-forms of different dimensional levels and different metric-function signatures), as well as time, where time now can become many-dimensional also (in the higher spatial dimensions, eg R(4,2) or C(4,2)).

There may be some value in attaching harmonic function-spaces to this structure, or allow transient non-linear structures to be a part of this descriptive structure (to be contained in the coordinate base-spaces), so as to fill in details of material dynamics and changing material geometry that can exist between the stable sets (of simple geometries), but are not a part of the math structures upon which the relatively stable properties of existence depend. That is the mechanisms which can allow for the descriptions of more details are mechanisms which are unstable and they change, so as to become the stable linear space-form structures of existence (though the time duration for the change might be long or short).

The stable, discrete, definitive math structures which are the basis for our relatively stable existence are extremely simple math structures, which the arrogant math and scientists feel they are too important (too arrogant) to consider. Yet these math structures lead into much more accessible, but very complex and interesting systems than any system they can describe with their vague basis in language, even though the math scientists are trying with their very complicated descriptive structures of math, to describe the smallest detail of everything which is a part of existence. Nonetheless it is only their classical physics which has any descriptive validity, and it is the linear, metric-invariant, and separable geometries which are the important part of classical physics and it is these properties which need to be considered more carefully (which they have not done) and thus they have missed the stable (simple) underpinnings of existence.

The basis of the descriptions of the experts is determined in relation to being able to describe all the details by using: (1) non-linearity is a belief which does not take into account quantitative consistency but only considers the process of identifying convergence as necessarily identifying a valid quantitative value, and (2) function space techniques which are not consistent with descriptive continuity, nor does it take into account the need for a random set (which is the object of description, or which is what is being described) to be related to a stable and well defined elementary event space.

That is, it is believed that "If a property can be distinguished, then one can count such identifiable sets of events, and those distinguished properties in turn identify a probability distribution."

But if these events are unstable then this is not a valid quantitative set which is consistent with elementary operations of numbers, ie counting, thus it cannot be a basis for a meaningful definition of probability, or function space structure, with which probabilities and function's spectral properties can be believed.

The experts "follow the masters" who are full of themselves and who are full of illusions and who follow very complicated structures, the experts (those who win authoritative contests, so that the society says that they should be "full of themselves") study complicated illusions and complicated irrelevant patterns which are studied by self-centered people in a system of wage-slavery run by self-centered people who follow illusions, but after heroic attempts to describe complicated patterns, one must ask "What wide range of great creative applications have come from the descriptive structure?"

Subsequently one might ask, "Are complicated patterns actually irrelevant illusions?"

In the new many-dimensional containment set . . . , whose stable bounding geometries, to which unstable non-linear properties (can) converge, (or stable material systems) are space-forms . . . , there exist some space-forms of odd-dimension which when occupied by charge . . . , ie the relatively stable space-forms (of the next lower dimension [than the dimension of the odd-dimensional, material, space-form]) . . . , have on themselves a charge imbalance, yet these space-forms remain relatively stable . . . , (similar to the relatively stable atom) . . . , and subsequently begin to oscillate so as to generate their own energy (by pushing together opposite-state charges (which fit onto the space-form, which in turn, [the space-form] should be interpreted as a metric-space)) so as to create (or give-off) low-level energy when compared to the binding energy of these (certain stable) oscillating, odd-dimensional space-forms. This energy-generating structure of certain (odd-dimensional) oscillating space-forms now also forms a mechanism for resonance between the spectra of different dimensional levels . . . , though geometric properties of a given dimensional level are independent of the geometries of other dimensional levels . . . , and the space-forms contained in a higher-dimensional, oscillating space-form structure.

When a system is placed in (or moves to, by intent) a "new" (many-dimensional) containing space then different geometric relations (in relation to size-scales of material systems) can exist between the different dimensional levels based on the conformal multiplying constants, ie the physical constants, which can exist between the different dimensional levels, and can be changed based on the selection of the over-all containing space, ie there are many such over-all high dimensional containing sets.

Furthermore, subsystems of one "containing set" can move between a certain range of different "over-all containing spaces" but when they move they conform to the "new" geometric-spectral properties of the set into which they moved, which can change as the conformal multiplying constants can change between dimensional levels.

Are we (humans) either 7-dimensional or 9-dimensional oscillating space-forms, which (who) create existence itself (if sufficiently knowledgeable)? That is, knowledgeable about the relation that the discrete, separable geometries of semi-simple Lie groups have to the properties of existence.

Unfortunately, we seem to be stuck in a space with 3-spatial dimensions, due to both our constant internal dialog between ourselves and the external social world, and due to the simple (radial, spherically symmetric) interaction structure of material at the 3-spatial-dimension level and a quite different interaction geometry at higher dimensions.

This paper (and all the others) can be summed-up in a short outline.

There exist stable definitive discrete material systems which exist at all size-scales: nuclei, atoms, molecules, crystals, and solar systems and possibly (orbital) motions of stars in galaxies. These stable discrete spectral-orbital systems do not have adequate descriptions within the context of "what are now (2011) considered to be the laws of physics" (eg there is, currently, no known reason as to why the solar system is stable, nor how a general nuclei comes to have stable spectral properties).

The existence of these stable systems implies that measuring makes sense in regard to these stable definitive systems which exist at all size-scales, and that they can be quantitatively described in a stable, measurable descriptive language.

However, one can only measure in a context of stable structures:

> stable systems,
> stable geometries,
> stable and consistent quantitative structures, and which are not too complicated so as to hide inconsistencies,
> within sets which are not too-big, and (also) not too-complicated, so that the set properties of descriptions are both consistent and comprehensible, and
> descriptions are to be placed in stable contexts where the values one measures are stable and consistent, so that measurable values fit into a description which is also stable and practically useful, and which actually accurately fits data (such as describing the stable discrete spectral structure of a general nucleus), and which provides information which, in turn, can be used in a practical manner because both the system and its descriptive structures are stable.

How is stability to be introduced into the descriptive language of math-physics?

Answer: The key is stable geometry, and how to introduce (or fit) stable geometric structures into a quantitatively based descriptive language, is now made clear by the following list of "eight" math properties:

(1) The geometrization theorem of Thurston-Perelman [Note: The Thurston-Perelman theorem is actually unnecessary, one can simply appeal to a need for measuring-stability (linear, separable, metric-invariant) and the stable discrete shapes exist within this stable context.], where Thurston-Perelman (essentially) show that the discrete hyperbolic shapes and discrete Euclidean shapes . . . , (as well as the sphere, but the sphere is non-linear, and is stable, in regard to a system which has only one spherically symmetric body composing the system) . . . , are the central shapes in regard to classifying geometry, in its general topological context, and all the other shapes (at least in 3-space) are unstable and they either disintegrate or evolve towards the stable shapes.

(2) The knowledge that "the most stable shapes" are the discrete hyperbolic shapes, and that the stability of the discrete hyperbolic shapes are robust and their properties (of shape) are unaffected by their (it) being near other geometries, whereas in the spherical-case, one sphere is a stable, while two spheres forms an indescribable (or unresolved-able) geometric structure.

(3) Coxeter showed what the fundamental properties of the discrete hyperbolic shapes are distinguished by: dimension, genus (number of holes in the shape), and stable geometric measures, which are measures on the faces of the discrete hyperbolic shapes' fundamental domains (ie polyhedron, or cubical-simplex structures [the square on a checker-board is a fundamental domain of a discrete Euclidean shape]), [the stable geometric measures on discrete hyperbolic shapes are the basis for stable spectral-orbital properties which are associated to the discrete hyperbolic shapes], and the discrete hyperbolic shapes of each dimensional level can include infinite-extent discrete hyperbolic shapes (so that these shapes have finite "volume" measures. *) associated to themselves, ie these infinite extent shapes model light, and perhaps neutrinos.

(4) Furthermore, there are many simultaneous (but different types of) metric-spaces involved in physical description, where each metric-space (identified by dimension and signature) is associated to a different fundamental physical property. These fundamental physical properties and their associated metric-spaces (identified by dimension and metric-function signature) are distinguished by the elementary displacement symmetries of E Noether, which are (also) related to conservation properties. Eg inertia-spatial-displacement is a property of Euclidean space and charge-energy are primarily properties of hyperbolic space (hyperbolic space is equivalent to space-time, it can be thought of as the velocity space of space-time)

(5) Opposite metric-space states can be defined on this structure. In turn, opposite metric-space states is the structure upon which "the spin-rotation of opposite metric-space states" can be defined. Furthermore, opposite metric-space states can be defined in complex-coordinates, and this leads to containment within complex-coordinates, ie "the real state" and its "pure-imaginary, opposite-state," which subsequently leads to unitary fiber groups (extending the description beyond the usual (real) rotation [or isometry] fiber groups).

(6) The structure is many-dimensional, since it is assumed that the structure of material is based on the same idea as the structure of a metric-space, namely, they (both material and metric-spaces) are both based on discrete hyperbolic shapes, and the discrete material shapes are contained within (adjacent higher-dimension) metric-spaces.

The discrete shapes in other (non-hyperbolic) metric-spaces are involved in dynamic processes, and the subsequent spatial displacements of these dynamic processes, where (both) opposite metric-space states ((7) as well as the fiber groups, associated to local coordinate transformations of the interaction geometries) are also involved in dynamic processes.

(7) the math containment set is a (set of) principle fiber bundles (where a principle fiber bundle has a base space, wherein a system would be contained, with its properties distributed between the different types of metric-spaces, and associated to each point of the base-space is an appropriate, allowable (isometry or unitary), local coordinate transformation group)

The base-space is composed on the many different dimensions, and many different types of metric-spaces up to hyperbolic dimension-11 (12-space-time), including coordinate base-spaces which have complex coordinate-systems, and the fiber groups are the set of classical groups which are associated to these different (real, isometric) metric-spaces as well as Hermitian spaces (with their complex-coordinates and unitary fiber groups).

(8) Material has a shape, and it also has a distinguished point (so interactions appear to be point-like), and metric-spaces also have stable shapes, so the coordinate metric-spaces now (also) possess spectral properties. That is, the containing space has spectral properties associated to itself.

Note: Stable systems can emerge out of interactions (if the energy of the interaction is within a proper range of values), and this emergence of stable systems out of material interactions would be due to the existence of resonances between the new system and the spectra of the (high-dimension) containing space.

Note: Quantum randomness can be derived from the structure of material interactions within the new descriptive language, but the basis for this "new" descriptive language is stable geometry.

The context of stability

When formulating how to describe a (physical) system, it is the stable geometric shapes, in regard to the descriptions of inertia and charge (and their geometries and motions in space), which are fundamental to the formulation and solution of a system's description (of a system's properties). The existence of a stable (geometrically separable) global coordinate geometry is a fundamental aspect of a stable and solvable system's description.

Because there are so many (many-component) systems which seem so independent of an external geometry, eg the neutrality of atoms, one also wants a model of a global, separable shape which is closed (sealed-in) and bounded.

(Note: There are many topological sets which can be both closed [contain all of their limit points] and open)

There are such bounded, separable shapes (which are not spheres) and they are (essentially) the discrete shapes of the Euclidean (tori) and hyperbolic spaces (at least two tori geometrically combined together).

These discrete shapes, especially the very stable discrete hyperbolic shapes, will be, mostly, unaffected by external conditions.

Another outline of the structure of math, whose patterns are organized, in an alternative descriptive structure

There is an alternative description of physical existence, which can be defined in a context of "linear geometric" properties and defined in a context of many-dimensions, where each of the dimensional levels (or each metric-space of a different dimension) is itself associated to a "linear geometric" shape,

> (or possibly as an "orbit" in a discrete hyperbolic shape, or equivalently [to an orbit, (would be)] a pair of opposing faces of the "cubical-simplex" associated to the discrete hyperbolic shape).

Furthermore, these same hyperbolic shapes define material systems when contained within the (an) adjacent higher dimensional level.

Where "linear geometry" means that one solves linear, separable, differential equations defined in a metric-invariant context (where the metric-functions have constant coefficients) in order to accurately describe the measurable properties of a (linear) geometric shape. The linear, separable shapes, defined by differential equations (or by differential-forms), are essentially both the discrete Euclidean shapes and the discrete hyperbolic shapes.

The discrete hyperbolic shapes have very stable geometric properties so they would be good models for the stable discrete quantum systems.

It is natural to define conformal factors between adjacent dimensional levels, where these factors would be equivalent to physical constants such as, h (Planck's constant), and G (the gravitational constant), or between different types of metric-spaces, where the speed of light, c, relates Euclidean 4-space to hyperbolic 3-space.

The discrete spherical geometries are non-linear and thus they are not considered.

Linear geometry implies the existence of stable discrete shapes in a many-metric-space context, and in a many-dimensional context.

The differential operators which determine material interactions, in the new descriptive language, are descriptions which are similar to the classical descriptions, but now they are explicitly (ie the operators are) discrete linear shapes (that is, they are harmonic structures) which inter-relate the different types of metric-spaces and the different materials which are part of the description (where stable material systems are actually (closed) metric-spaces).

Furthermore, the different metric-spaces also distinguish different physical properties.

The operators, which are discrete shapes, are essentially classical differential operators, but now they are defined in a math structure where the opposite metric-space states are spin-rotated from one to the other (the spin-rotation of metric-space states).

Nonetheless, the unstable non-linear operators are (also) still defined, in a manner similar to how the classical non-linear differential equations are defined, ie non-linear systems still exist, and these differential equations are related to actual geometric properties which are defined by the material interactions, to which these classical differential equation are related. Thus, the geometric relevance of the critical points and associated (geometric) limit cycles which are associated to these (most often) classically defined non-linear differential equations. Thus the question becomes, "When does any particular non-linear differential equations lose its relevance in regard to defining the material interactions of the system one is trying to define?"

However, the stable systems are the discrete hyperbolic shapes, and the material interactions in 3-space (where hyperbolic 3-space is equivalent to 4-space-time) are spherically symmetric . . . , as are both Newton's laws of gravity, and (as are) the basic assumptions of (one-body) general relativity . . . , where spherical symmetry of a material interaction in 3-space persists up to the existence of a bounding discrete hyperbolic shape, which will affect motion when the interacting system gets small enough (as small as the discrete shapes of material systems which define the material interaction), so that at a particular geometric size (in regard to a material interaction), "what was a spherically symmetric interaction would no longer be definable as a spherically symmetric shape," which defines the dynamics of interactions within the small distance-range of the interaction where the material interaction begins to take on the structure of a collision, where the collision gets defined around the distinguished points of the discrete shapes which are part of the interaction process.

How interactions transition to stable structures is a description which is based on the new "containment system's" relation to its many-dimensional containing space. During a material interaction, the interaction structure (if the interaction is within certain energy ranges) . . . ,

(as the material interaction approaches the distance-range which defines a material-collision as opposed to the material interaction defining an interaction being described in the context of a force-field interaction)

. . . , may begin to resonate with the spectra of the over-all high-dimension containing space so as to transition to a new stable discrete hyperbolic shape, due to both resonance and energetic consistency of the interacting system.

Furthermore, this new description has a structure which is entirely similar to the math structure E Nelson (Princeton, 1967) used to derive that "quantum-randomness is a result of the Brownian motions of the atomic structure of material." Thus quantum-randomness can be derived from a linear-geometric model of both material and space, wherein the stable quantum systems are discrete hyperbolic shapes, which in 3-space are also related (up to bounding discrete hyperbolic shapes) to spherical harmonics (due to interactions in 3-space being spherically symmetric (even) in the new theory, where in the atom it seems to be true that the size of the nucleus is (assumed to be) small compared to the size of the atom, where the atomic size determines the distance-range of the interaction within an atom).

However, now there is a math structure which allows for the description of the existence of a stable many-component orbital structure (since the discrete hyperbolic shape can have many holes in its geometric structure, where each hole can be associated to a separate orbit, or separate spectral-value).

New ideas

However, the new ideas, placed in a context of metric-invariance for metric-functions with constant coefficients, about how to describe existence provide material with a geometric structure. In doing this, the new ideas can also take into account quantum physics.

In the new description, material interactions . . . , which are related to the descriptions of inertial properties . . . , are described in detail within an entirely geometric context (one is not seeking to describe the many unstable observed geometric forms, rather only the stable geometries of material, namely discrete hyperbolic shapes, are related to inertia, where discrete hyperbolic shapes have an inertial-(mass)-energy relation to discrete Euclidean shapes by means of resonances). This interaction structure allows one to distinguish between fields and acceleration, whereas Einstein assumed that the differences between force-fields and acceleration was "not" distinguishable. However, the difference between force-fields and acceleration is (are) distinguishable within the new descriptive context. ** The acceleration is due to local coordinate transformations associated to each small time interval, which is a part of the interaction process, while the force fields are the structure of Euclidean interaction shapes which determine the spatial displacements of the local coordinate transformations which act on the structure of the Euclidean interaction shape.

Einstein believed in some absolute (or magical) property in regard to the existence for a force-field. But he assigned no form [(or no geometric property) was associated] to any atomic-component (or solid) object of material.

However, in the new descriptive context, a force-field (or a material interaction) depends on material having a relatively stable geometric structure (whereas general relativity is trying to account for the relatively stable geometric properties of material systems, eg the precession of the perihelion of Mercury, the new description says that relative stability depends on how discrete hyperbolic shapes are organized in a many-dimensional containing space, where each dimensional level, itself, possesses a stable discrete hyperbolic shape, where within these shapes the geometry is either with a constant curvature [in hyperbolic space] or is flat [in Euclidean space]).

In the new descriptive structure inertial forces form instantaneously in relation to the geometric properties of the interacting materials. However, in the new description inertial mass is equivalent to either gravitational mass or electromagnetic mass (etc) since all of these masses are expressed in a Euclidean context, ie expressed as discrete Euclidean shapes [which are in resonance with some defining aspect of the stable discrete hyperbolic shapes], upon which inertial description depends.

The mass equals energy formula of Einstein, $E=mc^2$, is easiest to believe (or to deduce) based on E Noether's idea that "time displacement symmetry is to be related to conservation of energy." Thus, assuming energy conservation for the universe, the "time slot" of the 4-momentum set of coordinates is to be associated to energy, ie $m/(1-(v/c)^2)^{1/2}=Energy$, etc.

Note: Remember that the spatial displacement symmetries of E Noether (in the 19th century) imply that inertia should be represented in Euclidean space (not local space-time space), since the "conservation of momentum" for systems with spatial displacement symmetries is a well established concept.

So it is also necessary to follow E Noether, in regard to the properties of coordinate frames.
Namely, that inertia (or spatial displacement related to small time intervals) is to be associated to Euclidean space, along with (or apparently) instantaneous "action-at-a-distance" models of fields "which affect inertia."
On the other hand, both electromagnetism and material's relation to energy (including, mass = Energy, which expresses the relation of Euclidean space to energy where this relation is caused by resonances between discrete hyperbolic and Euclidean shapes, which are needed to define discrete Euclidean shapes) are best described in a space-time metric-space, or (equivalently) in hyperbolic space.
That is, one wants the field properties of electromagnetism to be described in space-time so as to be consistent with Lorentz invariance. While material object's relation to energy is associated to the sectional curvature of the spatial faces of a Euclidean interaction shape. **

The new descriptive structures

Furthermore, (in the new descriptive structure) the dynamic process (in Euclidean space) is discrete, it is not continuous. However, the time intervals . . . ,
whose properties (or period) are determined by the "period of a spin-rotation between the metric-space states" (where "metric-space states" must be defined in complex coordinates, because they (the metric-space states) occur in pairs, so as to have a real part and a pure-imaginary part [where the pure imaginary part is the opposite of the real-part of metric-space states])
. . . , are of very short duration, so that dynamic properties appears to be continuous.

Euclidean discrete shapes (also called Euclidean space-forms) form (or come into being) due to: (1) their resonance with hyperbolic space-forms, which represent the charged (or energy) systems, and (2) during a material interaction. Material systems are always interacting, where "the resonance," in regard to the Euclidean discrete shape associated to the material components of the interaction, forms into an average value of the discrete hyperbolic shape's energy (the two interacting discrete shapes also instantaneously average their energies between themselves so that the two material components which are interacting can be identified with one another) then in the process of a material interaction the spatial faces form (instantaneously come into being) between the two interacting material shapes (or material space-forms) so as to form both a discrete Euclidean shape, and (which along with) an associated local coordinate transformation of the spatial faces . . . , where the existence of the local coordinate transformation (acting on the spatial faces) is due to the fiber (spin) group . . . , (which) together (local geometric properties and local coordinate transformation properties) determine the dynamics of the interaction.

The new discrete Euclidean shape is contained in the next higher dimension metric-space wherein it determines a 2-differential-form. The 2-forms have the same dimension as the spin fiber group (or, if metric-space states are considered, then a complexities [or unitary] fiber group) and thus the local coordinate transformations, which the discrete Euclidean shape (or Euclidean space-form) of interaction demands, are determined by the local geometric relations . . . , that the discrete Euclidean interaction space-form possesses . . . , which in turn, are related to the local geometric properties of the fiber group so as to select the local coordinate transformations of the fiber group (based on these geometric inter-relationships), ie the dynamic changes are related to either spatial changes caused by the relationship that the spatial geometry of the Euclidean interaction space-form has to the local geometry of the fiber group, or (they could be related) to (new) changes in frames which are associated to new and different energy structures for the interacting material systems, which can form from the interacting space-forms, where these new energy structures, ie new stable discrete hyperbolic shapes, result from resonances between the spectral properties of the interacting space-forms and (with) the spectral properties of the over-all high dimensional containing space, ie a quantum interaction (but also the form of a classical field-interaction).

This is an extremely simplified model of quantum interaction in comparison to the quantum and particle physic's model of quantum interactions, and it describes classical interactions in the same (descriptive) context, ie the quantum interaction is equivalent to a classical interaction (in the new descriptive language) but it is contained in a (new) context which allows both stable material geometry and new "spectral possibility" where this (new) "spectral possibility" is associated to the entire high-dimensional (many geometrically independent dimensional levels) containing space.

The new description and frames of reference

That is, the new context for dynamic interactions are systems which change with each discrete time interval (determined by the period of the rotation between the metric-space states) so that the changes are based on (instantaneous) geometric properties of discrete Euclidean (or unitary) subgroups (or discrete Euclidean shapes) and their geometric relation to the local transformation properties of the fiber group (which act on the spatial faces of the discrete Euclidean interaction geometry).

On the other hand, the properties of space-time metric-functions are related to charge and energy by tensor properties which are consistent with the Lorentz transformation, eg the 2-forms which define electromagnetism (see below).

The dynamic (or velocity) relations defined by the Lorentz transformations are the observed properties seen between hyperbolic frames which are linked by the light (through which the observations of properties are made).

However, the inertial relationship that the electromagnetic field has to material systems can be kept in a Euclidean context, yet (still) remain consistent with Lorentz-invariance due to the existence of an interaction Euclidean space-form which exists in the next higher dimension, Euclidean 4-space, where the (4th-Euclidean) dimension is modeled as time, while the other three dimensions determine the 3-Euclidean space where the inertia manifests, so that the (Euclidean 4-dimensional) interaction 2-form can be made Lorentz-invariant by assigning the 4th-dimension to time (in the proper way), yet the description can stay in the 3-space inertial context where the Euclidean 4-space is projected down to Euclidean 3-space.

Though the new (math) descriptive structure is consistent with "special relativity" it is not clear that "special relativity" (though applicable to light) is applicable to an entirely inertial context where light is no longer conveying the geometric (or field) information within the system (as is the case for electromagnetism in space-time), where now instantaneous discrete Euclidean geometric structures are now conveying that information, yet the formal structure of the field can remain Lorentz invariant.

Frames

(general relativity) The forms of physical laws should be the same in all frames either moving or accelerating. However, a force field can only be invariant to rotations if it is spherically symmetric, but magnetic forces exist which are not spherically symmetric. Thus, general coordinates (or general frames) are not possible, the coordinates (or frames) must be consistent with both metric invariance and the metric-functions must have constant coefficients, thus, it must be that the frames be moving, or accelerating, in a translational sense (in regard to the fixed stars) and that, if the motion is an acceleration then the acceleration needs to be uniform for all material contained in a (particular) region of space, in which case (or thus) a constant acceleration will be added to both sides of the F=ma Euclidean inertial law, since F depends on the geometry of material. That is, the principle of general relativity must be modified, but when modified it is consistent with Galilean frames (for Euclidean frames), and can be made consistent with Lorentz transformations for electromagnetism. However, uniform acceleration for all the material in (a large region of) space would be detectable in space-time due to the change of the speed of light (in regard to curved light-paths, but "Is this detectable?") which is not associated to light interacting with material. Thus, if the curved paths of light can be detected then uniform accelerating frames for all the material in (a large region of) space are not equivalent frames in space-time, and thus not equivalent frames in hyperbolic space.

Note: The two metric-space states of Euclidean space are the frames of the fixed stars as opposed to the frame with rotating stars, ie translational frames vs. rotating frames. It happens that the inertial force fields (ie gravity) in 3-space are spherically symmetric (and directed in the radial direction) due to the new structure for interactions.

The two metric-space states of hyperbolic space are forward progressing time or backward time.

Extra stuff about frames

The most agreeable property of equivalence between inertial frames, is "frames at different relative velocities," while Einstein claimed that frames with different relative motions of any kind are equivalent. But this leads to a non-linear idea about geometry's relation to inertial motion.

Perhaps if frames for material are uniformly accelerating, in regard to translational motion, then either different levels of translational acceleration or different relative translational velocities would be equivalent frames.

The force field can transform under a different (local coordinate) transformation laws than do the inertial properties.

But this way of thinking (about frames) is in relation to a continuous idea about how frames are defined. What if, for short time intervals, inertial frames are fixed in regard to spatial separation, and contained within sets of equivalent frames (where this would include frames with different translational velocities), when (where) the spatial separation changes, due to the local coordinate transformations, at the end of the short time interval, but during the short time interval the spatial separation, in any frame (in Euclidean space), is fixed.

It should be noted that time is a property of space-time, which, in turn, is associated to hyperbolic space.

The frames which are equivalent, in space-time (or equivalently hyperbolic) coordinates, are frames which are

> either
> attached to systems with fixed motions
> and (or)
> with fixed energies (within the short time intervals which are a part of the interaction structure),

and the interactions in these frames (at the end of the short time interval)
either
change the system's property of motion
or
change the properties of discrete hyperbolic shapes (and subsequent changes in energy), which are determined at the end of each time interval.

Note: (in the new description) The property of motion can only be changed in an apparently continuous manner within Euclidean frames.

Does one want to describe all geometric structures, or does one want to use geometry to understand stability?

Answer: One wants to understand the causes of stability, and to transcend the idea of materialism so as to describe an existence which contains both material and (it fits into) a many-dimensional context for geometric description.

Whereas, if "one" models physical reality in a high dimensional context which is at the level of macroscopic field interactions, ie a many dimensional descriptive structure which is consistent with classical physics, and its subsequent relation to inertia, is placed into a linear, separable, and metric-invariant context, where the metric-function has constant coefficients, (where this context is associated to discrete Euclidean and hyperbolic shapes) then the descriptive structure will be:

(1) stable (in the case of discrete hyperbolic shapes),
(2) quantitatively consistent, and
(3) can be related to practical useful creativeness.

Furthermore, the many dimensions can also be used so as to fit the data of particle physics, ie a macroscopic structure associated to Lie fiber groups modeled within a many-dimensional containing space, where the dimensions go beyond the 3-spatial dimensions which materialism defines

> (though the patterns of particle physics are identifies a fairly unimportant set of patterns, which one might [or might not] want to fit).

Rather, what is desirable is a geometric description of higher dimensions which has a relation to stable, definitive, discrete spectral-orbital structures, and describes how interactions can both determine dynamics and relate interactions to the observed stable spectral-orbital properties of stable (system) geometries (which depends on the spectral properties of the higher dimensional containment set).

Chapter 16

ABSTRACTS

Abstract I

A new context in which to apply geometry to: math, quantum physics, and the solar system, etc

Quantum physics assumes the global and descends to the local (ie random particle-spectral measures).

Is geometry a better vehicle to define the stability of quantum systems rather than function spaces?

Is the stable construct to be the very stable discrete hyperbolic shapes, in a many-dimensional context?

A geometrically stable and spectrally finite math construct, where, in adjacent dimensional levels, the bounding discrete hyperbolic and Euclidean shapes are defined, and then mixed as "metric-space states" in a Hermitian (or unitary) context, can provide a structure for stable properties.

Assume that math be consistent with (local) geometric-measures of stable shapes, which define finite spectral sets, contained in higher-dimensions.

The stable shapes in the different dimensional levels are con-formally similar, and resonate with a finite geometric-spectral set contained in a high-dimension space.

A new interaction type consists of a combination of hyperbolic and Euclidean components, but when in an "energy-size range" the system can resonate with the spectra of the containing space, and thus it can change to a new stable, discrete shape.

Abstract II

There are (moded-out) "cubical" simplexes in a many-dimensional context, whose structure is determined by hyperbolic metric-spaces, which can, themselves, be modeled as moded-out "cubical" simplexes. Transition between the different dimensions determine physical constants, and the value of these physical constants can imply that:

1. the different dimensional levels can be hidden from one another, ie the size of the interacting materials change from dimensional level to dimensional level and the geometry of the interaction can also change, ie material interactions are not usually spherically symmetric.
2. The over-all high-dimension containing space can be defined as having a finite spectral set.
3. The descriptions of both mathematical and the physical systems are (or can be) stable, because the "cubical" simplexes can define discrete hyperbolic shapes.

Metric-spaces have properties and subsequently an associated metric-space state, and this determines the dimensional distinction between Fermions and Bosons, as well as determining the unitary (invariant) mixing of (metric-space) states in subsets of complex-coordinates.

Unitary invariance implies continuity, or the conservation laws, eg the conservation of energy and material, etc.

Each dimensional level is a discrete hyperbolic shape, and this implies such a set can define a finite spectral set for the entire space.

A new interaction type consists of a combination of hyperbolic and Euclidean components which are one dimension less that the dimension of their containing metric-space. A 2-form construct emerges from this geometric context which is the same dimension as the adjacent (higher) dimension Euclidean base space of its fiber group which determines discrete spatial displacements. This interaction construct is either chaotic or it could begin to resonate during the interaction and, subsequently, to become a new stable spectral-orbital (discrete hyperbolic shape) structure, by means of its resonance with the spectra of the many-dimension containing space.

A new way to apply geometry to both math and to quantum physics

Tradition and authority vs. (a very educated) Equal free-inquiry

This is a careful expression of ideas about a geometrically measurable set of systems associated to spectral-material and high-dimension math constructs, which are built from the sophisticated math patterns uncovered by this culture.

It is based on measurability, math stability, quantitative consistency, and finite spectral sets which have a unitary-invariant structure. That is, the coherent, or connected, or consistently resonant systems which are described conserve a many-dimensional, geometric-spectral math construct, ie the continuity of material existence is tied to a finite spectral, high-dimensional, geometrically based math structure.

It can describe the precise and stable properties of general quantum systems, and because it is based in geometry it is practically useful.

New ideas

Why are carefully expressed, new and fundamental, ideas which "more than any other ideas being expressed" address the core problems which face our society concerning "the failure of the society's intellectual and technical development," are so easily ignored and/or marginalized?

Where modern physics, which is, essentially, based on quantum physics, has only really contributed

1. The bomb (based on the rate or reaction being related to the probabilities of particle-collisions, ie based on 19th century theories of chemistry),
2. The laser (clearly invented based on quantum principles), and possibly
3. The transistor

   (but really the transistor, like most technology associated to quantum properties of material, is mostly based on a classical thermal manipulation of material so as to allow for the existence of an observed quantum property in a semiconductor material)

That is, most technology related to quantum properties is about creating a classical system which can couple to a quantum property, ie the calculations and understandings associated to quantum physics is only very slightly associated to the technical development, and this is because the quantum description is mostly about systems composed of a relatively small numbers of components whose quantum properties can only be associated to random processes.

The quantum descriptions based on randomness, for systems built on relatively few components, do not provide a mechanism of system (or material) control, though the stable precise context of the quantum properties, which have been observed, suggests they existence in a controllable context.

Quantum physics has had the greatest impact on statistical physics, where the probabilities of a great number of system-components can be related to measurable and controllable properties as averages, but the quantum influence is essentially about partitioning a statistical system's energy structure and applying the statistical properties of Bosons and Fermions in a system's quantized energy structure.

That is, the impact of quantum physics, and its associated particle-physics, in regard to using knowledge to advance technical development has been very slight.

The marginalization of new and very important ideas is possible since we live in an unequal society, where the communication system (or propaganda system) only serves the interests (and the ideas) of society's "top few."

This control over institutional and public communication within society gives great credibility to the ideas, which the "top few" believe, and it is those same "top few" people, who control the resources, creativity (by investment), knowledge, and communication within all of society.

[Note: The education system is an arm of the propaganda system, and serves big-business interests. Education is not a ladder to high-culture leading-up to an authoritative truth, rather education is about challenging the authority of high-culture, especially if the high-culture is not developing the practical creative capacity of knowledge, but rather languishing in its relation to business interests. That is, education is not about wage-slavery, rather it is about free-inquiry, and the relation of knowledge to practical creativity]

The character traits which are desired and rewarded by the big business interests, the institution which most affects the nature of our society, are the traits of being: selfish, narrow, absolute, self-righteous, cruel, unfeeling, petty, and extremely violent (so as to be able to hold onto what one has in order to give to themselves social-value [the value which is expressed to us by big business interests and its (or the) media]).

Thus, it is easy to marginalize new ideas which are not consistent with the ideas of the propaganda-education system.

This is exactly the same social structure by means of which the Catholics controlled thought in their (Holy-Roman-) Empire, but now it is not popes but rather the "lords of the monopolistic businesses" . . . , eg oil, agri-oil-business, banking-oil, military, medicine-drugs . . . , who are in control the resources, creativity (by investment), knowledge, and communication within all of society. Thus, physics is about the interests of the big military businesses, etc.

However, if the discussion is about both new ideas and to show the substantial problems which permeate the traditional and authoritative ideas which today dominate thinking in the major US universities, and the discussion is at the level of assumption, interpretation, and (new) contexts, then the discussion can be very educational (in the strongest sense of what education means, namely, the relation that knowledge has to the practical creativity of individuals, as opposed to the creativity of large institutional businesses).

For example, the risk models used by the big banks failed because the math being used is based on ideas which, such as indefinable randomness (an arbitrary quantization math construct {or math scheme}) have fundamental flaws in their: assumptions, contexts, interpretations, the structure of set containment, and set size, ie its failure is based on elementary ideas, and thus it should be much more widely challenged.

But it is not widely challenged, thanks to the extreme fear, as well as extreme violence, by which the society is controlled, so that people are afraid to challenge authority since it leads to a one-sided contest in which the authorities get to destroy those who challenge their authority.

The social positions of those on top cannot be challenged in any context, especially in regard to how language is used.

Only a person outside the social institutions can express ideas about these topics and this is because such a person is so easy to marginalize, even the alternative media only give credence to the ideas of those people established within the big institutions, even if the new ideas being expressed are ideas derived from those same big (educational) institutions.

The only place where the public are allowed to express ideas, in the (so called) alternative media, is in regard to civil protests.

This is an idea of the legitimacy of public expression, which the authoritarian, traditionalist, and so called, progressive, N Chomsky, touts, ie only civil protest is a valid way to express ideas.

That is, one of the so called champions of the liberal cause, which is best defined as those who believe in equality, are, in fact, authoritarian traditionalists, and thus they are allowed onto the media, and in this position they are truly conservatives, who oppose change in regard to challenging the authority of traditional knowledge, and in fact, they support both tradition and authority.

Clearly the finer judgments of the likes of Chomsky lie with the W Lippmanns of the world, ie with those who support class warfare. Where early cultural sophistication, ie what is called intelligence in our society, should define one's social position, in an education system which filters for "cultural expression which will advance the authority of the culture."

There are

1. Socrates who supported equal free-inquiry, ie the inquirer is to be given the authority, and
2. Christ who expressed his belief in equality by the expression, love one's enemy, or "treat you neighbor as you want them to treat you," and then there are

3. those who populate the higher levels of the society, and who support the structures which determine and maintain their social positions, which are judged to be superior to the social positions of their neighbors.

The split is easy to determine, does one believe in equality, or does one believe in tradition and authority of the fixed way in which resources and knowledge is used, and thus such people are in opposition to new knowledge and in opposition to equal-creativity.

The evangelical religious people, the true protestant reformers, should believe that "each and everyone of us" is an equal creator, because this is the "best way" to interpret the golden rule [love one's enemy, ie we are all equal], and that equality is about the very "core nature of mankind" and our relation to creativity, where one cannot believe in the material world, wherein the value of human creativity is judged, there is a deeper relation that life has to creativity (than to material creations, or artistic creations marketed through material-based communication systems).

The new ideas place creativity into a "new way in which to consider" the "ideal" world, as opposed to the material world, and they provide a new context for human creativity, in the context of an ideal existence, in which the material context is a proper subset.

"What about ideas which are fundamental to the practical creativity of a culture (in regard to both its scientific and religious structures of belief)?" and not so much ideas which are concerned about "How power is organized within society (ie civil protest)?"

That is, even if people want knowledge to remain associated to practical creativity then the authority of traditional math ideas should be challenged in regard to the ideas of professional mathematicians (since practical creativity is still related to 19th century, materialistic (or scientific), view of existence) about:

1. Quantitative sets (set size, the "quantitative sets," eg the real numbers, are "too big")
2. Quantitative consistency (non-linearity does not work in a quantitatively consistent manner, and indefinable randomness is not capable of forming a valid math structure [this is why the big banks models of risk failed])
3. Logical inconsistency (self-contradictory, eg when probability based constructs depend on geometry [particle-collisions])
4. We have an academic-industrial-vocational education system, eg physics departments are nuclear weapons engineering departments, which adheres to tradition and authority, and to the sources by which they get funded, so as to develop data-fitting math-science constructs which support authority and tradition. These data-fitting math constructs are based on non-linearity and indefinable randomness (see below), and these are math constructs which are in opposition to the math properties of both stability and quantitative consistency.

5. Though the academic institutions developed the idea that:

   (a) precise language has clear limitations as to the patterns which it is capable of describing in a measurable context (Godel's incompleteness theorem), and
   (b) the example of Copernicus vs. Ptolemy is often discussed (but the discussion is framed as "science is correct and religious authority is wrong in regard to the material world"; rather than it being related to the idea that measurably verifiable constructs can be framed in the wrong context), and whereas
   (c) the usefulness of "classical physics descriptions" vs. "the technically un-useful quantum physics," where quantum physics is based on indefinable randomness and contributes very little to technical development; but on the other hand, most of the technical development today is still based on the classical physics of the 19th century, nonetheless, the stable preciseness as the observed properties of quantum systems suggest a very controllable context for their correct descriptive construct.

(this is a discussion which is fundamental to education, in regard to its relation to:

Creativity (the trait which best characterizes human life),
Knowledge, (knowledge is developed by means of equal free-inquiry),
Freedom (the freedom to inquire at an elementary level of language, and the freedom to use knowledge and create [gifts for the world]), and
Equality (we are all equal creators))

Abstract

A new context in which to apply geometry to: math, quantum physics, and the solar system, etc

Quantum physics assumes the global and descends to the local (ie random particle-spectral measures).

Is geometry a better vehicle to define the stability of quantum systems rather than function spaces?

Is the stable construct to be the very stable discrete hyperbolic shapes, in a many-dimensional context?

A geometrically stable and spectrally finite math construct, where, in adjacent dimensional levels, the bounding discrete hyperbolic and Euclidean shapes are defined, and then mixed as "metric-space states" in a Hermitian (or unitary) context, can provide a structure for stable properties.

Assume that math be consistent with (local) geometric-measures of stable shapes, which define finite spectral sets, contained in higher-dimensions.

The stable shapes in the different dimensional levels are con-formally similar, and resonate with a finite geometric-spectral set contained in a high-dimension space.

A new interaction type consists of a combination of hyperbolic and Euclidean components, but when in an "energy-size range" the system can resonate with the spectra of the containing space, and thus it can change to a new stable, discrete shape.

Introduction to the Abstract

If there are statements . . . , upon which this discussion of a new math context depend . . . ,

which people with expert status will be sure are wrong . . . , then this belief is related to the context and interpretations of a description which the expert person in opposition is following, eg those who followed Ptolemy would surely find the ideas of Copernicus completely wrong . . . , but according to the media we live in an age of science, where physical theories are measurably verified . . . , but in fact the theory of Ptolemy was also measurably verified . . . , but Ptolemy was in an incorrect context . . . , but today's physics theories are also clearly in an incorrect context

> . . . {since it is not possible to precisely describe the properties of general quantum systems using the laws of either quantum physics or particle physics . . . , and the candidates for the theories of everything describe much less that quantum physics (which can describe the H-atom) and particle physics (which can be used to describe the patterns of particle-collisions in particle-accelerators) . . . , can describe, . . . , but nonetheless . . . }

. . . , since fundamental physical systems, eg nuclei, general atoms, molecules, molecular shape, crystals and the properties of the stable solar system are stable and precisely discrete and observable [which means that these stable systems are continually emerging into their uniform stable structures, and thus they are emerging in an interaction context (ie a differential equation) which is linear, solvable, and controllable] yet the general types of the fundamental material systems, just listed, remain indescribable based on modern physical law (2012).

However, there is a new descriptive relation . . . related to the patterns of a "physical-form which has been observed" (or to the set of observed system properties, and whose properties {or their patterns} one is trying to describe) . . . . which is being identified in this discussion (in this article).

The new descriptive structure allows one to describe, within a new context, ie a new containment set, the stable properties of many-component systems, composed of charge and/or mass, whose geometry of interaction, in 3-spatial-dimensions is spherically symmetric . . . ,

> [but {in the new descriptive context} the micro-structure of material systems is discrete hyperbolic shapes, where, in turn, these shapes and the angles between the system's components are influenced by the angular relations which exist between the Weyl chambers to maximal tori of the fiber group and the relation of these angular properties to angular changes (or angular shifts) in regard to the "toral sub-geometries" of the discrete hyperbolic shapes, ie causing an angular bending between the hole-structure of a discrete hyperbolic shape],

. . . , but the new descriptive context also describes how control can be affected over the subsystems of both living and the mental subsystems of living systems.

Whereas in the currently accepted math constructs of the material interactions between mass or charge these interactions are assumed to be (absolutely) spherically symmetric in 3-space, and in all other dimensions. This assumption, concerning an unmeasured property (of material interactions being spherically symmetric in higher-dimensions), is used as a proof that higher dimensions do not exist, where it is assumed that the higher-dimensional structure would form a continuum and that all material interactions would be spherically symmetric, since there are not observed higher powers than the inverse square powers, when materials interact, thus proving the non-existence of higher-dimensions. This is a context built on assumptions in a similar manner as Ptolemy's system was based on assumptions.

By describing both the fundamental building blocks which fit into the new context, as well as the properties of global systems, such as the solar system as well as the spectral structures of living systems, one is educated about both the details and assumptions of the descriptive construct as well as the large context and range of applicability in the new context of containment, ie both its details and its distant reach are identified in close juxtaposition.

Some principles which will likely be contested by the authorities in the (unnecessarily competitive) world of academic intellectualism:
-1. The physical systems which are being described are very stable and precise systems . . . , nuclei, general atoms, molecules and their shapes, crystals, and solar system stability, [as well as dark matter, and universe expansion], but also life and mind . . . , and this implies that the

interaction structure, between the components which make-up these systems (ie the system's differential equations), are: linear, metric-invariant (for a fixed metric, with non-positive constant curvature), and separable, ie solvable and geometric, and thus these are controllable or determinable systems, and furthermore, their being modeled geometrically allows them to be both understandable (they can be pictured and measured {as opposed to "waiting for random events as a model of measuring" }) and practically useful.

That is, if the material interaction processes were non-linear and indefinably random then the physical properties of the resulting material systems would not be so uniform.

The uniform properties of so many fundamental systems implies that their interactions are in the realm of determinable mathematics, not in the realm of non-linear and indefinably random math structures.

How is this possible?
What context would allow for small material systems (quantum systems) to be described in a determinable manner?

This is possible within a context in which the very stable discrete hyperbolic shapes as well as the continuous (in shape) discrete Euclidean shapes are placed in a many-dimension context of containment [as well as the considerations of discrete shapes associated to other metric-spaces of various signatures] and these discrete shapes are used to model both material and metric-spaces (and these geometric shapes [as well as their determination of a large but fixed and finite spectral set] also contribute significantly to the "material" interaction process).

Furthermore the state-structure of both material (eg spin) and metric-spaces (eg metric-space states [eg positive and negative time directions associated to different subsets of complex coordinates]) requires that this description also be unitary.

One cannot simply build on the work of authorities, eg in quantum field theory this process (of honoring authority and its associated dogma and fixed interpretations) has led to a model of material interactions based on random point-particle-spectral (collision) interactions interpreted geometrically as particle-collisions which exist in a math structure where quantitative consistency and continuous geometry do not exist, yet these particle-collisions are modeled as if possessing a continuous geometry of object and path and identifiable collision, and this absurdity has led to other absurdities which show the plasticity of infinite-sets, indefinable randomness, and non-linearity, and a descriptive structure which logically proceeds from an assumed (but indefinable) global structure (of a function) to a local spectral measurement, ie a local particle-spectral (random) event in a 3-dimensional spatial subspace.

These assumptions have not worked, they are neither accurate for general systems, nor are they practically useful.

Mathematicians like to claim that they would consider any math pattern, ie in the math context of quantities and shapes, but this is apparently not true as they only consider peer reviewed literature, ie following authority and tradition within which a narrowly defined contest is constructed . . . , and apparently unknown to the professionals, their knowledge is channeled towards particular industrial (or big business) interests concerning engineering and production or about calculating risk (where it is clearly seen that top university PhD math and science people make calculations of risks which are clearly invalid and failed, the math of university PhD's are about math patterns which are not related to the world of our experience, rather the math experts are creating math patterns which describe a "world" which is an illusion, they are not math patterns which are practically useful, though they may be measurably verified, as Ptolemy's ideas were also measurably verified).

Some of the contested statements will be:

0. Stable physical systems must be based on a geometric descriptive construct (to be both measurable and usable).

1. Stable structures must be stable-geometric, ie based on the discrete hyperbolic shapes (and their associated discrete Euclidean shapes, which are related to material interactions as well as inertia).

2. Math structures must be quantitatively stable (metric-invariant, for metric-functions with constant coefficients), thus the only valid descriptions (quantitatively consistent descriptions) are diagonal, ie parallelizable and orthogonal at each point of the system's global geometry.

Note:
Only a linear model of either the derivative or of an equation which is defined in terms of derivatives acting on functions, ie such an equation must be linear, in order for these math constructs to be quantitatively consistent.

This means that "the derivative represented as a connection" identifies a non-linear context which is quantitatively inconsistent, since the local measures between the domain set and the set of function values do not relate the quantities in the domain set to the quantities which represent a function's values by a simple, diagonal, linear (matrix) relation.

3. Math structures must be quantitatively consistent, or linear and finite, and where the global structure is to be determined (from a local structure).

The new math context can give correct answers, and it can give consistent models of the organization of very complicated systems, eg living systems and their associated mental systems.

4. The fiber bundle becomes a part of the model (even more so than in quantum physics), not simply a "book-keeping set" used for the description of the properties of an external system, or an external math structure for a system, which is assumed to be non-linear, ie connections can be represented within "principle fiber bundles," and "function spaces" require that fundamental measurable properties be non-commutative, though quantum law wishes for commutativity. Thus the system [represented as a function space] {or represented as its set of physical operators, ie its measurable properties} is also non-linear.

Note: The commuting property for the "complete set of *commuting* Hermitan operators" which is supposed to define linear quantum systems has never been realizable for general quantum systems.

Thus, the linear superposition of spectral functions as a means by which to define a wave-function for a (general) quantum system "has been a wish" rather than a (true) principle of the laws of quantum physics.

Note:
The laws for a quantum system are (essentially):

(1) represented as a complex-number function space (but indefinably random), which is
(2) acted on by linear unitary operators (or in the context of local Lie algebras, function spaces are acted on as Hermitian operators) which represent physically measurable properties of the quantum system, (mostly the energy [or wave] operator and its decomposition into other Hermitian operators) and which is
(3) unitary invariant, ie conserves energy, and a
(4) linear quantum system is realized in regard to finding its "complete set of commuting Hermitan operators," etc.

Quantum physics is formally linear but in the context of

(1) a function space and its dual space being related to non-commutativity, ie the uncertainty principle, and
(2) in general quantum systems composed of 5 or more charged components, there does not exist simple symmetries, such as spherical symmetry, and subsequently the non-commuting pairs of properties (operators and their associated function spaces) become a permanent, non-linear, fixture of the descriptive construct.

A non-commutative structure for a differential equation (for a set of operators) can (does) imply curvature so the system's differential equations are not diagonal, rather (for example) upper tri-angular, and then the matrix can have many solutions, the system is under-determined, and there can be bifurcations of value,

or

The differential equation can be made diagonal, but the coordinated in which it is diagonal are not quantitatively consistent with the metric-function, the local coordinate structure is different for the two cases of diagonalization, ie the metric-function and the differential equation, where again this causes quantitative inconsistency.

(3) the charged components identify a set of independent singular points, which seem to be able to arrange themselves so as to create a neutral system, ie no singularities, but singularities lead to divergences.

Why do some systems composed of charged components have dipole and multi-pole (charge distribution) structure, and some are neutral?

The endless problems with math descriptions based on indefinable randomness (see below for definition of indefinable randomness)

A set of commuting (linear) Hermitian operators must be selected (found) in order to identify a quantum system's measurable properties, and so as to represent the system as a linear superposition of its spectral functions, but for general quantum systems, eg spherically symmetric quantum systems with five or more particle-components, this set of commuting operators cannot be found (has not yet been found).
They cannot be found for: nuclei, general atoms, molecules, nor for crystals.
Thus the set of a quantum system's measurable properties, eg measurable properties and their dual operators, together define a non-linear or curved space associated to the Lie algebra of operators, and their relation to a Lie group of operators which act on the infinite-dimension function space.
The description of a quantum system (in regard to its measurable properties, ie Hermitian operators) which has a linear structure (where the set of physical operators commute) is rare. Thus, there is an attempt to associate an approximate (linear) spectral structure to a non-linear context [non-commutative operator context] (for measuring [or modeling] quantum properties).

Problems with spherical symmetry

It should be noted that the singularities of the $1/r$ potential energy of a spherically symmetric structure for material interactions cause solutions to wave-equations to diverge.

Even for the H-atom the solution to the radial equation diverges, and thus it is truncated (the solution series cut-off) so as to fit data.

Furthermore, the singularity of the non-linear spherical geometry (associated to charged particles) is the focus for manipulating ways in which a spectral set, associated to the 1/r singularities of the system's various charges, can be manipulated to fit data.

Local symmetries of particle-physics

The (complex-number) function solutions of quantum wave-equations are energy invariant and thus they possess unitary invariance in regard to a global symmetry (ie global invariance).

On the other hand, the eigenvalues of the global functions are separated into (local) spin-eigenvalue-states, and thus in the model of "perturbation of wave-function spectral functions" by means of local eigenvalue-particle-state properties the global invariance (of the wave-function) is apparently assumed to be reduced to a local wave-function structure for the quantum system's eigenvalue structure.

These local particle-state invariances, which model material interactions as particle-collisions, which exist within a probabilistic construct (which should not allow the geometry of a particle-collision), which also models the random local spectral-particle event, ie the point about which the quantum description is trying to model, ie "how data is to be fit into a math construct."

Yet, how is this math construct of local small adjustments (due to particle-collisions in a context of indefinable randomness) to be related to the non-local properties of quantum systems, ie to the global wave-function properties of quantum systems?

The local gauge invariance ie how the wave-function is invariant to a local phase-transitions, is supposedly central to the construct of the math structure of physical law (based on the particle-collision model of material interactions). But a global wave-function needs a global relation to its phase.

These local symmetries, in regard to wave-phase, are related to the property of mass (and charge and field-particles etc) in the particle-collision models.

Thus, the question is about, "though the idea of gauge invariance" is supposed to be about the determination of the type of particle-collision interaction . . . ,

> since local gauge-invariance requires that the local particle-state matrix be introduced (as a connection term) so as to assure gauge-invariance, and this allows the local phase term (due to differentiation of products of functions) to be a part of the differential equation of the material interaction-system, so as to define local gauge-invariance,

..., nonetheless, "how does a quantum system still have the property of non-localness?"

However, because local symmetry is about "how the type (or which type) of particle-collision interaction is a part of the differential equation of the material interaction-system," this results in a discussion (concerning these differential equations) about the multiplicative constants which are part of the description, eg q (charge), and photons, and m (mass) and g (gluons), and "D" dark-matter, etc (as elementary particle keep getting added to the elementary event space, ie indefinable randomness).

That is, the relation in regard to "How force-fields are related to charge and mass, and g, etc".

Symmetry breaking is about pulling the energy spectra, ie the mass, of particle-collisions away from zero. That is, the graphs of quantum properties (graphs of wave-functions) are given in the scalar values of energy. That is, the perturbation of a wave-function is described in the context of the wave-function's energy. That is, charged and strong-force interactions take place about the zero energy point, and mass is about spectral (energy) cut-offs associated to the manifestation of mass.

This type of a model is only of interest to bomb engineering and the manipulation of measurable parameters in regard to finding probabilities of collisions in particle-collisions eg mass and charge related parameters, the only measurable properties in the bubble chambers of the particle-accelerators (?).

Whereas, mass (and charge) are a part of the spectral-energy properties of particle-physics, and these are about local symmetries of wave-phase, but the spectral-energy properties of the wave-function must be about global invariance of wave-phase, so that non-local properties can manifest (as a property of the wave-function).

Are these un-resolvable properties of quantum systems, or are they defined away, by claiming that "away from the local particle-collision the average structure of the harmonic wave-function, the non-local wave-function re-manifests itself."

But how is this done?

Though spin can be placed into the global structure of a wave-equation, the local particle-state structure cannot be placed in an energy-invariant (non-local) wave-equation, because particle-physics, eg including renormalization, is all about the loss of a geometric continuum for space's small structure, yet the continuum of a particle-collision is maintained within the non-geometric context of particle-states.

A local particle spectral-value event, is mediated by a math construct of local, non-linear connection terms (in the system's differential equation, or in the system's operator space) which preserve local wave-phase invariance, where the spectra is about mass (since mass equals energy), but mass is now a construct which is only global in a different context, than in the local context in which the material-interaction (particle-collision) model is defining mass (and charge etc). Whereas a global wave-function has a global energy-invariant construct, which has a different relation to its operator space, than does the local wave-phase adjustments to a waves energy.

The problem with such a description is that, the global wave-function for general quantum systems cannot be found, thus the local adjustments are meaningless.

Local adjustments have no meaning in an indefinably random context, a context which has come to be about data-fitting, but a context which seems to have no validity.

The lack of validity, for the math structure of indefinable randomness, is also shown by the failures of the risk calculations for business-risk, where the calculation of business-risks are also defined in a context of indefinable randomness, and are used to try to fit data, and thus, to identify risk within the structure of function spaces, and their operator (or differential equation) fiber bundles.

The failure of indefinable randomness is clear since there are no valid calculations for the spectra of general quantum systems, eg nuclei or atoms with five or more components, and the fact that the descriptive context of quantum physics has not led to any (robust) form of technical development, yet the stable spectral structures of general quantum systems suggest a controllable context for the descriptions of general quantum systems.

The obsession (in regard to social interest, of the military industry) with particle-collisions is about the fact that university physics has been turned into nuclear weapons engineering (for the military industry), and the way by which Asperger's syndrome, ie autism, is sought-out (by those who guide social institutions) and exploited (obsessive and narrow and competitive, ie the measure of intelligence etc), so as to serve the interests of the military industry (essentially, the only industry in the US which still makes things).

The new context claims that, there are unitary symmetries, but it provides a stable, many-dimensional, geometric context, which is also a stable finite spectral context, within which to interpret these unitary symmetries, and it is a context which is consistent with classical physics, but based on stable geometry. That is, it is not a description based on manipulating, in a very artificial way, the quantitative structures of function spaces and an associated set of fanciful convergences, and thus it is a math structure which is stable, as the properties of quantum systems are stable, and it is quantitatively consistent, ie the measuring structure means something, and it is geometric so one can picture the descriptive context so as to facilitate "practical" creativity, where practical is newly defined in terms of higher-dimensional stable geometries and their stable spectral properties.

Apparently

The non-linear structure, it is believed, allows for a greater variety of ways in which data can be fit with cut-offs and bounds on measurable values. It is believed that this process, about "how to manipulate the difficult context of (spectral) analysis for a set of non-linear operators" is considered OK by the experts, because the containing math set is "very big" and the non-linear and indefinably random patterns are sensitive to small changes in a system's bounding conditions either spectral or spatial. But the description is now non-linear, and thus attempting a "linear superposition" of spectral functions results in "new spectral properties (different from the linear spectra)" for the (new) sum. That is, if one changes a spectral set of functions, in the context of non-linearity [or non-commutative operator context] then the spectra associated to that (same) set of functions can be radically changed.

It is a process of going from the frying pan into the fire.

The new descriptive context

In the new descriptive construct the fiber group (of a principle fiber bundle) has an active part of the interaction and the new description's base space is also related directly to a geometry composed of various dimensions, because the fiber group (of a classical Lie group) has a spectral storage structure, ie a set of maximal tori, and the discrete hyperbolic shapes, the geometry upon which the descriptive context is based, are stable spectral structures, thus resonances can exist between the base space and the maximal tori of the fiber group.

In the current description of quantum physics the system is the function space while its physical properties are operators in its fiber group (there is a division between the system and its measurable properties)

In quantum physics the space of operators is a (book-keeping) mechanism for manipulating the structure of a function space around singularities and other critical points of a system's operator structure (or differential equation structure), with a focus on fitting data.

However, if one considers the physical systems themselves, these stable spectral quantum systems would not have the same spectral properties so consistently as they do have (from the various interactions in which these stable systems emerge) if the physical structure of these interactions was the same as the descriptive structure, which is so non-linear.

It should be noted that a non-linear math structure is quantitatively inconsistent (and logically compromised), and thus from such an inconsistent set of measurable properties an "infinite spectral variety" would emerge as the measurable properties for such systems, ie from non-linear and indefinably random structures of the material interactions there would not be uniformly identifiable quantum systems (based on their stable and uniform spectra).

High dimension context of the new description within which the set structure could be finite

Since, in the new descriptive structure, stable existence depends on a well defined spectra and the bounded spectral-set contained within an 11-dimensional hyperbolic metric-space...

> [resulting from the discrete hyperbolic shapes contained within this 11-dimensional hyperbolic metric-space (where the 11-dimensional hyperbolic metric-space models the containing metric-space)]

..., would be finite (or can be finite).

This is the basis for a (new) description based on a finite spectral set (though it is quite a large finite spectral set).

The nature of descriptive knowledge (for language which are precisely defined)

The point of Godel's incompleteness theorem is that the precise language of math (or of any specifically defined language) needs to be considered in terms of its (continual) re-organization at the level of assumption, context, interpretation, use, set-containment, and definition.

Tradition and authority are not valid determinants in regard to the validity of mathematics and science, rather these properties (of tradition and authority) define social class, and they define contests associated to using ideas in very narrow ways (where the contests are used to determine those who qualify for the upper social-intellectual class).

The current focus within the subject of math does not account for fundamental issues:

1. Quantitative consistency (eg it uses non-linearity as well as indefinable randomness as valid math constructs which are quantitatively inconsistent).

2. Set size (axiom of choice allows for "'too big' of a set," for quantities to be consistently defined). Sets which are "too big" cannot remain logically consistent.

3. Logical consistency which results both from sets which are "too big" as well as from quantitative inconsistency.

The data obtained from the local spectral-particle observations of general quantum systems

3b. There is another weird attribute, associated to the idea of logical consistency, which appears when contrasting the math of classical physics with the math techniques of quantum physics, where classical descriptions which are built upon local so as to determine a system's global properties, ie the laws of classical physics (Newton's laws), uses local model of linear measures (derivatives) to define differential equations, and the solutions of linear, separable differential equations (defined within Euclidean space) provide global information about the system.

Whereas "In the practice of quantum physics" the description assumes a global structure, ie it assumes a function space associated to a quantum system, which, in turn, is related to the local measuring of a system's spectral properties. The spectral functions identify locally-determined (or observed) properties.

The descriptions of micro-material properties seem to require that there be a fundamental change of assumptions wherein a stable global context is assumed.
When one is trying to find a measurable attribute which exists then it seems better to assume that such a description of measurable attributes be a result of a local linear measurements and their associated differential equations whose form is guided by an assumption of some stable bounding structure (or construct) [in classical physics this would be the material geometry surrounding a system's material component], but for micro-material properties "Is this stable structure a function-space along with a set of measurable properties (or set of Hermitian operators)?" or "Is it a stable geometry, eg a set of discrete hyperbolic shapes?"
It is a set of discrete hyperbolic shapes.

However, local, random particle-spectral events in space and time, ie the data one is trying to fit, and the set of global properties, ie the system's set of operators applied to the system's function space as well as the function-space itself, cannot be made (or used) to match the observed local spectra (in regard to some general quantum system).
The local spectral measures cannot be made to match-up with (or to fit) a general quantum system's calculated spectra which identify local particle-spectral measurements (or observations) [a general quantum system is assumed to be random, but the measurable nature of its randomness (its spectra and its spectral functions) is not determinable by physical law, ie it is an "indefinably random" system (see below, too).

4. Quantitative stability, the need for a descriptive and mathematical context of quantitative stability.

5. Accepting that creativity as well as observing a system's stable properties demand a stable geometric basis within which measuring is both reliable and useful for practical creativity.

The crux of the matter, in regard to basing descriptive language on tradition and authority

Simply because a math technique, or math process, or a math construct, can be made to fit (a set of cherry-picked) data does not mean that such a math construct is valid, the construct also needs to have a wide range of applicability, so as to (also) provide a framework within which practical creative development can take place.

Ptolemy's models fit data quite well.

The need for a descriptive structure to be widely applicable

The best example of wide applicability and great usefulness is classical physics, in regard to the linear, solvable physical systems (of a classical system's differential equations).

But similar context of linear, separable, and metric-invariant structure, within which the math techniques of quantum physics are applied, are not valid for systems with spherical symmetry and more than five-components composing the system (or geometric symmetries as properties of the quantum system's potential energy term), where quantum physics is framed in the context of function spaces and probabilities associated to spectral values, but the spectral values (that which one is trying to calculate) are local properties [not global properties]).

The failings of function spaces (as a basis for physical description)

Why is the function space context (for differential equations) not as reliable of a math construct for quantum systems as is the geometric context for the classical context?

The interpretation of materialism in classical physics means relatively local stable geometric structure while for quantum physics materialism means spectral discreteness, randomness, and mysterious spectral cut-offs which exist for quantum systems, cut-offs and spectral approximations which are needed to solve the differential equations.

These spectral cut-offs, which are needed to solve for the spectral structure of quantum wave-systems, exist for (all the) quantum systems (and are a characteristic of function space math-techniques), from the H-atom, ie the truncation of the radial equation's infinite series (diverging) solution, ... to particle-physics, ie re-normalization due to (or based upon the idea of) "symmetry breaking," where the breaks in (mass-less) symmetry (the existence of mass) are supposed to identify an underlying spectral structure (spectral cut-offs, or spectral approximations) about which quantum systems are (supposed to eventually be) organized.

This might be true, except for the fact that symmetry breaking (and particle-physics) and the identification of elementary particles is only related to perturbations of spectral properties of quantum systems, where (it is assumed) the main spectral approximations have already been found from physical considerations (a complete set of commuting Hermitian operators) . . . ,

> [which (without a sufficient amount of highly ordered geometric structure) are not sufficient constructs (ie sets of measurable properties [Hermitian operators]

have not been found for quantum systems in general) by which to find a close approximation to a general quantum system's spectral properties]

..., where these (assumed to be) well defined approximations to the system's spectra can then be perturbed by using particle-physics,

[furthermore, the process of adjusting the spectra to be related to the singularity of a spherically symmetric energy {or interaction} structure (or set of such singularities) has not worked].

The lack of geometric structure within quantum description means that the function space techniques do not work for quantum systems which are (assumed to be) spherical symmetric but which have five or more components (particles), ie a nuclei and a set of electrons.

However, in such a system the spherical symmetry is lost, and instead the system has a set of $1/r$ singularities associated to each charged component.

The non-linear structure of function spaces

The reason for such a break-down of these math techniques should be clear (elementary) it is because function space methods are non-linear (and non-linear means quantitatively inconsistent), a property for differential equations which emerges (when solved by means of function spaces and their associated "spectral cut-off [or spectral approximation]" techniques) from the uncertainty principle related to a function space and its dual space [and a probability density-function's relation to standard deviations (or variances) {where the spectral functions of the function space are probability density-functions}].

The non-communitivity of, say, "momentum and position" [which identify dual sets of functions (or dual operators) under Fourier transformations] means that these properties define a non-linear structure in the Lie group of unitary operators (with its Hermitain Lie algebra) upon which (or within which) the wave-functions are defined to be invariant. However, these properties (eg momentum and position) are always physically present (it is assumed) [certainly for a system which does not possess spherical symmetry].

Furthermore, the spectral functions possess geometries which are not linearly compatible with one another, ie they are different shapes. Momentum functions and position functions have shapes which are incompatible with commutativity.

Solvability and commutativity

Linearity requires consistently parallel geometries, eg congruent or similar shapes, but similar shapes are not metric-invariant shapes (so similar shapes might exist in different dimensional levels so that the conformal structures defined by physical constants can intervene in the stable math relations which need to be defined for quantitative consistency).

That is, the physical properties, which are assumed to always be related to a measuring process, are quantitatively incompatible as "measured properties" and such physical properties would only be consistent in a commutative, linear description, but such descriptions cannot be found, (perhaps, mostly due to singularities associated to spherically symmetric of charged material interactions in systems with five or more charged components).

Thus, the quantum structures are indefinably random, since (1) the spectra of general quantum systems cannot be approximated in a valid manner and (2) the linear theory is non-linear in its formulation "based on sets of operators which identify physical properties," ie the math process is non-linear in regard to general operators acting on function spaces.

Indefinable randomness (defined)

Indefinable randomness is about elementary event spaces (the sets upon which probabilities are defined) are composed of: unstable events, un-define-able (or incalculable) events, the elementary event space is not fully identified, ie new events are added to the elementary event space at will . . . , eg dark-energy particles added to the set of elementary-particles etc . . . , where each (or any) of these properties for elementary-events makes these event spaces uncountable, ie one cannot rely on the counting process to determine probabilities. Thus, the probabilities cannot be determined for such elementary events spaces.

That is, even though it might be claimed that certain types of events are distinguishable, and thus countable (so as to be related to probability), however, if these events actually fit into a space which is "indefinably random" then one cannot rely on the probabilities which are determined from any counting process applied to such a probability space.

Where indefinable randomness is defined to be a set of elementary events of a random process which cannot be counted in a consistent manner, and thus it is not a quantitatively consistent structure, ie counting is not reliable within such a set of distinguishable (but unstable and undefined) elementary events, ie the structure is not a valid math structure.

Measuring (or defining measurements) on (or in) such a math structure is neither reliable nor conceptually useful at the level of practical creativity.

Geometry vs. sets of Hermitian operators (and their dual structure [which result form Fourier transforms])

eg The "position vs. motion" dual pair

The problem seems to be about material geometry being limited in its range, and material geometry deals with a finite range (or context) of locality (ie consistently relatable to local linear measures) (closely associated to the idea of closed and bounded (in a metric-space)), ie the system is (thought to be) a bounded object (when one considers a physical system in a geometric context).

The "too big" set structure of function spaces

The implicit infinity of function spaces, ie infinite dimensional, is not mathematically manageable in a quantitatively consistent manner (or in a consistent set containment structure) so that quantum descriptions cannot be relatable to a context of relatively stable local geometric properties, or local geometric stability, ie the quantum-foam of particle-physics.

Function space methods are non-linear due both to the uncertainty principle associated to dual spaces, and to the general context in which the operators are applied (or used), ie the geometric symmetries are not known for systems composed of five or more charged "particles."

Furthermore, the geometries of the spectral functions in function spaces which are discretely distinguished from one another, and thus the shapes of the different spectral functions cannot (in general) be continuously related to either shape or to topology ie holes in the space. Their local vector field geometries do not commute.

Classical physics assumes materialism and local linear measuring contained within a geometric and continuous context. However, classical inertial properties are non-local, while the properties of charged material possess "positive" and "negative" time-states associated to such physical systems.

Quantum physics assumes materialism, non-localness (ie action-at-a-distance), and a fundamental indefinable randomness which is contained in a non-linear context due to either the uncertainty principle (of a function space and its dual space) or the non-linear and indefinably random structure (of quantum description) is due to the indefinably random and non-linear model of particle-collision models of material interactions, as defined within quantum physics.

Preface

The new (alternative) model for mathematical description of existence is based on material and space both being composed of stable "cubical" simplexes (and their associated moded-out shapes) for both discrete Euclidean and (the very stable) discrete hyperbolic shapes, where these cubical simplexes are placed in a many-dimensional context, and where the properties of material and space are separated from one another by one dimensional level, ie n-dimensional space contains (n-1)-dimensional materials.

This simple math structure is needed so as to have mathematical stability, and to be able to describe over a wide range of general physical systems the observed stable set of physical properties associated to so many fundamental physical systems. Stable spectral-orbital properties are possessed by fundamental physical systems such as: nuclei, general atoms, molecules, crystals, as well as the mysterious stability of the solar system, which are all systems characterized by their stable spectral-orbital properties.

The stable basis for our experience, exists within a geometric context, where geometric descriptions can be used in practical ways when these descriptions are associated to measuring.

This math structure identifies a linear, metric-invariant (for fixed metric-functions with constant coefficients) math context, and it also provides a context for geometry which is based on parallelizable and orthogonal vector fields (which are associated to these simple geometries [of the cubical simplexes] which now model material systems as well as modeling the containment space for these material systems).

This metric-space structure has a clear relation to the classical Lie group isometries as well as the unitary fiber groups defined over the above mentioned metric-spaces and Hermitian-spaces, where complex coordinates are needed to describe the state-structure of both material (spin properties etc) and metric-spaces (which now possess "metric-space states" which are an important part of the dynamic interaction process of material interactions) where, in complex space, the real isometries are divided into the two subsets (real and pure imaginary) associated to positive and negative time directions, which represent opposite metric-space states of dynamical systems separated into subsets of the complex coordinates, thus leading to a unitary fiber group, and a subsequent mixing of these metric-space states.

The new context of containment is restricted to the spaces of non-positive constant curvature, ie Euclidean and hyperbolic spaces, as well as spaces of various divisions of spatial and temporal subspaces associated to a metric-function's signature (but not spherical space, which is a non-linear space).

However, the model for material interactions in 3-space (or in 3-dimensional spatial subspaces) is spherically symmetric, until the interaction approaches the size of the interacting material components, which have the geometry of discrete hyperbolic shapes.

The geometry of higher-dimensional interaction structures depend on the geometry of the fiber group, ie material interaction are not universally spherically symmetric. For example, SO(3) has the shape of a 3-sphere, but SO(4) has the shape of two 3-spheres, ie the inertia of the material interactions in hyperbolic 4-space is not spherically symmetric.

This math structure is both simple and very constrained, except for its higher-dimensional structure, which creates a context of great variety.

The higher-dimensional structure allows for descriptions of coherent higher-dimensional models of living systems which can control lower dimensions of their own set composition, and thus they can control chemistry, and such systems can also coordinate the functioning of the organs which compose their own living system.

Minds are also describable in such a context.

Rather than using infinity to ensure that the gaps of measuring structures are filled, might it be better "for the mathematical context to be stable" and quantitatively and logically and consistent, so that the descriptive structure (the stable geometric construct) is contained within higher-dimensions, and within which a finite spectral set can be defined (which is contained within a higher-dimensional [hyperbolic 11-dimensional metric-space] containment set), where the fixed finite spectral set is based on the stable, bounded, discrete hyperbolic shapes which compose the structure of the metric-spaces of the various dimensions.

That is, "gaps" (or incompleteness) in the quantitative structure can exist as long as the spectral set structure, and its associated stable geometric structure, is maintained within the over-all high-dimension containment set.

The new context is a stable geometric context so this will allow such a construct to be applicable to practical technical development, but it (also) seems to signal that human creativity is better focused outside of the realm of materialism, based on new higher-dimensional and "global" descriptions of life.

The new material interaction structure accounts for both

(1) the apparent indefinable randomness of small components in space, as well as
(2) the hidden properties of the higher-dimensions which compose existence, where physical constants also help hide the properties of higher-dimensional space and its material components.

How to build stable sets?

Inclusion in a set also provides the elements of the set all the properties which have already been attributed to the set.

But should the set be built from the ground upward to greater complication or should the set be based on known stable structures and set containment of elements be more about stability and consistency in regard to the models of physical measuring.

It could be noted that functions "model simultaneously" both measuring and set containment.

This is the main issue of quantum physics which assumes global structures and descends to the local (random particle-spectral measures), while the interpretation of the classical differential equation assumes a local measuring structure and finds a global relation.

Should the structure of math also be re-organized from the stable down to the measurable.

But is geometry a better vehicles to define stability rather than function spaces, and/or their operators?

> That is, is the stable math construct to remain function spaces and/or their sets of operators (local Hermitian operators of physically measurable properties)?

or

> Is the stable construct about which the math descriptions of physical existence are to "pivot (or revolve) around" the very stable discrete hyperbolic shapes, in a many-dimensional context, up to a dimension-11 hyperbolic metric-space (as identified by D Coxeter), (where it should be noted that hyperbolic 3-space is equivalent to space-time [one-to-one and onto]).

The Abstract:

Can a geometrically stable and spectrally finite math construct . . . ,

> which is contained in a many-dimensional context, associated to an open-closed, but non-local (ie action-at-a-distance applies), set of bounding stable discrete hyperbolic shapes wherein adjacent dimensional levels are geometrically significant in regard to both containment and for a set of (partially) bounded dynamic-geometric math processes, which are defined within a bounding (containment set) structure of discrete hyperbolic, Euclidean, and unitary subgroups (or simply, linear discrete shapes),

. . . , provide a better math structure within which [or to be used] to frame the descriptions of observed physical properties?

Yes.

Rather than using infinity, might it be better "for the mathematical context to be stable" and quantitatively and logically and consistent, so that the descriptive structure is contained within higher-dimensions, wherein measurement is consistent with geometric measures of stable geometric shapes, and within which a finite spectral set can be defined (which is contained within a higher-dimensional [hyperbolic 11-dimensional metric-space] containment set), where the fixed finite spectral set is based on the stable, bounded, discrete hyperbolic shapes which compose the structure of the metric-spaces of the various dimensions. The geometries of the different dimensions have similar or consistent discrete structures, ie the geometries are compatible with one another, ie geometries can con-formally fit into one another, or as "free stable geometric structures" they can be resonant with a subset of a fixed, finite geometric-spectral containment-set structure.

When a metric-space modeled as a discrete hyperbolic shape is placed in (or contained within) an adjacent higher-dimensional metric-space then the original metric-space becomes a closed shape (which is easiest to think of as being a closed bounded shape in the higher dimensional metric-space within which it is contained). (This would be a model of a "free geometry" contained within a metric-space of some given dimension.)

That is, each dimensional level can contain within itself a variety of lower dimensional (lower than the dimension of the containing set itself) stable discrete hyperbolic shapes as well as a set of associated resonating Euclidean shapes, (which are resonating to a discrete hyperbolic geometries dominant {or occupied} energy structure [of the hyperbolic shapes various hole-spectral structures).

However, these stable lower dimensional shapes [of dimension (n-1) or less, when they are contained within a metric-space of some given dimension, n) can always be placed within (or can occupy) a higher-dimensional shape, up to shapes of dimension-(n-1). Thus, for an n-dimensional containment metric-space (which also has a shape) the main focus of interactions would be in regard to the (n-1)-dimensional shapes (it contains) which (naturally) interact with one another, by means of action-at-a-distance discrete Euclidean shapes, which fit into the geometric interaction construct in a geometrically consistent manner.

Note:

The interaction structure accounts for both random properties, as well as for the hidden properties of higher dimensions, (metric-spaces with spatial subspaces greater than three, seem to be hidden from our senses) where the higher dimensional spaces are hidden by both the interacting structure and by the interacting system's size which is (partly) determined by the physical constants (which are defined between the different dimensional levels, where physical constants determine the sizes of the interacting shapes (or systems)) within a metric-space.

It needs to be noted that both material and metric-spaces are discrete hyperbolic shapes, but with different dimensions, ie metric-spaces contain the discrete shapes of material systems.

The central focus of this new descriptive construct is spectra, as is also the case of function space descriptions

Since the new theory is a finite spectral theory, based on stable discrete (hyperbolic) geometries, it is a description which depends on spectral cut-offs (or spectral approximations) which are associated to physical constants.

In the new description Bosons are Euclidean structures attached to "infinite extent" discrete hyperbolic shapes, which are a part of the descriptive structure of a "material" system (the dimension of these Bosons is one less than the dimension of the metric-space which contains the given infinite-extent discrete hyperbolic shape), (or perhaps the same dimension as the dimension of the metric-space which contains the given infinite-extent discrete hyperbolic shape) . . .

. . . . However, there exist "interaction structures" which are higher-dimensional, and the spectral geometry of hyperbolic space (ie the discrete hyperbolic shapes) which are 6-dimensional and beyond . . . ,

(Note: up to 10-dimensional hyperbolic space, but (apparently, according to D Coxeter) 11-dimensional discrete hyperbolic shapes do not exist)

. . . , are determined by infinite-extent shapes, and the spectra of such infinite-extent discrete hyperbolic shapes depends either on intent or on the lower dimension compact space-material (discrete) geometric-spectral structures, of five-dimensions or less, which are a finite set (and contained within an over-all containing, hyperbolic, 11-dimensional metric-space), and any of the spectra (of the appropriate dimension) of this "finite spectral set" of the over-all containing space can be carried (or defined) upon an infinite extent shape.

Old and new

The new theory is not at odds with the observed properties of particle-physics it is unitary invariant and focuses of spectra (other than being at odds with the fact that particle-physics cannot be used describe any stable physical systems and its descriptive context is practically useless), but it is geometric and the construct of its interaction processes are simpler, and it provides a better defined context in which to identify "what the . . . 'mass resonances' or the 'broken symmetries' . . . . actually are," in a geometric-spectral context.

Conjecture: A scalar boson would be a physical constant defined between dimensional levels (of higher dimensions than the spatial dimension defined by materialism) [and/or between different signature metric-spaces, of (say) the same spatial subspace dimension], ie it would define a different material size-scale in regard to significant (or easily observable) interaction forces of material within a certain dimension metric-space [and/or a metric-space which may have "varied" properties for the signature of its metric-function].

In the new description there is no need for scalar Bosons since mass is assumed to be an intrinsic part of the descriptive construct, namely the existence of discrete Euclidean shapes resonating to the sub-toral components of the stable discrete hyperbolic shapes

If physical description has a geometric context and global description can be based on local measures to find global structure then the description can be based on differential equations motivated by geometry where local linear measures allow the description to be quantitatively consistent but being contained within a metric-space the differential equation needs to be metric-invariant where the metric-functions have constant coefficients, but metric 2-forms are symmetric and thus in local coordinates can be found which diagonalize the metric 2-form, thus the linear structure will only be quantitatively consistent with the metric-function in these local coordinates but "are these local coordinates global?" but if the differential equations is separable, ie the geometry upon which it is defined is parallelizable and orthogonal, then the linear structure would be globally (on the "separable" shape) diagonal and thus both consistent with the geometric measures and invertible, and thus solvable.

In the new math context, if the description is determined by the a stable material system's shape, ie a discrete hyperbolic or Euclidean shape, then the metric-invariant, linear differential equation will be quantitatively consistent and solvable.

However, most interaction material systems are not defined on these stable discrete hyperbolic shapes, but rather they are defined between these shapes, and they are mostly non-linear interaction structures. However, if the interaction is defined within a "correct" range of energy and the material (or geometric) components of the interaction comes "close enough" to one another then the system can resonate with the spectra of the (over-all high-dimension) containing space (where the containing space is [also] composed of stable discrete hyperbolic shapes) and the interacting system can "deform" into a stable discrete hyperbolic shape (or adopt its new shape of its being) so as to form a stable system [of hyperbolic and Euclidean shapes together form a connected shape similar to some discrete hyperbolic shape, and thus a subset of this geometry could begin resonating, and, subsequently, instantaneously change to a discrete hyperbolic shape which does [exactly] resonate to the containing space's spectral properties] whose differential equation also changes into a linear, metric-invariant, separable differential equation associated to the newly emerged stable material system, (which emerged from the material interactions).

If these "correct" energy properties or the necessary property of "closeness" are not realized by the interacting system then the material interaction is non-linear and chaotic in its nature.

High-dimension discrete shapes and the relation of interactions to finite spectral sets

Inert material systems interact, and sometimes resonate and subsequently "jump" to a new stable, higher-dimensional material spectral-geometry.

The Mathematical Structure of Stable Physical Systems

Living systems can (now) be modeled to be some of the odd-dimensional, discrete hyperbolic shapes, which possess an odd-numbers of holes in their shapes, and all spectral flows are occupied within this shape, so as to form an unbalanced charged geometry within a stable hyperbolic shape. Thus, the shape possessing such an unbalanced charged distribution begins to oscillate, and to thus generate its own energy.

This is an elementary model of both life (and of a radioactive nucleus).

How do compact, living, oscillating, higher-dimensional "material" systems interact?

What is the spectral structure of an infinite-extent discrete hyperbolic shape?
Answer: It can take-on the spectral properties of the finite spectral set within which it is contained, ie it is contained in an over-all high-dimension containing space.

Can an infinite-extent living system "jump" between large finite spectral sets, where each spectral set is contained within an over-all high-dimension containing set (space), ie contained within a 10- or 11-dimensional hyperbolic space.

What is the relation between lower dimensional discrete shapes and the higher-dimensional metric-spaces within which they are contained?

How do lower-dimensional shapes relate geometrically to the higher dimensional discrete shapes, where these discrete shapes are based on "cubical" simplexes. Do they move about freely in the cube, or do they get confined to its faces?

Think of a given metric-space as an n-cube (or cubical simplex) . . . ,

> {which can be moded-out to form an intrinsically n-dimensional shape whose extrinsic (n+1)-dimensional shape is contained in an (n+1)-dimensional metric-space (this is true for metric-spaces with non-positive constant curvature), eg the shape of the 2-torus is contained in Euclidean 3-space}

. . . , then the material system would be an intrinsic (n-1)-dimensional cube in the n-metric-space.

But the n-cube has (n-1)-faces, thus the (n-1)-cube, which defines a material system, can be one of the (n-1)-faces of the metric-space's n-cube structure, or it (the (n-1)-dimensional cube) can be contained in the n-cube so as to interact with another (n-1)-cube which is also contained in the n-cube (model of a metric-space). But the interaction of (n-1)-cubes (in a containing n-cube metric-space) in turn, also forms an (interaction) n-cube whose extrinsic (or moded-out) shape is contained in an (n+1)-dimensional cube. If the conformal constants defined between adjacent dimensional levels are such that only very large extrinsically (n+1)-dimensional shapes interact (by means of a detectable force) then the (n+1)-shape defined by (interacting) (n-1)-components can be stable, and in such a case would appear to move about as (n-1)-components in n-space, ie the n-cube model of the metric-space. Thus the (n-1)-dimensional shapes can either determine the (n-1)-flows of the moded-out n-cube within which it is contained, or they can move "freely" about the n-cube, ie the freely moving (n-1)-components are within the intrinsic (closed) geometry of the n-cube {whether the n-cube is moded-out or not}.

It can be noted that the n-cube (of the given n-dimensional metric-space) is part of the flow geometry of a moded-out (n+1)-cube, modeling an (n+1)-metric-space, when the (n+1)-metric-space is moded-out.

Now if infinite-extent discrete hyperbolic shapes are contained within a closed and bounded shape (ie a shape which is compact, in regard to metric-spaces, and also a moded-out shape) then such infinite-extent shapes would structurally prefer the spectra of the moded-out discrete hyperbolic shape.

Is the difference between light and a neutrino (both of which [in the new descriptive context] can have infinite-extent geometric properties) that of a difference of dimension, where the neutrino can be contained in a compact shape, while the "infinite extent geometric structure of light" is such that it is contained within the closed set structure of the next (or adjacent) dimensional level, and thus appears to be infinite-extent?

The finite spectral set which is built upon discrete hyperbolic shapes depends on the compact shapes of discrete hyperbolic shapes which are only defined up-to and including the 5-dimensional discrete hyperbolic shapes. (This is due to a Theorem by D Coxeter.)

Thus, if an over-all high-dimension containing space is 11-dimensional then the spectra of all the infinite-extent discrete hyperbolic shapes, 6-dimensions and above, would have to carry a given finite spectral set defined by the set's compact discrete hyperbolic shapes of dimension 1, 2, 3, 4, and 5.

Because the infinite extent discrete hyperbolic shape can carry on its geometric structure many different spectral values, since an infinite structure has no natural values for geometric measures, then what can hold such geometric structures to a particular finite spectral set?

or

How can such infinite extent geometries change to new spectral sets?

Can the infinite-extent discrete hyperbolic shapes of an (or contained within an) over-all 11-dimensional containing space, which contains a finite spectral set, "jump" to other, different, 11-dimensional containing spaces, defined by different (finite) spectral sets?

or

Is the structure to be considered the "jumping" of 10-dimensional infinite-extent shapes to other 10-dimensional infinite-extent shapes both of which are contained within an 11-dimensional hyperbolic metric-space?

Is this mostly a question of concern for the living infinite-extent systems in an 11-dimensional containing space, an 11-dimensional space defined by a finite spectral set. That is, for the infinite-extent 7-dimensional oscillating hyperbolic structures, and the infinite-extent 9-dimensional oscillating discrete hyperbolic shapes.

By the interactions of 9-dimensional discrete hyperbolic shapes the other 10-dimensional finite spectral structures can be contacted in the 11-dimensional hyperbolic metric-space, where the 11-dimensional hyperbolic metric-space has no spectral structure of its own. Thus in 11-dimensions all the different high-dimension, 10-dimensional spaces defined by finite spectral sets, can be accessed, and thus transitions between high-dimensional spectral sets is a possibility.

Thus, for high-dimension (infinite-extent) living systems, some of these other spectral sets could sustain the experience of the living system (survivability and being able to perceive), and others might not have the correct organization to sustain experience.

Are these issues of survivability about the "types" of infinite-extent 10-dimensional discrete hyperbolic shapes, which were identified by Coxeter?

What are the main issues concerning creativity for such a structure of existence?

Chapter 17

NEW BASIS FOR PHYSICS

The explicit alternative

On the other hand if one assumes that there is a many-dimensional, discrete, separable, metric-invariant (where the metric-functions have constant coefficients) geometries . . . , which are defined on the coordinate base spaces . . . , as fundamental to existence, then this gives a new descriptive structure for all material system interactions (where material is now a compact space-form defined in a particular dimensional level, contained in a metric-space, which in turn, has a discrete separable shape associated to itself), so that this structure of material interactions is consistent with classical physics (when defined on 3-dimensional spatial subspaces, eg R(3,0) and R(3,1)), and this new geometric structure for . . . :

(1) metric-spaces,
(2) material systems, and
(3) material interactions,

. . . . accounts for:

(1) point-like spectral events, and it also accounts for
(2) quantum randomness, and it also accounts for
(3) unitary-ness (each dimensional level of a metric-space has "pairs" of metric-space states associated to itself, and these "pairs" of metric-space states identify complex-coordinates) as well as

(4) spin-rotation of metric-space states,

(where the spin-groups are related to the geometry of the dynamics of material interactions, where it is only in 3-spatial-dimensions where the interaction geometry is spherically symmetric, yet even in 3-spatial-dimensions the interaction does not depend on 1/r singularities).

The interaction structure is based on differential-forms, as in the case of classical physics (where, now, the differential forms identify space-forms), but now the interactions depend on new sets of dimensional levels.

Basically it is both the structure of interactions and the values of physical constants, eg h, G, c, etc, which keep the different dimensional levels from being (geometrically) inter-dimensionally dependent, though spectrally a system is dependent on the spectral structures of all the different dimensional levels.

The spectral properties of the different dimensional "material" space-forms contained in the different dimensional metric-spaces (which also have space-form shapes associated to them) are those spectral values which are in resonance with all the spectral space-form structures (of many different dimensions) which are contained in the over-all high-dimensional containing space.

This geometric structure provides the stable, discrete spectral structures which are needed to model the stable discrete physical systems related to: nuclei, atoms, molecules, crystals, and solar systems.

These shapes for material systems and the interaction structures are consistent with properties which are approximately spherically symmetric related to 3-dimensional spatial subspaces of metric-spaces, and the associated rotational symmetries of these systems.

Consider the table:

Dim of material	Interaction space-form dim	Dim of containing space	fiber group	phys const
1-dimensional	2-dimensional	3-Euclidean space	SU(2)	G
1-dimensional	2-dimensional	3-hyperbolic space	SU(2) + iSU(2)	c
2-dimensional	3-dimensional	4-Euclidean space	SU(2) + SU(2)	h
2-dimensional	3-dimensional	4-hyperbolic space,	Spin-rep [R(4,1) + R(4,2)],	?

The last row can represent either ... two interacting charged space-forms where the subsequent (relatively) stable nuclei are related to either R(4,1) or representing a radioactive nucleus related to R(4,2), or ... two or more interacting atoms where the subsequent stable inert molecules would be related to R(4,1), and the biologically-active molecules related to R(4,2), eg chlorophyll would be related to R(4,2). Note: R(4,2) is related to SO(4,2), while C(4,2) is related to U(4,2), and SO(4,2) will have a spin representation etc.

The chart can continue:

Dim of material	Interaction space-form dim	Dim of containing space	fiber group	phys const
3-dimensional	4-dimensional	5-Euclidean space	spin rep of SO(5),	(?)
3-dimensional	4-dimensional	5-hyperbolic space,	spin rep of [SO(5,1)+SO(5,2)],	(?)

Etc, etc

This chart is presented in a "real" coordinate context, but the natural coordinates are complex coordinates so as to take into account the two (opposite) metric-space states which exist at each dimensional level. Thus the fiber groups which one would also have to consider are the SU(n) and SU(s,t), where s+t=n, unitary fiber groups, and their relation to spin representation groups of the SO(n) and SO(s,t) real subgroups of SU(n) and SU(s,t) respectively.

Note: Stable systems emerge from both the system of interaction structures and the resonances that all the space-forms (which exist in an over-all high-dimensional containing space) . . . , where the containing spaces themselves have discrete separable shapes associated to themselves, though this is hidden from our experience by the structure of interactions and the values of the physical constants, but nonetheless is related to the physical properties of the existence of stable planetary orbits . . . , have with all the other space-form spectral properties in all the different dimensions and all the different subspaces (of the same dimension as some given level). As the material space-form interaction proceeds, the interacting space-form can start to experience resonances with the containing set, if the energy of interaction is of the correct level then a stable space-form can emerge from both the interaction and the phenomenon of resonance, thus ending the interaction process with the formation of a new stable space-form. Note: spectral stability and spectral discreteness is a property of hyperbolic space-forms.

Atoms are described using 3-dimensional hyperbolic interaction space-forms defined on a 4-dimensional hyperbolic space, while atoms and molecules (may) interact as 4-dimensional interaction space-forms on a 5-dimensional hyperbolic space, so that these interactions are very weak, ie the van-der-Waal's forces, they are very weak forces because the interaction's corresponding* conformal constant . . , analogous (or similar) to c . . . , requires that the interacting "charges" be very large or very close (to have noticeable changes of motion)* so that interactions between small systems of the correct dimension, ie in the case of van-der Waals forces it is dimension 4, interact only slightly.

Note: These van-der-Waal's forces are (could be) the forces which hold crystals together, but they are perhaps off-set (or altered) by small multi-pole electric charge distributions.

while h may be the 3-dimensional Euclidean interaction space-forms (or differential forms) of harmonic functions associated to 2-dimensional inertial properties (mass) related to 4-Euclidean space. That is, these physical constants are the conformal factors which exist between dimensional levels in regard to models of material interactions.

G is related to 2-dimensional interaction structures which are contained in 3-Euclidean-space

c is related to 2-dimensional interaction structures which are contained in 3-hyperbolic-space

c is related to 3-dimensional interaction structures which are contained in 4-hyperbolic-space (?)

h is related to 3-dimensional interaction structures which are contained in 4-Euclidean-space

To relate the physical constants to interaction structures
G may be the 2-dimensional interaction space-forms of gravity related to 3-Euclidean space,

while

h may be the conformal constant between the harmonic function space structures of 2-dimensional charged structures and 3-Euclidean space.

while c may be the 3-dimensional interaction space-forms (or differential forms) of 2-dimensional charged material of electromagnetism interactions related to 4-hyperbolic space.

Note:
One wants free speech, and one wants many ideas about how to organize, and structure the math in relation to the descriptions of a measurable or couple-able (or a useful) existence, so that many paths to creative developments exist.

Unless there is a math structure which integrates complex systems into a unique system structure upon which hierarchies of control can manifest then there is no means of describing the precise and integrated functioning of living systems and their complicated subsystems based only on the idea of materialism.

Science made religion irrelevant but science has in turn been made fixed authoritative and dogmatic, just as a religion, so that there is a battle between two dogmas, but (ideally) science is not dogmatic. In other words religion is irrelevant unless its abstract truths are associated to a descriptive language which is associated to practical creativity, not simply related to social hierarchy and the manipulation of social forces, ie manipulation of the icons of high social value.

Probability based descriptions are not useful for direct creative use, unless the measured values are averages defined over large reservoirs which contain many components, while only descriptions based on geometry have been related to creative use as in the case of classical physics, but in classical physics it is only the separable geometries which are useful.

* (Referencing the Copernicus vs. Ptolemy descriptive structures) In fact, it is probably possible to enhance the epicycle structures of Ptolemy so as to still provide a more precise description of planetary motion than other (currently accepted) descriptions of planetary motions associated to authoritative physical theories about planetary motions, either perturbed Newtonian motions, or other types of (non-linear) descriptions (artificially adjusted by boundary conditions).

Considering the simple quantitative nature of math, it is difficult to find simple descriptions of math patterns.

Math reconsidered

Godel's Incompleteness theorem implies that there are limitations as to the set of measurable patterns which a precise language can describe. Thus there is the question, in regard to descriptive knowledge, as to "What is true?" An answer to this question can be that, "A described pattern is 'true' if it is widely applicable and its (descriptive) information can be used in regard to many practical creations."

Math is about description based on measurable quantities and shapes upon which measurement (or coupling, or control) occurs.

Functions define measurable properties on sets.

Math descriptions begin with identifying the sets within which the descriptions of measurable patterns are contained.

Numbers (quantitative sets) have assumptions associated to them concerning arithmetic operations, where (for example) identity elements and inverse operators are central to solving equations defined on quantitative set structures.

On these (algebraic) sets of numbers one can define sequences, ie a sequentially ordered set of numbers. If sequences are bounded then it has a convergent subsequence. Thus, "Is the value to which convergence is identified (actually) a number (value) in the same set, ie does the number defined by convergence also follow the same arithmetic laws as the numbers which compose the sequence?"

If yes, then sets of numbers, which obey the laws of arithmetic, can be very large (because the numbers to which the sequences converge depend for their definition on the axiom of choice).

Can quantitative sets be "too large?"

This (property of the real-numbers being a very large set) may be the core reason as to why algebraic structures defined on large quantitative sets, and on large function spaces, are not yielding useful information, ie the realm (the set structure) of description is not sufficiently restricted.

It has been further found that almost all real quantities require an infinite amount of information to identify the number's value. (G Chaitin)

It has been noted that trying to use algebraic patterns which are defined on sets which are extremely large (or extremely complex) has not been all that effective at identifying ordered structures (or solution sets, which are practically useful in a creative context), eg using sets of commuting operators to diagonalize function spaces is quite often not successful.

Fundamental randomness destroys the logical context upon which geometric restrictions can be defined for quantum systems, due to the uncertainty principle of probability based descriptive structures, but it is the geometric properties of wave-operators which have been the types of operators which are most related to being able to define a spectral structure, which is (at all) close to a quantum system's spectral properties. H-atom, crystal energy levels, etc.

In regard to convergences defined on functions in the form of a ratio, ie [(rise, function values)/(run, domain values)], there are two interpretations of the existing convergences in regard to this ratio:

(1) geometric and measurable property of the local slope of a tangent line to a function's graph (this defines a local linear scale-relationship between the domain set and the function's values), and this tangent line can also be given a direction, ie it also defines a local vector structure related to the domain's geometric properties.

or

(2) since the convergence defines (or happens to also define) a new function, the "convergent dependent" derivative, and its almost inverse integral-operator, becomes an [(sort of) an approximate] algebra of operators acting on function spaces so that solutions to differential equations are related to the algebraic properties of operators (and their inverse operators) acting on function spaces.

This determines the context of a spectral description of either the domain set of the functions, or of the function space itself.

Though this relation of algebras of operators to function spaces suggests a relation to spectral sets of functions (which are) defined on the function space, so that the spectral functions can also be associated to spectral values, where each operator, in the set of commuting operators, should be associated to its own eigenvalue equation for each eigenfunction.

However, this structure is difficult to correlate with (for example) quantum systems' observed spectral values (though these techniques work better for the more limited geometric context of classical waves of electromagnetism and sound where the waves have physical properties and the wave-equation is a hyperbolic second order equation).

Furthermore, the techniques of "function spaces being associated to sets of operators" can be applied to geometries, but most often this would be in relation to non-linear (or non-solvable) geometries, and this has not led to anything of practical value in regard to creativity.

When local linear coordinates are defined on "shaped sets," the interpretation of the local directions, which derivatives define, is geometric on the domain space, which in turn, is most often a non-linear general geometric shape. However, non-linear geometries continuously (smoothly) deform so as to either disintegrate (ie the non-linear shape disintegrates), or into stable, discrete, separable, linear, hyperbolic (and Euclidean) geometries (but mostly hyperbolic).

A property identified by Thurston and Perelman, whereas knowledge of the geometry of hyperbolic manifolds was greatly expanded by Coxeter.

This deformation property of general geometries is described in a very clear way by C McMullen, in AMS Bulletin April 2011, in the article, (The evolution of geometric structures on 3-manifolds). McMullen's article is quite clear until the final section on "Open Problems" which pulls the discussion of geometry and its limitations back into an traditional (and illusionary) context, which has shown itself to be a fairly useless (in regard to practical creativity) context.

Note: Math people should want 8 year olds to be able to understand their discussions about math, since the math language is such a primitive and elementary language, yet it can be so closely related to practical creative actions. (But math people talk either in riddles or they discuss fine details in an overly complicated model, usually a model which has very limited practical uses, and subsequently they assume that their descriptions should only be addressed to their fellow experts [in an, essentially, autistic context])

Furthermore, the conclusion of Godel's incompleteness theorem is that "it is at the very simple level of: assumption, context, interpretation, containment, and shape from which useful knowledge most directly emerges," and this level of assumption is the level which is most consistent with young people, who are developing language skills. Thus, young people might be best suited to acquire a capacity to change the assumptions of a (traditional) language which has lost any relation to practical creativity.

The fact that discrete hyperbolic geometry is central to understanding stable geometric properties makes it also related to studying stable spectral properties, but the descriptive context needs to be changed.

The stable and limited context of geometry, where non-linear and other unstable geometries smoothly deform into either being dissipated or into stable hyperbolic geometries, as Perelman's analysis determines (or so shows), where the hyperbolic shapes are ridged and stable so as to make them good models for the underlying geometries upon which the stable and limited spectra of quantum systems (would) depend for their stable discrete structures.

Hyperbolic geometry is central to describing stable geometry and to describing the stable and limited quantitative context where spectral properties and linear, geometrically separable, and thus solvable and controllable, properties of measurable physical systems. The dynamics of material interactions can be formed by relating hyperbolic geometry to Euclidean geometry (through resonance and an approximate geometric structure), where spatial displacements are most naturally described in Euclidean space, eg Noether's symmetries (or invariants).

Why push the descriptions of physical systems into "an undefined fundamentally random context?" where there is too much quantitative information which exists in a context which does not have enough (or without enough) limiting structure.

This is what quantum physics and non-linear (based) physics now (both) do?

Instead place quantitative description in the more limited, and stable, context of an underlying hyperbolic geometry.

Containment of measurable descriptions within metric-spaces of "coordinate spaces of constant curvature," ie metric-invariance with metric functions which have constant coefficients, (so as to be contained within a containing space of [various signature] higher-dimensions) has a natural geometric relation to the space's discrete isometry subgroups, and they are shapes which are (also) linear, and geometrically separable (shapes), and thus form the natural geometric basis of a coordinate space's (descriptive) structure, which is needed for describing the observed stability of spectral orbital structures at all size scales.

The point of precise description is to change . . . , especially . . . , if the current descriptions have few relations to practical creativity. However, this is the case for virtually all math-science today, despite the incessant "hype" of dominant scientific and mathematical knowledge associated to today's assumptions concerning math and science.

When sets of differential equations (or sets of symmetries, or invariants) are interpreted as (complete) sets of commuting (Hermitian) operators acting on function spaces, then this is interpreted either harmonically and/or related to fundamental randomness of a spectral event space. This descriptive structure is too diverse, and unstable, and undefined, because the underlying geometry (which are to be contained in the structure of the operators which act on the function space) is not sufficiently defined so as to be able to identify the spectral sets to which the function space is "supposed to be" related.

However, if there are strong geometric restrictions (in spaces of constant curvature) (or finite, discrete, spectral restrictions) imposed on the spectral structure, then the spectral sets may be identifiable. That is, the new, more geometrically restricted, description of both geometry and spectral properties (both) seem to be better described in a geometric context in which there are several different dimensional levels to consider, as opposed to a function-space context whose domain spaces are determined by a failing assumption, namely, that material determines a description's dimensional structure.

Note: The stable set of hyperbolic geometries to which unstable geometries (continuously) deform are characterized by their holes in their shapes, and their subsequent (hole-encircling) spectral lengths. Thus if metric-space also have discrete hyperbolic geometric shapes then such metric-space structures would also possess, within themselves, spectral properties. Thus, discrete material hyperbolic shapes could be in resonance with discrete metric-space hyperbolic shapes, or resonating with hyperbolic shapes of the same dimension but which exist in different subspaces, which in turn, exist in the higher dimensional containing space.

That is, most of the defined properties of convergence and of associated algebraic structures, form into a very large set, ie into "too large of a set," which is related to very little (if any) order.

Yet both spectral sets, and geometric sets can be placed in a solvable and controllable context by allowing measurable sets (ie sets of quantities) to be organized by the geometry of linear, separable, discrete hyperbolic and Euclidean geometries, which exist in coordinate spaces of constant curvature, since the hyperbolic shapes are both stable geometries and have stable definitive and discrete spectral properties, and they are the geometric shapes to which virtually all general geometries deform in a smooth context of deformation (as envisioned by Thurston and Perelman).

Spectral flows defined on discrete hyperbolic geometries (or space-forms), and metric-space states defined on space-forms, which are related to the spin rotations of metric-space states defined on space-forms (which in turn model metric-spaces), and variety in regard to metric-functions of material (or sub-geometry) containing sets, can (all together) be used to describe both geometric properties of dynamics associated to interactions as well as finite spectral sets.

This can be accomplished by placing metric-spaces in a many dimensional set where both material and metric-spaces have discrete hyperbolic shapes associated to themselves, so that hyperbolic space-form shapes which model stable material systems can be related to Euclidean dynamical changes within a particular dimensional level.

There are conformal factors which can exist between dimensional levels, where values of these constants have the practical effect of limiting spectral sets defined on the over-all containing space, and thus allowing for spectral and geometric stability.

Now consider physics:

There are stable definitive spectral-orbital properties for material systems at all size scales: Nuclear, atomic, eg Fe-atoms, molecular, crystalline, planetary orbits (why the apparent great stability of the solar system in what seems to be properly described by non-linear equations?), galactic motions (dark matter), and universal dynamics of (apparent) divergence of all material in the universe, so that the shape of the universe is (interpreted) to be (Euclidean) flat, where all of these (above mentioned) material systems have observed (stable) properties which go without valid descriptions . . . , ie without descriptions which are widely applicable and creatively related to great practical usefulness.

Only stable discrete spectral properties of quantum systems and the Fermi or Bose components . . . , with distinct statistical properties in regard to the occupation of these discrete quantum structures . . . , is capable of being used in a practical manner, where most often macroscopic systems are coupled to discrete quantum properties, eg crystal energy levels are "coupled to" within transistors which, in turn, fit within electric circuits, etc.

Whereas particle-collision models, which are central to the physical models of quantum interaction in particle physics, only have practical use in bomb engineering, where an explosion depends on a reaction, which in turn, depends on probabilities of particle-collisions as a system transitions between two relatively stable initial and final states.

Perhaps the above mentioned "newly" described math structure "should be used" to model physical systems so as to have a linear separable geometry, thus solvable and controllable descriptive structure, associated to physical description.

Yes, it "should be so used."

The above math-based containment set, can account for quantum randomness, which is due to Brownian motion, which in turn, results from micro-interactions (described within the theoretical structure which exists for this math-science model) which in turn, E Nelson (Princeton, 1957) has shown is equivalent to quantum randomness.

This leads to two more considerations (1) The only successes that quantum physics has, in regard to getting energy levels as "valid" approximations to observed quantum systems, are those quantum models associated to precise geometric potential energy terms in the wave-equation, yet there is no aspect of quantum physics which is supposed to allow such an explicit geometric structure (due to its fundamental randomness), ie quantum interactions are modeled as random particle-collisions, [although (point-like) field sources (are supposed to) identify a spherically symmetric randomness of field-particles emanating from a point source]. But the particle-collision model of quantum interactions has no relation to a useful description of quantum properties, but nonetheless it needs to be pointed-out again, that quantum properties are stable and definitive, so that this property of stability does imply that there is some underlying geometry associated to these stable, definitive quantum systems, but such a geometric idea opposes the fundamental randomness of quantum physics. [That is, the new math ideas about physical description need to be taken seriously.] (2) In this new descriptive context, geometric deformations are not always continuous. In fact, essentially smooth geometric interaction structures (or geometric deformations) can suddenly jump to a new stable, definitive geometric (space-form) structure (due to new resonances [which are part of the interaction] with the over-all [many-dimensional] containing space). This jump to a new form of stability can occur if the energies of interaction are within (or fit into) a particular range of energies.

In this new descriptive structure, all of the stable definitive material geometries are (already) associated to discrete hyperbolic shapes (whereas smooth geometric deformations were needed to prove that general (or non-linear) geometries deform into (mostly) discrete hyperbolic shapes).

That is, as an alternative "set of assumptions" one can begin with metric-invariant spaces of constant curvature and within this context the natural shapes are the discrete isometry (or unitary) subgroups (or space-form shapes) and these shapes can be either bounded or not bounded, and these shapes are linear, and geometrically separable, and they can be defined as differential-forms, and the hyperbolic space-forms are stable and definitive in their geometric and spectral properties.

Combinations of these geometrically separable and linear space-form shapes from both the hyperbolic and Euclidean spaces can be used to describe the observed physical properties of material systems, and these descriptive structures are constructs which generalize to higher dimensions.

It should be noted that apparently it is only in 3-Euclidean space that spherical symmetry is a part of the structure of material interaction, and it may not be because the fields are themselves are spherically symmetric, rather the spherical symmetry of material interactions in 3-Euclidean-space is due to the relation ship between the geometry of the spatial differential-forms and a corresponding geometry of the Euclidean fiber group, which is spherically-symmetric, and it is the spherical symmetry of the fiber group and how it is involved in the structure of the interaction (the interaction construct) which causes material geometry and its dynamic relations in Euclidean 3-space to be related (so much) to spherical symmetry.

Note: According to math, a bounded (closed) simply connected shape, as our universe is supposedly identified to be, must be a sphere, but measurements (if they can be believed) imply that the universe is flat, ie Euclidean, not spherical.

Because the physics community cannot put-forth realistic models . . . (which actually describe the observed structures) and so that their models are, in fact, related to practically creative technical development, the physics community only puts forth physical models which have fundamental problems . . . , there is no reason to believe any of their claims in regard to "what is true" about both physical systems, and about the nature of physical law.

The sub-oligarchy of math and science is based on materialism, indefinable randomness (ie there is no valid way in which to count random events), where this indefinable randomness is quite often built upon a foundation of non-linearity, but non-linearity (as well as indefinable randomness) are quantitatively inconsistent, and the math processes of indefinably random and non-linear systems must be defined (or are placed) on sets which are "too big," so that even the rules of algebra can be questioned on such big sets.

Thus, one must ask "How can the stability of a number system be ensured within the math descriptions of the observed patterns (of physical systems), where these observed patterns appear to be both indefinably random as well as being stable, definitive, discrete structures (or properties), where these stable patterns are observed at all size scales of existence?"

To answer the question in regard to ensuring a stable basis for math descriptions (a question which is outside of the interests of the controlling oligarchs of society) one might first consider the very useful descriptions of classical physics, which assume the existence of the stable geometries which are a part of the descriptive structures. The usable aspects of classical physics are based on:

1. linearity of local measuring (linear differential equations),
2. geometrically separable shapes,
3. contained within metric-spaces whose metric-functions can only have constant coefficients, where
4. in this descriptive context, the main math structure is the differential-form and exterior differentiating.

This is the basis of stable (and practically useful) math descriptions for the observed physical patterns.

These are the properties of the discrete shapes derived from "cubical" simplexes so as to exclude the sphere, where the sphere is geometrically separable but it is non-linear, ie its metric-function does not have constant coefficients.

[Note: The discrete hyperbolic shapes are both very numerous and very stable.]

These discrete shapes are defined on many dimensional levels, so the descriptions can extend into higher dimensions, yet the interaction structure is dimensionally dependent so as to make the different dimensional levels geometrically distinct and separate, ie it is not easy to observe the higher dimensions. However, the stable spectral properties of the over-all high-dimension containing space do affect the allowed spectral structure in each dimensional level (see below).

There are also the observed physical properties of:

0. what seems to be is that, "the spectral-particle events" (of a quantum system) in space appear to be random,
1. stable discreteness of quantum systems,
2. spin,
3. unitary,
4. higher-dimensional (is one interpretation of the elementary-particle data), while
5. the inertial or Euclidean spaces possess the property of action-at-a-distance (or having non-local properties), while
6. the space-time spaces, or equivalently the hyperbolic spaces, define
   a. charge
   b. they seem to have the property of causality, and
   c. waves (which exist in hyperbolic space) are (can be) based on a field structures, as well as
   d. hyperbolic metric-spaces possessing the property of having stable systems defined in the context of energy.

Thus one can change one's set of assumptions so that instead of basing descriptions on

(1) the very general context of: indefinable randomness, non-linearity, and on math sets which are "too big," where it is believed that the functions which are associated to the structures of: non-linearity and indefinable randomness are vaguely related to holes and knots defined on the domain space, where these abstract geometric structures, in turn, are related to the math structures of homotopy (continuous deformations), co-homology (integration around holes in the shapes of spaces), and cobordism (manifolds as boundaries of one-higher-dimension manifolds), along with some very complicated algebraic structures, in turn, related to vector fields and (associated solution) functions (to non-linear differential equations),

> [but they are also related to the geometrization properties of shapes, wherein the discrete hyperbolic shapes and the discrete Euclidean shapes are the (most) stable shapes],

and instead (or and thus) base the math descriptions of the observed patterns . . . , of very stable spectral-orbital properties (which go without valid descriptions) . . . , on

(2) the properties of the discrete (isometry) shapes derived from "cubical" simplexes, which exist within spaces whose metric-function only have constant coefficients. These discrete shapes are linear, and geometrically separable (or parallelizable and always locally orthogonal), thus all the vague abstractness goes away, and the algebra of these shapes is based on diagonal matrices,

since the geometry of these discrete shapes is based on the (fundamental domain of) "cubical" simplex.

This eliminates the complications of:

1. homotopy,
2. co-homology, and
3. knot theory, and perhaps even
4. cobordism (since the math structure is specific, wherein it seems that cobordism would reduce to properties of "cubical" simplexes), as well as
5. making the algebra of these abstract constructions very simple (always working with diagonal matrices greatly simplifies the algebra).

But
 If the geometric basis for a descriptive math structure is not based on "cubical" simplexes (which are very specific and very simple geometric structures) then the underlying number-system will "not" be stable, and the math descriptions (of such a math structure) do not have a valid (basis for) meaning.

> This can be seen because in . . . :
> general relativity,
> quantum theory,
> particle-physics (except related to randomly based particle-collision reactions),
> string-theory,
> grand unification,
> supergravity,
> Etc, which are all based on indefinable randomness and non-linearity,

. . . , there are no (or very few) practical applications which come from these descriptive structures, and the descriptions cannot be used to identify the stable spectra of a general quantum system.
 Nonetheless, the general quantum systems, which these theories are trying to describe, do have very stable and very specific measurable properties, and this means that there is a descriptive math structure which can account for these stable properties, and that the logical foundations of indefinably random and non-linearity (or instability and inconsistency) are not correct, rather the fundamental assumption of materialism needs to be changed to an assumption that the logical foundations of physical description is to be "cubical" simplexes (for both the structures of stable (material) geometry and the stable containing metric-spaces which are to be placed in a many-dimensional context).

Reconsidering the foundations of math and physics

Essentially, t1he math simplification (which challenges the foundations of math) requires that higher dimensional geometry (or topology) be considered in the context of "spaces of constant curvature," where these are the spaces of:

1. the sphere,
2. Euclidean space and
3. hyperbolic space,

but the sphere is non-linear, so it is excluded as a basis for a descriptive structure in math, if one wants the number system, upon which the math description is based, to be stable so that the math description can be valid.

However, now the description is many-dimensional, and spaces such as [R(s,t)] need to be considered and related to the descriptions of existence, including relating these spaces to physical properties (defining types of material which correspond to the types of R(s,t)-discrete-shapes). These spaces have metric-functions which have constant coefficients, so this leads "to these spaces also having discrete geometries," similar to discrete Euclidean and discrete hyperbolic geometries.

Furthermore, for physical systems, the stability problem and the interaction problem is solved (as given below) by this simpler math structure for physical descriptions.

The triumph of this new descriptive structure [of only allowing geometries which are discrete shapes which are associated to metric-spaces whose metric-functions have constant coefficients (to enter into the math methods)], is that it "does solve" the problem of stability and material interaction, but it also extends into higher-dimensions, and it provides (new) models of life and mind based on a connected single organizing geometric structure, about which these very complicated structures . . . , yet unified and relatively stable (living) structures (such as life and mind) . . . , can be organized in a causal manner.

One can base the descriptions of existence so that both material and space have the structures of discrete hyperbolic and Euclidean shapes, where material interactions are described as the probing (by discrete Euclidean shapes) of the stable discrete hyperbolic shapes . . . , which represent the stable boundary properties of the "material containing" metric-space . . . , where this probing of boundary conditions is being done by the action-at-a-distance discrete Euclidean shapes (with each time period of a spin-rotation), whose faces are averages of the properties (shapes) of the hyperbolic faces, which form the stable boundaries which the discrete Euclidean shapes are probing.

Furthermore, since the containing metric-spaces also have stable shapes, this means that the over-all high-dimensional containing space can determine the spectra of any newly emerging interacting material systems, where the spectra of the new discrete hyperbolic shapes are determined by the "emerging systems" being in resonance with the spectra of the over-all high-dimension containing metric-space.

* The new math structure of material interactions uses the vector-field structure of the related interacting space-forms (discrete isometry shapes) and its relation to the geometry of the fiber group, and is similar to a connection (ie a connection is about using the properties of local coordinate transformations of the fiber group acting on the above mentioned vector-fields) but this new group action is in relation to the 2-forms defined on the above mentioned interacting (very simple) space-forms and acts on the Euclidean interaction space-form's vector-field, where the Euclidean 2-form and the spin-fiber group have the same dimension, and thus are natural to relate to one another.

This is not a connection, where a connection is about a derivative, which is supposed to be a linear relation, but when the shape (or manifold) upon which the derivative is defined has curvature then the linear part (of the derivative) needs to be adjusted so the coordinates of the manifold's coordinate patch fit onto the local coordinates of the shape (or manifold). But in the new description the shapes are all linear and defined on separable coordinates (or on a separable shape), thus no adjustments need to be considered. However, the local structure of the instantaneous Euclidean shape changes with each (time) period of the spin rotation of metric-space states, which also causes the associated (force-field) 2-form to (also) change. Thus, the changes of position of the stable hyperbolic shapes (which determine the bounds of the instantaneous (discrete) Euclidean shape) caused by the action of the fiber group (in relation to the vector structure of the 2-forms) also changes as the instantaneous Euclidean 2-form changes in each discrete time period. These changes in position (in Euclidean space) are defined (as oppositely directed changes of position) in each of the two states of time for the hyperbolic metric-space. When stable discrete hyperbolic shapes are interacting so that both the energy of the interaction is within an acceptable range of energy of a (possible) new system and the interacting structure of the discrete hyperbolic shapes and the instantaneous discrete Euclidean shape come close enough to the spectra of the containing space so that a combination of conditions "of acceptable energy values and spectral resonances of the interacting shape with the spectral of the high-dimension containing space" allow the resonances to settle into a new stable discrete hyperbolic shape (but in a higher-dimensional hyperbolic metric-space) so as to form a new stable discrete hyperbolic shape from the interaction. In such a deformation of the interacting stable discrete hyperbolic shapes, the details of the deformation are not significant, rather the resonances and energy structure, ie the energy of the interaction (ie essentially a collision of discrete hyperbolic shapes) and the existence of some resonances between the interaction shape and the containing space's spectral set (and the rearrangement of the other energy states of the original discrete hyperbolic shapes) determine if the new stable discrete hyperbolic shape can emerge from the interaction.

A discrete hyperbolic shape will have an average-value discrete Euclidean shape associated to itself, which determines both the discrete Euclidean interaction shape and the inertial properties of the discrete hyperbolic shape during an interaction, where it is the Euclidean inertial properties of the discrete hyperbolic shape whose position is changed during the interaction.

This paper (and all the others) can be summed-up in a short outline.

There exist stable definitive discrete material systems which exist at all size-scales: nuclei, atoms, molecules, crystals, and solar systems and possibly (orbital) motions of stars in galaxies. These stable discrete spectral-orbital systems do not have adequate descriptions within the context of "what are now (2011) considered to be the laws of physics" (eg there is, currently, no known reason as to why the solar system is stable, nor how a general nuclei comes to have stable spectral properties).

The existence of these stable systems implies that measuring makes sense in regard to these stable definitive systems which exist at all size-scales, and that they can be quantitatively described in a stable, measurable descriptive language.

However, one can only measure in a context of stable structures:

> stable systems,
> stable geometries,
> stable and consistent quantitative structures, and which are not too complicated so as to hide inconsistencies,
> within sets which are not too-big, and (also) not too-complicated, so that the set properties of descriptions are both consistent and comprehensible, and
> descriptions are to be placed in stable contexts where the values one measures are stable and consistent, so that measurable values fit into a description which is also stable and practically useful, and which actually accurately fits data (such as describing the stable discrete spectral structure of a general nucleus), and which provides information which, in turn, can be used in a practical manner because both the system and its descriptive structures are stable.

How is stability to be introduced into the descriptive language of math-physics?

Answer: The key is stable geometry, and how to introduce (or fit) stable geometric structures into a quantitatively based descriptive language, is now made clear by the following list of "eight" math properties:

(1) The geometrization theorem of Thurston-Perelman [Note: The Thurston-Perelman theorem is actually unnecessary, one can simply appeal to a need for measuring-stability (linear, separable, metric-invariant) and the stable discrete shapes exist within this stable context.],

where Thurston-Perelman (essentially) show that the discrete hyperbolic shapes and discrete Euclidean shapes . . . , (as well as the sphere, but the sphere is non-linear, and is stable, in regard to a system which has only one spherically symmetric body composing the system) . . . , are the central shapes in regard to classifying geometry, in its general topological context, and all the other shapes (at least in 3-space) are unstable and they either disintegrate or evolve towards the stable shapes.

(2) The knowledge that "the most stable shapes" are the discrete hyperbolic shapes, and that the stability of the discrete hyperbolic shapes are robust and their properties (of shape) are unaffected by their (it) being near other geometries, whereas in the spherical-case, one sphere is a stable, while two spheres forms an indescribable (or unresolved-able) geometric structure.

(3) Coxeter showed what the fundamental properties of the discrete hyperbolic shapes are distinguished by: dimension, genus (number of holes in the shape), and stable geometric measures, which are measures on the faces of the discrete hyperbolic shapes' fundamental domains (ie polyhedron, or cubical-simplex structures [the square on a checker-board is a fundamental domain of a discrete Euclidean shape]), [the stable geometric measures on discrete hyperbolic shapes are the basis for stable spectral-orbital properties which are associated to the discrete hyperbolic shapes], and the discrete hyperbolic shapes of each dimensional level can include infinite-extent discrete hyperbolic shapes (so that these shapes have finite "volume" measures. *) associated to themselves, ie these infinite extent shapes model light, and perhaps neutrinos.

(4) Furthermore, there are many simultaneous (but different types of) metric-spaces involved in physical description, where each metric-space (identified by dimension and signature) is associated to a different fundamental physical property. These fundamental physical properties and their associated metric-spaces (identified by dimension and metric-function signature) are distinguished by the elementary displacement symmetries of E Noether, which are (also) related to conservation properties. Eg inertia-spatial-displacement is a property of Euclidean space and charge-energy are primarily properties of hyperbolic space (hyperbolic space is equivalent to space-time, it can be thought of as the velocity space of space-time)

(5) Opposite metric-space states can be defined on this structure. In turn, opposite metric-space states is the structure upon which "the spin-rotation of opposite metric-space states" can be defined. Furthermore, opposite metric-space states can be defined in complex-coordinates, and this leads to containment within complex-coordinates, ie "the real state" and its "pure-imaginary, opposite-state," which subsequently leads to unitary fiber groups (extending the description beyond the usual (real) rotation [or isometry] fiber groups).

(6) The structure is many-dimensional, since it is assumed that the structure of material is based on the same idea as the structure of a metric-space, namely, they (both material and metric-spaces) are both based on discrete hyperbolic shapes, and the discrete material shapes are contained within (adjacent higher-dimension) metric-spaces.

The discrete shapes in other (non-hyperbolic) metric-spaces are involved in dynamic processes, and the subsequent spatial displacements of these dynamic processes, where (both) opposite metric-space states ((7) as well as the fiber groups, associated to local coordinate transformations of the interaction geometries) are also involved in dynamic processes.

(7) the math containment set is a (set of) principle fiber bundles (where a principle fiber bundle has a base space, wherein a system would be contained, with its properties distributed between the different types of metric-spaces, and associated to each point of the base-space is an appropriate, allowable (isometry or unitary), local coordinate transformation group)

The base-space is composed on the many different dimensions, and many different types of metric-spaces up to hyperbolic dimension-11 (12-space-time), including coordinate base-spaces which have complex coordinate-systems, and the fiber groups are the set of classical groups which are associated to these different (real, isometric) metric-spaces as well as Hermitian spaces (with their complex-coordinates and unitary fiber groups).

(8) Material has a shape, and it also has a distinguished point (so interactions appear to be point-like), and metric-spaces also have stable shapes, so the coordinate metric-spaces now (also) possess spectral properties. That is, the containing space has spectral properties associated to itself.

Note: Stable systems can emerge out of interactions (if the energy of the interaction is within a proper range of values), and this emergence of stable systems out of material interactions would be due to the existence of resonances between the new system and the spectra of the (high-dimension) containing space.

Note: Quantum randomness can be derived from the structure of material interactions within the new descriptive language, but the basis for this "new" descriptive language is stable geometry.

The context of stability

When formulating how to describe a (physical) system, it is the stable geometric shapes, in regard to the descriptions of inertia and charge (and their geometries and motions in space), which are fundamental to the formulation and solution of a system's description (of a system's properties). The existence of a stable (geometrically separable) global coordinate geometry is a fundamental aspect of a stable and solvable system's description.

Because there are so many (many-component) systems which seem so independent of an external geometry, eg the neutrality of atoms, one also wants a model of a global, separable shape which is closed (sealed-in) and bounded.

(Note: There are many topological sets which can be both closed [contain all of their limit points] and open)

There are such bounded, separable shapes (which are not spheres) and they are (essentially) the discrete shapes of the Euclidean (tori) and hyperbolic spaces (at least two tori geometrically combined together).

These discrete shapes, especially the very stable discrete hyperbolic shapes, will be, mostly, unaffected by external conditions.

Another outline of the structure of math, whose patterns are organized, in an alternative descriptive structure

There is an alternative description of physical existence, which can be defined in a context of "linear geometric" properties and defined in a context of many-dimensions, where each of the dimensional levels (or each metric-space of a different dimension) is itself associated to a "linear geometric" shape,

> (or possibly as an "orbit" in a discrete hyperbolic shape, or equivalently [to an orbit, (would be)] a pair of opposing faces of the "cubical-simplex" associated to the discrete hyperbolic shape).

Furthermore, these same hyperbolic shapes define material systems when contained within the (an) adjacent higher dimensional level.

Where "linear geometry" means that one solves linear, separable, differential equations defined in a metric-invariant context (where the metric-functions have constant coefficients) in order to accurately describe the measurable properties of a (linear) geometric shape. The linear, separable shapes, defined by differential equations (or by differential-forms), are essentially both the discrete Euclidean shapes and the discrete hyperbolic shapes.

The discrete hyperbolic shapes have very stable geometric properties so they would be good models for the stable discrete quantum systems.

It is natural to define conformal factors between adjacent dimensional levels, where these factors would be equivalent to physical constants such as, h (Planck's constant), and G (the gravitational constant), or between different types of metric-spaces, where the speed of light, c, relates Euclidean 4-space to hyperbolic 3-space.

The discrete spherical geometries are non-linear and thus they are not considered.

Linear geometry implies the existence of stable discrete shapes in a many-metric-space context, and in a many-dimensional context.

The differential operators which determine material interactions, in the new descriptive language, are descriptions which are similar to the classical descriptions, but now they are explicitly (ie the operators are) discrete linear shapes (that is, they are harmonic structures) which inter-relate the different types of metric-spaces and the different materials which are part of the description (where stable material systems are actually (closed) metric-spaces).

Furthermore, the different metric-spaces also distinguish different physical properties.

The operators, which are discrete shapes, are essentially classical differential operators, but now they are defined in a math structure where the opposite metric-space states are spin-rotated from one to the other (the spin-rotation of metric-space states).

Nonetheless, the unstable non-linear operators are (also) still defined, in a manner similar to how the classical non-linear differential equations are defined, ie non-linear systems still exist, and these differential equations are related to actual geometric properties which are defined by the material interactions, to which these classical differential equation are related. Thus, the geometric relevance of the critical points and associated (geometric) limit cycles which are associated to these (most often) classically defined non-linear differential equations. Thus the question becomes, "When does any particular non-linear differential equations lose its relevance in regard to defining the material interactions of the system one is trying to define?"

However, the stable systems are the discrete hyperbolic shapes, and the material interactions in 3-space (where hyperbolic 3-space is equivalent to 4-space-time) are spherically symmetric . . . , as are both Newton's laws of gravity, and (as are) the basic assumptions of (one-body) general relativity . . . , where spherical symmetry of a material interaction in 3-space persists up to the existence of a bounding discrete hyperbolic shape, which will affect motion when the interacting system gets small enough (as small as the discrete shapes of material systems which define the material interaction), so that at a particular geometric size (in regard to a material interaction), "what was a spherically symmetric interaction would no longer be definable as a spherically symmetric shape," which defines the dynamics of interactions within the small distance-range of the interaction where the material interaction begins to take on the structure of a collision, where the collision gets defined around the distinguished points of the discrete shapes which are part of the interaction process.

How interactions transition to stable structures is a description which is based on the new "containment system's" relation to its many-dimensional containing space. During a material interaction, the interaction structure (if the interaction is within certain energy ranges) . . . ,

> (as the material interaction approaches the distance-range which defines a material-collision as opposed to the material interaction defining an interaction being described in the context of a force-field interaction)

. . . , may begin to resonate with the spectra of the over-all high-dimension containing space so as to transition to a new stable discrete hyperbolic shape, due to both resonance and energetic consistency of the interacting system.

Furthermore, this new description has a structure which is entirely similar to the math structure E Nelson (Princeton, 1967) used to derive that "quantum-randomness is a result of the Brownian motions of the atomic structure of material." Thus quantum-randomness can be derived from a linear-geometric model of both material and space, wherein the stable quantum systems are discrete hyperbolic shapes, which in 3-space are also related (up to bounding discrete hyperbolic shapes) to spherical harmonics (due to interactions in 3-space being spherically symmetric (even) in the new theory, where in the atom it seems to be true that the size of the nucleus is (assumed to be) small compared to the size of the atom, where the atomic size determines the distance-range of the interaction within an atom).

However, now there is a math structure which allows for the description of the existence of a stable many-component orbital structure (since the discrete hyperbolic shape can have many holes in its geometric structure, where each hole can be associated to a separate orbit, or separate spectral-value).

The new knowledge (which is following the true ["free inquiry"] intent of science, as expressed by Copernicus) is based on identifying a descriptive structure (an alternative descriptive structure) which contains both the existing physical systems as well as their containing (metric) space, and this new (alternative) structure [about which the uppity sorcerers should consider] is based on intrinsic (stable) shapes which are associated to both material properties and to the properties of "material containing metric-spaces."

A containing space contains the measurable quantities (physical properties) which are limited by the geometric-spectral properties of stable definitive discrete hyperbolic shapes which model (in the new alternative model) the fundamental spectral systems, where such stable spectral properties have been observed in general quantum systems (the new alternative descriptive structure also allows for the properties of rotational symmetry for material interactions in 3-space).

Existence is to (now) be described so as to be fundamentally based on the stable shapes of discrete hyperbolic shapes defined on many dimensions so that both material systems contained in metric-spaces and the metric-spaces (both) have discrete hyperbolic shapes. This means that on the next higher dimensional metric-space the (now) lower dimensional metric-space becomes a material system [(within the next higher dimensional level metric-space)] with a (relatively) stable discrete hyperbolic shape. Thus, a metric-space can have spectral properties in the same way that a material system modeled as a discrete hyperbolic shape also has a definitive stable discrete set of spectral properties (associated to either the material system's discrete hyperbolic shape or the metric-space's discrete hyperbolic shape).

Note: The two shapes (1) a material system's shape and the (2) the shape of the "system containing metric-space," are defined at different dimension values, ie adjacent dimensional levels (a containment property which discrete shapes possess).

That is, the (stable) discrete hyperbolic shapes along with some associated discrete Euclidean shapes (associated by resonance) can, together, in a (pair of) principle fiber bundle(s) math-structure . . . ,

> [where a principle fiber bundle is a metric-space wherein each coordinate point has associated to itself a model of a local coordinate transformation space (or a local [fiber] Lie group), where the local transformations leave the local geometric measures invariant from point to point in the coordinate space]

. . . , can (within a many dimensional context) determine the spectral, and geometric, and interactive-dynamic properties of both the material and the containment space (though these two shapes will be of different dimensions).

These discrete hyperbolic shapes determine the describable, measurable, useable . . . , (in regard to information needed for creativity and control) . . . , properties of both "physical systems" and the containment spaces within which the measurable properties of a "physical system" are being described.

Thus the "place" for creating existence is in higher dimensions and whereas the correct model of (human) life is also a relatively stable geometric shape which is higher dimensional.

Discrete hyperbolic shapes defined at various dimensional levels can be used to describe the physical systems as well as the metric-spaces (which also have these discrete hyperbolic shapes) and thus the metric-spaces also have spectral properties, and the spectral set of the over-all containing space determines the type of spectral systems (of material and metric-space sub-systems) which can exist within the over-all (high dimensional containing) metric-space (which would be a hyperbolic metric-space [which does not have a discrete hyperbolic shape] of hyperbolic dimension-11 (where there do exist 10-dimensional discrete hyperbolic shapes)).

This is the basis for the description of existence and such descriptions are (will be) about using the fundamental stable geometric shapes (ie discrete hyperbolic shapes) to describe the properties of existence where the fundamental-ness of the discrete hyperbolic shapes has been identified by the geometrization theorem of Thurston-Perelman (to which [according to the proof of the geometrization theorem] all non-stable geometries evolve, namely they evolve to the discrete hyperbolic shapes).

By turning this theorem on its head, so that one identifies a new descriptive structure in which it is assumed that discrete hyperbolic shapes are the fundamental properties of the descriptive structure. These shapes can be used to describe both material and the dimensional (and spectral) properties within which the existence of both "material and its 'containing spaces'" are (together) contained, in a hierarchy of dimensions (along with their associated stable shapes and spectra) which determine a hierarchy of stable spectral and geometric (including dynamic) properties for both material systems and for the metric-spaces which contain the (material) lower dimensional discrete hyperbolic shapes, (all) within an over-all, high dimension containing space wherein interactions, based on spectral properties as well as the geometric shapes of interaction, can be described, ie the shapes each dimensional level contains determine the stable spectra which each dimensional level possesses.

Some of these (discrete hyperbolic shapes), shapes are infinite-extent and determine the "lines of the world," and some of these shapes can determine simple (higher dimensional) models of life.

Nonetheless:

The point of a measurable descriptive language is to realize what brings about stability, "what are the natural limits" that a set of descriptive patterns can have, and how coupling can occur so that the description is useful, and how changes in the fundamental patterns of existence (How displacements in space) can be realized.

The new assertion is that:

Ultimately the core of existence is geometric, and it is based on:

(1) the stable discrete hyperbolic shapes,
(2) the many-dimensional structure and geometric structure within which they are (or can be) contained (and these stable geometric and spectral structures are inter-related to one another), and
(3) an idea about controlling change, and "what properties of existence are open to change, and when doing this (controlled) change, one might also consider, "do other properties need to be undone (or unlocked)?"

There are many properties which seem to be a stable part of existence, but which there is no apparent reason that these properties "are the way they are."

For example:

Why does the existence of discrete hyperbolic shapes stop after hyperbolic dimension-10?

Why is the 5-dimensional hyperbolic shape the last dimension for a discrete hyperbolic shape for (at) which bounded discrete hyperbolic shapes exist?

How can these dimensional constraints be changed?

How can one use the fact that the quantitative structure of existence can be placed in a finite set of spectral (or geometric) quantities so there is the idea to extend (or make bigger) these quantitative structures, and thus extend existence?

There already is (within this geometrically based model of existence) a simple model of life, but how life can control the flow of energy within itself is still an open question (what is this process of change?) Is it an intended geometric shape which can cause changes? What changes does one want to accomplish?

These are questions about interaction, and how changes take place.

One question is about the nature of quantitative description, and what the model of a derivative . . . , and (in classical physics) its associated geometry of change . . , is?

Measuring is related to coupling a description of a system to a further structure, so that the further structure is designed to use the information (of the description) in a creative way. So that a similar measurable geometric cause for change can be realized, so as to best describe intentional changes related to changing, or expanding, the properties of existence.

There are fundamental questions about measurability and precise description.

What is the relation between quantities and measuring?

Is it best to assume that a stable system is a result of a finite containment set structure? (This, perhaps, might be true)

What is the relationship between quantitative sets and the quantitative containment sets within which it is assumed a system is contained?

How is a measured quantity to be modeled? In quantitative structures, eg a coordinate system, there need to be defined functions which, in turn, identify geometric measures within the quantitative (containing) structures, eg the metric-function defined on coordinates.

Must a measured quantity be a locally linear math structure? (Yes)

Are measured quantities consistent with a model of quantitative set containment? For example, are non-linear models consistent with the measurable properties of a containment set? (No)

Is measurement a much more limiting property when placed within a consistent quantitative containment space than is presently assumed? For example, when many sequences which identify convergences to values (or convergence to points in a containing space) are defined on quantitative structures [whose cardinality (measure of the set's size) is very large], where the convergent value (or point) whose property (type) is forced onto the descriptive structure, are these quantitative values (or points in the containment space) consistent with measured values within the containment set?

It turns out that (for measured values defined inside quantitative structures) when a derivative (or local linear model of measuring) is used to define a measured property, this measured value is only consistent with the measured values which can be defined within the (containing) quantitative structure if the derivative defines:

a (1) linear relationship, and
if (2) the measuring structure, eg the differential operator (or the differential equation), is measurement invariant to the measured properties which are defined on the quantitative

structure by means of "measuring functions," that is the differential operator (or equation) must be invariant to certain types of local coordinate transformations, and

if (3) the linear differential operator (or linear differential equation) is geometrically separable, ie the set of local directions for the coordinate curves, (defined by solution functions) at every point, always define a local orthogonal coordinate system (or vector space) eg the circle is a separable shape. The property of geometric separability allows the linear, metric-invariant, differential equations defined on separable geometries to be solved and controlled by initial and boundary conditions.

What are the geometric models of measurement? For example, the real number line.

Must measuring be related to a stable geometric shape?

Apparently, (due to the Geometrization Theorem of Thurston-Perelman) the only very stable shapes are the discrete hyperbolic shapes, and the discrete hyperbolic shapes are most often geometrically separable.

Conditions (1), (2) and, (3) (just mentioned) suggest that measurement must be related to both (the) particular quantitative structures and related to certain types of geometrically separable shapes (most notably the discrete hyperbolic shapes, which also have the property of being very stable).

The vast majority of separable geometries are:

(1) related to discrete isometry subgroups, ie paralizable manifolds defined on Lie groups, eg SO(s,t), [and these are most often the discrete subgroups of the isometry or unitary Lie groups],
(2) such as Euclidean tori (where tori are related to the discrete subgroups of SO(n)), and
(3) discrete hyperbolic shapes (ie related to the discrete subgroups of SO(n,1)), as well as
(4) rectangular boxes (or in a word [general] cubes, where a cube can be extended in a general way to all dimensions) as well as the separable shapes of
(5) circles,
(6) cylinders and
(7) spheres etc, (but spheres have non-linear metric-functions and thus lead to an inconsistent relation between measured values [defined by local linear measurements, ie defined by derivatives] and the containment set).

The above are the natural questions with which one can begin free inquiry concerning the structure and valid uses of precise (or quantitative, or measurable) descriptive languages.

One needs to be cautioned that . . . , both non-linear descriptions and fundamentally random descriptions . . . , have great limitations as to their usefulness, the inductive proof (of this) is that, if one simply considers that Faraday's geometrically based description of electromagnetism has consistently led to ever greater technical innovation and invention. However, the very limited technical development which quantum physics has fostered is almost always related to the way in which Fermions and Bosons occupy quantum states, ie quantum physics has essentially led to the development of the laser. Otherwise classical systems can couple to quantum properties, so as to have technical development related to quantum properties, but they are quantum properties found by experiment. The quantum properties are found by experiment because quantum calculations (based on the laws of quantum physics) are not capable of describing any of the fundamental stable spectral properties of quantum systems:

(1) nuclei,
(2) general atoms,
(3) molecules,
(4) molecular shapes, and
(5) crystal properties (the crystal theory of superconductivity, ie BCS, predicted a cut-off temperature which high temperature superconductivity has exceeded) cannot be described based on the laws of quantum physics.

That is, all of quantum physics (including particle physics, and all the other derivative-theories (eg string-theory)) predict none of the fundamental stable spectral-orbital properties of quantum systems.

Thus, quantum physics and particle physics have not led to any new means for technical development.

Quantitative descriptions based on fundamental randomness, at best can follow data, but since their calculations cannot be used to describe the stable spectral properties of general quantum systems, it is clear that such descriptions (based on randomness) have no descriptive value, they have no meaning in regard to useful creativity. That is, descriptions of random systems should only be considered if one has a random system based on a finite set of stable, well defined set of (random) events which compose the random structure whose probabilities for the random system's events one wants to calculate, ie it is only good for determining probabilities in gambling.

Otherwise, one should not trust the math of a fundamentally random system whose event space is only vaguely defined, no matter how rigorous the quantitative description might appear to be. One cannot claim to possess a well defined elementary event space when one cannot use the laws of quantum physics to calculate the spectral values which are supposed to compose the well defined elementary event space.

It is not logically consistent to say that one can calculate one or two of the quantum system's spectral values so "one can get rid of the calculated spectral values which are not seen" and then "add the spectral values that are observed" and to still claim that one has a well defined elementary event space.

Nor is it valid to say we do not know what dark matter is but we will add a new "dark matter particle" to our elementary particle space. Rather a fundamentally random structure needs a well defined elementary event space from the beginning, eg from experimentally measuring the quantum system's spectral structure but then that structure needs to be fixed (though one might have doubts about elementary event spaces where the number of elementary events is not finite).

It is like calculating the risk of nuclear meltdown for a nuclear reactor, but where (after the melt-down) the cause of the melt-down was not on the list upon which the (so called) risk was calculated, and then for the authority who calculated to risk to say, oh-yeah by the way, there are a new sets of events which are (should have been a) part of the elementary event space upon which the risks of nuclear melt-down were supposed to be calculated.

Models of fundamentally random systems, whose elementary event spaces are not identifiable, are not models of anything, it is a descriptive structure which has no meaning.

On the other hand some statistical constructs seem quite believable, eg scurvy is caused by a lack of vitamin C (note this relation is identified experimentally, or from experience). If one did develop a quantitative model about scurvy it would have to be something about how the molecule [or molecules] of vitamin C interact with (?) What aspect of the chemistry of the body? (?). This is difficult to answer since the chemistry of life is so very complicated, so any chemical relationship (and chemical correlation) may have many different pathways.

On the other hand global warming has already been measured and thus there will be a higher probability of higher energy weather events, since the weather system now contains more energy, but measuring how much energy is contained in a weather event is difficult to do. (Do weather events have energy cut-offs? It is likely that they do eg the total energy contained in the atmosphere would have to be such a cut-off. However, a good measure for the total energy contained in the atmosphere is $E=kT$, where T is the average global temperature [assuming that the total amount of material and the total volume of the atmosphere stay constant, thus the values of the pressure stay within a certain range of values, in an ideal gas].).

That is, though it is interesting to talk about correlations which sometimes appear to exist in a random, many variable descriptive structures [but perhaps the variable structure can be made simple], nonetheless it is futile to base the laws of description on a property of randomness.

That is, that quantum systems always possess the properties of being stable, definitive, discrete measurable spectral values and this can be interpreted to imply that these spectral properties are emanating from some stable, discrete, geometric structure.

That is, the description of quantum systems does not have to be based on fundamental randomness, it can be based on geometry (as long as the new geometric model can derive the properties of randomness for quantum systems which are observed, [where the new descriptive structure can do this]).

Gödel's incompleteness theorem (as well as the example of Copernicus vs. Ptolemy) says that such categories (language which depends on such fixed and limited narrowness) identified by (precise, or meaningful) language, eventually cannot be used to describe the observed patterns that are considered to be a part of the existence, (where each one of the given categories is supposed to be (so) related).

For example:

The energy levels of the: nuclei, general atoms, molecules, and crystals, ie fundamental systems of the physical world, are not being identified by using the laws of either quantum physics or particle physics. The risks of markets or nuclear reactors are not being properly modeled using math based on fundamental randomness and/or non-linearity.

Chapter 18

NEW MATH CATEGORIES

The context of the geometric-spectral category

Material 2-space-forms exist in a 3-metric-space so that the material 2-space-forms interact. Geometric force fields are defined on interaction 3-space-forms, whose 1-dimensional facial structure of the spatial toral-component (of the interaction space-form) identifies three directions associated to an interaction geometry, which in turn is related to inertial motion (geometries) projected (down) into the 3-metric-space, so that this inertial motion is also related to local Lie algebra transformations from the fiber group (or a related spin fiber group) where the group geometry, and the geometry of the interaction space-form in the base space, are inter-related in regard to determining the inertial changes based on spatial geometry. That is, inertial motion is also associated to the geometry of the fiber group, so that this motion can be directly related to the geometry on the interaction 3-space-form (projected down to the 3-metric-space). It should be noted that the interaction 3-space-form is contained in a 4-metric-space.

It should be noted that, spatial 2-forms defined on an interaction space-form, which is contained in the base space, are isomorphic (same dimension) to the spin (fiber) group's Lie algebra.

For any dimensional level, an n-space-form is contained in an (n+1)-metric-space, so that in the (n+1)-metric-space a 1-form is isomorphic (same dimension as) to the n-forms, and 2-forms are isomorphic to (n−1)-forms, where (n−1)-forms, determine the spectral currents (or the (n−1)-faces) on the n-space-form.

There is a natural geometric relation between the 2-forms defined on an interaction space-form . . , which identify the n-spectral-currents on the interaction (n+1)-space-form . . , and the Lie algebra of the base-space's spin fiber group (where the base-space is an (n+2)-metric-space, in which case the 2-forms are associated to the spatial and material n-faces of the (n+1)-interaction-space-form), so that the properties of inertia are related to the geometry of the interaction space-form, and due to isomorphism (the properties of inertia) are also related to the local Lie algebra transformations which act on the spatial face of the interaction space-form, thus identifying the inertial properties of material space-form interactions on any dimensional level.

If an interaction results in a stable orbital (or stable space-form) system it would be because the interaction space-form began resonating (during the interaction) with a corresponding spectral set (or spectral property) in the containing space, and the energy of the interaction was within an energy range which allowed the resonance to manifest as a (new) stable ("material") space-form,

> (this stable space-form can be a hidden structure in regard to the material containing metric-space, ie the interaction space-form is defined in the next higher dimension metric-space [of the same (or analogous) signature] and usually in that dimension the new is not of the correct size to manifest as material, and/or it is rigidly stuck on the boundary of the metric-space which contains the interacting materials, and thus it does not manifest as material in the next higher dimension metric-space, whereas the space-form shape of the lower dimension metric-space is that structure which does manifest as a material space-form in the next higher dimension metric-space).

Interactions and their energy ranges determine both the dynamics and the stable systems which exist in a given dimensional level, where the new stable systems are based on the spectral-geometric properties of the containing space and the (range of) energy of the interacting space-form.

Separable geometries lend themselves to very simple descriptions of both geometric and spectral systems.

The new descriptive structure is: simpler, it fits data better, and it is based on geometry (thus it will be related to creative development).

Introduction

A theory, or a descriptive language, gives a context within which to both interpret data and to couple together systems in relation to the measured values of a (described) system's observed (or described, or predicted) properties, so as to develop new systems (or creative ideas) by means of either control of a system's properties, or by the use of information in regard to a system's described properties.

Knowledge based on measurable (or precise, or geometric) description, used to create new systems.

A descriptive language is based on: mathematical and physical assumptions, data interpretations, and new contexts within which to imagine new ideas.

Science is fundamentally about useful descriptions. (and it is the heritage of everyone, in an equal way, to consider how precise measurable descriptions can be changed so as to be made more useful and more widely applicable)

Consider the properties of "quantity and shape" (the fundamental category of mathematics) in relation to the many types of patterns and the many sets of possibilities which "sets of measurable properties" can be used describe.

There are the distinctions between:

Geometry (classical physics, based on materialism)
Confinement (bounds on a system, thermal properties, statistical physics)

Note: Where, in statistical physics, probabilities are derived from [assumed probability distributions of the] many components of slightly (or non-) interacting components in relation to measured properties, where average values of the different (thermal, statistical) properties are measured.

and

Probability (quantum physics, which is a fundamentally random theory)

The atomic hypothesis has led to a great deal of emphasis on small unseen material-particle components which compose material systems, and the subsequent quantum ideas related to observed random particle-spectral (localized) events in space and time.

Description begins with several general contexts (or concepts) . . . . :
eg materialism, the atomic hypothesis, etc, (the categories mentioned in this paper are some fundamental categories from both physics and math)

. . . . and then at a particular point the descriptive language leaves the general category so as to follow (or to become) either another general category, or a specific aspect of description, ie the description becomes specific and highly conditioned and detailed.

The general question is, "Which set of patterns?" and in, "What sequence 'in which they (the different sets or the different patterns) are put-together' results in the greatest usefulness?"

To say that one "wide ranging category" is more fundamental than some other "wide ranging category" (or idea, or pattern) is to claim knowledge which one does not (or can not) possess.

One needs to consider a variety of descriptive languages, and the fundamental assumptions upon which those languages are built, and then see which languages become the more useful.

Some categories and related subcategories of both physics and mathematics:

1. Materialism

    I. Geometry

        (1) Continuous
        (2) Smooth
        (3) Linear
        (4) Discrete, separable geometries
        (5) Non-linear

    II. Probability [Random particle-spectral events in space, and/or large sets of non-interacting particles and the use of (assumed) average values]

        (1) Fixed, finite elementary event spaces ([supposedly] for a quantum system, this is the basis for probability descriptions which correctly identify betting probabilities)
        (2) Harmonic functions, or Hilbert function spaces (the random structure of the quantum system, this is also the source of non-localness, where (non-local) global wave-functions which [instantaneously] collapse to a point when an event is observed),
        (3) Hermitian operators (the quantum system's defining spectral set is associated to these operators), and
        (4) Unitary invariance (conservation of energy in the context of both a complex function space and Hermitian operators),

Note: Unfortunately, these function-space; operator, techniques of quantum physics have not been developed so as to be able to identify a general quantum system's observed fixed elementary event space.

(5) Unstable and undefined elementary event spaces (particle physics, and regular quantum physics, finance also uses this invalid basis for their "probability" models of risk, [and all of these efforts have failed]).

2. Non-materialism: Discrete and/or separable geometries defined on a many-dimensional containment set.

(Particle physics and string theory etc use many-dimensions in their descriptive language but they also maintain the idea of materialism by identifying a special 3-space (or special 4-space-time) for both macroscopic material existence and the place of random particle-spectral events, so that the higher dimensions are hidden [exactly like the hidden structure of God in authoritative religions, where only authorities can talk to God, or where only the authorities of [a very dogmatic] physics can interpret the incomprehensible language of hidden dimensions] this is very artificial and quite unnecessary, because the patterns of math can be used in many ways so as to model material and quantum properties.)

The key to describing (the apparently) random and discrete quantum events in the same space where macroscopic material is continuous (as well as conserved) is the way in which a new containing space can model new ideas about continuity in relation to both the containing space's dimensional structure and a (the) new containing space's geometry.

Quantitative types, the elementary properties of quantity, and models of measurement

Geometry and probability have two quite different constructions for measurement . . . . (one the local measurement of the derivative and the subsequent differential equation in relation to the system's material geometry, and the other a complete set of commuting Hermitian operators associated to the quantum system's spectral set, where the spectral set is the quantum system's defining elementary event space)

. . . . yet Math, itself, has its own relation to quantity:

(1) If one is dealing with an elementary event space which is not a fixed set, then the probability structure does not properly relate to quantitative sets. One is no longer

counting in a fixed set, rather one is creating a new subjective number system. Thus, the description oscillates between different and incompatible quantitative sets.

(2) Only a linear representation of measurement, as in a differential equation, identifies a valid inter-relationship between a measured function's value and the measured properties in the domain space, ie only a, y=mx, linear relation identifies a valid inter-relationship between two quantitative sets. A differential equation must be linear so as to maintain a valid measuring relationship between the function's values and the domain's measured values.

Using the information which comes from a description

In order to use descriptive information, the information about measurable properties must be placed in a descriptive context where a described property (of a system) is inter-related with other (independently) measurable properties, so that the system can be coupled to a subsystem, where the subsystem is associated to a . . . given property of the system's . . . . set of inter-related (and often independent) measurable properties, ... so that a subsystem coupling to the system can be made, ... so that the properties can be used in some further creative process, where the new process or the new system is coupled to the descriptive properties either through the system or through the subsystem. This is a description of either a system and its geometric subsystem properties, or a component (Boson or Fermion component) of a system and the energetic (or geometric) subsystem properties of the system.

Classical physics deals with the measurements of material (system) properties caused by the geometry of material, while quantum description deals with the very stable spectral properties of (small) quantum systems, whose spectral properties are measured during the observation of random particle-spectral events in space and time. If the techniques of quantum physics could actually identify the spectral structure of general quantum systems, then there would be a greater possibility that the information provided by quantum physics could (perhaps) be useful, but quantum physics cannot provide this type of information about general quantum systems. On the other hand even if quantum physics could provide valid spectral information about a general quantum system (which it has not been able to do), because the description is a probability description there exist no causal relationships, and there exist no geometric relationships between the description and the system which exists in both space and in a random event space. However, one might be able to select light waves with particular properties which can (possibly) lift an electron to a particular energy level (but no guarantees, only a probability for such a transition, ie very limited control). Though in regard to the same transition, the new theory may not be able to add any causal relation, but it is likely that the energy levels would be found within the new theory, and furthermore, the new description provides a new context through which the containment of the general quantum system is defined so that properties of this greater containment set can be controlled, eg by modifying the spectral properties of the quantum system's containing space (eg by introducing oscillating energy generating systems into the context of containment).

Note: Quantum description claims that randomness is fundamental, whereas such "random structure" of small component events can be derivable from a geometric context.

How does one build a new descriptive language?

Does one consider (dance through) the various categories which have been erected (built) by tradition and by a tyranny..., of an assumed superior intellectual structure of institutions associated to both the fixed set of traditional ideas (or categories) and to authority (in the traditional descriptive set of structures bogged down by their own irrelevance)..., or does one develop one's own new categories?

One does not have to stay within a set of categories because the properties which identify a category might be describing a pattern which, in the world which we experience the category, is an illusion.

A new set of categories to consider

Geometry needs to be placed into the categories of:
 linear,
 separable, and
 non-linear but
 non-linear descriptions make no sense in regard to elementary quantitative structure.

Due to elementary considerations, both non-linear and probabilistic descriptive languages need to be discarded, because they are not consistent with the elementary properties of quantitative sets (quite true),

or (alternatively)

Due to induction, which is based on the very limited applications to which both non-linear and probabilistic descriptive languages are associated, these (two) descriptive languages need to be discarded.

Thus, geometry is divided into the categories of linear and metric-invariant, or linear, separable, and metric-invariant, but linear and metric invariant is either in regard to linear differential equations with constant coefficients or in regard to one-variable (since multi-variable differential equations in a geometric context are only consistent with the diagonal metric-function [because the metric-function is symmetric and thus diagonalizable] if the geometry is separable).

That is separable geometries are fundamental to physical description. The largest set of separable geometries can be identified by the discrete isometry (or unitary) subgroups, or space-forms of various dimensional metric-spaces. Space-form have properties of both geometric shapes as well as spectral properties associated to the bounded space-form shapes.

In classical physics, there are the categories of continuous or smooth.
Thus, questions about, "What are the limits of continuity or smoothness?" Are questions about separable geometries.

A new category

Thus one (can) needs to identify a "new category," the "geometry-spectral" category.
This can (will) be studied by means of the properties of separable geometries and discrete isometry subgroups and in relation to real geometries which also posses the distinct properties of being in metric-space states. Thus, one should also consider discrete unitary geometries (and "other" discrete geometries). Note: There could also be (other) 4-state geometries and 8-state geometries etc.

The category of continuity

In the context of discrete geometries the property of continuity implies conservation. The conservation of the number of holes in a separable geometry, as well as the conservation of energy, mass, and charge. Thus separable geometries with holes in their geometric structure are good models for material systems characterized by conserved properties. The discrete isometry (and unitary) subgroups are characterized by the number of holes that are a part of their geometric structure, and they also define separable geometries.

The category of many-dimensions

This category transcends the idea of materialism, the category of many-dimensions where each dimensional level can be associated to discrete isometry (or unitary) subgroups which can have various size relations to one another (ie no need to role [or fold] this geometry into an invisible-small size, as in string theory).

There is also the "new category" of many-dimensional separable (or discrete) geometries, which has both the property of discontinuity between dimensional levels (when going from a lower dimensional level to an adjacent higher dimension) and it can also have the property of inter-dimensional spectral continuities, or spectral properties which are inter-dimensionally continuous (or inter-dimensionally dependent on one another).

Conjecture: Material is made of space-forms of dimension-(n-1) contained in a metric-space of dimension-n.

Material interactions can be modeled as separable geometric (or space-form) structures which bridge the gap between adjacent dimensional-levels (which are discontinuous, eg the structure of both material systems and the material interactions cause the discontinuity in regard to geometric properties).

Spectral properties imply an inter-dimensional continuity, eg conservation of energy

While geometry implies continuity at one (given) dimensional level, as well as for all lower dimensional levels contained within the given dimensional level, the spectral structure of material (space-form) systems implies continuity of spectral properties between the different dimensional levels, ie the space-form material systems which a given dimensional level contains are conserved if they resonate with the full containing (many-dimensional) space. That is, the material space-forms of the given dimensional level (with their particular set of spectral properties) are confined in a continuous manner (or in relation to the property of conservation) to that given dimensional level.

Separable and/or discrete geometries, eg discrete isometry (or unitary) subgroups (or space-forms), defined on a multi-dimensional, and many metric-space state, structure as (for) a containment space for all the separable and/or discrete geometric properties which fit within the confines of the many-dimensional structure, provides a basis for a descriptive language which is capable of giving an actual geometric and spectral structure to the observed properties of both classical and quantum physics, so that this description has several relations in regard to the idea of continuity, and conservation, in regard to the description's full containing space and its new and various inter-dimensional properties.

Information about physical (or separable geometric) systems is not primarily contained in differential equations, rather it is primarily contained in the geometric structure of discrete and/or separable geometries, subsequently time structure as well as the inter-connections between systems (or energy structures) are often clarified in higher dimensional, material-spectral inter-dimensional, multi-metric-space state, system structures. Higher-dimensional spaces can have new properties in regard to both time and material.

There exist rules, in regard to resonances, about how lower-dimensional discrete separable geometries (or space-forms) can form together, in the higher-dimensional containing space of spectral properties, within which they are contained.

However, if resonances during interactions cannot define a new stable system, then the interaction follows the rules of interactions which are followed in classical physics.

That is, the differential equation is still a means through which the details of a material-space-form interaction (or some other relevant space-form structure) can be found (and this is found in regard to differential forms and inertia, defined on space-forms and in metric-spaces of various dimensions [and with various metric-function signatures]), but now the containing set is full of new information about the spectral relations between a space-form system and its many dimensional-levels which compose the (new) containing space.

Particle-collision experiments seem to show that the property of unitary containment (or unitary invariance, the analog of metric-invariant transformations) is an observed physical property.

This is consistent with a descriptive structure in which metric-spaces must have two metric-space states associated to themselves.

Thus, two dimensional-levels would imply four different metric-space states, while three dimensional-levels would imply eight different metric-space states. This could be the metric-space states which exist between the 3-dimensional level and the 5-dimensional level, where the 5-dimensional level is the last level of hyperbolic space which has bounded separable-discrete (space-form) geometries (Due to a theorem by Coxeter). That is, hyperbolic space-forms of dimension-6 and above (up to hyperbolic dimension-10) are infinite extent space-forms, and thus without a definite position in space and time and without a definite spectra, ie their spectra must be determined by the lower dimensional space-forms which are contained within themselves. Infinite space-forms in some ways can be quite similar to light and thus they could vibrate along its infinite extent arms so as to identify for itself a spectral property.

This property of an infinite-extent space-form being spectrally dependent on the lower dimensional space-forms it contains, might define the condition of the "freedom of life," where when life can change the types of spectra "which a life-form's infinite extent space-form contains within itself," then the life-form can become free to create in many new contexts.

Such living space-forms would have to be either of dimension-7 or of dimension-9, since it is these dimensions in which such space-forms can have the property of being oscillating, energy generating space-forms, and thus simple models of life. In fact, an oscillating, energy generating, infinite-extent space-form may be associated to a very complicated model of life, including cell structures, and system's of organs and enzymes (or hormones), and inter-relations to DNA.

How do the spectral properties manifest on an infinite extent separable (space-form) geometry?

The infinite extent space-form can oscillate [or vibrate] . . . . (along its geometric structures out to infinity) with various spectral "wave-lengths" (or more generally, "wave-areas" or "wave-volumes" etc) on the various dimensional faces (or natural spectral structures of various dimensions on the space-form) . . . . out to infinity, along the space-form's geometric sub-structures. Thus the "wave-lengths," going out to infinity, would be in resonance with the spectra of the lower dimensional space-forms which it (the higher dimensional space-form) contains.

There is another structure upon which the spectral properties of the lower dimensional space-form structures can be maintained upon the higher-dimensional, infinite extent, space-form . . . , and that is through a maximal torus which exists in the (unitary) fiber group. The space-form is defined on the base space and there is a fiber group for each point of the space-form, and this structure is called a principle fiber bundle. The different spectral properties of the lower dimensional space-forms can resonate with the spectra which can be defined on corresponding dimensional structures which exist on a maximal torus, which is a bounded geometry, defined in the fiber group of the higher dimensional oscillating space-form. This spectral structure, existing on a maximal torus, is a model of a mind. That is, the spectral properties associated to an infinite extent, oscillating, energy generating, space-form can be held in the (its) mind. Thus, the spectral properties associated to the oscillating infinite extent space-form can be related to an intent held in the (its) mind. This can be a model of both perception and memory and thus it is related to (the idea of) knowledge.

The stable spectra in the base space of the principle fiber bundle come mostly from the hyperbolic space-forms. The Euclidean space-forms have spectral structures which can adjust to the stable hyperbolic spectra. However, there are also spectra which can come from metric-spaces with various metric-function signatures for each dimensional level (where the signatures are related to rules about the existence of new material types min the higher dimensional levels), and even the possibility of space-forms defined in subspaces of purely time dimensions (which might also be related to spectral properties of intent, in relation to infinite extent oscillating space-forms). Nonetheless the spectral structures defined on maximal tori in the fiber group of a principle fiber bundle are assumed to be stable and effectively in resonance (or in agreement) with the stable spectra defined in the various dimensional levels of the base space. This spectral set on a maximal torus can even be thought of as the spectral set which holds an entire spectral structure defined on a many-dimensional base space together, thus moving between such spectral sets within the fiber group would be like moving between "different worlds," each distinguished by a characteristic spectral set on the sets of maximal tori in a (high dimensional) fiber group.

New mathematical context for physical descriptions

In fact, the whole issue about spectral properties and their relation to local measuring structures (such as differential equations) can be resolved by re-framing physics, not as a set of laws used to define a physical system's differential equations, but rather by re-defining (or re-framing) the context within which material and space are defined.

That is, physics is fundamentally about the stable geometry of the containing set structure, and it is within these stable bounding structures that measurable physical processes, ie differential equations, are to be defined.

This new metric-space structure has dimensions which are defined beyond the idea of materialism, so that materialism is but a subset of the new descriptive structure (where materialism is based on three spatial dimensions).

The proof that such a stable context is needed to understand the physical processes of physical description is that:

(1) classical solvable systems are very useful in regard to practical creativity, whereas its non-linear equations provide very limited (useful) quantitative information, essentially relegated to the limit cycles associated to a (non-linear) differential equation's critical points, and
(2) that quantum systems are observed to have such very stable properties associated to themselves and these properties are measured in a metric-invariant context.

Furthermore, there is evidence that action-at-a-distance is a verifiable physical property (though apparently remote from physical manipulation [unless perhaps it is given its proper descriptive context]), and thus the existence of absolute frames in relation to the fixed stars, as identified by Newton, also seem to be a part of the true aspects of physical description.

Whereas special relativistic frames seem to hold for hyperbolic space, but hyperbolic space is primarily characterized by its relation to energy, eg the conservation of energy, so that the sectional curvature of hyperbolic-space's discrete hyperbolic shapes is related to both potential energy and kinetic energy. However, its relation to potential energy is best modeled in the discrete Euclidean geometry, in turn, associated to the material interactions of discrete hyperbolic shapes, where discrete hyperbolic shapes model stable charged material systems, where the discrete Euclidean shapes model the spatial separation which exists between the discrete hyperbolic shapes which are interacting.

However, general relativistic frames, which are characterized by non-linearity, have no basis in the observed properties of existence (eg the solar system is stable), though in 3-space the material interactions are spherically symmetric in the new descriptive context, but the prevalence of orbital planes in regard to gravitational systems seems to imply that the gravitational interaction is best described in two-spatial-dimensions, so that the basic model of "mass" in the new geometric context would be a 1-dimensional loop (or a circle) [or perhaps several loops tangent to one another in a plane, eg a figure-eight (on its side) {in resonance with an underlying stable 2-dimensional discrete hyperbolic shape}].

Thus, the new idea is that a [stable geometric] containing structure (for math descriptions, which accurately [to an acceptable level of precision] and usefully describe material properties) must have a relation to stable geometric shapes, ie discrete hyperbolic and discrete Euclidean shapes, and it must be many-dimensional.

That is, both material and metric-spaces are based-on the very stable discrete hyperbolic shapes where n-dimensional (or lower) metric-space are contained in (n+1)-dimensional metric-spaces, so that these metric-spaces have discrete hyperbolic shapes and this containment structure continues up to 11-dimensional hyperbolic space, ie 12-dimensional space-time. These spaces, viewed as shapes are open-closed. Furthermore, material would be the bounded shapes which are 1-dimension less than their containing metric-space (which also possesses a shape, but it is higher dimensional than the material shapes, which is actually a closed metric-space). That is, a closed, bounded hyperbolic metric-space (which has a discrete hyperbolic shape) of dimension-n identifies "charged-material" in a hyperbolic metric-space of dimension-(n+1), while its mass is associated to a closed bounded (interaction) Euclidean metric-space of dimension-n (or of dimension-(n+1)) which is in resonance with the given (closed, bounded) hyperbolic metric-space.

In this new geometric context material interactions are determined by differential forms defined around discrete Euclidean shapes which are defined as averages, ie center of mass coordinates, between pairs of holes in the material space-form's discrete hyperbolic shapes, and this is done in a higher dimension so that the material containing metric-space so that a dominating property which determines the interactions in the material containing metric-space (in relation to the differential-form's geometric relations with the fiber group of the interaction metric-space, where the differential-forms are defined around discrete Euclidean shapes) while the averages over other holes (identified by the discrete Euclidean shapes of material interactions) contribute to the envelope of stability (related to the material interaction in the material containing metric-space), where this is possible in the higher-dimensional context of the description of the material interaction processes (that is, the interactions which contribute to the envelope of stability are in a direction [in the higher-dimensional interaction-space] which is tangent to the material containing metric-space).

This is all geometric, thus, if it is essentially correct, then it is useable and controllable information [since (if it is a stable system then) it is: linear, metric-invariant, and geometrically separable].

This interaction structure has a structure similar to a connection . . . , that is if the fiber group geometry and the system's geometry are not aligned (or not consistent with one another, so that geometric separability is not maintained) . . . , and thus it allows for the very prevalent non-linear structures observed for material interactions, but these non-linear interactions fit between a stable containing framework for physical description of discrete hyperbolic shapes which are related through material interaction by discrete Euclidean shapes.

The material systems have stable discrete hyperbolic shapes upon which stable spectra can be defined while also the metric-spaces have stable discrete hyperbolic shapes and it is the set of spectra . . . defined by "all the different stable discrete hyperbolic shapes which, in turn, define all of the metric-spaces" which are contained in the hyperbolic 11-dimensional over-all containing metric-space . . . of which the material spectral structures must be in resonance (so as to come into existence in a stable form during material interactions).

Ignoring the failed ideas about how to describe quantum physics as well as particle physics . . . ,

> ie sets of operators and sets of random particle-collisions [which adjust the (unfound) solution to the wave-equation],

. . . , the useful laws of physics ie the laws of classical physics, are about "how discrete Euclidean shapes guide the structure of differential-forms" and the relation these differential forms have to discrete spatial displacements determined by a geometric relation (of the differential-forms) to the Euclidean fiber group (ie a geometric structure realized through the relation that the geometry of the differential-forms have to the geometry of the Euclidean fiber groups), within the new context of physical existence.

This paper (and all the others) can be summed-up in a short outline.

There exist stable definitive discrete material systems which exist at all size-scales: nuclei, atoms, molecules, crystals, and solar systems and possibly (orbital) motions of stars in galaxies. These stable discrete spectral-orbital systems do not have adequate descriptions within the context of "what are now (2011) considered to be the laws of physics" (eg there is, currently, no known reason as to why the solar system is stable, nor how a general nuclei comes to have stable spectral properties).

The existence of these stable systems implies that measuring makes sense in regard to these stable definitive systems which exist at all size-scales, and that they can be quantitatively described in a stable, measurable descriptive language.

However, one can only measure in a context of stable structures:

> stable systems,
> stable geometries,
> stable and consistent quantitative structures, and which are not too complicated
> so as to hide inconsistencies,

within sets which are not too-big, and (also) not too-complicated, so that the set properties of descriptions are both consistent and comprehensible, and
descriptions are to be placed in stable contexts where the values one measures are stable and consistent, so that measurable values fit into a description which is also stable and practically useful, and which actually accurately fits data (such as describing the stable discrete spectral structure of a general nucleus), and which provides information which, in turn, can be used in a practical manner because both the system and its descriptive structures are stable.

How is stability to be introduced into the descriptive language of math-physics?

Answer: The key is stable geometry, and how to introduce (or fit) stable geometric structures into a quantitatively based descriptive language, is now made clear by the following list of "eight" math properties:

(1) The geometrization theorem of Thurston-Perelman [Note: The Thurston-Perelman theorem is actually unnecessary, one can simply appeal to a need for measuring-stability (linear, separable, metric-invariant) and the stable discrete shapes exist within this stable context.], where Thurston-Perelman (essentially) show that the discrete hyperbolic shapes and discrete Euclidean shapes . . . , (as well as the sphere, but the sphere is non-linear, and is stable, in regard to a system which has only one spherically symmetric body composing the system) . . . , are the central shapes in regard to classifying geometry, in its general topological context, and all the other shapes (at least in 3-space) are unstable and they either disintegrate or evolve towards the stable shapes.

(2) The knowledge that "the most stable shapes" are the discrete hyperbolic shapes, and that the stability of the discrete hyperbolic shapes are robust and their properties (of shape) are unaffected by their (it) being near other geometries, whereas in the spherical-case, one sphere is a stable, while two spheres forms an indescribable (or unresolved-able) geometric structure.

(3) Coxeter showed what the fundamental properties of the discrete hyperbolic shapes are distinguished by: dimension, genus (number of holes in the shape), and stable geometric measures, which are measures on the faces of the discrete hyperbolic shapes' fundamental domains (ie polyhedron, or cubical-simplex structures [the square on a checker-board is a fundamental domain of a discrete Euclidean shape]), [the stable geometric measures on discrete hyperbolic shapes are the basis for stable spectral-orbital properties which are associated to the discrete hyperbolic shapes], and the discrete hyperbolic shapes of each dimensional level can include infinite-extent discrete hyperbolic shapes (so that these shapes have finite "volume" measures. *) associated to themselves, ie these infinite extent shapes model light, and perhaps neutrinos.

(4) Furthermore, there are many simultaneous (but different types of) metric-spaces involved in physical description, where each metric-space (identified by dimension and signature) is

associated to a different fundamental physical property. These fundamental physical properties and their associated metric-spaces (identified by dimension and metric-function signature) are distinguished by the elementary displacement symmetries of E Noether, which are (also) related to conservation properties. Eg inertia-spatial-displacement is a property of Euclidean space and charge-energy are primarily properties of hyperbolic space (hyperbolic space is equivalent to space-time, it can be thought of as the velocity space of space-time)

(5) Opposite metric-space states can be defined on this structure. In turn, opposite metric-space states is the structure upon which "the spin-rotation of opposite metric-space states" can be defined. Furthermore, opposite metric-space states can be defined in complex-coordinates, and this leads to containment within complex-coordinates, ie "the real state" and its "pure-imaginary, opposite-state," which subsequently leads to unitary fiber groups (extending the description beyond the usual (real) rotation [or isometry] fiber groups).

(6) The structure is many-dimensional, since it is assumed that the structure of material is based on the same idea as the structure of a metric-space, namely, they (both material and metric-spaces) are both based on discrete hyperbolic shapes, and the discrete material shapes are contained within (adjacent higher-dimension) metric-spaces.

The discrete shapes in other (non-hyperbolic) metric-spaces are involved in dynamic processes, and the subsequent spatial displacements of these dynamic processes, where (both) opposite metric-space states ((7) as well as the fiber groups, associated to local coordinate transformations of the interaction geometries) are also involved in dynamic processes.

(7) the math containment set is a (set of) principle fiber bundles (where a principle fiber bundle has a base space, wherein a system would be contained, with its properties distributed between the different types of metric-spaces, and associated to each point of the base-space is an appropriate, allowable (isometry or unitary), local coordinate transformation group)

The base-space is composed on the many different dimensions, and many different types of metric-spaces up to hyperbolic dimension-11 (12-space-time), including coordinate base-spaces which have complex coordinate-systems, and the fiber groups are the set of classical groups which are associated to these different (real, isometric) metric-spaces as well as Hermitian spaces (with their complex-coordinates and unitary fiber groups).

(8) Material has a shape, and it also has a distinguished point (so interactions appear to be point-like), and metric-spaces also have stable shapes, so the coordinate metric-spaces now (also) possess spectral properties. That is, the containing space has spectral properties associated to itself.

Note: Stable systems can emerge out of interactions (if the energy of the interaction is within a proper range of values), and this emergence of stable systems out of material interactions would be due to the existence of resonances between the new system and the spectra of the (high-dimension) containing space.

Note: Quantum randomness can be derived from the structure of material interactions within the new descriptive language, but the basis for this "new" descriptive language is stable geometry.

The structure of the discrete hyperbolic shapes are related to the different dimensions of hyperbolic metric-spaces is as follows (as determined by Coxeter):

> 1-dimensional, fixed "shortest periodic lengths," genus properties (?),
> 2-dimensional, shapes identified by genus (number of holes in the shape) and the lengths of 1-faces of the discrete hyperbolic shape's (cubical) fundamental domain, as well as the angles defined between adjacent 1-faces on the fundamental domain,
> 3-dimensional, shapes identified by genus and the areas of 2-faces of the discrete hyperbolic shape's fundamental domain, as well as the solid angles defined between adjacent 2-faces on the fundamental domain,
> {Perelman showed in 3-space (perhaps in general spaces, up to and including hyperbolic dimension-5) all shapes which are either not discrete Euclidean shapes (or closed metric-spaces) or are not discrete hyperbolic shapes (or closed metric-spaces), then they are unstable, and they either eventually disintegrate or they evolve towards the stable discrete hyperbolic shapes.},
> 4-dimensional, shapes identified by genus and the areas of 3-faces of the discrete hyperbolic shape's (cubical) fundamental domain, as well as the solid angles defined between adjacent 3-faces fundamental domain,
> 5-dimensional, shapes identified by genus and the areas of 4-faces of the discrete hyperbolic shape's (cubical) fundamental domain, as well as the solid angles defined between adjacent 4-faces fundamental domain,
> 6-dimensional, shapes identified by genus, but only infinite extent discrete hyperbolic shapes exist, but the fundamental domain has (can have) finite volume
> *

The same type of infinite-extent structures exist up to 10-dimensional hyperbolic metric-spaces

As for the 11-dimensional level, there are no discrete hyperbolic shapes which seem to exist at this dimension, or above, the hyperbolic dimension-11.

Thus the over-all containing space for this descriptive structure is an 11-dimensional hyperbolic metric-space (12-space-time), which is not associated to a geometric shape, either bounded or unbounded, but all the other lower dimension coordinate metric-spaces are associated to very stable discrete hyperbolic shapes, where these shapes may be either bounded (if the hyperbolic dimension is 5 or less) or unbounded.

Note: The reason why higher-dimensional levels (than 3-space) are not "seen" is:

(1) the values of physical constants (defined between dimensional levels [not all physical constants are now known (2011)]), which, in turn, affect,
(2) the relative sizes of the interacting material systems in the different dimensional levels, (eg the size of planetary orbits in 4-space vs. the size of atoms in 3-space)
(3) the geometric nature and dimensional structure of the interaction process, and
(4) the closed nature of the containing space of a particular dimensional level, where each dimensional level is associated to a discrete hyperbolic shape.
(5) perhaps the data form particle accelerators is best interpreted as evidence for the existence higher-dimensions beyond a space-time dimensional space (which is defined by the assumption of materialism, ie material determines existence).

This above geometric context is the essential context of an existing underlying geometric stability which is (can be) a part of the descriptive structure for all of existence.

Thus, one wants to use these above listed properties to construct a stable measurable descriptive language to be used to describe the stable spectral-orbital physical systems of all size-scales, and in a high-dimensional context.

However, one should be unsure of the interpretations concerning the properties of galaxies and the "universe," provided by main-stream authoritarian science and math.

The nuclei, atoms, molecules, crystals and solar systems all have stable, discrete, definitive spectral-orbital properties, which in the new theory have an existing stable underlying geometric structure within which (or into which) these stable systems can form. But galaxies have different shapes: spiral, elliptical, etc, and the spiral galaxies seem not to be comprised of clear orbits, but (rather) the "orbits" appear "smeared-out," where star motions can be interpreted as "almost orbital motions" (of stars). This ambiguity of orbital-shape for the stars within galaxies makes one wonder if an underlying geometry and/or and underlying geometric structure which guide material interactions within the large structure of galaxies, has some very different geometric properties than do the smaller sized stable orbital systems [Note: In the new description the geometry of the "fiber groups" affects the geometric properties of material interactions, thus affecting the shapes of orbits (in higher-dimensions).].

On-the-other-hand, the interpretation of the "universe's" increased rate of expansion, ie an interpretation which assumes material has (geometric) properties beyond its interactive-relation to other material so that material is not passively fitting within (or defining) a containment set. Whereas there may be a geometric structure (in, say, hyperbolic-11 space) which includes both:

(1) a simultaneous sharing of spatial position of separate independent spectral sets, where a possible access between the spectral sets may exist by means of some shared set of resonances (between different and distinct spectral sets), where the separate spectral sets are modeled as a superposition of independent sets of spectra, and
(2) a math pattern of internal spectral separation which exists between the different maximal tori within the fiber group, where maximal tori can be reservoirs of the different base space's spectral sets (where one of the maximal tori in SU(11) might be the "spectral reservoir" for one galaxy), and what causes the apparent external motion of "drifting apart" in space is a result of the 1st and 2nd derivatives of the conjugations (which define transitions between these different maximal tori in the fiber group) placed in a locally continuous-time context, and which result in an apparent spatial drifting apart of the galaxies (where the spectra of one galaxy fills one of the maximal tori within SU(11)), and where conjugation defines the property of these different spectral sets simultaneously sharing the same hyperbolic-11 space.

The structure identified by (1) [in the previous paragraph] also allows for a model of action-at-a-distance. Where action-at-a-distance is a property which seems to exist in both,

(1) a few simple (entangled) quantum systems (though entanglement for more complicated quantum systems would be more easily changed (or disrupted), due to the many types of entanglements which a larger quantum system can have), and
(2) in Newtonian gravity, where Newton's gravity is used in guiding satellites in the solar system.

The new context

Thus a new context related to the linear, separable geometric (properties), so as to also be contained within a metric-invariant context, might be considered in relation to considering both discrete (isometry or unitary) subgroups (defined in a metric-invariant or Hermitian-invariant context, respectively), so that the containing space has many dimensions, and where the discrete geometries "contained within a metric-space (of some given dimension)" are models of material systems, and furthermore, that very same (metric-invariant) metric-space has (can have) a discrete geometry associated to itself, when observed from a higher dimensional context.

In this new context the idea of materialism is easily defined in terms of the material space-forms (or discrete separable geometries), and the system-defining differential equation can be associated to an interaction space-form, which is defined in the dimensional space just above the dimension of the material containing metric-space.

The relation that this new interaction can have to quantum physics is that these same interaction structures, just described, can be placed in a context of a "many-dimensional containing space" so that each dimensional level is related to discrete, separable geometric structures, which in turn, are related to the properties of "cubical" simplexes, and thus within a structure of many independent simplicial faces (of the containing high-dimension "cubical" simplex), there can be (or there can exist) many spectral properties associated directly to the containing space, and in this context the above described interactions can be suddenly disrupted (and if energy bounds of the interaction are satisfied) [and if] so that "the" new transitioning space-form begins to resonate with the spectra of the containing space, so as to become a new stable material space-form physical system, so that the structure of the interaction can lead to new stable material space-form systems within the (new) spectral context of the containing space.

The Hermitian structure is needed since the new description incorporates an idea of a metric-space-state, where each metric-space has a pair of opposite metric-space states associated to itself, and these states are discretely spin-rotated into one another in regard to very small discrete time intervals of a metric-space state where the state-property is determined by the idea of time, where there exist states of opposite time, which exist (in their pure opposite states) in the subsets of real and "pure imaginary" numbers in the Hermitian-invariant complex coordinate containing space.

If one partitions the descriptive set structure (which is related to the transitions between relatively stable systems) by means of a discrete set which is composed of subsets, which in turn, belong to (or are a part of) the (new) linear, metric-invariant, separable, many-dimensional category, where the partitions are determined by the (time) period of the (discrete) spin-rotations between metric-space states, so that in these (periods of spin-rotation of metric-space state, time intervals) time intervals, there are formed separate and discrete spatial displacement faces, which are faces of a high dimensional interaction "cubical" simplex of the interaction space-form, which in turn, identify the dynamical relations between the interacting material (space-form) components, which in turn, are related to the both original relatively stable systems and the end relatively stable systems but this new discrete context can re-arrange its material components so as to form a new relatively stable physical structure which is distinct from the originally interacting material (space-form) systems.

By making dynamics discrete, (this) it allows for a new stable discrete structure to emerge from the dynamics.

CHAPTER 19

NEW MATH IDEAS

An alternative idea

Consider a containment set construct in which:
The containment set has many-dimensions, up to hyperbolic-dimension-11, and each dimensional level (or separate and different subspaces of the same dimension) are determined by a stable discrete hyperbolic shape, where the containment of a lower dimension discrete hyperbolic shape, in the adjacent higher-dimensional containing (hyperbolic) metric-space, can be conceived-of as either flows, or as free-shapes.

Physical properties are associated to the different metric-spaces. Spatial-position, spatial displacement, and inertia are associated to Euclidean space, while time, energy, and charge are associated to hyperbolic space, where the stability of an energetic system is based on the discrete hyperbolic shapes. The mechanical energy associated to mass results from the adherence of mass to the stable discrete hyperbolic shapes which is due to a resonance between natural spectral-flows of discrete hyperbolic shapes and discrete Euclidean shapes (which also manifest as maximal tori in the fiber Lie groups of the natural principle fiber bundles [for each physical property] used to contain and describe physical systems and their interactions).

Within this new set construct:

1 (a). What is material?
1 (b). What is a metric-space?

Are metric-spaces and (stable) material systems essentially the same type of geometric pattern but separated by one (or more) dimensions? Yes.

2 (a). What is a material interaction?
2 (b). What are the interaction geometries? and "How does one model action-at-a-distance?" (see other source)
3. What are physical constants? Constant factors which affect the relative sizes of the material systems which are needed to cause a detectable force.

For example:

One can conjecture:

That gravity, and the gravitational constant, G, and the measure of inertia, m, is about the relation from 1-dimensional material to 2-dimensional containment space, ie the plane upon which the planetary orbits are defined.
Electromagnetism, c and j, is the relation between 3-dimensions and 4-dimensions.
Discrete energies, h and Boltzmann's constant k and q, is the relation between 2-dimensions and 3-dimensions, etc.

4. What causes material systems to be stable? They are composed of very stable discrete hyperbolic shapes.
5. Where do the stable spectra of physical systems come from? Material systems resonate with the spectra of the metric-spaces, which also possess the stable shapes of discrete hyperbolic shapes.

In a fundamentally geometric model, which includes a "causal" material-interaction structure,

6. "What causes point-like (spectrally identified) randomness?"

Answer: The material interaction construct defined for small stable discrete hyperbolic shapes, which possess distinguished points, leads to local Brownian motions for the small components, and thus it leads to an apparent (fundamentally) random appearance characterized by point-like interaction events, but the points are associated to discrete geometries which possess spectral properties and this spectral property is revealed at the site of the interacting material system's distinguished point.

Unitary invariance (of the solution function and the wave-equation) implies conservation of energy, but does this also mean continuity of dimension?

The unitary structures of particle-physics are related to a discrete particle-collision material-interaction process, but it can also be related to a many-dimensional discrete shape structure, ie a set of discrete shapes ranging over many-dimensions, so that (or that is) one needs to relate finite-dimensional Hermitian geometry (ie geometry contained in a finite number of complex-coordinates) to real geometry (ie geometry measured by real-numbers in finite dimensions) associated to the base space of an isometry group, ie and associated discrete shapes (as well as to spin group geometries, ie rotation within a discrete shape represented as a complex rotation (the shape must be even-dimensional in order to ensure that the shape will be a very stable shape)).

That is, real-number discrete geometries exists-in (are contained within) both isometry groups and within spin-groups (or the even-dimensional real geometries can be easily related to spin groups).

A more general context for material, where material is associated with certain physical properties which are also associated to certain types of metric-spaces.

Is life about existence in higher-dimensions, and when written history began, did humans lose this knowledge?

There are the discrete geometries in the base space of R(3,0), ie Euclidean 3-space (namely 2-dimensional discrete Euclidean shapes), which is associated to an isometry fiber group SO(3), as well as the unitary fiber group SU(3), which mixes the metric-space states of Euclidean space (ie matter and anti-matter, or fixed-stars vs. rotating-stars, Euclidean metric-space frames).

A similar situation occurs for R(4,0) and SO(4), as well as associated spin-groups, which exist for classical isometry groups. But, whereas the 2-shapes contained in 3-space are easily related to spin groups the 3-shapes in 4-space are more difficult to relate to spin groups, [where the spin-rotations are defined on the 3-dimensional shapes] where 3-shapes can be charge-unbalanced, and subsequently they can oscillate (as semi-stable discrete hyperbolic geometries), thus identifying a new material, which in turn must be associated to a 2-dimensional time subspace, ie a 2-dimensional metric-space structure for time, eg R(4,2) {time1, is associated to charged material, while, time2, is associated to oscillating charged material}.

How to interpret the new material and the new dimensional structure to a higher-dimensional time-subspace?

Nonetheless: How can mathematical descriptions of physical systems be changed so that the descriptive language is more widely applicable and much more useful?

What are some mathematical properties of physical description?

In what context are differential equations (the most) useful? Answer: Both in the context of the differential equations being a model for the measurable properties of a physical system (ie classical physics), and in the context of their being linear, or being both linear and separable, where the property of being separable is a property of (the tangent directions to the system's natural containing and curved coordinates) being locally orthogonal at all points of the containing coordinate space. It is through solutions, to (these classical) differential equations which model physical systems, that this context has provided the information (of solution functions whose properties depend on initial conditions (or boundary conditions)) needed to control these systems, and subsequently to allow the greatest control and use of physical system's, whose differential equations have these properties.

This also happens to be the context of the differential equations which are related to space-forms, where space-forms lead to a useful geometric context for both the relations between differential forms and physical description (classical systems), and for the geometric context of stable, definitive spectral (or orbital) properties, ie the context of quantum systems as well as stable macroscopic orbital structures.

The context in which physical description is most useful, is where differential equations of physical systems are linear and/or separable, and this context is given its most productive general setting in the realm of space-forms, especially if the space-form's spectral properties are related to higher dimensional metric spaces, which in turn, are modeled as space-forms with spectral properties.

This space-form, higher dimensional context is a better, more useful, mathematical context for physical description than either classical physics (with its known limitations) and quantum physics which has failed as yet to provide a truly useful context, and thus the "scientific truth" of quantum physics is to be questioned in a serious manner, ie other descriptive languages which challenge quantum physic should be taken seriously.

Note: General relativity is only applicable to the 1-body problem which has spherical symmetry, ie the mass of the planet Mercury when the progression of its perihelion is calculated in general relativity is zero. That is, general relativity is neither widely applicable nor useful, it is only a questionable speculation because it is so limited in what it has so far described.

The new descriptive language, based on space-forms in a multi-dimensional context, causes re-examination of the meaning of the mathematical structures and their multi-dimensional properties . . . , which by the properties of space-forms . . . , and subsequently the properties of:

(1) the shape of space,
(2) the fundamental properties of frames, and these frames relation to space-forms, and
(3) subsequently to large scale structure of physical systems

.... so that these three properties of space-forms cause these higher dimensions to be mostly undetectable, with one exception being detection through the determination of atomic and nuclear spectral properties (which are now undeterminable by the methods of quantum physics) ..., and this new space-form context also allows the concept of materialism to (also) be challenged in a fundamental way. Namely, the idea of materialism can be discarded.

A descriptive language whose mathematical structure contains the classical description of material properties and which also supports the descriptive properties of a probability wave but which nonetheless is both fundamentally geometric in its basis and transcends the idea of materialism should be given serious consideration by both "professional" scientists and by religious people (although religion is really about that which one believes is true).

However, it must be realized that the academics and professionals within our society have a "trained incapacity to determine truth," a property of professional and academic people which has been previously observed and noted by M Parenti, ie professionals have a narrow viewpoint about what is true, and that viewpoint is authoritative and domineering, and such a narrow viewpoint of truth helps the owners of society keep their power. This is similar, in nature, to what characterizes the few leaders of monotheistic religions.

The new language of physical description based on space-forms is a superior descriptive language than what now exists, so that the current language will eventually give way to a new language, which in turn, either leads to the above proposed new language, or leads to a new language within which the above proposed language is immersed.

Superior knowledge is about a pattern which organizes descriptive language so the new description is more useful and/or more easily applied.

Namely, discrete geometries based on "cubical" simplexes (or space-forms) are more fundamental in regard to describing existence than is both calculus (and/or local measurability) and operators acting on function spaces (in order to find the spectral sets of physical systems), so that material does not reduce to elementary particles (rather both material and space reduce to space-forms, where material and space are essentially space-forms of adjacent dimensional levels), though the complete description of dynamics (contained within a particular metric-space state, at a particular dimensional level) is best done in terms of metric-space states and unitary fiber groups, whereas isometry groups apply to individual (or particular) metric-space states (ie within one of the two metric-space states), so that this new description is multi-dimensional and moves past the philosophy of materialism.

Perhaps it is the discrete properties of separable geometry, in relation to "cubical" simplexes which is the dominant property of physical description and it is orbital (or discrete) geometry from which descends the smooth description of classical physics, and so that the discrete structure (discrete interaction structure) can account for Brownian motion at the quantum level, and thus according to E Nelson these discrete geometric structures do not oppose quantum properties of randomness for small (space-form) systems.

Should one consider:

1. Holes in space

or

the number of toral components of a linear shape?

2. Order of operations, where the commutative property is not satisfied for non-linear shapes (and thus non-linearity cannot be placed in a quantitatively consistent structure), and inverse operators, which can only exist in very small regions if the quantitative structures of the description of several dimensions, essentially, do not commute,

Or

Should one consider linear, geometrically separable shapes, where differential operators on functions are (almost entirely) globally diagonal, ie they (nearly) always commute?

3. For non-linear, non-commuting systems, which are either contained in coordinates or placed in a (harmonic) function space,

Should one consider:

Going to a limited, non-linear, fixed spectral structure which increases the dimensional structure of the spectra, ie the random collisions of elementary particles used to perturb the spectra of a wave-function, where the original spectral wave-function (which one wants to perturb) cannot be made to even come close to approximating the system's spectral values, but in this case the higher dimensions which contain a fixed mass spectra, are irrelevant as perturbing agents, but nonetheless, this model conforms to the idea of materialism, and it mostly relates probabilities of particle-collisions to rates of (nuclear) reactions,

Or

Are oscillating functions, based on an indefinable spectral structure, being used to try to recover the system's descriptive measurable structure, which has (requires) a stable uniform unit of measuring (which the indefinable spectra as well as the property of non-linearity destroy), so as to model an approximation to a flow of a vector field (or sets of differential operators) which is supposedly defined by a geometric shape (but "is the geometric shape stable or unstable?").

Answer: Neither case seems to be usable, thus the non-linear, non-commuting context seems to mostly be irrelevant.

4. Non-linear metric-structures, where one can be assured that chaotic vector fields, identifying the local "linear" structure of a geometric shape, cannot be placed in a continuous (or smooth) measurable context which determines a quantitatively consistent measurable pattern, since there are properties of both space and set-containment which cannot be formulated in a non-linear (quantitatively consistent) context.

Or

Place the descriptive context in metric-invariant metric-spaces with metric-functions which have constant coefficients in metric-space which have non-positive constant curvature (where a spherical shape is only valid if it remains unperturbed, and thus the sphere is only about identifying the bounding regions where stable structures of math description can be defined). Note: The sphere is about defining regions of opposite metric-space states, in regard to material interactions, eg related to the advanced and retarded wave solutions (or advanced and retarded potential functions) in regard to electromagnetism.

5. Are limit cycles a result of defining measurements within a geometric context wherein vector fields are chaotic, thus requiring a relation to an assumed geometry, ie to some bounding, stable geometry (a necessary region which has some stability [continuity] needed for measuring, ie is the system truly in a measurable context, ie the material is stable but the interacting system is transitioning in a non-linear context, so its measurable properties are unstable and quantitatively inconsistent. That is, there is a valid context for measuring (the material is conserved, or identifies a continuous geometric [or spectral] structure) but it is non-linear, so there can be (can exist) stable shapes in the descriptive context (associated to the stable geometric [circle] structures of the fiber Lie group).

6. For a bounded system should one consider a harmonic function space if neither geometry nor spectral cut-offs [or spectral estimates] can be used to determine the system's observed stable discrete spectra,

Or

Should the system be placed in many dimensions, with the dimensional levels modeled as stable shapes, and lower dimension stable shapes are contained in each (independent) dimensional level, thus forming a new source for a system (at a particular dimensional level) to have (possess) particular spectral properties, ie the properties which already exist within the higher dimensional containment set, composed of different dimensional levels, which in turn, are modeled to have stable (spectral) shapes.

Spectra can be used in a discrete spectral context where the spectra of a stable (spectral) system can be enumerated, but how can the eigenfunctions be interpreted?
Are they related to random distributions of the spectral event to which they are associated?

or

Are they oscillations about a relatively stable geometric continuum?
That is, spectra can also be used in a geometric context, but what set of spectra can be related to any particular geometry?

and

By what principle are spectra to be related to any particular geometry?
One obvious relation would be spectral properties associated to some of the geometric measures of any particular shape.
Is the geometric relation to spectra the only valid inter-relationship which exists between spectra and a system's description?
Must physical system description, even spectral systems, always be about geometry?

An alternative to the experts (Space-forms)

If there is an alternative descriptive language for quantum physics, it might be a geometric language.
The geometry of what? or What geometric properties does existence have?
Metric invariant metric spaces are associated to particular types of quantities.
This assumption places the isometry group structure of a coordinate space (and the differential equations defined on these spaces), as well as its spectral-geometric structures, into the complex number system and unitary fiber group structure.

Note: In the real context some isometry fiber group structures (or differential structures of the manifolds) determine the geometry of the manifold which composes (or which determines) the coordinate space, eg R^3 becomes $S^2 \times R$, due to the isometry group of $SO(3) = S^3$.

However, mostly the geometry of a coordinate containing space (or coordinate, metric-invariant, manifold) depends on a fiber group structure determined by discrete subgroups (holes in space) which are related to fundamental domains which are based on "cubical" simplex structures.

The mathematical structure of discrete isometry (or unitary) subgroups leads to linear separable differential equations, the type of differential equations whose solutions give the most useful information about physical systems, as well as determining (in the hyperbolic space-form type) stable definitive spectral-geometric properties, the type of properties which quantum systems possess.

The space-form manifolds which discrete isometry subgroups define determine both material systems as well as metric-invariant metric-spaces so that these metric-spaces also have a complex structure associated to themselves. This determines the geometric-spectral structures (or properties) of existence (as well as the metric-space state structures). A spectral connection can be defined on the complete set of higher dimensional space-forms which determine the full containing metric-space structure of a physical system, a system which is modeled as a space-form of a particular dimension, and the complete set of spectral properties of the physical system.

If homotopies (holes in space) are the math structures which are most related to the definition of material, then it is clear, from the stable definitive properties of quantum systems and planetary orbits, that homotopies are really about the properties of discrete subgroups whose fundamental domains are built from "cubical" simplexes, thus leading to linear, separable differential equations and to a (material) system which has stable, definitive (spectral and orbital) properties.

The type of quantity (physical quantity) is of central importance in regard to the correct type of mathematical structures needed to describe existence in a widely applicable and useful manner.

For example, inertia is related to spatial displacement which is also related to mass which is naturally a part of Euclidean space, while energy is related to time displacement and to charge which is all related to hyperbolic metric spaces (or equivalently the space-time type metric spaces).

Furthermore, metric spaces have pairs of opposite states associated to themselves. This is the basis for a metric space to have a complex structure associated to itself, and it is a property which is related to the spin-rotation of states, eg the spin of an electron.

In the space-form formulation of material and space, the interaction structure for quantum systems is an analogous structure as the interaction structure of classical systems, where both material and space define faces (of the same dimension, where the "edges" of the material space-forms define a subset of the edges of the spatial faces of an interaction space-form) of an interaction space-form in the next higher dimension, which can become a stable space-form in the higher dimension, if the (energy) conditions of interaction as well as the spectral connection (applicable to stable space-forms) allow for a new stable space-form to form. The spectral connection is related to the space-form, multidimensional structure of existence.

For quantum systems: 1-dimensional material space-form systems interact to form 2-dimensional material space-form systems (contained in 3-space), while for classical systems: 2-dimensional material space-form systems interact to form 3-dimensional material space-form systems (contained in 4-space, but projected down to 3-space), whereas stable planetary orbits are 3-flows on a 4-space-form contained in a 5-space, etc.

New ideas

In science one must reflect on what a measurable description should do. A practical answer to this question is that it (a true description of existence) should be useful. If it is not useful, then other mathematically based (or specific) descriptive languages should be invented, their context identified, and such new speculative descriptions must be considered in terms of their usefulness, where the (current) orthodoxed description is just another speculative description which has not shown itself to be useful.

All such (speculative) descriptions (or descriptive languages) must be a part of scientific inquiry.

That is, there is a "real" existence which is measurable, so that if a specific quantitative description is not practically useful then, "what is such a description's value?" Answer: Not much.

The type of description is also of importance. Should it be a probability based description or should it be a geometric descriptive language? Answer: It should be a geometric based description because a geometric description, if correct, is more useful than a probability based description.

Quantum systems' spectra are precise and stable, ie quantum systems settle into (or exist in) an array of precise spectral values (or precise and stable orbits). Since quantum physics is not providing the information about the spectral structure of quantum systems, by using sets of Hermitian operators acting on a quantum system's function space, then one can speculate, using an alternative descriptive language. Namely, one can model the array of precise spectral values of quantum systems by (or result from) the geometry of discrete isometry subgroups (or space-forms). The harmonic properties of these spectral geometries [solution functions, and/or related function spaces] mean that resonances, between (space-form) material and their (space-form) containing metric space, have a precise (geometric) context. This means that geometric (or space-form) properties are related to both interaction properties and spectral properties of quantum systems, but there are also "hidden" geometric properties which identify a range of precise resonance values for the higher dimensional metric-spaces (modeled as space-forms) within which the spectral (or quantum) system and its interaction structures are contained.

The "hidden" spectral-geometric structure is related to a higher dimensional structure of existence due to modeling both material and space as space-forms.

This is not the same higher dimensional structure of string-membrane theory, though it is similar to it. In string-membrane theory it is not clear if higher dimensions are small or if membranes allow a macroscopic higher dimensional geometric structure.

The new model (or relation), of both a higher and lower dimensional (space-form) structure to existence, depends on modeling all of existence: space, material, and interactions: as discrete isometry subgroups or space-forms. Space-forms have a "cubical" simplex structure. Independent dimensional levels are allowed (or caused) by the space-form (or simplex) structure of: light, space, and interactions of material in space, which are all modeled as space-forms. This set of both microscopic and macroscopic space-form structures define independent higher dimensional structures of space (and material interactions) and these higher dimensional structures are related to the independent facial properties of "cubical" simplexes.

This independence has a similar logical basis as the dimensional independence of macroscopic membranes of string-membrane theory, ie in terms of how interactions and material (as well as space) are defined within a dimensional structure of (material and interactions and) space, so that for space-forms space has independent facial (or dimensional) properties related to their (its) "cubical" simplex structure.

Consider the question: "What is the shape of space?" One answer is related to the ideas of general relativity for inertia in (spherically symmetric) Euclidean space. However, it is only 3-dimensional Euclidean space which has the differential geometric properties of spherical symmetry (its differential structure is SO(3), which, in turn, is a spherically symmetric space), while most evidence points to space having the property of being flat.

That is, both classical physics [Newton's gravity, electromagnetism, and thermal physics] as well as the evidence from the 3 degree Kelvin background radiation (and other astronomical evidence), together imply that space is macroscopically flat (though the interpretation of background radiation should be questioned in a fundamental way, ie particular geometric properties are being assumed from shifts of electromagnetic spectra (this is only science if the assumed geometric properties can be confirmed in another way)).

If one considers flatness and differential structures of "flat" spaces (or differential manifolds modeled as principle fiber bundles), then the structure of space is most naturally to be modeled as discrete isomerty subgroups (or space-forms), so that inertial properties in 3-Euclidean space have spherical symmery.

It should be noted that only 3-dimensional Euclidean space has a natural differential structure of spherical symmetry, ie the fiber group SO(3) for 3-dimensional flat Euclidean space has the properties of a sphere. The spherical symmetry of the differential structure of Euclidean 3-space could be the origins of the spherical symmetry of the inertial properties of material systems contained in 3-space. It might be noted that SO(4), the differential structure for 4-dimensional Euclidean space, has the property of being a product of two 3-spheres.

Note 1: Conjecture: Dark matter has a large scale geometric interaction (space-form) structure which extends beyond the geometric range of the solar system. This is an example of subtle energy structure with a larger (or higher energy) macroscopic system which has stable, large scale, interaction (or orbital) properties. This is determined by the spectral properties of a containing metric space, which only makes sense if metric spaces have a geometric (or space-form) structure of their own.

Note 2: Macroscopic systems modeled as space-forms, eg planetary orbits, which have stable orbital properties, will have inertial (interaction) properties which are "tangent" to the radial direction which connects the two interacting space-form systems, ie similar to (or exactly related to) the precession of Hg's perihelion.

Conjecture: Spherical symmetry is only a property of 3-dimensional Euclidean space and it is this spherical symmetry of inertial properties in the material experience which determines the radial nature of interaction forces in 3-Euclidean space. In higher dimensions interaction inertial forces are tangent, not radial, thus causing, in spatially extended metric-spaces, dynamics to have spiral inertial properties. If the interaction is restricted to a closed space-form then the tangential forces define distinct orbital properties confined to flows on the orbital geometry of the space-form.

Descriptions of quantum systems need to be geometrically based

Note: Though probability claims wide applicability, if there are not general "widely applicable" rules for applying a probability based description to a system, so that instead each system is its own independent statistical study, then the probability based rules of description are not widely applicable. [Note: This is the problem which (regular) quantum physics is not resolving, ie each quantum system seems to be its own independent statistical study (ie complete sets of commuting Hermitian operators cannot be found which determine a general quantum system's spectral-statistical structure).] Furthermore, a probability based description is not all that useful. Expected values (or means) and allowable variations (or ranges) of errors for the expected value is the language of (passive) laboratory measurement, it is not the language required to describe precise stable spectral properties of both macroscopic and microscopic quantum systems (precise descriptions of micro-spectral systems requires geometric definiteness). Probability is not the language of active control, unless one is trying to control the "expected value" of a measurable property of a system, eg change the temperature of a thermal system, by causal relations, eg heating up a thermal system.

However, if a system is simply a (probability) distributions of random events which cannot be altered, then the value one can "expect" to measure for the system is its main (measurable) attribute (its expected value) within a probability based description. Nonetheless, a system which has stable, highly organized spectral properties does not suggest that such a system should be described with probabilities. Each new event is not simply a "roll of a dice," rather the interactions that occur (can) have precise relations to the highly organized system, eg a system with highly organized energy levels (which are changed in precise ways with interactions [within certain ranges of energy]).

Can a current (of charges) both flow in a crystal and (simultaneously, ie in the same system) interact with spin properties at the crystal's lattice sites? The probability based description of this system (or possible system) cannot be described (at all) with quantum physics, nor can it be described with particle interaction (collision) physics [which depends on there existing an approximate wave function for the system of "a current flowing in a crystal's energy levels and the current also interacting with spin properties of the crystal systems lattice sites" (yet such a wave function does not exist)], nonetheless certain crystals have both very well defined current-voltage properties and, very well defined spin properties at their lattice sites. The probability descriptions of quantum physics are failing at describing such a system, so a geometric description should be considered.

Yet the "aristocracy of intellect" (the oligarchs of academia) only allow variations in the description which are consistent with (or based on) its own authoritative dogmas, but in a probability based description eventually the description can be adjusted until its methods can be used to match the observed data.

This is (also) the problem with the probability models of the economy and risk, its claim to truth are that it can always be adjusted to fit with the truth, but it is a little late to get the right answer after the fact of an economic collapse.

New ideas

A property of existence which is such a mystery is about the fact that there are many, many examples of very stable systems which have very distinct and measurable properties, ie well defined spectra, as well as many very complicated systems which have a great deal of control over themselves, such as life forms.

If the probability based descriptions have failed to describe these stable systems then perhaps these stable systems have the basis for their stability (and control) in the properties of geometry. Perhaps a probability based descriptive language cannot describe the cause of such very stable systems which have such very precise measurable properties.

The useful part of classical physics is mostly about the linear, solvable differential equations of physical systems which are contained in flat metric spaces, and which are uniquely related to the system's boundary conditions (or initial conditions).

The classical description of physical systems is about materialism. Material is allowed to define space and time and it is assumed that such coordinates are of infinite extent (and smooth, or always locally measurable to second order, which implies a relation to a metric function [or a particular flat metric space]) so that the material geometry within such coordinate systems identify differential equations as well as the boundary conditions for the system's differential equation. A preferred coordinate system, the system's symmetries (or conserved motions) and the system's boundary conditions with a preferred metric structure allow the system to be solved and then measured (in a metric space). Most boundary conditions for physical systems, which are composed of mass and charge (where the rest frame of the charge identifies a well defined geometry), and whose inertial properties are contained in 3-Euclidean space, are related to spherically symmetric boundary conditions, whereas physical systems characterized by electric currents, ie electric currents move through a fixed geometry, are often related to cylindrical (or toral) symmetry. Note: Toral symmetry could (also) be related to the geometry of hyperbolic (or space-time) space-forms.

On the other hand spectral systems can be modeled as either function spaces or as hyperbolic space-forms. The function space model of quantum systems has not worked (ie it has not been useful). Hyperbolic space-forms have spectral properties which are very fixed and very stable. Space-forms are also differential forms, consistent with the geometric structure of classical physics.

Note: Quantum physics is all based on a particular interpretation of the observed patterns of random spectral-particle events (ie material systems are reducible to elementary [or discrete] particles), and a subsequent model of particle-collision interactions. Particle collisions occur between independent (usually macroscopic) systems, whereas within stable spectral systems, particle collisions might, in fact, be quite rare events (or [perhaps] non-existent). The observed micro-spectral-particle events could be interpreted in terms of special, physically distinguishable, points of space-form geometries, where space-forms model stable micro-spectral systems. Furthermore, the micro-interactions between independent micro-system space-forms result in stochastic processes which preserve the random, wave-like, properties described in quantum physics, ie wave-particle duality (where stochastic processes [of interacting space-forms] define the wave properties and the space-forms and their distinguished points identify particles).

"What can the alternative geometric description of material systems, which are both geometric and spectral systems, be?"
"Is there an alternative to the number of dimensions that space and time can have?" and
"Must there always be an infinite extent flat coordinate system which has the proper number of dimensions and, as well, it contains the (material) physical system?"

The natural structures in the mathematics of physical systems to consider are the different dimensional metric-spaces and different "signature" metrics with their associated metric spaces. Furthermore, for physical systems it is natural to consider their being contained in fiber bundles and in particular fiber bundles with fiber groups which are isometry groups, ie groups of local coordinate transformations which leave the metric (of the particular signature and dimension metric space) invariant (This property identifies a flat metric-space). Included in these flat spaces are the discrete isometry subgroups, or spaces of constant curvature, or (equivalently) space-forms.

"Measurability," and "what differential equations do," need a new context. Quantum physic's understanding of what . . . "Measurability," and "differential equations" . . . are, has not been useful. The idea of (a property of) local measurability, or the derivative of a function (or property, or quantity), set equal to a geometric model of a property (or quantity) of the same type as (the number type which) the derivative identifies, determines classical physic's idea about (of) "what a system's differential equation is," as well as defining the classical idea of measurability (of what measurability is). Local measurability of a (particular) property contained in a coordinate system is relatable to geometric measurability of a property which is quantitatively a number of the same type (as the locally measurable property). Classical physics is about smooth (quantitative) properties (of material in space) used to define linear differential equations of classical physical systems within flat metric-spaces, so that these properties (or functions) are most often defined in the context of differential-forms acted on by the generalized (metric invariant) Laplace differential operator. A Laplace operator is the (metric-space) square of a first order differential.

Note, however, that the square root of the generalized Laplace operator is the Dirac operator (not necessarily a first order differential, but related to a differential operator which is a representation of a simply connected covering group of the real metric space's differential structure, ie the metric space's isometry group). This in turn implies a mixing of (pairs) metric-space states.

For a particular dimensional level, a system's geometry must be (mostly) independent of higher dimensions (or its dependence is hidden), as well as being independent of the material that a higher dimensional space contains. The properties of faces on simplexes allow this to occur, where "cubical" simplexes are the fundamental domains of space-forms. This allows space, which is modeled as a space-form, to have the required properties in regard to both (the dimension of) the material that is contained, and the dimension of the containing space (or the dimension of the material's containing metric-space) as well as the independence of the material geometry (and the metric-containing-space) to higher dimensional material-geometric effects.

Because space-forms (and interactions between space-forms) are, themselves, differential-forms which are always solutions to differential equations (see L Eisenhart, Riemannian Geometry) this means that physical systems modeled as space-forms are (or can be) consistent with classical physics (whose systems are often modeled as differential forms).

However, the interactions of between micro-space-forms (which model atoms and molecules) identify a stochastic (or random) process (ie Brownian motion), and thus (as pointed out by E Nelson) the dynamic behavior of such micro-material (or space-form) systems would be equivalent to the behavior of probability based wave functions of quantum systems defined for micro-material particles (or equivalently space-forms and their distinguished points) in space.

In the case of space-forms identifying both material systems and metric-space (containment sets), the geometry and spectral properties of (hyperbolic) space-forms in a higher dimensional context (than it is believed materialism implies) become the dominant aspect of the description of a physical system's properties (mostly its spectral properties). That is, a system's properties are still solutions to differential equations, but the differential equations might be defined on higher dimensions (than they are traditionally defined), and on (both) smaller (and larger) well defined space-form geometries, so that these higher dimensional geometries define a new context for spectral resonances, than it has been customary to define resonances of material systems, so that the solutions to these defining differential equations for space-form systems, are also differential forms whose (stable spectral and geometric) properties can project down to a lower dimensional space.

These space-forms can be both macroscopic and microscopic so as to conform to the geometry of classical physics (as the language of classical physics can be adapted to the language of space-forms), they have random dynamic properties so as to conform to matter-waves, they can be microscopic and have well defined spectral-orbital properties, thus they also conform to the mathematics (as well as the observed properties) of quantum, or spectral, physics. Furthermore, space-forms can extend into higher dimensions in a geometrically consistent manner (the higher dimensions appear hidden), and it can be done in a manner which is analogous to (or consistent with) particle physics unitary (particle-state) vector transformation properties.

In the new physical description it is not the identification and solution of a system's differential equation which is fundamental, rather it is the geometry of the space-forms which compose the system in a multi-dimensional multi-metric-space context (where the definition of resonances are fundamental to both dynamics and stability) which is fundamental to the physical (or existing) system's description (both geometric and spectral and stability).

When the relevant space-form geometry is identified then a differential equation for the space-form of interest can always be identified. That is, the differential equation of an existing (or physical) system is dependent on (or a subset [or subordinate] to) the geometric structure (or geometric context) of the space-forms [which are defined on the different dimensional levels of a containing space-form structure] that define the (existing, or the lower 3-dimensional physical) system.

It should be noted that chemical models of molecular spectra have already been made with 2-dimensional hyperbolic space-forms.

Although the new description, which is based on space-forms, is closely allied with the mathematics of classical physics, it is very different from the mathematical properties and ideas of quantum physics. This is basically because it is a description based in geometry not probability. Nonetheless the new descriptive language (or new descriptive context) accounts for both the probability (or wave-like) structure of micro-material systems, and it accounts for the spectral properties of quantum systems with a geometric space-form structure. Furthermore, it accounts for the higher dimensional structure of internal particle states of particle physics, but it does this in a way in which the higher dimensional structure (of what was internal particle-states) refers to true geometric vectors of both macroscopic and microscopic (interaction) space-forms in a higher dimensional containing space, when the system is fully defined in its (higher dimensional) space-form structure.

Flat spaces

Physics is about materialism and solution functions to differential equations which represent the measurable properties of physical systems.

When the differential equation represents a locally measurable property within a geometric context, the solution function is about a geometric/dynamic property of the physical system (classical physics). In classical physics the physical systems which either correspond to linear differential equations or are defined in flat metric spaces so as to have separable coordinates which are solvable so that the solutions give very useful geometric information about the system.

When the differential equations represent (Hermitian) operators acting on a quantum system's function space then the spectral functions of the function space represent a particular spectral-particle event and the probabilities of it occurring in space (quantum physics).

However, in quantum physics, the spectral functions give some information about expected values, but in most cases the spectral functions cannot be found.

Yet quantum systems have very precise spectral properties.

When physical systems are modeled as space-forms, it should first be noted that space-forms are always solutions to differential equations, whose equation structure is similar to classical physics, but whose geometric (and spin rotation of metric-space states) properties encompass the spectral properties of quantum systems. However, in this space-form centered descriptive format one is mostly concerned with the geometric relations between space-forms which exist between adjacent dimensional levels, as well as considerations concerning coherent and consistent higher dimensional space-form (or spectral) structures which can control, through resonance properties between metric-spaces and material space-forms, an entire system (whose properties can project down to lower dimensions in a concrete manner). There can be various spectral-space-form relations that can exist between different dimensional levels, while the material geometry of each level is mostly independent of other levels, except interactions which are defined between adjacent dimensional levels. The space-form metric-space structure of our experience would be dominated by the 3-flow on a 4-space-form so that the 3-flow would identify the envelope of stability for the earth's orbit. The boundary of this 3-flow could interact in its containing 4-space, but such an interaction would involve a similar sized 3-flow in the containing 4-space, and the effects of such an interaction would be the movement of the 3-flow, this acceleration would not be detectable within the 3-flow.

This geometric-spectral space-form based description of material and space and spectra can be used to place concrete spectral and geometric (higher dimensional space-form) relations between worlds of experience with different and independent spectral structures. Moving between such all encompassing spectral structures, between different and independent worlds, could be an important function of life, ie a new context for our existence.

Life seems to be able to move between these different dimensional and independent spectral structures in order to construct geometric models of worlds (ie in order to think) and to move.

Compare particle physics with space-form physics

There are two fundamental statements of particle physics: (1) The global wave function of a quantum system, whose energy structure needs to be corrected, is invariant to unitary transformations, (2) In quantum systems reduced to elementary particles, particle-collision interactions are determined by unitary Lie algebra transformations of the elementary particle-states.

It is not clear that the first statement (1) has any content (or any meaning) when a global wave function is both made discrete . . .

(How is one discrete point related to another discrete point in a global wave function? and How is convergence determined at each point of the global wave function? and Is the convergence being properly defined? That is, does the convergence defined for convergence to the world we experience? How could this be determined if infinities are being subtracted to realize a finite result [for the wave function's defining infinite series]?)

. . . . and placed in the context of particle-states, so that these particle states are only defined at a single point (Which point of the global wave function should be selected?) and points in the domain space (or function space of harmonic oscillators) where similar particle collision processes are defined must be independent of one another. Note: Each point in the domain space represents a harmonic oscillator.

And in the second statement (2), it is not clear that stable quantum systems are reducible to elementary particles (instead they might be reducible to space-forms) and subsequently it is not clear that (if) there exists any elementary particle collisions in stable quantum states (or stable quantum orbits), eg particle physics cannot describe the stable orbital properties of a general nucleus.

There are two fundamental statements about material space-form systems and space-form interactions: (1) material space-form systems are invariant to unitary (spin rotation of metric-space states) transformations, and (2) space-form interactions are determined by (local) unitary Lie algebra transformations of an interaction space-form's deformations so that this description exists in a context of spectral resonances between the interaction space-form and its higher dimensional containing metric-space. (Ricci flows can be defined in such a context, or perhaps Ricci flows better named as the more elementary "harmonic flows").

> [The simplex (or space-form) structure of space allows fixed dimensional levels to be geometrically independent (materially independent) of higher dimensional levels (or equivalently, each dimensional level is geometrically independent of other dimensional levels).]

**The math structure of the newly developed containing-space for all existence

This new descriptive structure places a stable, linear, geometric (and geometrically-separable) model for (of) physical description (where these properties of stability and geometry are attributed to both material systems and to the spaces which contain these material systems) into a context of feedback, which in turn, is based on a fixed geometric-spectral set which, through resonance, upholds the stable properties of a verifiable and practically useful "geometric description" of existence. This new model has a many-dimensional base-space which is defined within a principle fiber bundle, where the fiber groups are both orthogonal and unitary (ranging over both the various dimensions and over various types of metric-function signatures [see below for definition] of the metric-spaces and Hermitian spaces [for the complex-coordinates associated to the metric-spaces' dual-state structures] which comprise the base space).

This new math-physics construction, whose purpose is to be a containing-space of all existence, is (also) a data fitting device . . . ,

> since all of science is about following observable data which is described in a quantitatively consistent descriptive context, (thus, one should always be critical of such descriptive structures, and that is why one must relate precise descriptions to practical creativity)

. . . , the process of fitting data is (will always be) based on a set of assumptions about "what the context for (of) existence actually is." The new rule which this new descriptive structure follows is that, science is about following observable data which is described in both a geometric and quantitatively consistent descriptive context. This is because, if a description is not geometric or its defining measurements not based on averages over many components which fit into large geometrically measurable reservoirs then the description will not have any relation to practical creativity.

For example, (1) materialism and a sun-centered solar system (identifies that from which classical physics emerged, and provided the [coordinate] context within which the differential equation was defined) (2) currently (2011) physical law is about fundamental indefinable randomness defined in terms of sets of (differential) operators acting on function-spaces . . . , which identify the random properties of the system's components . . . , whose spectral properties are supposed to determine the spectral properties of observed physical systems, where these systems are composed of a relatively few number of components, (the current basis for physical description [which is opposed to geometry]).

Any new alternative context for description should provide the solution space whose problem is implied in the new model (or new context) of existence, ie the containing space within which the problem is to be formulated. For example-(1), in the previous paragraph, in regard to the ideas of Copernicus, the natural problem is: What determines the orbital-motion of planets in a sun centered solar system?

The new alternative context of existence, provides a vision about both what the important properties of a data-fitting containing-space would be (a many-dimensional space which contains, primarily, discrete hyperbolic shapes, upon which the stability of existence depends), as well as "what the process of fitting the data to the new containing space would be," (the mathematical structure for the interactions between the discrete hyperbolic shapes, which is to be based on discrete Euclidean shapes).

In the new context, one aligns the observed data (of the stable definitive discrete spectral-orbital properties of material systems, which exist at all size scales) with the (spectral-geometric) properties of an "over-all higher-dimensional containing space" whose formal properties are based on the set of very stable discrete hyperbolic shapes which are defined on an array of many different dimensional levels (which model existence). In fact, the new assumptions give information about the spectral-geometry, at each dimensional level, of the high-dimensional structure of an "over-all containing-space."

This stable spectral-geometry, defined on many different dimensional levels, is the basis for the relatively stable, definitive, discrete spectral-orbital properties of material systems which exist at all size scales (yet go without adequate descriptions). Furthermore, this array of many-dimensional discrete hyperbolic shapes, form the basis upon which can be defined a material-interaction process which is consistent with the existence of the (these) many stable discrete spectral-orbital properties of material systems.

The observed data of physical systems is related to a stable geometric-spectral structure which is associated to each dimensional level.

Furthermore, this new descriptive context might define an "over-all higher-dimensional containing space" whose fundamental descriptive properties of spectral-values (namely, the characteristic geometric-spectral values of the discrete hyperbolic shapes), might form into a finite set, though this spectral-geometric set would be large (see below, where elementary models of these spectral sets are counted).

Thus, the problem to solve is:

(1) identify the spectral set of the elements of the periodic table of the molecules, the elements, and their nuclei, as well as the spectra of crystals, ie identify the existing observed spectral-set related to relatively stable material systems, and this would include the orbital geometry of the (apparently) stable solar system, as well as the orbits of stars in galaxies, etc, then
(2) to relate this set to both

(I) a set of conformal factors (which exist between dimensional levels), ie physical constants, and
(II) a set of discrete hyperbolic shapes, characterized by their sizes, which define both the set of independent metric-spaces at each dimensional level, as well as the material which is contained in these metric-spaces, ie the set of ([closed and] bounded) independent, discrete hyperbolic shapes which compose both the metric-spaces and the "material" systems at each dimensional level. Then to describe how these different spectral-geometric systems can be organized in regard to their interactions, and how they form into their (final) stable geometric-spectral-structures as a result of these (material) interactions.

Brief examination of the currently accepted physical description's failures to describe the stable spectral properties of general quantum systems

The descriptive structures of quantum physics and particle-physics can only provide bad approximations of the observed patterns of physical systems.

The random math patterns seem to simply not be able to describe the stable, definitive, discrete spectral properties of general quantum systems, (with a few exceptions, eg the H-atom).

However, the example of the H-atom is used to provide the energy levels of all the other atoms from the periodic table of the elements, but this is not a valid description of the energy spectra of a general atom picked from the periodic-table.

Currently (2011) atoms are all modeled as H-atoms, so the spectra of atoms is considered to have about 140 (for n=7) spectral states, or 280 spectral states (including spin states), but the spectra of atoms do not fit into this very limited spectral set. The relation that the (principle quantum) number, n=7, has to the number of 140 energy-levels (for the atoms which compose the periodic table) is obtained from the energy degeneracy of quantized angular-momentum properties of an H-atom system, where the energy operator is assumed to have the property of spherical symmetry. That is, the energy-operator is based on geometry, but geometry is a property which a fundamentally random descriptive structures are not supposed to possess. Furthermore, the (power series) solution to the radial equation of the H-atom diverges, unless artificially cut-off.

One sees that, at the start, quantum physics is both invalid and incoherent, and this incoherence quickly leads to problems in regard to the descriptions of slightly more general quantum systems, ie only the H-atom is describable. Quantum theory cannot describe the more general atomic-structures, other than in terms of statistical data fitting (or data fitting through epicycle structures). Such an ad hoc data fitting process (using randomness to fit data) has not shown itself to be related to practical, useful, wide-ranging, technical development, a problem which Ptolemy's model of the planets also possessed (the problem of its failing to be a useful description, though it might have been used for navigation, in a similar way in which particle-physics is useful to bomb engineering)

On the other hand particle-physics only describes the observed unitary patterns of the "internal" (higher-dimensional) symmetries of unstable particles which emerge from high-energy particle-collision experiments. These high-dimension so called "inner-symmetries" of elementary-particles would be better interpreted in a more geometric model of existing higher-dimensions provided by the new descriptive structure for existence.

Particle-physics cannot be taken seriously as a way in which to adjust the values of the energy levels of quantum systems, but that is how its math-structure is organized.

Nonetheless the biggest impetus for particle-physics was its application to the energy levels of crystals, and subsequently its relation to the property of superconductivity in crystals, but this phonon-electron-interaction based method predicted a critical temperature above which superconductivity cannot exist, but this critical temperature has been exceeded by "high-temperature superconductivity," and the theory has never been adjusted (or remedied) in a satisfactory way.

Apparently, when a crystalline system is cooled sufficiently then there are energy-orbital structures . . . , whose size-scale is the same as the size of the crystal itself . . . , into which charges can be both fit and become organized. Such a discrete linear shape for a crystal can be related in a simple manner to the discrete hyperbolic and Euclidean shapes of the new descriptive structure of many dimensions, where each dimensional level is (or can be) associated to a stable, discrete hyperbolic shape. That is, a crystal's discrete structure can related to various hyperbolic shapes of various sizes, defined over a (small) number of dimensional levels.

Randomness seems to simply be the wrong pattern to try to use in order to describe the stable discrete properties of stable general quantum systems.

While non-linearity simply adds more fundamental randomness to a system, due to the quantitative inconsistencies (and incompatibilities) which non-linearity brings to the descriptions.

PART IV

PART IV

Chapter 20

GEOMETRY

Science is always a (relatively) precise (or measurably verified) data fitting description, but is the description useful?

or

Most troubling is a discussion about the continuum-hypothesis and it being put into relationship with Godel's incompleteness theorem (New Scientist august 1, 2011, #2823, R Elwes, Ultimate Logic:) [a typically very poorly written quasi-technical propaganda article]

An article about resolving the continuum hypothesis but which includes an unnecessarily (overly) generalized context of Godel's incompleteness theorem (a discussion about extending descriptions to have infinite complexity so as to imply completeness rather than incompleteness) . . . ,

> (where the incompleteness theorem essentially states, that there are always precise patterns which descriptive languages based on quantitative structures cannot describe, ie in order to describe the elusive non-describable patterns one needs to provide a new set of assumptions as well as new contexts upon which to base ones new descriptive language)

. . . , namely (the article argues), that the axioms "of the quantitative structures" of math are sufficient to describe all math patterns if infinite sets are properly organized within the quantitative sets.

It should be noted (it was reported in this article) that K Godel, himself, believed quantities could be used to describe all math patterns and that this could be resolved in regard to using very large sets of quantities (using infinities and greater complexity to make the axioms complete, eg using the equivalent of Ptolemy's complex epicycle structures to fit data, but not focusing on the need for a very-stable "unit" of measuring upon which numbers depend for their definition (in regard to their definition by counting), an important idea which math community seems to not want to consider in regard to fundamental assumptions about describing measurable patterns, and yet the stable unit of measuring is a simple geometric structure) [Perhaps Godel was too easily influenced by (the inept) J Von Neumann].

This idea of making the structure of quantitative sets more complicated is in opposition to a more geometric interpretation of the "incompleteness theorem," and thus it seems that it should be asked, if the very large infinite sets of quantities which are used to describe a greater range of measurable patterns, "Whether the descriptions are consistent with either the quantitative structures (namely, consistent with the number system's uniform unit of measuring) or with quantitative structures of the measurable properties "of the system" . . . ,

> [where an actual "system" is again an example of a very geometric part of measurable descriptions, even if the system-geometry reduces to the geometry of collisions (which might be given a random structure so as to fit into the complex quantitative structures which are being used to fit data)]

. . . , which one is describing with the sets of numbers, which define very large sets?"

Should one solve a logical problem by making quantitative structures more complicated, more inconsistent, and less capable of being related to a useful description, in regard to practical uses of measurable descriptions.

If almost all real numbers require an infinite amount of information in order to identify each real value (G Chaitin) (so that these real numbers are associated to either an unknown event space, or an unknown eigenvalue set, or to a non-linear math structure) then quantities can be defined by means of sequences in arbitrary ways so as to fit data.

Can a quantitative structure describe all measurable patterns, or is geometry also needed? The fundamental aspect of a uniform stable unit of measurement (in the structure of sets of numbers) requires that geometry enter the structure of the descriptive sets.

Are non-linear geometries valid? No, they are fundamentally random systems, where their randomness does not have a stable, well-defined elementary event space, and they are not properly related to the unit of measuring, where a number's relation to a stable unit is used to identify the values of numbers.

However, "Can a description which is described by means of a quantitative structure, successfully describe a measurable structure in a consistent and useable manner?" It seems that this is possible only if the description is placed in some geometric context which is consistent with a stable unit of measuring.

How is the continuum of real values constructed? By applying an absolute idea about defining a quantity's value and then constructing sequences of value in arbitrary ways and then using this to fit data.

This can be done with both non-linear properties and from difficult to define elementary events spaces which have an infinite number of elementary events?

How can one know that such sets of real numbers are properly defined so as to be consistent with the "stable unit" (upon which quantities depend for their definition)?

and

How can one know if these real-values are consistent with the measurable properties "of the system" being described?

However, in regard to very big quantitative sets there is the problem of whether the defined quantitative patterns are consistent with a system's (or a geometry's) description. Are the newly defined quantities . . . , whose definitions depend on both sequences and the axiom of choice . . . , properly associated to the "stable uniform unit" upon which a quantitative set depends for its definition (where the stable unit is associated to the system whose properties are being measured and described).

It is clear that rather than considering abstractions of very large sets . . . , which may or may not be consistent with the measurable properties of a system being described . . . , one should consider whether the description is practically useful, that is one should consider "finite descriptive structures for stable, measurable systems." What geometric context is needed to describe a system's measurable properties, and can it be related to a "finite" quantitative set? Apparently, this is (may be) possible in the context of discrete hyperbolic shapes.

What exists so as to have stable properties? Discrete hyperbolic shapes, and some stable (and finite) elementary event spaces, but only random systems with a very large number of components so that physical measurements are averages of such a system's statistical properties (defined on the many component system) have descriptions which can be controlled and (measurable information) used. Thus the focus should be on discrete hyperbolic shapes of various dimensions. That is, descriptions based on randomness for a system with fewer than (say) 100 components are descriptions which cannot be used in direct creative ways (they are not useful descriptions).

What allows for a stable unit of measurement, (which in turn) can be associated to quantitative sets so that the elements of these quantitative sets can have any-size at all (or sizes which require an infinite amount of information to identify)?

Is there some stable set which makes measured values valid?

For example,

> a stable, finite, well defined elementary event space used for describing random systems,

or

> a set of controllable, stable solutions related to some stable system of geometry.

But it is only the geometric description which will be directly related to the creative usefulness of a measurable system's description, and consistently relatable to the stable "unit of measuring" upon which the quantitative sets are defined.

Is it clear that for (other) calculable numbers within a system's descriptive structure (or differential equation based on physical law) to have any (or a) valid relation to the system being described, [or to be related to "that in the description" which is stable], then it is the property of stability . . . , of either the unit of measurement or of the system being described . . . , which is fundamental to the validity and usefulness of the description (or the property of usability of the measurable description).

Furthermore, it is stability which (or is what) allows numbers (or measures) to be defined in relation to the (stable) system's description.

Its not clear if the (real number, or measurable) values come from a formula or from a stable geometry, etc.

Though calculus is defined by sequences defined on both function spaces and domain spaces, where a function's value is defined on (in) the domain space, nonetheless the calculus operators (the result of the sequentially defined operator values) are linear models of geometric measures defined on the domain space (but multiplied by a function's value).

That is, (this can be interpreted to mean that) the values of a system's descriptive properties based on solutions to differential equations must be associated to a consistent geometry of the domain space.

This can be interpreted to mean that the eigenvalue equations (of physical systems) need to be related to stable geometries, namely, the discrete hyperbolic geometries.

The problem with large sets (and set theory) is the question, "To what geometric set are the quantities quantitatively consistent (to what geometric set is the stable unit attached)?"

Are the different values (defined by [arbitrary] sequential processes) consistent with different geometries or to different elementary event spaces?

That is, the values of large quantitative sets must be limited as to what they can describe so as to be consistent with a particular geometry (or well defined elementary event spaces) which they are being used to describe.

Creating existence
(the knowledge which apparently D Juan did not possess)

The ideas of quantitative descriptions based on discrete hyperbolic shapes placed in a many-dimensional context both transcend the idea of materialism and provide a geometric blue-print upon which create ideas can be developed. One can explore higher dimensions by considering the attachment of line segments to the vertices of a lower dimensional cube (they could be rectangles). That is, to the line segment attach in a perpendicular manner to line segments to the line-segments end-points. the same technique can be used to construct a cube from a square. So try to continue this construct up to higher dimensions (this could be considered to be an example of not-doing). Both discrete Euclidean shapes and discrete hyperbolic shapes are built out of their relationship to cubes, where opposite faces of these cubes are identified with one another so that the line-segments defined at the vertexes of opposite faces are turned into circles, so as to form a discrete shape. It should be noted that if the cube is in 3-dimensions then the discrete shape exists in 4-dimensions.

It is within this higher dimensional very simple geometric structures that one can consider the prospect of creating existence itself.

Schaflie (19th century) and Coxeter (20th century) have examined these simple Euclidean and hyperbolic geometries at higher dimensions, so many higher dimensional properties of discrete hyperbolic shapes are "known." Thus, there can be listed within the context of "known" properties of these shapes which one might like to try to change (with one's higher dimensional intent) in order to create new properties for existence.

The discrete hyperbolic shapes are very stable so it is the discrete hyperbolic shapes which one might want to focus their creative interests in regard to creating existence, itself. In the different higher dimensions there are the "existence properties" of these discrete shapes, and there are issues about whether these shapes at different dimensional levels are "bounded" or not.

Bounded (hyperbolic) shapes have definite spectral properties.

How to change properties of existence and change properties of bounded-ness in the different dimensional levels?

This new description depends on the relation that the geometric shapes have to their classical Lie fiber groups, and Lie groups have distinctive geometric properties, ie SO(3) has the shape of a 3-sphere and this is related to the spherical symmetry of material interactions in 3-space, where SO(3) is the fiber group to $R^3(3,0)$. Lie groups always possess a maximal tori geometry within themselves and discrete shapes are related to discrete subgroups of the fiber Lie groups. Thus, considering different lattice structures, ie structures which define discrete subgroups, could be of value, though the simplicity of the regularly shaped cubes has value in understanding the relation of discrete shapes to physical description.

The relation that SO and SU have to one another, especially in the context of spin-covering groups, where in spin groups opposite metric-space states are united (or put together) in one coordinate space (a coordinate space which is often a complex-coordinate space).

The further relations that can be determined in regard to SO, SU, and Sp, can be considered as possible points where existence can be expanded.

Central to the descriptive structure is the relation that the maximal tori of the fiber groups can have to the spectral and geometric properties of discrete shapes in the metric-spaces of the different dimensional levels. The tori can be used to resonate with the spectral sets of the metric-spaces, thus they can be used as a form of memory of geometries (which also depend on these spectral properties) which exist in the different dimensional levels, and are thus related to different dimensional aspects of maximal tori.

Always focusing on the memory of 3-space results in useless knowledge in regard to our true context of existence, thus we are left with limited knowledge which is essentially useless (C Castaneda referred to this capacity for memories of higher dimensions, or fixedness to one dimensional level, as being related to each person's ring-of-power through which we hold our idea of existence as being fixed [if we only maintain our memories of 3-space]).

Certain odd-dimensional discrete hyperbolic shapes (dimensions: 3,5, 7, 9) can have an odd-number of holes in their shapes, if their spectral-flows around these holes, ie atomic energy levels (in 3-space these are radial energies), are filled with charge then there would be a charge imbalance and thus these charge unbalanced discrete hyperbolic shapes would begin to oscillate and thus they would also push metric-space states together so as to generate energy when they oscillate. Thus along with these oscillating shapes being associated to a memory structure through which and informed intent can move or act. This is a simple model of life and how a 3-dimensional or 5-dimensional such oscillating hyperbolic shape can be projected down to 3-space can be very complicated and thus can have a deep relation to the chemistry of life in 3-space. But one being able to move in higher dimensional space might be the more over-whelming type of knowledge through which one might want to deal with the creative aspects of existence.

This knowledge of higher dimensions results in gaining a lot of power but one needs to focus on the intent of creating existence itself which is most important to existence, ie personal gain corrupts and lessens one's creative power in the true context of our existence.

If By using and maintaining memories of different dimensional levels then energy can be allowed (or controlled so as) to flow into different spectral sub-structures of oscillating discrete hyperbolic shapes so as to affect both changes and motions in different dimensional levels, but the motions (and its geometries) must be consistent with the local coordinate transformations of displacements of that dimensional level which define geometric inter-relationships with the geometry of the local coordinate transformations of the fiber group.

Life

In some odd-dimensional discrete hyperbolic shapes, which have an odd-genus (odd number of holes in their shapes), which (if their spectral flows [or spectral faces] are filled with charge) are unbalanced in regard to charge, and thus naturally oscillate, and in doing this generate their own energy. Furthermore, spectral structures can be represented on maximal-tori (associated to each [metric-space] point's fiber Lie [isometry] group) so as to give the particular oscillating, energy generating, discrete hyperbolic shape a means through which a memory of spectral properties can be realized, and thus these certain types of naturally oscillating (energy generating) discrete hyperbolic shapes form a "model of life."

If life is a multi-dimensional shape then knowledge which is only low-dimensional is mostly unrelated to the high-dimension wherein a life-form exists, though these higher dimensional functions can descend to the lower (material) dimensions in the form of muscles causing movement in the life-form. Thus, if one possesses only low dimensional knowledge then one cannot create to life's full capacity to create.

Geometry

Should one consider the main properties of geometry to be those properties which are related to local diffeomorphic maps between geometric containing spaces of general curvilinear coordinates (or between equivalent manifolds) upon which geometric measurement can be introduced,

or

Is geometry primarily about fixed measurements related to metric-invariant coordinates which can be used to identify geometric shapes?

Geometry which is not integrable (or not separable), ie geometry with curvature which is not constant, is not stable in regard to both shape and dynamics. Such non-integrable geometry will not have stable properties particularly in relation to dynamics.

Should the fundamental measuring functions (or metric functions and subsequent alternating forms) be contained within a coordinate space which fixes measurements with metric-invariant transformation groups for local coordinate transformations, or should one seek a coordinate space which measures the geometry in its "most natural manner," ie along curvilinear coordinates, upon which it is also most natural to define a metric function, ie introduce measurement as a property between two (diffeomorphic) coordinate sets, with which to measure the locally diffeomorphic geometric (or coordinate) maps, rather than beginning with a well defined measuring function (or metric function).

Assuming the two geometries are the same, and using the same path between two fixed positions on the geometry, one can measure the lengths of these paths based on different local metric functions. How can one know which of these measures is the correct measure?

Answer: Metric invariant measures can be compared with the measurable physical properties that physical systems (or the material geometries) which are contained in such metric-invariant coordinates, ie the systems that such metric-invariant coordinates are used to model. Metric invariant coordinate spaces which contain the material geometries of physical systems are used to great practical value. Namely, in the theories of: Newtonian gravity, electromagnetism, thermal physics, and statistical physics.

Most of our technology depends on these metric-invariant theories (or descriptions).

The less useful quantum physics is also formulated in the metric-invariant context, while general relativity is the only physical description dependent on the locally invertible properties. However, general relativity has only described one approximate system which (perhaps) has been measurably validated, the one-body system with spherical symmetry and a zero-test-mass. Since general relativity is a non-linear theory, slight differences in how one approximates geometric properties of more than one-body can result in quite different dynamic descriptions. Thus general relativity has not been able to describe a two-body system which has more geometric diversity than spherical symmetry in a useful manner.

It appears that local invertibility is more general than metric-invariance, but the fundamental distinguishing properties of different geometric types is their discrete group (or space-form) structures, not delicate and/or sudden (but smooth) curvature changes in the general curved coordinates. The reason for this is that subtle changes in curvature (or metric function) properties for a physical system are not stable, since they identify non-linear differential equations, whereas the separable coordinates of space-forms identify linear differential equations which are easy to solve.

Separable linear systems are solvable and stable while systems with curvature are (usually) not solvable and (usually) not stable. There exist many stable systems which the currently accepted descriptive language is not capable of describing. That is, the description is saying that such observed stable systems should not be stable. Thus a new descriptive language is needed.

In order to relate geometric measures to physical description:

(1) Should one map between metric functions in a metric-function "function space" and in this way determine if the true measure (or length) of a given geometry is being determined?

or

(2) Should one remain in a metric invariant context and use stable "hyperbolic discrete group" manifolds to determine both (1) stable (material) geometry, eg atoms, planetary

orbits etc., whose geometric coordinates are separable, and (2) stable metric invariant metric-spaces whose coordinates are fixed, but the geometric tangents (of curves on a manifold, or on a geometry) can be locally transformed by the isometry fiber group to align with the containing metric-space's fixed coordinate structure (directly relatable to the metric-invariant metric-function for the set of local coordinate transformations which are determined by the actions on the local coordinates by isometry fiber group) so as to establish geometric measurements of the stable geometry which is contained in the metric-invariant metric-space?

However, metric-invariant spaces can, themselves, be modeled, with the help of discrete (isometry) subgroups, as space-form shapes, so that within these metric-spaces (modeled as space-forms whose definition depends on discrete groups) their metric-function remains metric-invariant in the same differential structure of a metric-invariant fiber group for a manifold, or principle fiber bundle, that is contained in the metric-space.

Local invertibility (or locally invertible maps, ie a diffeomorphism fiber group) of a map between pairs of coordinate spaces (or sets) of the same geometric entity, eg local (or global) coordinates defined on a manifold, results in a relative representation of measuring length, that is one can ask, "Which of the two different length values is the true length?" However, if the motion (inertia) of a zero-test-mass only depends on the shortest path on the geometry (or geodesic), then ultimately the exact (or correct) geometric measure of length (and subsequently all geometric measures (eg area) can be determined from experience. Unfortunately curvilinear coordinates and geodesics on manifolds only work, in actual experience for the one-body spherically symmetric system (with very slight measuring differences from the Newtonian Euclidean model). Thus no general data exists on how to resolve the issue of "actual lengths" (or actual geometric measures) for the curvilinear coordinate case.

This could (alternatively) be considered in relation to the principle fiber bundle model, of a geometric manifold, with connections (or derivatives) based on Lie algebra (of the fiber group) actions on local coordinates. It should be noted that particle physics is modeled within a principle fiber bundle with a metric-invariant context, which includes a unitary fiber group, upon which a local Lie algebra acts on internal symmetries of point particles (as opposed to acting on geometrically determined vectors).

Geometric shapes and the holes in the geometric shape

In the metric-invariant model, a geometric measure is well defined, so that when applied to physical models it is consistent with measurement in experience (ie in the lab). Such fixed metric functions are conformal with the three types of coordinates which have constant curvature of 1,

0, and (-1), for the coordinate metric-spaces of the sphere, Euclidean space, and hyperbolic space respectively (and their associated discrete isometry subgroup geometries, or space-form shapes). These metric spaces are coordinate containing spaces for geometries.

Note: A general n-manifold can be contained in at most a 2n-Riemannian manifold, where a Riemannian manifold (usually) has Local Euclidean coordinates.

The constant curvature metric-invariant spaces have discrete isometry subgroup structures, or space-form geometries, associated to themselves, where the "holes" in the space-form shapes (contained in these metric-invariant spaces) most characterize stable geometry-types (as well as stable spectral properties).

Furthermore, it is the metric-invariant spaces which are related to the most useful descriptions of physical existence, eg Newton's gravity, electromagnetism, sound and light waves, thermal physics, and statistical physics, are all descriptions which are contained within metric-invariant spaces. These descriptions are all (though wave descriptions might be somewhat exceptional) based on differential forms which are alternating matrices which are local geometric measures, and which ultimately depend on a metric-invariant context.

If a metric is not fixed (or it is fixed but for a particular space of curvilinear coordinates) then there is (emerges from the calculation) the Ricci term in the general Laplace operator and the description is then placed in the locally invertible context, so that valid geometric measures are difficult to identify (in experience).

It seems that the fundamental aspects of (stable) geometry is most related to the number of holes that exist in a geometry so as to determine within which metric-invariant space the geometry most naturally exists. On the other hand, geometry does not depend so much on the "surface" molding of curvilinear shapes, which seems to be unstable, and curvilinear shapes are most related to the properties of locally invertible maps defined between equivalent (but not necessarily of the same geometric measure) coordinate curves for the ("surface" molded) geometry.

Note: The global geometry of Euclidean 3-space is spherical symmetry. This is due to the spherical structure of the isometry fiber group, $SO(3)$, for Euclidean 3-space. However, Euclidean space-forms for material (geometric) systems can have stable shapes which are in resonance with certain spectra of the spectrally stable hyperbolic space-forms which also compose a given physical material system.

As for diffeomorphism: The shape of the coordinates contained in a particular dimensional space allows a functional relation between the dimensional variables of two containing coordinate sets of the given manifold (or geometry), so that this allows a metric to be determined in relation to an agreed measure of one of these two coordinate sets. This allows a measure of the relative separation of sets on a chosen path connecting two points on the manifold. The metric depends on a property of continuity of local measurability of mutually defined coordinate functions for the two coordinate sets defined on the manifold (or defined on the geometry). However, if the geometry has physical existence then the defined geometric measures may not be consistent with physical geometric measures.

Alternating forms and differential forms

The measures of geometric properties, eg length, area, etc., are determined by local measures of local vectors, (usually) referenced to Euclidean space, related to coordinate functions of the geometry, so that the tangent vectors of the intersecting coordinate curves (where they intersect) are placed in to a matrix and the geometric measures are calculated using the rules of alternating forms (or determinants) so as to measure the geometric properties of area, volume, etc., and "for length" the local vectors (or their Euclidean coordinate functions) are parameterized by a single variable, eg time, and used to measure length integrals.

Local alternating forms are also called differential forms, and it is upon differential forms which is defined the "differential" operator, d, and subsequently the general Laplace operator, dd, which act on differential forms, whose vector structures are local measures of geometry related to coordinate functions (or local coordinate directions) so as to be put into a very systematic geometric structure which can (be used to) measure geometric properties, eg length, area, volume, etc. However, the Laplace operator is most naturally applied within a metric-invariant setting, and it has been proven to be more useful than in the metric-invariant setting than rather than within a "general metric" setting [or (equivalently) within the local diffeomorphic setting].

What is geometry?

What are the main aspects (or properties) of geometry?
Local invertibility and subsequent development of the assumed measurable relations between coordinates defined on a manifold (But does the manifold naturally fit into a particular type of coordinate space?),

or

Placing a manifold in a metric invariant space so that the maps on the manifold have definite geometric measures (of the metric-invariant space) which are properly related to the type of local coordinate structures which measure a particular physical property which characterizes a geometry made out of a particular type of material. That is, material and its associated physical property are attached to metric-invariant spaces with metric functions that have particular signatures, ie where a metric's signature is related to the number of spatial sub-space dimensions and number of time sub-space dimensions of the metric-space and the metric function.

However, not all manifolds (or geometries) can be contained within a particular metric-invariant space in relation to both coordinate containment and metric-function inter-relations, eg spherically symmetric spaces naturally belong in Euclidean space, but spherical symmetry does not naturally belong in a hyperbolic (or equivalently a space-time) space.

Riemannian geometry versus local Lorentzian geometry. Einstein formulated general relativity in Riemannian geometry, because inertia is related to displacement in Euclidean space (translational or rotational displacement), yet the Schwarzschild solution does not define a Riemannian manifold, where the Schwarzschild solution is locally space-time (or equivalently hyperbolic space).

The geometry (or geometric measures) of metric-invariant spaces can be related to local vector (or tangent) properties of coordinate functions (and slightly deformable geometric [or curved] coordinates so that the local invertible transformations of these curved coordinate functions are (or remain within) the more restricted metric-invariant fiber group [or isometry group], as opposed to the curved coordinates conforming to the (material) geometry's more natural curvilinear coordinates). However, the stability of a geometry in such metric-invariant spaces is not related to its deformations (limited to metric-invariant transformations), but rather it is most related to its discrete isometry subgroup space-form structures. This is because the metric-invariant coordinates of a space-form are separable coordinates, thus the differential equation identifying a space-form is (linear and) solvable.

Is the property of local inversion (second order measurable, inversion), enough of a property to insure (or allow) that a geometric shape is fixed and the distances between two fixed points (or between two fixed positions) along specific paths, is the same value when two different coordinate structures are used to describe the manifold's geometry.

or

Can lengths be different when the geometry and its separated positions (and given connecting paths) remain fixed, but different metric-functions or different coordinate functions are used to determine length? (Yes).
Are there metric functions which are (best) associated to particular coordinate functions on a particular dimensional space and which identify a "true" measure of length for a geometry made of a particular type of material? So that all other measures of length, of the (a) geometry composed of a particular material space-form, should be referenced locally to these "favored" metric-invariant coordinate metric-spaces of the particular material type? (It seems that this is true, certain invariant metrics are naturally associated with certain material types and certain physical properties.)

How to distinguish geometric invariance (locally invertible) from geometric-measure invariance (if one is only dealing with functions between coordinates [or functions of geometric-measure defined on local coordinates])?

Geometry is the geometry of material within space, where space has (can have) a coordinate (or geometric, ie curvilinear) measuring system.
What are the main attributes of geometry, in particular the geometry which material identifies?
Is the material geometry contained within metric-invariant coordinates which are related (in there most useful form) to separable (or integrable) coordinates which can be related to isometry transformations acting on local coordinate properties,

or

Is geometry primarily about diffeomorphism (local inversion) in relation to measurement defined between (1) curved coordinates, and (2) an arbitrary local measuring space, so that these curved coordinates are mostly not integrable?

The main attribute of geometry is its "hole in space (or space-form)" linear structure, rather than the clay-like molding of a sphere's surface geometry (or a bounded region of a general manifold's surface), which is non-linear and not stable.

The separable geometries that are related to holes in space (which discrete isometry subgroups define) . . . are geometric properties which are central to the stability of material properties, eg (1) the stability of small atoms, which when present in very large numbers can be molded (like damp sand can be molded), as well as (2) the stability of planetary orbits.

Where discrete isometry subgroups define the holes in space in a space-form, which is a geometry which can be contained in the different metric-invariant types of metric-spaces such as:
Euclidean space [zero curvature] (not well defined spectra associated to its toral structure),
Hyperbolic space [constant curvature of (-1)](well defined stable spectral structures defined on each of the holes which the hyperbolic space-forms identify),

The other important stable, simple geometry is the Spherical geometry [with constant curvature of (+1)].

The sphere is the stable aspect of a ball of clay which can be molded (but without handles), but spherical geometry is only central to the geometry of the inertial properties of material, and subsequently important in regard to material interactions, when those interactions are defined on of Euclidean 3-space. This is because the isometry group of Euclidean 3-space is SO(3) which has the geometry of a 3-sphere. It is this local coordinate, metric-invariant transformation group which is shaped as a 3-sphere which causes the global spherical symmetry properties . . . .

for the inertial material properties
(of inertial material [toral] 2-space-forms which exists in Euclidean 3-space)

.... by identifying a radial direction in relation to an interaction (approximately spherical) tangent surface so that this interaction tangent surface is identified by an interaction space-form which forms between interacting (inertial) material 2-space-forms in Euclidean 3-space so as to form an interaction 3-space-form.

The discrete subgroups of the metric invariant isometry fiber groups, of either Euclidean or spherical or hyperbolic coordinate spaces, can be used to identify the stable and intergrable (or separable) geometries of manifolds which are (or can be) contained in a metric-invariant coordinate spaces. This discrete geometry (or space-form geometry) is very stable (hyperbolic) geometry which is also separable (or integrable).

Geometry which is curved (or has curvature) is not integrable (or not separable), ie curved geometry is geometry which has curvature which is not constant curvature (or zero curvature). Such curved geometry, or non-integrable geometry applied to the dynamics of zero-test particle masses, will not have stable properties in space, particularly in relation to dynamics.

The theory of general relativity is about curvature, thus it is about non-linearity and instability, yet the solar system seems to have been stable for billions of years. Thus general relativity is not a valid model for the gravitational properties of the solar system.

Geometry which is not integrable (or not separable), ie geometry with curvature which is not constant, is not stable in regard to both shape and dynamics. Such non-integrable geometry will not have stable properties particularly in relation to dynamics.

Is geometry about general properties of curvature within mappings between pairs of measuring (or coordinate) spaces or is it about the holes in a containing metric-space (coordinate) space which has a definite metric measure of length?

Answer: The distinguishing property of geometry is most related to the holes within the geometry, while subtle molding properties of geometry (related to the Ricci curvature of curvilinear coordinates) is non-linear, and thus it is unstable.

Stability of material (atoms, nuclei, etc.) and of planetary orbits, and subsequently the description of (relatively) stable material geometry, are fundamental to the systematic description of material geometry, and thus is relevant to the description of all existence.

There may be some exceptions to the following statement, but the most important examples of non-linear properties have to do with limit cycles of solutions to non-linear differential equations, and these limit cycles are related to the singularities (or holes) in the solution function's domain space, and the fundamental properties of solutions to non-linear differential equations which model physical systems is not about subtle geometries of curvilinear coordinates.

Can the holes in a space, due to discrete group properties, be related to the stable spectral properties of physical systems? (Yes).

or

Must stable spectral properties of micro-systems be described using operators acting on function spaces within the context of probability? (No, space-forms can be used to describe quantum systems which have stable spectral (or orbital) properties).

Does quantum physics have any validity, or has it been validated in an absolute manner for all measurable physical systems because it is based on probability, or does it possess (or is this) an illusion of validation because it is based on the probability properties of an elementary event space (the set of a quantum system's measurable spectral properties) which can be changed as it is deemed convenient (or as an observer deems convenient) for measuring purposes, ie changing the description to fit the data?

How should spectral properties of physical systems be represented? Either as probabilities or as stable geometric properties of the discrete isometry subgroups of the fiber groups of metric-invariant spaces? (Answer: As discrete isometry subgroups.)

Experts

When experts talk, it is only a limited viewpoint which is expressed.
Can any part of general relativity or quantum physics be taken seriously? Or are these simply complicated expressions of useless ideas which the experts try to extend (but which are not useful), but which are only relevant to bomb engineering, or more limited than that, perhaps only applicable (or relevant) to particle accelerators, because particle collisions do not actually occur (or rarely occur) in stable quantum systems.

(The data about particle collisions within particle accelerators is properly interpreted to mean that the fiber groups of micro-spectral systems are unitary, though this is also be true for macro-systems.

The containing coordinates of physical systems are complex because the two metric-space states with which each metric-space are associated [namely, metric-spaces are associated to particular physical properties which have two opposite states, eg (+t) and (-t),] requires that the natural coordinate space be a complex number coordinate space, but our time and material perception cause us to observe material geometry and its dynamics in a real space so as to not observe the mixture of metric-space states upon which dynamics depends).

Nonetheless, there remains the idea that complication can be extended so as to form some specialized useful description, ie a great truth built from the rigor and complication of useless ideas. In other words, it is believed that there exists a truth so complicated that this truth only allows experts to understand and speak about this truth with their expert knowledge, so that only within that particular authoritative language structure (for description) can truth be described. [Sort of like trying to prove that capitalism can be made to work by making contracts ever more complicated, so the elites can dominate and there still exists social equity.] This is a futile hope (or futile belief), and this limiting reliance on expert truth and expert authority, causes the descriptions of these authorities to become more limited and more and more useless.

It is a process of trying to extend a limited descriptive language, however, this process ignores Godel's incompleteness theorem, which states that precise language has limits to the pattern which it can describe, and thus this limits the usefulness that a precisely formulated descriptive language can have.

Fundamental observed patterns of the physical world cannot be described in the physical descriptive language of today's expert physicist-mathematician:

>   the solar system's stability,
>   the orbital properties of a general nucleus,
>   the energy structure of a general atom (not the H-atom),
>   spin-current interactions within a crystal
>   etc.

Yet the claim is that we have come to the edge of what we can know, the edge of truth, rather than that we have come to the edge of what patterns that the currently accepted authoritative, fixed, descriptive language can describe. The second interpretation, that we have come to the limits of a descriptive language capacity to describe the observed patterns of existence, implies that a new descriptive language must be sought.

New languages are not found by assuming the presently accepted structure of things, and then deducing the new truth. Unfortunately, this seems to be the attitude that the experts (and their adoring [or dependent] public) have adopted.

Instead new languages are found by associating new patterns to fundamental ideas (or fundamental properties). For example: Modeling an elementary particle as a distinguished point of a space-form upon which a material property manifests in relation to stable spectral flows on a (hyperbolic) space-form. Or identifying a metric-invariant space of a certain dimension and of a certain metric signature with particular (set of) physical properties and particular material types, etc.

To arbitrarily identify a descriptive language and its associated interpretive and deductive processes as having high value and to claim that, it is only from such high value descriptive sources that (from which) truth will come, is similar to the guarded absolute truth which Copernicus faced in his age.

For example, materialism is the basis for classical physics and then in quantum physics material reduces to random elementary particle events in space, so that high energy systems are either classical or general relativistic in nature (where general relativity focuses on the more limited geometric descriptive capacities of Ricci curvature and its geodesic relation to dynamics [or inertia], than the more wide ranging descriptive geometry of space-forms and metric-invariant spaces, where space-forms deal more directly with the properties of material and dynamical stability than do geodesics defined on curved coordinates), otherwise material (or particle) interactions are to be determined by (internal) particle-state unitary transformations which occur during random collisions of elementary particles. This is the same arbitrary character of truth within which Copernicus presented an alternative to the academic and fixed authoritative truth the centers of learning espoused in his day.

The currently used authoritative descriptive language is not able to describe:

>the stable general nucleus,
>the stable general atom,
>the stable solar system, and
>the interactions of current and lattice spin properties in a crystal.

These are fundamental systems which the current descriptive language claims are too complex to ever describe, yet they are relatively stable and well organized systems, and this does not suggest complex systems based on probability which are too difficult to describe. That is, if the laws of probability cannot describe these systems then the unsuccessful application of those laws should suggest that these systems are not stable (based on the laws of probability and operators), yet they are stable. Thus the laws of quantum physics are to be questioned.

Geometry

In regard to geometry, it is the integrable holes in space which range over the greatest amount of geometrically diverse types of geometries, and when such space-forms are applicable to modeling the geometric properties of physical systems it is this space-form geometry which is the most useful. This is partly (or mostly) because space-form geometries identify diverse properties of stable geometries, including the microscopic quantum systems.

What is geometry?
Is it locally invertible maps between a manifold and its curved coordinate functions with a diffeomorphism fiber group and a curvilinear coordinate base space?

or

Is it metric-invariant spaces with isometry fiber groups and local maps between some curved coordinates and particular metric-invariant coordinates which contain the manifold (the manifold is immersed in the metric-invariant coordinate space)?

It seems that the second type of geometric representations (or metric-invariant spaces) with their discrete subgroups and their subsequent space-form geometries which is the most useful geometric viewpoint, and covers a very diverse set of fundamental geometric possibilities so as to be presented in their most useful context.

Groups

The isometry and unitary groups, that are naturally associated to metric invariant based coordinate descriptions, are semi-simple Lie groups and have discrete subgroups which identify space-form geometries which (when hyperbolic) are the stable geometric structures of a metric-space of a particular dimension and of a particular metric signature (see above).

From the discrete isometry subgroup structures a spectral-dimensional (spatial) hierarchy of inter-related spectral properties in resonance with the containing metric-space (or containing space-form) determine the spectral properties of the space-form systems contained within the over-all containing space, ie a set of spectral inter-relationships determine existence and what is perceived within these inter-related spectral-dimensional sets.

A general diffeomorphism group has a much more complicated group structure than semi-simple Lie groups. However, though the complication of such a locally invertible fiber group might allow detailed descriptions of the geometry of complicated systems, but what is the range of stable geometric structures which the diffeomorphism group is supposed to describe where it must be noted that most of these geometric-coordinate structures will be unstable (due to their non-linearity). Despite the wide range of geometric possibility for general diffeomorphic groups, nonetheless it is better to use the semi-simple Lie groups to outline (or approximate) the basic stable geometric structures which (might) exist within a spectral-dimensional hierarchy, of some highest dimension, which in turn, defines the observable (and non-observable) material and other geometry of existence, where spectral properties of one dimensional level can be dependent on the space-form (or spectral) properties which depend on other dimensional levels.

However, the geometry of a particular dimensional level is independent of the geometry of higher dimensional levels, while material in a particular dimensional level is composed of the space-forms which are lower dimensional than the given dimensional level, ie the given dimension of the containing metric-space, though the material space-forms of a given dimension metric space are mostly composed of the space-forms of one dimensional lower than the given metric-space dimension.

Palladium-deuterium energy generation

Recently there has been a new set of reports about the Palladium-deuterium energy generation phenomenon. This story demonstrates the trouble with the media, it is so authority driven that it is not capable of presenting observed patterns, the observed patterns must always be interpreted by the authorities.

It has been reported that some labs see energy generation by the palladium-deuterium system, while others have not.

Since it has not been used as an energy source it might be guessed that things are not working, but then new reports say that many labs do see energy generation, but it is not explainable.

There are 3-dimensional polyhedron fundamental domains for the hyperbolic space-forms. The (bounded) 3-dimensional hyperbolic polyhedron which have: 6, 18, 30, 42, 54, . . . , etc number of faces, have a "magical" number of faces, and naturally oscillate and generate energy. Such (bounded) oscillating hyperbolic polyhedrons can be used to model a radioactive nucleus.

This is a better model of a nucleus (as well as a better model of radioactivity) than the "nucleus made of colliding particles" model which quantum physics uses (which has never been an explicit model).

The natural oscillations, of the space-forms of these (above mentioned) particular polyhedrons, push together occupied flows of material which exists in opposite metric-space states. The oscillation seems to cause these opposite material states to come together and annihilate each other so as to give off energy, ie exactly the same as matter and anti-matter coming in contact.

The contacts between material in opposite states is caused by oscillation, and seems to be governed by the probability of contact events (or energy manifestation between metric-space states of opposite sign).

The deuterium-palladium crystal interaction can be modeled as a 3-dimensional polyhedron (or space-form), and because of resonances of this polyhedron with the containing 4-space (as well as the energy level of the interaction), is turning this interaction space-form into a stable oscillating space-form which is related to one (or more) of the above polyhedron (which have a magic number of faces, for the space-form's polyhedral fundamental domain).

Note: D Hunter has described these oscillating polyhedral systems in the book, "The Authority of Material vs. The Spirit."

> The palladium-deuterium system is an oscillating 3-space-forms in a 4-dimensional hyperbolic space . . . (which we do not naturally observe, since the material we observe [or focus on] is contained on a 3-face which has a well defined facial dimension, partly due to the space-form structure of light)

. . . . thus the physical constants involved in such a physical processes may be affected by the different dimension containing spaces.

The basic model for usage is to use E = h/T, where h is Planck's constant, and where T is the period of oscillation and E would be (related to) the excess energy generated in the palladium-deuterium system. Use this frequency, 1/T, to design a resonating circuit.

D Hunter discusses this, and his comments can be enhanced or modified a bit.

The engineering of forming resonating systems with the palladium-deuterium system will depend on the geometric structures of 3-dimensional interaction space-forms which are contained in 4-space. Interacting material systems in 3-dimensional space naturally have a 3-dimensional interaction space-form, so to build the device it might require that the device (the circuit) be composed of particular stable interacting systems, ie composed of the right materials with the right spectral properties put together in the right order and with the right size to achieve a resonating system. If it is an electric resonance circuit it must resonate with and couple to the oscillating space-form, which seems to be related to the palladium-deuterium crystal.

The geometric and spectral properties of 4-space can be determined from the atomic spectral properties of material, since it appears that the H-atom is the only atom which has a spectral structure which are naturally resonant with the spectral properties of our containing 3-dimensional hyperbolic metric-space.

Note: The speed of light is also a spectral (or space-form geometry) property of our containing 3-dimensional hyperbolic metric-space modeled as a (large) space-form.

The exact conformal (or constant) curvature of our space-time metric space is related to the size of its containing space-form. This containing space-form (or something like it) may also be related to the stability of the solar system.

Separable geometries and Lie groups

Note: Many separable geometries can be found (or constructed, the parallelizable sub-manifolds of Lie groups) and such separable geometries can be defined in the context of the classical Lie groups, eg:

SO (orthogonal, symmetric bilinear forms),
SU (unitary, Hermitian bilinear forms [always diagonalizable]),
Sp (symplectic, where the bilinear form of Sp is anti-symmetric).

[A stellar paragraph]

Consider the relation between the spectral ranges of a system and geometric properties of the containment set, where the containment set is a high-dimensional system-containing space, where the external and internal structures are both stable, discrete separable-geometries, but of different dimensions, so that the higher dimensional space also has spectral properties of the same dimension but in different (spatial) subspaces than the containing space of the low dimension system. That is, let the vertices, which are associated to particle-collisions, be extended to a geometric structure so as to be the vertices of a high dimensional "cubical" simplex and its associated separable discrete high dimensional shape. Now the spectra of the system, and the spectra of the high dimensional containing space can be related (through spectral resonance).

So what are the basic ideas upon which these above descriptive languages..., which are getting funded by business interests..., are based?

In particular, consider quantum physics, it is based on:

(1) fundamental randomness of physical (spectral-particle) events (2) all physical systems reduce to elementary particles (3) quantum systems are stable, discrete, definitive systems (4) quantum interactions (and thus all physical interactions) are based on random particle-collisions,

> (this might be quite wrong even at the level of thermal systems where a random particle collision model seems to work, but on the other hand the properties of clouds (and thus thermal systems in general) may not be understandable unless a global geometric structure can be associated to the entire cloud)

(5) quantum properties are described on complex coordinate systems, and thus have unitary local transformation groups (6) modeling the structure of particles so as to have particle-states associated to themselves, suggests a many-dimensional structure of the base (coordinate) space of the descriptive structure.

Alternatively, and claim that discrete isometry (or unitary) subgroups defined by means of "cubical" simplexes, with distinguished points (the source of the apparent particle properties) and (through interaction structures) associated to Brownian motions (thus accounting for quantum randomness), with many metric-space states (thus accounting for the unitary structure of physical description), and many metric-function signatures or many material types (eg mass, charge, etc), and defined on many-dimensions.

It can be used to account for both discrete stable quantum (and orbital) structures and classical structures (without the idea of materialism associated to the new descriptive structure.

Furthermore it is a geometric description and thus it will be useful for technical development though understanding the geometry of higher dimensions will be central to being able to use the description for practical things.

This brings almost the entire math description of existence into a geometric context.

In the context of higher dimensions the idea of materialism can be modeled as being related to material (discrete structure) interactions so that such interaction models (the description of space-form interactions are dependent on specific dimensional properties) are independent of higher dimensional properties within the over-all "high-dimensional" containing space.

It explains the stable discrete spectral-orbital properties of both quantum systems and classical systems, eg the stability of the solar system.

The new ideas are about placing the useful part of classical description, ie separable geometries, into a higher dimensional context so that the new description contains the stable discrete spectral structures of "discrete separable geometries" (or space-forms), but does not separate the system's measurable properties (as the spectral set [or elementary event space] of a quantum description) from the domain's (or system containing space's) properties (such as local geometric measures) as is done in quantum physics.

However, in quantum physics, it is only the geometric properties incorporated into the energy operator which can limit spectral structures of quantum systems, eg $1/r$ term in the H-atom's energy operator, but due to the separation of the spectral set and the system containing space (in the quantum physics descriptive language), the spectral structures of quantum systems never have a basis for their distinct formation, that is the spectra do not have any relation to geometry. This is why representing quantum interactions as a geometric model of random particle-state transformations (of a non-linear connection term) determined by energy levels at random particle-collision points in space will never provide the structure needed to determine a discrete spectral set . . . , since it is energy-level (or spectrally) dependent . . . , and the structure in the energy operator . . . , which determines the spectra of a quantum system's wave-function . . . , is geometric. That is, the particle-collision model of quantum interaction must affect the, $1/r$, term in the energy operator, otherwise energy and space are being put together in a quantitatively inconsistent manner. However, if the, $1/r$, term were adjusted by the non-linear interaction operator then the linear energy operator would become non-linear and thus the quantum system would not have a stable spectra.

The math of classical physics is (relatively) easy, ie the calculus of differential forms (which also implies metric-invariance,) and it is solvable when it is both linear and defined on separable geometry. The higher dimensional context shows how geometry affects motion, although motion also depends on the local coordinate transformations of the fiber group.

The relation between "natural coordinates of a physical system" and local coordinate changes of components of the physical system is that, in stable systems the local coordinate changes are linear (can be solved) and can be represented as diagonal matrices, and this occurs for separable geometries, which are the stable, discrete, definitive geometric structures of both the solvable part of classical and (it now seems) the geometries which carry the stable spectral properties in quantum physics.

Geometry is about measurable, stable shapes.

If a system has coherence which is associated to a position in space and the system is composed of components which are also identifiable, then this descriptive construct must be related to geometry which in turn is intimately related to a stable math construct for measuring (shapes, spatial-regions, and system-associated measurable properties).

Stable geometric systems and stable spectral properties of systems seem to be about solvable differential equations, ie linear solvable differential equations. Perhaps separable geometry, [ie curved coordinates whose tangent directions at each point in the domain space always form an orthogonal set (of real vectors)], which is related to linear differential equations is more important in the description of material-spectral-geometric-spatial existence (or a material system) than are (usually non-linear) differential equations, (in general).

The separable geometric properties of material-space are more related to holes in space (or holes in geometric shapes) for metric invariant spaces, ie topological (or the continuous) properties of space-material. Issues of diffeomorphisms between differential structures (defined on a manifold) are more related to non-linear differential equations which (may [or may not]) have definitive geometries associated to themselves.

Perhaps differential equations are super-ceded (for physical description) by the geometry of separable spatial structures such as discrete groups. That is, perhaps linear differential equations which are associated to separable geometries in E, S, and H metric-spaces are sufficient to describe the material-spectral-geometric-spatial properties of existence, where E is Euclidean space, S is spherical space, ie Euclidean space restricted to a sphere, and H is hyperbolic (or space-time) space, where E, S, and H are the metric-spaces of constant curvature with curvatures 0, 1, and -1, respectively.

Perhaps the idea that has been overlooked is the idea that stable, definitive, separable geometries are fundamental to stable definitive observable properties at all size scales, however, for small systems such geometries (as well as their discrete interaction geometries acting on adjacent geometries) leads to Brownian motion, and these random Brownian motions are (related to) the source of quantum wave-particle properties. However, the particle is actually a distinguished point of a small space-form geometry used to model atoms and nuclei as well as stable electrons and protons, though geonium shows that size and charge may vary for leptons. That is, space-forms and the differential structures of isometry groups determine material, space, and the interaction process between (space-form) material, where in 3-space the higher dimensional interaction space-forms are unseen (the interaction space-forms [which mediate material interactions] do not exist in 3-space).

These same interaction space-forms, when acting on an atomic scale, have many types of local (or adjacent) interactions which change from instant to instant, and thus result in Brownian motion for small (or quantum) systems. However, when an interaction results in a stable new quantum system that system is best modeled as a space-form.

Can curvature be related to energy?
Yes, sectional curvature can be related to kinetic energy.
But, within the (fixed, ridged) shape "what would move?"
Can sectional curvature be related to kinetic energy within the patterns of mathematics, or is this a physical assumption? It would be a physical assumptions.
Do metric-space need to be identified with physical properties?

The formulation of geometry is virtually always represented (by professional mathematicians) in a context of non-linear geometric properties, and in physical problems there is an absolute idea about material geometry, eg material interactions are always spherically symmetric (and the belief that material existence would always "define a set of continuous geometric relations" on the "domain" space of a function, which defines a geometric property, [for whatever fixed dimension that material defines]), where the dimension within which material is contained, is always given special consideration, ie 3-Euclidean space or 4-dimensional space-time, and geometry is most often related to an abstract manifold, whose description is non-linear.

Shape is almost always considered in its most general contexts, namely that shape is mostly distinguished by the set of holes which exist in a space, and about these holes there are many "folding" geometric structures, which can fold back on themselves and which twist and identify discontinuities and other types of critical points also exist.

This generality is unnecessary and mis-leading.

How is geometry related to local patterns of local measuring?

Geometry (in the communities of professional mathematicians) is usually about putting together local coordinate functions (within small neighborhoods into which the shape has been partitioned) which are solutions to non-linear differential equations, related to either curvilinear coordinates (defined on the shape) or to general metric-functions defined on the shape (where by means of determining the shape's metric-function local geodesics can be found on the shape). However, these non-linear patterns cannot be put together in a quantitatively consistent pattern, ie a uniform unit of measuring cannot be consistently associated to the non-linear patterns which are being described on the shape (or space, or manifold).

This general context (of non-linear manifolds) is assumed to depend on abstract concepts such as holes in a space (or holes in a shape) ie non-zero cyclic integrals of functions defined on these shapes can only exist about holes in the space, and apparently these holes can be found by the continuous deformations of simple closed geometric shapes, or rather by the inability to continuously deform simple closed geometric shapes because the shapes get "hung-up" by the holes in the space.

The general context depends on "holes in a space" as well as, strange folds, turns, and twists about holes in space, and changes in a region's orientation, all defined in a non-linear unstable context of a fleeting existing geometric pattern.

But do holes have a stable structure? (One would expect that holes are not stable.)

If they are stable, What is the structure of their stable properties (within what context can one be assured that they remain stable)?

Answer: The basic math structure for stable shapes which contain holes so that the holes remain stable, and which carry stable spectral properties, are the discrete hyperbolic shapes, while discrete Euclidean shapes can have properties of stability when they are associated to discrete hyperbolic shapes (eg by resonance).

These discrete hyperbolic and Euclidean shapes are much simpler to describe than are the abstract "holes in space."

What does the simplification of math and physics imply for society? That is, great math generality reduced to simple linear, diagonal, metric-invariant (concrete) shapes of discrete hyperbolic shapes. These shapes can be related to both large stable orbital properties as well as to stable spectral properties, and the properties of sectional curvature of the discrete hyperbolic shapes, $1/r$, exists only for certain discrete values of r.

It means that knowledge including math knowledge's relation to creativity can be simple stable and geometric (but now in a higher-dimensional context) and this context can give to the public very sophisticated creative skills, ie it equalizes society.

But a new implication in regard to the new math-physics descriptive context is that human creativity is not confined to the material world. This is because the patterns of discrete shapes extend up to higher dimensions but the structure of their interactions makes detection of higher dimensions very difficult (higher dimensions are hidden from any particular dimensional level), or at least it requires a new context for exploration.

The great generality of "holes in space" is not needed, since the most important examples of holes in space are the holes in the closed shapes which are discrete isometry subgroups (mostly discrete hyperbolic and discrete Euclidean subgroups) which are the cornerstones of the math properties of stable math patterns (both geometric and spectral).

Sets of commuting operators cannot capture the spectral properties of general quantum systems, since operators (differential, or their dual) deal with either (the "currently" assumed) geometries of quantum systems, or with spectral cut-offs.

Apparently this is because the spectral properties (of quantum systems, or of spectral systems) come from outside of the space within which the quantum systems exist, and this also has little to do with the intrinsic (local) geometry (or particular states, or energy ranges within which a spectral system can be defined (or contained)) of a general quantum system, though it might seem that an energy term with a 1/r factor could contain spectral information, but it seems that it would only be accessible if the assumptions about the geometry of quantum systems changes from spherical symmetry to the shapes of discrete hyperbolic shapes.

Indeed . . . , that

> "the spectral properties of quantum systems come from a structure which is outside of the metric-space within which the material systems (which are being measured) are being contained,"

. . . , is the main idea of particle physics, as well as all the other "physical" theories derived from particle-physics, eg string theory.

But particle-physics is simply the statement that "there is a hidden geometry, but the geometry (when associated to particle-collisions) is only saying that there is a set of 'spectral bounds' or 'spectral cut-offs' (where the cut-offs are determined by a method of data fitting), which determine the spectral properties of a quantum system."

String theory is trying to say that geometry is significant in quantum description, but the circular argument concerning the idea of materialism associated to the process of measuring "which necessitates that measurements be related to material attributes," requires that "the geometric properties only be related to a statistical, non-linear, (supposedly) measurable descriptive structure," which particle-physics defines.

But this descriptive context is useless, "data cannot be fit" and the "information (of randomness) cannot be used."

Furthermore, "more and more" elementary-particles are being added to the elementary-particle structure of particle-physics, eg the dark matter particle, the dark energy particle, the Higgs particle (where the Higgs particle was sort-of predicted) etc, thus, the probability structure is not well defined, ie the elementary-event space of elementary-particles seems to be unknown.

This undefined probability structure has the property of being an "epicycle structure" with which the authorities are trying to fit data by means of complicating the descriptive structures to the point where their descriptions have become useless, but nonetheless "their authority" cannot be questioned (by the public).

While, classically, geometry can be related to locally measurable properties so that the resulting description is accurate and useful in a practical sense, (but "how does this geometry manifest so that the differential equation is valid?"). On the other hand, in general, energy cannot be related to a set of spectra for "charged and inertial" material systems contained in a metric-space. Though the cornerstone of modern (2012) physical law, is that, energy can be (so) related to a set of spectra for "charged and inertial" material systems contained in a metric-space.

Rather the spectral properties of a quantum system are (can be) related to the spectral properties of an over-all high-dimension containing space, where a high-dimensional over-all containing space can have very significant geometric-spectral properties associated to itself. This can be made mathematically possible by considering a new math context, in a new containing space, with new types of models of material interactions. That is, the high-dimensional structures are macroscopic, yet they remain hidden. However, reconsidering one's own living context may allow one to have a direct experience with the higher-dimensions, so as to allow a new context for measuring, when the patterns one is trying to see (expecting to experience) are properly (accurately) described.

However energy can be related to a set of spectra if the energetic signal exists in a geometric context of [an energy signal which has (graphically geometric)] energy oscillations in time, eg voltage signals in electric circuits. But this is part of the geometric context of (classical) electromagnetism, which, in turn, is contained within the geometric context of electric circuits. In a circuit, geometric measures, eg length of a circuit's loop, can be related to the angular phase of a wave's period, as in the cyclic period of a sine-wave, ie geometric properties of the circuit can be related to the intrinsic phase properties of a voltage signal's wave-function.

However, the loop structure of an electric circuit can also define a hole-structure in a space, eg a space of loops.

In the fact the different differential structures of 4-Euclidean-space on equivalent topological manifolds can be interpreted in terms of different spectral structures on the same topological set, eg $SU(2) \times SU(2)$ can act on (a,b) and (c,d) related to two spectral areas on one of the manifolds, ie a cross product of 3-tori in 4-space, while $SU(2) \times SU(2)$ can act on different, (e,f) and (g,h) again related to different spectral areas on a cross product of 3-tori in 4-space, ie different sized 3-tori, are topologically equivalent, but the metric functions on the two manifolds can be (quite) different, requiring a diffeomorphism which deforms a pair of in-equivalent metric functions from one manifold to the other, so that each separate 3-tori cross product manifolds has a metric-invariant differential structure over each separate tori.

Both classical and quantum descriptions of material depend on the operators of calculus, where the operators of calculus define what measurable properties are. That is, both classical and quantum descriptions of material depend on the derivative and integral operators of calculus, where the derivative operator of calculus defines the nature of measurable properties and the integral can sum over the locally measurable properties defined by the derivative so that the values obtained from integration are related to a quantitative set with a uniform definition of unity (or with a uniform definition of its unit, or measurement is consistent with the domain set).

If the derivative is the derivative of a position function then the integral of the "derivative of the position function" will again be the position function. Thus, the integral is the inverse operator of the derivative. This is an important property which might be used for solving differential equations.

Exploring derivatives

Are derivatives tangents to a function's graph . . . .

> [and thus they are lines in a coordinate space and thus derivatives (or tangent lines) are about the relationship that uniform measuring units (of various [or of two] quantitative types) have to one another],

. . . . or are derivatives mostly about their relationship to function spaces, eg converting position functions to motion functions?

So what are differential equations?

Are differential equations relations between local measurability and geometry (expressed as a formula), or are they about the relation that locally measurable properties (of motion) have to function spaces and subsequently to their Fourier transformed dual function spaces, so that differential equations can be Fourier transformed to (algebraic) polynomial equations?

Are function spaces about (1) probability (harmonic functions) or about (2) solving smooth systems of (differential) equations, that is, are function spaces in the (2)-category so that the function space is being used to describe:

> geometry,
> determining holes in space,
> non-linearity, or
> chaos?

Note: Some smooth systems are naturally harmonic functions eg some electromagnetic wave systems.

Another viewpoint, which compares space-form properties with the differential equations of classical physics:

What are differential equations?

Answer: Differential equations identify the relation that an interaction space-form in 4-dimensions has to 2-dimensional material space-forms in 3-space. Three-dimensional interaction space-forms (can) identify the (natural) coordinate curves on a 3-space-form (in 4-space), which in turn, are tangent to (a point on) the material 2-space-forms in 3-space, so that these tangents are either Euclidean, and thus determine tangents in regard to position and displacement (or momentum), or they are hyperbolic, and thus determine tangents in regard to motion and time (or energy). The geometry of the interaction space-form determines the geometry of the local tangents, which in turn, determines the locally measurable properties of position and motion. Furthermore, the geometry of the interaction space-form (also) determines the geometry of the geometric side of a physical system's classical differential equation.

The tangent properties of the interaction space-form change with each spin rotation-of-state [rotation (time)] period, so that both the subsequent displacements in space and the changes in motion are determined by local coordinate transformations caused by metric-invariant transformation-groups, and subsequently these coordinate transformations are determined by the geometry of the local coordinate transformations, ie the geometry of the transformation group (space). That is, the geometry of the local coordinate transformations is determined by the shape of the transformation-group space, so that this (coordinate transformation space) geometry is transforming local coordinate properties (of the space-forms) in relation to the tangents of the natural curves (or 1-faces) of the interaction space-form at the sites of the material space-forms (where the material space-forms compose the faces of the interaction space-form), so that these local coordinate transformations identify a global geometry on the coordinate space which corresponds to the geometry of the transformation group space.

Comparing the properties of space-forms in a multi-dimensional context with the spectral properties of quantum systems

In the new descriptive language . . . . of discrete isometry subgroups in a multi-dimensional multi-metric-function signature context . . . . quantum physics is the spectral relation (or resonance) of a space-form (newly formed from an interaction space-form) to its full multi-dimensional (space-form structured) containing space.

Within this spectral context the material and the interaction space-forms at a particular dimensional level can (or must) resonate with the spectral structure of its containing set . . . , a containing set whose structure is that of higher dimensional space-forms (that is, space-forms which have stable definitive spectral properties) . . . , and if an interacting system is within certain energy ranges then the resulting interaction can settle into a new stable space-form structure, which is in resonance with its containing set (or containing space composed of higher dimensional space-forms).

How is material to be modeled?

Is material essentially a hole in space, a single point, which has internal properties (as in particle physics) or is material a space-form in which its particle-like properties are caused by the distinguished point on the space-form, the distinguished point where all the vertices of the space-form's "cubical" simplex all come together, ie the particle-like properties observed are really the (observed) property of the distinguished point of a material space-form. Furthermore, the spectral properties of a material space-form are related by resonance to the higher dimensional spectral structure of higher dimensional containing metric-spaces which also are modeled as space-forms (which have spectral properties).

Holes in space compared to discrete isometry subgroups

Not only do discrete isometry subgroups identify holes in space but their quotient topologies surround these holes with separable, or locally "cubical" sets of coordinates.

Is it the holes in space which are fundamental (to geometric properties of space), or is it the locally "cubical" coordinates which are fundamental (to physical description)?

Do the sets of locally "cubical" coordinates (or space-forms) identify (especially in hyperbolic space) the set of stable definitive "material" (or space-form) systems which can exist? Answer: Yes (in hyperbolic space).

Are the holes in space . . . , [and simultaneously the separable coordinates defined by discrete isometry subgroups (or equivalently space-forms) which determine the holes in the space], ... actually the discrete isometry subgroup structures of metric invariant spaces (or coordinate principle fiber bundles) which determine (or define) material?

Answer: Yes.

By "material," it is meant, a space-form contained in an adjacent-dimension, higher dimensional metric-space, where the metric-space may, itself, be a (spectral) flow (on a space-form) or a space-form or both flows and (at the same time) space-forms.

However, due to a theorem about discrete hyperbolic reflective groups, eg discrete isometry subgroups of hyperbolic space, by D Coxeter there are no discrete hyperbolic reflective groups, ie no space-forms, in hyperbolic dimension 11, and above. Thus the 10-dimensional hyperbolic space-forms are the highest dimension hyperbolic space-forms and they are contained in an 11-dimensional hyperbolic metric-space.

Space-forms and classical physics

Are the two ideas ... that either

(1) local measurability of motion be set equal to a geometry (of the same quantitative type),
or
(2) local measurability of position (squared) added to a geometric potential energy term so as to identify the value of a system's energy,

.... fundamental in identifying the properties of physical existence which can be described, measured, and used?

Note: The above two relationships between local measurability and geometry define how the differential equations of classical physics are determined.

If "local measurability of certain properties are related by equality to geometric properties" is fundamental to physical description, then these fundamental tangent and geometric properties can be modeled as sets of material space-forms and (higher dimensional) interaction space-forms, thus space-forms also become fundamental to physical description.

That is, by means of the structures of, (1) tangents to material space-forms related to the natural interaction space-form geometry (where the interaction space-forms are contained in the next higher dimension metric-space than the dimension of the metric-space which contains the material space-forms), and (2) material dynamics of the interacting material systems also being related to the geometric properties of the interaction space-forms, it becomes clear that physical description based on space-forms becomes equivalent to the descriptive structure of classical physics.

The prevalence of space-forms in all aspects of the physical description of material (space-form) systems also implies that the natural coordinates, for describing an interacting material (or space-form) system, fit into a locally "cubical" coordinate system (or separable coordinate system) defined by space-forms, so that the space-form geometry of any system is "universal" (for the system), ie is always applicable to the description of a material, interacting system.

These universal space-form coordinate geometries can be associated to both geometric and local measurability properties of either position or motion, (as mentioned above).

Space-forms and quantum systems

Both classical and quantum physics, are about finding boundary conditions in relation to nested sets of containing space-forms, of different dimensions, so that the boundary conditions (or initial conditions of a physical system) are related to physical constants as well as related to both the spectra (or stable, definitive orbital properties) of the higher dimensional containing set, and the (energetic) dynamics of the physical system, where the dynamics of material (at a particular dimensional level) depends on the material geometry in a (that) particular dimension metric-space.

Some questions dealing with separable coordinates:

Is this dependence, of physical description on space-form, separable geometries, analogous to a parallelize-able vector field (on a manifold [which is a base space] in a principle fiber bundle) with respect to a Lie group of local coordinate transformations. That is, for separable coordinates one is always able to diagonalize the group transformation matrices of the tangent vectors of the (natural) coordinate curves of the system at each point in the containing coordinate space, eg a parallel vector field in a Lie group, so as to generate the curved separable coordinates. Are these parallel coordinates always associated to the maximal torus of the coordinate transformation Lie group, through the locally diagonalizable coordinate transformation matrices? Are the coordinate transformation groups of material systems always metric invariant transformations?

Which is more fundamental:

(1) Discrete isometry subgroups
(2) parallel vector fields on Lie groups, eg isometry groups,
(3) the properties of a maximal torus of a Lie (fiber) group, eg a unitary group, and their effects on:

   (a) the properties of metric invariant local transformation properties of the system's (natural) coordinates
   (b) the spectral properties on one particular dimensional level, due to the space-form spectral properties of higher dimensional levels?

Space-forms describe both stable, definitive quantum properties, as well as the linear, separable properties of highly controllable classical systems.

Measurements belong to metric-invariant metric-spaces, and material properties associated to particular material (or metric function) types also belong to particular metric-invariant metric-spaces, within which discrete isometry subgroups (or space-forms) identify material, as well as material interactions, though the structure of an interaction space-form exists in the next higher dimension metric-space than the dimension of the metric-space within which the material exists, whereas quantum systems are the material space-forms, within the given dimensional metric-space, whose spectral properties resonate with the spectral properties of the space-form structure of the higher dimensional containing set.

[separable geometry and stable definiteness]

* A separable geometry might be the only mathematical structure which can allow (or carry) the stable definitive spectral and orbital properties of material systems of all ranges in size. There is no other model of quantum systems which implies their stable, definitive spectral-orbital properties.

Note: the currently accepted function space techniques of quantum physics cannot be used to determine the actual spectral properties which are observed for quantum systems.

The shape and properties of:

(1) space,
(2) material, and
(3) material interactions,

are determined by space-forms (as well as the global geometry of local dynamic coordinate transformation groups) defined on metric invariant metric-spaces, and these properties are consistent with the above classical relations between local measurability and geometry as well as consistent with the stable definitive spectral properties of space-forms whose spectra is related to the spectral properties of the higher dimensional containing space (of the material space-form system).

That is, the distinction between the diffeomorphism groups and the isometry and unitary groups in regard to (coordinate) principle fiber bundles can only be compared in relation to higher dimensional immersions of (a lower dimensional) curved geometry into (higher dimensional) metric invariant spaces (a general curved n-manifold is contained in a 2n-dimensional metric invariant space) and in this context there exist new spectral properties in the higher dimensional metric-invariant context, which might account for metric-function deformations related to the diffeomorphism properties of the n-manfold.

Do holes in space, (or more directly) do space-forms, identify material with both geometric and spectral and dynamic interaction properties (or structures)?
Answer: Yes.

Diffeomorphism does not identify material, and it seems to only be related to the "shape of space" in the one case of spherical symmetry in Euclidean 3-space, and this spherical symmetry results from the geometry of the group SO(3), the fiber (isometry) group of Euclidean 3-space, which affects the geometric inertial (or dynamic) properties of (material) space-form interactions in 3-space.

It might be that the metric invariant spectral diversity at a particular dimensional level, which is allowed in a multi-dimensional context, is the property which defines the allowed sets (or classes) of diffeomorphism types of geometries (or manifolds, ie differential structures [which are related to fiber groups]) which are distinct from C^0 [or topological] types of geometries.

[Note: A function space model of a quantum system along with the sets of operators related . . . , by analogy with classical descriptive physical properties . . . , through geometry and dynamic (or momentum) operators to measurable properties of quantum systems (finding a quantum system's complete set of commuting operators) so as to determine the quantum system's spectral set, does not work within the process of finding a general quantum system's (or the function space's) spectral set, where the complete set of commuting operators act on the function space so as to "diagonalize" the function space (or to find one of the spectral "set of functions" for the function space of a particular quantum system, ie this process does not work for a general quantum system)]

Because quantum physics, general relativity, particle physics, non-linearity, and string-theory etc do not realize the (descriptive) properties of wide applicability and great usefulness (each case in which these descriptions are useful is always a special case, not related to general principles (of the very few cases where these descriptions lead to usefulness)) then classical physics is the only known scientific truth, since it is both widely applicable and greatly useful, though it has known limitations.

Appendix

This paper (and all the others) can be summed-up in a short outline.

There exist stable definitive discrete material systems which exist at all size-scales: nuclei, atoms, molecules, crystals, and solar systems and possibly (orbital) motions of stars in galaxies. These stable discrete spectral-orbital systems do not have adequate descriptions within the context of "what are now (2011) considered to be the laws of physics" (eg there is, currently, no known reason as to why the solar system is stable, nor how a general nuclei comes to have stable spectral properties).

The existence of these stable systems implies that measuring makes sense in regard to these stable definitive systems which exist at all size-scales, and that they can be quantitatively described in a stable, measurable descriptive language.

However, one can only measure in a context of stable structures:

> stable systems,
> stable geometries,
> stable and consistent quantitative structures, and which are not too complicated so as to hide inconsistencies,
> within sets which are not too-big, and (also) not too-complicated, so that the set properties of descriptions are both consistent and comprehensible, and
> descriptions are to be placed in stable contexts where the values one measures are stable and consistent, so that measurable values fit into a description which is also stable and practically useful, and which actually accurately fits data (such as describing the stable discrete spectral structure of a general nucleus), and which provides information which, in turn, can be used in a practical manner because both the system and its descriptive structures are stable.

How is stability to be introduced into the descriptive language of math-physics?

Answer: The key is stable geometry, and how to introduce (or fit) stable geometric structures into a quantitatively based descriptive language, is now made clear by the following list of "eight" math properties:

(1) The geometrization theorem of Thurston-Perelman [Note: The Thurston-Perelman theorem is actually unnecessary, one can simply appeal to a need for measuring-stability (linear, separable, metric-invariant) and the stable discrete shapes exist within this stable context.], where Thurston-Perelman (essentially) show that the discrete hyperbolic shapes and discrete Euclidean shapes . . . , (as well as the sphere, but the sphere is non-linear, and is stable, in regard to a system which has only one spherically symmetric body composing the system) . . . , are the central shapes in regard to classifying geometry, in its general topological context, and all the other shapes (at least in 3-space) are unstable and they either disintegrate or evolve towards the stable shapes.

(2) The knowledge that "the most stable shapes" are the discrete hyperbolic shapes, and that the stability of the discrete hyperbolic shapes are robust and their properties (of shape) are unaffected by their (it) being near other geometries, whereas in the spherical-case, one sphere is a stable, while two spheres forms an indescribable (or unresolved-able) geometric structure.

(3) Coxeter showed what the fundamental properties of the discrete hyperbolic shapes are distinguished by: dimension, genus (number of holes in the shape), and stable geometric

measures, which are measures on the faces of the discrete hyperbolic shapes' fundamental domains (ie polyhedron, or cubical-simplex structures [the square on a checker-board is a fundamental domain of a discrete Euclidean shape]), [the stable geometric measures on discrete hyperbolic shapes are the basis for stable spectral-orbital properties which are associated to the discrete hyperbolic shapes], and the discrete hyperbolic shapes of each dimensional level can include infinite-extent discrete hyperbolic shapes (so that these shapes have finite "volume" measures. *) associated to themselves, ie these infinite extent shapes model light, and perhaps neutrinos.

(4) Furthermore, there are many simultaneous (but different types of) metric-spaces involved in physical description, where each metric-space (identified by dimension and signature) is associated to a different fundamental physical property. These fundamental physical properties and their associated metric-spaces (identified by dimension and metric-function signature) are distinguished by the elementary displacement symmetries of E Noether, which are (also) related to conservation properties. Eg inertia-spatial-displacement is a property of Euclidean space and charge-energy are primarily properties of hyperbolic space (hyperbolic space is equivalent to space-time, it can be thought of as the velocity space of space-time)

(5) Opposite metric-space states can be defined on this structure. In turn, opposite metric-space states is the structure upon which "the spin-rotation of opposite metric-space states" can be defined. Furthermore, opposite metric-space states can be defined in complex-coordinates, and this leads to containment within complex-coordinates, ie "the real state" and its "pure-imaginary, opposite-state," which subsequently leads to unitary fiber groups (extending the description beyond the usual (real) rotation [or isometry] fiber groups).

(6) The structure is many-dimensional, since it is assumed that the structure of material is based on the same idea as the structure of a metric-space, namely, they (both material and metric-spaces) are both based on discrete hyperbolic shapes, and the discrete material shapes are contained within (adjacent higher-dimension) metric-spaces.

The discrete shapes in other (non-hyperbolic) metric-spaces are involved in dynamic processes, and the subsequent spatial displacements of these dynamic processes, where (both) opposite metric-space states ((7) as well as the fiber groups, associated to local coordinate transformations of the interaction geometries) are also involved in dynamic processes.

(7) the math containment set is a (set of) principle fiber bundles (where a principle fiber bundle has a base space, wherein a system would be contained, with its properties distributed between the different types of metric-spaces, and associated to each point of the base-space is an appropriate, allowable (isometry or unitary), local coordinate transformation group)

The Mathematical Structure of Stable Physical Systems

The base-space is composed on the many different dimensions, and many different types of metric-spaces up to hyperbolic dimension-11 (12-space-time), including coordinate base-spaces which have complex coordinate-systems, and the fiber groups are the set of classical groups which are associated to these different (real, isometric) metric-spaces as well as Hermitian spaces (with their complex-coordinates and unitary fiber groups).

(8) Material has a shape, and it also has a distinguished point (so interactions appear to be point-like), and metric-spaces also have stable shapes, so the coordinate metric-spaces now (also) possess spectral properties. That is, the containing space has spectral properties associated to itself.

Note: Stable systems can emerge out of interactions (if the energy of the interaction is within a proper range of values), and this emergence of stable systems out of material interactions would be due to the existence of resonances between the new system and the spectra of the (high-dimension) containing space.

Note: Quantum randomness can be derived from the structure of material interactions within the new descriptive language, but the basis for this "new" descriptive language is stable geometry.

Chapter 21

INTERACTIONS

The electric and gravitational fields of classical physics are expressing the geometric properties of an interaction space-form (in R^4) acting on material (space-forms) which are in R^3, and thus the geometry of the interaction has a spherically symmetric structure, so that one of the three tangent directions (of an interaction 3-dimensional Euclidean space-form upon which the structure of the material interaction depends) in R^4 becomes a radial direction in R^3.

A deformation process (in the new descriptive structure) so as to either result in decay or for the unstable system to deform into stable structures can take place by means of a non-local, discrete partition of the interaction-dynamic process into linear separable (Euclidean) geometries which (in regard to time) are periodically deformed in a process involving:

(1) spin-rotation of metric-space states, which determines time intervals between discrete processes,
(2) small time-interval periods (associated to the spin-rotation of metric-space states) and
(3) spin-group local transformations of either material-space-form positions or spatial space-form shapes (the spatial component [or face] of an interaction space-form),
(4) where the geometry of the dynamic structure is related to the 2-forms defined on the interaction space-form,
(5) where the dimension of the 2-forms is equal to the dimension of the spin-group (thus the 2-forms and the spin-group's local tangent structure can be geometrically inter-related).

As the dynamical process continues at some point the interaction space-form (during a "spin-rotation of state" time period) will begin to resonate with the spectra of the over-all containing space. If the energy values are "correct" then the deforming interaction space-form can enter into a stable space-form state, or if the energy properties are not correct then there will be the usual dynamics, eg it will be the system of colliding material space-forms which stay separate, where the collision can either let the interacting space-forms remain stable, or the collision can cause the interacting material space-forms (one or the other, or both) to break apart (depending on the energy of the collision). This is consistent with classical physics and it actually describes a quantum interaction process which determine stable quantum systems (so as to not be full of huge logical gaps in the descriptive structure, eg no global wave-functions yet there exist non-local processes, and such as divergences of the solution to the radial equation of the H-atom etc).

Note: The point-like properties of observed random particle-spectral quantum measured properties (interactions) are related to the distinguished point of the material space-forms.

Note: This model of geometric interactions can be related, at the atomic level, "to random behavior" (Brownian motion) from which the idea (or property) of quantum randomness emerges, and both the interaction structure and the physical constants (ie relative sizes of interacting space-forms of a given dimensional level) cause the interaction structure to (appear to) make the adjacent dimensional levels to be geometrically independent of one another.

Interactions are described using discrete separable shapes (space-forms) defined over different dimensional levels

In regard to interactions of material space-forms, the hyperbolic space-forms model the material, while the inertial interaction space-forms are the Euclidean spatial space-forms . . . , which forms a component of an interaction which define action-at-a-distance . . . , where one would expect action-at-a-distance to be related to an infinite-extent hyperbolic space-form, where and infinite-extent hyperbolic space-form is a space-form which represents (or models) light or (possibly) a neutrino. However, this relation would seem to be about spatial (Euclidean) action-at-a-distance and (hyperbolic) temporal states which reach to infinity.

However, hyperbolic space [the (traditional) interaction field the electromagnetic field as well as the space of energy] a light-signal requires a finite transition in time, due to an assumed bounded geometry of charged space-forms and a finite propagation speed of light in the bounded geometric context, where wave-propagation (electromagnetic waves transport momentum) occurs along the infinite extent hyperbolic space-form in the (hyperbolic) interaction context (moving the field between positions in space [where position is best modeled in Euclidean space]), subsequently, the force-field (carried by light, in the bounded geometric context of charge distribution and motions), leads to inertial changes of the interacting material system. In the case of electromagnetism, or the "hyperbolic space-form" case, the bounded hyperbolic space-forms (associated to stable charged systems) are also associated to resonating Euclidean space-forms, as well as a spatial interaction component, thus leading into the simultaneous inertial case (action-at-a-distance), where there is both a transitioning field through space and a spatial space-form component (which has the property of an instantaneous action at a distance) of the (charged) material interaction. That is inertia is ridged while electromagnetic fields have both advanced and retarded dynamic states, represented as opposite time-states in hyperbolic space, ie hyperbolic space is given a complex-coordinate structure which is associated to SU(3,1), which rotates the opposite time states so that at either (+t) or (-t) states in a complex coordinate representation the two time states are exactly opposite to one another. Where this becomes an action-at-a-distance structure.

To re-iterate,
In the inertial case, there is (instantaneous) action at a distance, as a property of inertial interactions. Thus, in the electromagnetism case these inertial changes (of charged systems) are about describing the mixtures of states of opposite time states (metric-space states in hyperbolic space) related to inertial changes which have an action at a distance property. Thus there is both an action at a distance spatial displacement component of the description of the material interaction and a field component of the description but the force-field determines the inertial changes thus there is a need for a mixture of time states in regard to the model of the force-field and this can be modeled either by a spatial component on the next higher dimension metric-space, or by a time component in 3-space, ie the differential forms of the electromagnetic field.

That is, in the new descriptive structure of space-forms (or equivalently differential-forms) defined on various higher dimensional metric-spaces, have geometric shapes so that opposite states of time are related to opposite geometric flows on a space-form (or on an interaction space-form).

Note: These mixtures of opposite metric-space (time) states (in hyperbolic space) is analogous to (but better related to geometry, than) Feynman (particle-collision) diagrams, which is also a mixture of "states of time," (or mixtures of the advanced and retarded electromagnetic fields, which emerge directly from electromagnetic field equations) but the mixtures of time-states in the Feynman diagrams are not so well defined in geometry, (but) they are associated to particle-types, in the (so called) particle-collision model of quantum interactions, and thus they are associated to irreducible group representations of the Poincare group, which (is a local coordinate transformation group which) improperly mixes together space-time (or hyperbolic space) and Euclidean spatial displacements, where these irreducible group representations . . .

> (of space-time and Euclidean space, ie mixtures of time and dynamic states [or dynamic displacements in Euclidean space])

. . . . are put together to build bounded Feynman diagrams of particle collisions (an unlikely model of existence properties when one is within the context of a stable system), so as to construct the (bounded) geometry of point-particle collisions, but they provide little clarification about time, since

(1) they are being used to build mixtures of both time states and spatial positions to create an unrealistic model of a particle-collision, the basis for the models of quantum interactions (such mixtures of time states might be better used to model spatial displacements of charge in the more meaningful geometric context of electromagnetic differential-forms), and
(2) the global linear wave-function of the quantum system, which is to be perturbed by the non-linear quantum-collision-interactions, is defined on a Euclidean domain space, yet (on the other hand) quantum electrodynamics (or other models of quantum particle-collision interactions of light particles) are described in a metric-invariant space-time context, ie hyperbolic space. That is, metric-invariant spaces which are also spaces of constant curvature are mixed in an inconsistent manner (where, even the Dirac equation must be squared in order to be solved, thus such solutions still separate spatial properties from time properties). Furthermore, the particle-collision has a non-linear unitary connection (thus it is a math model which is inconsistent with the metric-invariant structures of both space-time and Euclidean space).

On the other hand, the new descriptive structure of space-forms (or equivalently differential-forms) defined on various higher dimensional metric-spaces, have geometric shapes so that opposite states of time are related to opposite geometric flows on a space-form (or on an interaction space-form).

1. Review of the descriptions of material interactions based on "spin-group" geometric structures, in relation to the geometry of interaction space-forms, on both a many-dimensional and discrete subgroup structures for the base space of a principle fiber bundle whose fiber group is the spin-group.

Structure of material interactions

The principle fiber bundle has isometry (or spin), or unitary, fiber groups . . . , and a base space of both high and low dimensional metric-invariant metric-spaces of various dimensions and of various metric-function signatures, so that the lower dimensional metric-spaces are defined as faces of the "cubical" simplex "fundamental domains" of the high dimensional discrete isometry (or unitary) subgroups (or equivalently, defined as the spectral flows of high dimensional space-forms), where it should be noted that the hyperbolic space-forms have very stable spectral properties and their geometries are characterized by their genus (or the number of holes in the geometric space-form shape) so that these geometries can be either bounded or unbounded (but with finite volume) and (according to a theorem of Coxeter's . . . , the last of these space-forms exist in hyperbolic dimension-10, thus the highest dimensional hyperbolic space of hyperbolic dimension-11 has no spectral structure associated to itself.

For material space-forms contained in a particular dimensional-level metric-space . . . .

> (where the material space-forms and their containing metric-spaces have the same "type of signature," eg hyperbolic space has the subspace structure of R(n,1) where R is the real numbers (or rational numbers), n is the spatial subspace dimension, and 1 is the dimension of the time subspace . . . , together, for various values of n, these identify a set of metric-spaces which have the same "type of signature," ie a set of hyperbolic metric-spaces [or set of spectral-flows on hyperbolic space-forms])

. . . . there is also defined an interaction space-form whose dimension is the same as the material-containing metric-space, but which is contained, as a geometric shape, in the adjacent, next-higher, dimension metric-space.

The interaction space-form has a set of material faces, and a (set of) spatial-faces whose dimensions are the same as the dimension of the material faces. The spatial-face manifests as a space-form shape (or as a spectral flow within the interaction space-form) in the metric-space which contains the material space-forms.

The spatial displacements of a material interaction which characterize changes in motion (or which characterize inertial properties) take place in Euclidean metric-spaces, ie R(n,0) spaces, ie metric-spaces with no time subspaces.

** The force fields are defined in relation to either the exterior derivative of a 1-differential-forms or as 2-differential-forms, which in turn, are defined on the interaction space-form so that the dynamics which these force fields identify is (are) determined by the geometric relation that the 1-forms and 2-forms have to the spherical geometry which is defined in the fiber spin group. This "interaction space-form"—"fiber spin group" geometric relation identifies a geometric property between the spin-group's 2-dimensional and 3-dimensional spherical geometric properties (which can be identified in the fiber spin-group) and the 1-forms and 2-forms defined on the interaction space-form, and (but) the spin group is (of course) a group whose elements identify local coordinate transformations which act on the spatial face of the interaction space-form, which is a face which is one-dimension less than the dimension of the interaction space-form (or one-dimension less than the dimension of the [material containing] metric-space). Thus, if a particular direction is identified on a sphere (in the fiber spin-group) by a differential-form, then this direction identifies the direction in which the local coordinate transformations are defined by the elements in the spin group whose local coordinate transformations transform the coordinates in the base space in that well identified spatial (or Euclidean metric-space) direction.

That is, if a particular direction on the fiber group's geometry can be identified, then this, in turn, identifies a geometric direction of local coordinate transformations for the interacting materials in relation to the coordinates of the base space within which the material is contained. Namely, the coordinates which are transformed are the coordinates associated to the spatial-face of the interaction space-form, where the spatial-face determines the spatial relation that the material components of the interaction have to one another. Such a local coordinate transformation will change the geometric properties of the spatial part of the interaction space-form, thus changing the geometric relation (or changing the properties of separation) that the interacting materials have to one another in their containing metric-spaces (ie both Euclidean and hyperbolic, and in higher dimensions the higher oscillating-charge containing metric-spaces, which are characterized by the signature of their metric-function and the dimension [of their oscillating-charged-material containing] metric-space).

That is, a force-field is a Lie-algebra valued 2-form (though occasionally a 1-form) defined on the spatial face of the interaction space-form over a spin fiber-group.

This discrete process of change (of the spatial part of the interaction space-form) is related to the time period of a spin-rotation of metric-space states, in regard to the property of time (which is defined on hyperbolic metric-spaces, where the two states of time are forward moving time and backward moving time [the mixture of time states, which characterizes local coordinate transformations, are defined on complex coordinates; related to unitary fiber groups]).

The issue of higher dimensional time-subspaces, other than the one-dimensional time-subspaces which are related to hyperbolic space. These different time dimensions can either be related to the motion of a particular material-type, eg oscillating-charges of particular (but different) dimensions, or related to a velocity space similar to hyperbolic space, and thus in this hyperbolic-space, time can have a space-form structure, and this geometric structure can be related to directions of spatial displacements. This may, or may not, be consistent with identifying time-directions with the changes in motion of particular-types of oscillating-charges. Because time is related to energy, the relation that time might have to material displacement (different from constant motion, relative to some "near-by" center of mass) can be affected by the (relative) energy of an oscillating-charge.

** The dynamic "spatial displacement" of the spatial-face of the interaction space-form, is defined by local coordinate transformations of the spin-group which act locally on the metric-space coordinates. The directions of the local coordinate transformations is being determined in relation to how the geometry of the 1-forms and 2-forms . . . , which are defined on the interaction space-form . . . , are related to the 2-sphere and 3-sphere geometric structures of the spin group, respectively, so that the action of the local coordinate transformations will be on the spatial face of the interaction, ie down in the metric-space which contains the material which is interacting.

** Note: The local (or tangent) geometry of a spin group is analogous to the local geometry of the differential-forms, which are defined on the base space over (upon) which the spin group acts. This is because the spin group is a subset of the Clifford algebra, where the Clifford algebra has the same dimension as the set of differential-forms defined on the same metric-space over which the Clifford algebra is also defined.

The above construction solidifies the idea of favoring metric-invariance, since this identifies dynamics . . . .

> not in regard to a general-metric (and the associated non-linearity of a general metric) and the subsequent relation that the metric-function has to determining geodesics on a manifold,

. . . . but rather with local coordinate transformations related to isometry (or unitary) groups which identify spatial displacements in an interacting system (in which the two time-states "of a metric-invariant metric-space" are mixed).

2. What is spin?

or

What properties of spin groups are (most) useful in regard to physical description?

Spin-(n), is a low-dimension representation of a simply-connected covering group of the Euclidean isometry groups of dimension-n,

or

Is it the square-root of the Laplacian operator defined on each metric-space, ie the Dirac operator, which is related to both the spin-½ properties of components of quantum systems and to the positive and negative energies of these components, ie matter and anti-matter?

R Penrose defines "twisters" as geometric structures which are related to fiber groups which are spin-groups, in order to define coordinate properties (or coordinate shapes) in general relativity, but this is suspicious, since general relativity is defined with diffeomorphism fiber groups. Penrose's idea seems to be a "change of variables" which is (usually) only useful if it identifies separable coordinates. That is, in the only problem associated to general relativity, "the 1-body problem in spherical symmetry" the spherical geometric decomposition of the spin groups might help in the context of spherical symmetry, ie the sphere is identified with a separable coordinate geometry.

So

(A) How do spin groups relate to existence?

Is it a math property which is correlated with spin-½ properties (and the fairly rare, matter and anti-matter states) of components in statistical systems, or of components in quantum systems which have spherical symmetry?

Or

Is it a math property which is related to spherical symmetry in the very limited descriptive range of general relativity?

Answer to (A): It seems that the strongest relation of spin groups to existence is in relation to both the descriptions of force fields in material interactions, and the spin-rotation of pairs of metric-space states.

And

(B) Why are linear, separable differential equations defined on metric-invariant coordinate spaces so useful?

And

(C) Why is the derivative, and their subsequent differential equations, so useful, ie what causes the differential equation to be so effective in identifying (or describing) physical properties in the context of classical physics?

In regard to today (2010), question (C) is given two contexts:
Today the derivative is placed into two contexts:

(1) that of classical physics where the derivative models a local, linear, change of scale,

and

(2) that of quantum physics where the derivative is an operator which represents the classically measurable property of motion, and it (along with other operators, such as

the 1/r potential energy function, which represent classically measurable properties) act on a function-space which is Hermitian-invariant to unitary operators which represent classically measurable properties, such as the energy operator, ie Schrodinger's wave-equation, so that subsequently the "quantum interaction," related to the 1/r potential energy, is modeled, in an incoherent manner, by extending the idea of unitary invariance to both . . . particle-collisions modeled as unitary particle-state transformations at a point (to model the interaction), ... and Poincare (spin-)representations of these unitary particle-state transformations on the function space of the quantum system, then renormalizing, and then extending the values of the (global) functions . . . , which had been collapsed to the collision-point . . . , back out to infinity so as to reform the function-space of the global wave-function of a quantum system in its Poincare representation, which assumes the spherical symmetry of both the quantum system and the particle-collision interactions, which in turn, assumes the collision point as being a singularity point of a 1/r energy expression of the particle-collision interaction.

Essentially, the structure of (2) is not been sufficient so as to be able to solve for the radial equation in quantum systems which are assumed to have spherical (or Lorentz) symmetry (usually associated to the observed bounded-ness of the quantum system), ie the principle-energy value of general quantum systems cannot be found. Furthermore, the quantum description is an incoherent descriptive language, as well as being without practical value.

The precision proclaimed (for two or three systems) as proof of the validity of both quantum physics and particle physics . . . , in reality cannot be considered to be anything other than an unbelievable epicycle structure used to fit a few data points. That is, this descriptive method fits a very limited amount of data.

It is really the great limitations of an incoherent descriptive language which leads one, who is really a scientist . . . , ie one who questions authority and who presents alternative descriptive languages . . . , to <u>not-take</u> the context of (2) as a valid, useful, widely applicable, descriptive language.

Measurable verification is really to be related to the usefulness of the quantitative description, ie how much control and creativity emerges from the description, and not in relation to its (very few) measurable verifications, since based on its measurable descriptions quantum physics has "not gotten the job done," ie most general quantum systems do not have a valid measurable description in relation to fitting the description (the prediction) with the observed measured properties of general quantum systems.

New descriptive language

The point of the new description is that it gives a model for interaction which (1) allows for quantum properties, (2) in 3-dimensional Euclidean space it is in agreement with the properties of spherical symmetry, and (3) it provides the geometric structure upon which the principle quantum number, the fundamental energy of the system, can be realized and (4) this is done in relation to the quantum system being in resonance with the spectral properties of its high dimensional containing space, which has many spectral properties with which any low dimensional quantum system in the containing space can be in resonance, and (5) the stable space-form geometry . . . , into which the correct energy of the quantum system can exist . . . , is given.

Thus, the act of fitting data in the new description, is really an act of determining the geometric properties which are fundamental to the "complete containment set" of existence (or of some existence), and this information is related to a more complete set of information about how different material (or more general) systems can be related to one another, and thus it is a new descriptive language which is related to more control, and more capacity to create, within a new "broader" descriptive context of existence.

(A), (B), and (C), above are three questions which have been given a vastly expanded new set of contexts by the: many dimensional, discrete isometry (or spin) or unitary subgroups, where the descriptions of material interactions are placed in the context of the spin-groups.

The second question (B) is answered by the assumption that the geometry of material and the geometry of interactions are determined by the geometry of space-forms, ie separable, and in the context of metric-invariance. Furthermore, the properties of space-forms can be defined by solutions to linear differential equations defined on differential forms.

Linearity is simply the idea that two measuring scales must be related to one another in a consistent (or in a coherent) way, and that way is defined by linear relations. That is, there is only defined a linearity in regard to scale change.

Thus, non-linearity means that a function's values and the values of its domain space cannot be consistently related to one another. Thus, in a non-linear context, new values discontinuously emerge between the two scales of measurement due to this incompatibility of the two number systems, so as to create chaos for systems defined with non-linear differential equations.

And

The first question (A) is answered above in (section) 1.

Furthermore, (C) is best answered in relation to the relation that (or association of) both the derivative and the differential equation have to interaction space-forms. That is, the derivative has been so useful because there exist discrete geometric structures which determine the properties of material existence, and the tangent structures of these space-forms (associated to differential-forms) determine the useful differential equations for both classical physics and this new more broadly expanded context within which to describe existence and its useful properties.

It is best to assume that particle-collisions are rare occurrence's rather than the basis for fundamental interactions in material systems.

That is, the properties of existence are best placed in a context of continuity for each dimensional level, but in living systems this continuity might be best described in relation to a higher-dimensional oscillating-charge space-forms.

3. A new definition for a manifold

The interaction space-form takes the descriptive structure of the classical system down to the level of the quantum system where it defines Brownian motion.

This Brownian motion is the basis for apparent quantum-randomness of physically observed properties.

However, the interaction space-form in a many-dimensional context, which is characterized by its discrete properties, also re-defines the manifold structure (which has traditionally been associated to geometry), so as to form a new local (topological) structure (ie conservation properties [based on either continuity, or invariance to displacements of the system (or its properties) in space or time] are maintained within any given dimensional level).

In the new local (topological) structure, which is placed within the spin-group context, there can be a description of both interactions (of force-field spatial displacements) and the formation of stable quantum systems (associated to resonances and if the interaction is in an "allowable" energy range, within the many-dimensional and discrete structure). The local linear structure (associated to local geometric measures, ie associated to differential forms) of a traditional manifold is placed in the new context of interacting space-forms which span dimensional levels.

Furthermore, the energy-structures of oscillating-charges of a given dimensional level (ie associated to lower-dimensional material systems which the oscillating-charge contains) can be temporary structures, due to a constant energy value (for an oscillating-charge) being re-distributed (or allowed to flow) between the oscillating space-form's faces, so that within, say, living systems, both the dynamics and the geometry of the molecular systems, which are affected by the structure and energy of higher-dimensional oscillating-charges, can change from time to time (including molecular shape).

During such an energy re-arrangement, energy is conserved on a higher-dimension oscillating space-form structure (which contains the [living] system which is being described), so that energy can be transferred from one face (or from one spectral-flow) to another, due to resonances and/

or pushed energy flows in relation to a higher-dimensional oscillating-charge space-form (where a given energy is allowed to flow between different faces of the oscillating-charge of some given dimension, eg quasi-spatial faces of an oscillating-charge can change so as to change properties of energy resonance within the oscillating space-form, or the energy flow can be related to a time period defined by the spin-rotation of a time-state at a particular dimensional level, for a particular oscillating-charge space-form (which is associated to its own independent time-dimension), and of a particular metric-function signature).

Note: Professional science is about identifying an "in crowd" of "top scientists" who work on the projects which are of most interest to the owners of society, the owners want to control this knowledge so as to limit any competition, so what these "scientists" (or engineers) do is treated in the media as that which defines "legitimate" science, so that the outsiders, those not working on engineering projects, or those not in academia (academics are funded based on the projects upon which they work), do science which is amateurish and it does not measure-up to the "science" of the "master scientists" of real science's "in crowd," which is defined by salaries and access to equipment. That is, only people who are capable of understanding "science," as defined by the "in crowd," are let into the "inner circle" of "true science."

This is the same as a small clique of people defining an "in crowd," as occurs in high schools.

Indeed it is an organization of "science" which depends on an immature and arrogant behavior which opposes both reflection and careful analysis of the descriptive language which "scientists" use. The "scientists" of the "inner circle" have become "too important" for outside criticism, so that they maintain their descriptive structure in a fixed language, but the observed properties and the language which they use have become inconsistent. That is, the descriptive languages of: quantum physics, particle physics, general relativity, string theory, quantum gravity, are all languages which have never been widely applicable, and they have only been applicable to a very limited number of uses (if they have any relation to measurement at all, remember measurement is not about verification [Ptolemy's system was verified], rather it is about coupling to the system so as to use the properties of the system for some further creative uses). Defining science as an "in crowd" which can only use language in an orthodoxed manner results in a professional discussion about irrelevant illusions. It is all about making "knowledge" appear to be unobtainable, except for those few who can prove they belong within the "inner circle," and it is this definition of "science" as identifying social class which is now the true property which defines "science" today.

(see I M Benn and R W Tucker, in, An Introduction to Spinors and Geometry with Applications in Physics, 1987, Chapter 2)

Chapter 22

HIGHER DIMENSIONAL INTERACTION SPACE-FORM

The spectra of ... atoms and molecules etc ... can be found from a geometric description of the spectral-geometric properties of space-forms (discrete isometry subgroups) which can be related by a Lie algebra valued differential-form connection relating (or between) metric-spaces of different (adjacent) dimensions.

This is in relation to (or development of) the ideas of, Diagonalizing the function spaces of quantum systems (to find their spectra)

Setting a context

* Metric-spaces are (must be) associated to physical properties, eg properties such as to time (forward and backward) or spatial displacements (translation and rotation). These properties (or metric-space states) come in pairs which are opposite states to one another, and they define opposite frames of reference. Local (group, often a matrix group) transformations of the local structure of metric-space coordinates (or [isometry] group representations) must always account for the rotation between these opposite metric-space states, that is, spin rotation (between these opposite states).

Dynamics depends on small spatial displacements (for small intervals of time) which are dependent on mixtures of (opposite) metric-space states being present at the point of displacement. Different spectral flows on a space-form represent metric-spaces. These metric-space spectral-flows are paired together on the space-form as opposite metric-space-states. A spin rotation, in fact, rotates between these opposite metric-space states. This spin rotation of metric-space states is needed for any dynamic processes of materials that are contained in a metric-space.

Interaction space-forms

Interacting 2-space-forms form into (or naturally create) an interaction 3-space-form. The space between a pair of 2-space-forms can be modeled as another spatial rectangular component in an (hyperbolic approximation of an) interaction 2-space-form's fundamental domain. This new (approximation to a) 2-fundamental domain can be turned into a 3-fundamental domain by (1) first considering the corresponding (or similar) interaction space-form system in the metric-space which represents the opposite state for the (interacting) system, eg in the metric-space state in which time flows backwards (or the rotational frame which is opposite to (its) a translational frame), and (2) second (or then) connecting the corresponding vertices of these two (2-dimensional) fundamental domains (which exist in opposite metric-space states) by (1-dimensional) edges whose lengths are determined by the time period of the spin-rotation of metric-space states multiplied by the speed of light. This "distance" between metric-space states is determined by a Hermitian form (as opposed to being determined by a metric function in a real coordinate space) within the complex coordinate (or local vector) space. This complex coordinate space, which contains the above mentioned space-form, is used to model the mixture of different metric-space states which compose the interaction space-form system [ie a pair of real dynamic physical systems, which exist in opposite metric-space states]. Note: The velocity of each interacting 2-space-form can be multiplied by the time period of the spin-rotation of metric-space states to give the change in spatial displacement between the two interacting space-forms, for any given spin-rotation time period, in each of the different (and independent) real metric-space states (ie real and pure imaginary metric-space subsets of the complex coordinates), so that the direction of spatial displacement is opposite in each of the different metric-space states. This process results in the formation of an interaction 3-space-form which is contained in either a 4-complex space or it is a pair of (real) 3-space-forms each of which exists in one or the other of the opposite-state real 4-metric-spaces (ie projected into these real and pure imaginary subsets of the complex coordinate space), so that the interaction space-form is re-made with different geometric properties (due to spatial displacement) after each "spin-rotation of state" time interval.

That is, it should be noted that with each "spin-rotation time period" the spatial component of the interaction space-form changes its size.

Interaction 3-space-forms are on the boundary of a 3-dimensional metric-space which is modeled as a 3-flow (or 3-spectra) on a 4-space-form. The 3-flow (metric-space) would be the region which determines the stable orbital region of a (our) planet (earth). The interaction 3-space-forms are part of the boundary of the 3-flow metric-space, so only some of these 3-space-forms are stable. Namely, either the ones which are resonant with the 3-flow metric-space spectra, or those resonant with the spectra of the set of 3-faces which together compose the 4-space-form's 3-face "cubical" simplex structure. The sub-simplex elements (representing the interacting material space-forms) of an interaction space-form's "cubical" simplex (or polyhedral) fundamental domain do not participate in dynamics. It should be noted that 3-space-forms which do not identify a boundary of a 3-metric-space can interact and move in a 4-metric-space. That is, such interacting and dynamic 3-space-forms need to be within a 4-flow of a 5-space-form and not on the boundary of a 3-flow. Note: If such boundary dependent 3-space-forms did interact such acceleration would be uniform on their containing 4-flow and thus according to general relativity not be noticeable.

Conjecture: The boundary of fixed 3-space-forms within a 4-space-form might be analogous to neutrally charged hyperbolic space-forms, eg neutrons and neutrinos.

Thus to solve the space-form system for interacting 2-space-forms, each of which carries the property of charge in the form of currents (which fit into the 1-flow structures) on the interacting hyperbolic 2-space-forms, one must form the above mentioned (described) 3-space-form so that the dynamic energy of interaction does not exceed the binding energy of a (final) stable 3-space-form, eg the binding energy of an electron to an atom. Thus as the two material space-forms (which contain [opposite] charge) approach in a dynamical manner (placed in a center of mass frame) so that one of the 2-space-forms has a positive charge (which is a 1-flow [or current] on the 2-space-form) and the other 2-space-form is negatively charged. At some time (or [geometric] point of spatial separation) in the interaction, the two interacting charged 2-space-forms are spatially separated (in their containing 3-metric-space) by a 2-sub-space-form (or 2-face) which is in resonance with the (or some corresponding) largest allowed spectral or facial structure of the spectral 2-faces of a (or some) stable fundamental domain of a stable 3-space-form (or 3-face, or 3-spectra) which is contained in the particular 4-metric-space (or 4-space-form). That is, the spectra of the set of 2-faces of the interaction space-form, including the spatial 2-face, must be in resonance with the 2-face spectra of the 3-faces of the 4-metric-space (or 4-space-form), or equivalently the 2-spatial-face of an interaction 3-simplex (or interaction 3-space-form) must be in resonance with the 2-faces which form the spectra of the different 3-faces which, in turn, form the boundary of the 4-metric-space's 4-space-form (or simplex) structure. That is, the resonance can be with any of the 2-sub-faces which compose the 3-faces of the 4-fundamental domain, or simplex, which models the containing 4-metric-space (or 4-space-form, or 4-"cubical"-simplex, or 4-polyhedra), this can be contrasted with the 2-faces not being in resonance simply (or only) with the 2-faces of a particular 3-face of the 4-fundamental domain. Note: Our metric space (the metric space within which we are contained [or within which we fix our attention]) is a particular 3-face of the 4-fundamental domain (or a particular 3-flow of a 4-space-form).

The new stable 3-space-form can be determined as a (differential form) solution to the generalized non-homogeneous generalized Laplace equation in 4-hyperbolic space, or equivalently defined on a 5-dimensional space-time metric-space, ie a 5-dimensional space-time metric-space has 4-spatial dimensions and one time dimension.

Note: This could be placed in the context of a containing 5-hyperbolic-metric-space and the (natural) resonance (of the spectra of the 4-material-space-forms) would be determined with any of the 3-faces resonating with any of the 3-faces of the containing metric-space's 5-fundamental-domain, in which case the interaction space-form would be 5-dimensional whose spectra are 4-flows. In this case, when solving for the interaction 5-space-form the generalized Laplace operator would be in a 6-dimensional space-time metric-space.

The interaction 3-space-form (in a 4-metric-space) will be a 3-differential form (or a 3-form), F, so that the system (or space-form) would be defined on a simply connected component of the space-form, ie the space-form's simply connected fundamental domain, so there would be a 2-form G so that d*G=F. Then the (similar Maxwell) equation for the interaction 3-space-form would be dF=j, where j are the charged 2-flows (or 1-flows on 2-space-forms) of the positive and negative charges (on the two separate interacting 2-space-forms) which are dynamically interacting as 2-space-forms (in a 3-metric-space) to form a 3-interaction-space-form in a 4-metric-space. Furthermore, F^j=(force, between the current and the space-form, F). The 2-flows are 2-forms in a 5-space-time metric-space, so (5C2)=10, means that, in a 5-space-time metric-space, the 2-forms are contained within a 10-dimensional vector space (of 2-forms), and the subspace of 3-forms in the space of all differential forms on a 5-dimensional metric-space is also 10-dimensional so the 2-forms and the 3-forms on a 5-metric-space are isomorphic (1-1 and onto) to one another. Thus the 3-form, F, can be set equal to an equivalent 2-form F' and similarly for the 2-form, j, set equal to (or in an exact correspondence to) a 3-form, j', so dF'=j', can be solved for F' which can be made equal to F.

It should also be noted that, dd*G=j, is an equivalent equation to dF=j.

A connection (used in relation to calculus) is a mathematical mechanism which organizes the process of taking a derivative on curvilinear coordinates, which can be equivalent to taking a derivative when going between two different coordinate systems, eg the derivative of a coordinate transformation. Does the coordinate transformation leave the metric invariant (based on rotations and translations) or does it also stretch and/or shrink the (new) coordinate functions? The organization of taking a derivative can be done in terms of partial derivatives (and/or auxiliary [higher dimension] coordinates) or it can be done in terms of Lie algebras of coordinate transformation groups. In both cases the tangent structure to the original space (or the space which contains our experience) is central to what is being described.

However, "What is the purpose of a connection of a coordinate transformation in the context of space-forms, where the coordinate transformation is about a dimensional jump from one simply connected and flat fundamental domain region to another simply connected and flat fundamental domain region of a different dimension (so that each metric function is flat)?" For example, the description of interacting material space-forms takes place in a particular dimension metric-space, but an interaction space-form exists in a metric-space of one higher dimension. (Answer) In this case, the connection is about the spectral properties of the two different space-forms, where each different space-form is in a different dimension metric-space. That is, in this case the connection should be about a transformation between spectral properties of the two flat (but geometric) space-forms in their respective different dimensional metric-spaces. Each of the space-forms can be deformed. In particular, the lower dimensional space-form can be deformed (when it manifests as a stable higher dimensional space-form) based on deforming into the (new) spectral set into which the higher dimensional interaction space-form can be in resonance [or be represented (or can be realized)].

The focus of the problem is changing from the allowed spectral 2-flows (or 2-space-forms) in a particular 3-metric-space to the allowed spectral 2-flows in a particular 4-metric-space (or from a 4-space-time-metric-space to a 5-space-time-metric-space).

To do this (eg in space-time) we can move from the generalized Laplace operator, d^*d, to the Lie algebra valued #-form connection, so that the differential (or exterior derivative), d, becomes a connection, d+A, where A is a Lie algebra valued #-form (where the #-form is the same order form as j, but in a higher-dimension metric-space than the original j).

Now $dd^*G=j$ becomes $(d+A)(d+A)^*G=j$, which is equivalent to

$dd^*G + (dA)G + A\, d^*G + (A^2)G = j$, or
$dd^*G + d(AG) - A\, dG + A\, d^*G + (A^2)G = j$, or $dd^*G + d(A^\wedge G) + A(F-F^*) + (A^2)G = j$, and here A is a Lie algebra valued (differential) form, $A = L^\wedge j'$ . . . .

whose Lie algebra part (or element), L, acts on the space-form F, as well as both of its spectral-flow structures G and j.

. . . . [of the same type and order as the order of j, where j and G are (most often will be) of the same type (ie same order)].

The second term in the last equation, $d(A^\wedge G) = dA^\wedge G + [(-1)^{\wedge *}]A^\wedge dG = dAG + [(-1)^{\wedge *}]A^\wedge F^*$, and $A^\wedge F^* = (A[j^\wedge G]) = (AF^*)$ so $d(A^\wedge G) = d(AG) = (dA)G + [(-1)^{\wedge *}]A\, dG = (dA)G + [(-1)^{\wedge *}]AF$.

If, however, one allows A to be Lie algebra element, L, combined with the differential form j' so as to be $A = L^\wedge j'$, such that $A^\wedge G = L^\wedge j'^\wedge G = L^\wedge F$ then $d(A^\wedge G) = d(L^\wedge j'^\wedge G) = d(L^\wedge F) = (dL)F + L(dF) = (dL)F + Lj'$, and therefore

$dd^*G + (dL)F + A(F-F^*) + (A^2)G = j - Lj'$. (in 5-space-time (or equivalently 4-hyperbolic space) it is OK for G=1-form, and where $A(F-F^*)$=volume change, or an inertial force acting on the two real 3-space-forms in the two metric-space states), where, for example, F and its dual F^* can represent the geometric measures of the (spectral) faces of the corresponding spectral properties, which can change with a change in dimension of the shape, eg 2-dimensional shapes to 3-dimensional shapes [in 5-space], where both spectral properties are measured in the higher-dimensional metric-space [5-space], where the correspondence (when the dimensions change) is a constant of proportion (defined between dimensional levels). {Note: In 5-space the 2-forms and the 3-forms form dual vector-spaces which possess the same dimension, thus interacting 3-shapes allows for the 2-faces of the lower-dimension shape and the 3-faces of the higher-dimension shape to be treated as equal-dimension constructs which can be involved in a transformation of both shape and dimension.}

Thus the space-form F and its integral G, as well as the spectral currents j, are all Lie algebra transformed in the space of faces (or the space of space-forms) of the interaction space-form's containing metric-space. For example, 2-shapes interact as 3-shapes in 4-space and 3-shapes interact as 4-shapes in 5-space, etc. This equation defines an interaction space-form which exists in a metric-space which is one dimension higher than the metric space which contains the actual interacting material 2-space-forms, so that the interaction space-form, and its lower dimensional projections, depends on the spectral context defined by the Lie algebra transformations, which are associated with a proportionality constant defined between dimensional levels.

The Lie algebra transformations can be represented as a set [series (or sum)] of allowed, stable spectra. That is, the faces of fundamental domains with the correct dimensional properties and the allowed, finite-set of spectral-values, or the allowed set of "spectral geometric measures of the faces of the stable shapes." That is, the Lie algebra transformations can be used to determine the spectral set of a particular dimension metric-space within which all the spectral-flows of a stable interaction space-form must be in resonance with that spectral set, however, the spectra in this stable spectral set can (also) be multiplied by conformal factors, ie the conformal factors and the differential forms (or spectral-faces), so that the conformal factors multiplying the elements of the set of differential forms (or space-forms) determine the (complete) spectral set of a metric-space (which is itself a space-form). The dimension of the space of differential forms of a particular order provides the minimal size of a spectral set on a metric-space, where each different (linearly independent) differential form of a particular order can carry a specific (and possibly) different spectral property (or spectral measure). Thus when moving from a metric-space which has a minimal number of [combinations]* (4C2)=6 spectral possibilities of 2-faces for a 3-space-form, to a metric space which has a minimal number of (5C2)=10 spectral possibilities of 2-faces for a 3-space-form, then when projecting down to the lower dimensional metric-space this gives (10C6)=210 minimal number of elements (terms) in the spectral set (or series) of the new allowed stable and bigger spectral set, along with the set of conformal factors, which identify the Lie algebra transformation . . .

(between the different spectral sets which identify the different metric-space spectra of the two different dimensional levels) . . . . involved in the connection which relates the one coordinate (or spectral) set of one dimension to another coordinate (or spectral) set of another dimension.

That is as the dimensions of the system containing metric-space increase this allows a larger spectral set, yet these spectral properties are not dependent on particular dimensional directions, that is each of the new spectral possibilities can be placed into any fixed lower dimensional metric-space, that is a metric-space of a particular dimension can have as wide a variety of space-forms in that metric-space so that these space-forms can carry the new (or different) spectral properties.

The new space-form (in the original metric-space) will depend on its projected down spectra being resonant with the spectral set of the higher dimensional metric space, within which it, F, is contained, so that the spectra of the higher dimensional metric space is identified by the Lie algebra transformations of the Lie algebra valued connection form.

The Lie algebra transformation between the spectra of the two metric-spaces of different dimensions can be represented in terms of the Lie algebra matrix transformations acting on the different dimensional spaces of differential-forms, of a particular order, between each of the two different dimensional metric-spaces.

Given an interacting electron whose current occupied 2-flow has both a 1-flow boundary of about 10^{-5} m (or greater) [this is obtained from H Dehmelt's description of geonium (ie an orbiting isolated, solitary electron)] and its corresponding opposite metric-space state 1-flow is also of this radius, so that this electron on a 2-space-form is interacting with a proton whose current occupied 2-flow (or 2-space-form) has both a 1-flow boundary of about 10^{-15} m and its corresponding opposite metric-space state 1-flow is also of this radius. Another relevant spectral radius of this system is the Bohr radius of 10^{-10} meters, the radius of a 1-flow for an electron in a stable H-atom, where the H-atom is being modeled as a stable 2-space-form.

> The equation $dd*G + (dL)F + A(F-F*) + (A^2)G = j - Lj$,
> (G=1-form, and $A(F-F*)$=volume change [or an inertial force] of the space-form)

can be used to describe the interaction between a 2-form proton and a 2-form electron, where

> the term $A(F-F*) = L^j \wedge (F-F*) = L[j \wedge (F-F*)]$ is the Lie algebra transformed inertial force which acts on the pair of 2-forms so as to deform the two spatial faces of the opposite metric-space states of the interaction space-form after each spin rotation time period, and

the term $(A^2)G = (L^2) \wedge j \wedge (j \wedge G) = (L^2)(j \wedge F) = Lj \wedge LF$ is the new inertial force term between the two interacting 2-space-forms, but this force can also act on the new 3-space-form, and

> the $(dL)F$ term represents the space-form, F, being transformed by the Lie algebra elements in the (spectra of the) new metric-space so as to transform the interaction space-form with its old spatial faces into a new stable space-form so

that the new 3-space-form will be built out a new set of spectra, (after the spin rotation time period) and

the Lj, term represents the current, j, being Lie algebra transformed (after [before] the spin rotation time period) in the new spectral set of the new metric-space.

The Lie algebra transformations in the case of the equation for the H-atom, it will be assumed that the spectral properties of the 3-space-form in the 4-metric-space are the same as the spectral properties of the original interacting 2-space-forms in their containing 3-metric-space. Thus if the kinetic energy and the potential energy of the interacting 2-space-forms is low enough, (ie below the binding energy of an electron to an H-atom), and if the spatial face of the interaction 3-space-form is of the size or diameter of 10^{-10} meters, then the term dealing with the forces on (or changes in size of) the spatial faces of the interaction 3-space-form, ie $L[j\wedge(F-F^*)]$, is no longer relevant because the new 3-space-form now depends on being resonant to the spectral set of the containing 4-metric-space (of the new stable 3-space-form).

The other force term, which acts on the pair of space-forms, $Lj\wedge LF$, now acts to deform the new 3-space-form so that the new space-form can have spectral-flows which are resonant with the spectra of its containing 4-metric-space. It is a term which can also be thought to act in concert with the space-form deformation term, $L[j\wedge(F-F^*)]$, but that would be just as the interaction 2-space-forms are transitioning to a single stable 3-space-form.

The dL(F) and Lj terms are both now changed (or transformed) so as to be the means through which the new system can realize the new spectral range of the 3-space-form's that are contained in the 4-metric-space, (which in the case of the H-atom, is the same as the spectra of the beginning 3-metric-space).

The total energy of the new interaction system is related to both the new spectral range (determined by the 4-metric-space) and the deformation energy of the new space-form, so that if the values of the new and old systems' energies are "OK" (within the correct ranges), then the new 3-space-form deforms (due to the forcing $(A^2)G$ [or $L(j\wedge LF)$] term) into a new stable 3-space-form related to the spectra determined by the dL(F) and Lj terms of the generalized Laplacian. Where the generalized Laplacian was built from the generalized differential connection forms. This is a connection which goes between two flat metric-spaces, which have both different dimensions, and different spectral properties (or different spectral structures).

That is, the different geometric properties (of the different dimensional flat metric spaces) are determined by the spectral properties (or relationships) which exist in the two spaces, but ("new") spectra which are determined by the higher dimensional metric-space can be projected down to the lower dimensional space so as to be stable in the lower dimensional space. This means that the spectra of the space-forms which exist in a metric space can be different (or more varied) than the spectra of the metric-space itself. However, there must exist some spectral element of a space-form in a particular metric space that is still resonant with its containing metric-space's spectra.

Thus the energy of the original pair of interacting 2-space-forms (Kinetic, potential, and space-form formation energy) and the energy of deformation on the (new) stable (interaction) 3-space-form must be considered, as well as the energy of the new spectra of the new stable 3-space-form [and/or its projection as a 2-space-form down into the lower dimensional (or original) metric space which contained the original interacting material 2-space-forms], when considering the stability of a new 3-space-form which developed out of two interacting 2-space-forms, so that this consideration is done in both the (original) containing 3-metric space and the (spectrally relevant) 4-metric-space. That is, Can a new stable space-form be created [or result] from such an interaction?

This means that this particular new H-atom space-form can be solved by solving dF=j so that j has the values of the proton and the electron and the H-atom or Bohr's radius (whose radial values are mentioned above). This gives a space-form model of the H-atom whose 1-flows (or spectral 1-currents) can be used to model the geometric orbits on the space-form which are similar to Bohr's orbits (or more precisely A Somerfeld's elliptic orbits). However, the inertial structure of this stable H-atom also has spherical symmetry, thus the spectral levels of this H-atom (in a 3-metric-space) will have energy levels adjusted by inertial angular momentum properties.

In this particular case of the H-atom the higher dimensional metric space is assumed to have the same spectral properties as the original metric space. However, in most space-form (or quantum) systems this will not be the case, and the other inertial symmetries which can adjust the energy levels of these higher dimensional space-forms will depend on the structure of the metric-space's isometry group, ie an isometry group different from SO(3).

Solving dF=j where F is a 3-space-form (contained in a 4-metric-space) which has 2-faces which are equal to the 2-faces which model the "current," j, (thus thinking of F as a fundamental domain), and because of the properties of differential forms on 5-space-time (where 2-forms are isomorphic to 3-forms) and that integration of differential forms on ("cubical") simplexes must be consistent with the type (or order) of the differential forms, then F for the H-atom is simply determined by the properties of its faces which are j.

Thus the system for the H-atom can be solved by knowing the allowed spectral range of the current j, for the H-atom, and then writing down the answer which must also be consistent with the energy properties of the before and after space-forms, where the before system has kinetic energy properties.

** That is:

A metric space has two boundaries, (1) the lower-dimensional (material) space-forms which the metric space contains (or which exist within the metric-space), and (2) the boundary of the current-flow (or spectral-flow) which identifies a metric-space that exists as a current flow on a higher dimension space-form upon which the metric-space (or spectral-flow structure) exists. In the description of an interaction space-form some of the faces** of the interaction space-form may be identified with the boundary of the metric-space, [** namely, the faces which exist within the subspace defined by the infinite extent light space-forms which identify the subspace within which the material spaces-forms exist], the interaction space-form extends past the boundary of the space-form and thus is spectrally related to a higher dimensional metric-space than the metric space that contains the interacting material space-forms (which compose some of the faces of the interaction space-form). When the interaction space-form is represented as a differential form with respect to being acted on by an exterior derivative which, in turn, can be used to define a differential equation whose solution is the interaction space-form which exists in the higher dimensional metric-space, then some of its faces naturally project down to the lower dimensional metric space which contains the original interacting material space-forms. These faces which project into the lower dimensional metric-space are related to the orbital geometry and sets of conformal factors (ie spectral structure) of the higher dimensional metric-space and its (spectral) faces which contain the interaction space-form.

The new space-forms of the lower dimensional metric-space (which are the new faces of the interaction space-form) identify the dual (homological) structure to the higher dimensional interaction space-form in the projection (down) of the interaction space-form into the lower dimensional metric-space, ie the homology exists in the lower dimensional metric-space so as to be dual to the differential form structure (co-homology) of the stable interaction space-form which is part of the boundary of the lower dimensional metric-space, ie the intersection of the interaction space-form and the lower dimensional metric-space is not the empty set.

The homology (in the lower dimensional metric-space) of the new stable higher dimensional interaction space-form projected into the lower dimensional metric-space acts on the differential form (co-homology) which represents the new stable space-form in the differential equation defined on the interaction space-form (or differential form) which exists in the higher dimensional metric-space as an interaction space-form which has deformed into a stable space-form in the higher dimensional metric-space (resonant with the spectral properties of the higher dimensional metric-space), in a deformation process which is identified above based on the connection, d+A, and its square differential operator.

Conjecture: The energy of a space-form whose spectral flows are occupied with charge is determined by the energy (force x distance) in the containing metric-space, that is the energy needed to form the charge occupied current-flows of the space-form, where the charge starts out at infinity.

** considering a case of a specific dimensional level, eg the case of three spatial dimensions for material systems which are 2-space-forms. Is there only the case of an interaction 3-space-form contained in a 4-metric-space, ie four spatial dimensions, to consider for the spectral properties of a material 2-space-form contained in a 3-metric-space (three spatial dimensions), or does the observer actually exist in a yet higher dimensional metric-space so that higher dimensional metric-spaces than dimension four (spatial dimensions) [as well as higher dimensional space-form structures] need to be considered in the spectral structure of material (space-forms) which the observer can observe, for material 2-space-forms projected down to a 3-metric-space (for a material [and/or metric-space] subspace which intersects with the observer's space-form structure, so the intersection is not an empty set)?

This forms a well developed outline of how the geometry (and inertial properties) of space-forms can be used to describe the properties of quantum systems, and this is done in a way in which the act of fitting data can be used to develop a geometric-spectral picture of the properties of higher dimensional containing flat metric-spaces. This would provide new avenues (or methods) for controlling physical systems as well as develop a new understanding (or description) of existence.

The geometric nature of classical physics, that is, the geometric properties of the solutions of a classical system's differential equation is the focus of classical physics, is placed in a descriptive context in which the geometry of the space-forms which form the geometry of physical systems super-cedes the differential equation in defining a physical system, where the properties of the system, both geometric and spectral, depend on both the spectral properties of space-forms and the geometric properties of space-forms, of various types, and in various different dimensions.

By using the spectral properties of material space-form systems in the metric space in which we measure things, the geometric properties of higher dimensional metric spaces (related to the material systems in our experience) can be determined.

Consider, (4C2)=6 is the number of 2-faces (or 2-holes) on a 4-cubical-simplex, while (5C2)=10 is the number of 2-holes on a 5-cubical-simplex. This can be hypothesized to be the number of distinct principle quantum numbers, ie n, that there are in each of these respective metric spaces.

However, in 5-space-time the shape of space, ie the geometry of the differential structure or the geometry of the isometry group in regard to the spatial subspace (or the inertial [Euclidean] space), is that of two 3-spheres, whereas 4-space-time's spatial part (or the inertial [Euclidean] space), can be thought to have an isometry group which is a single 3-sphere. Thus in 5-space-time the principle numbers, n, and the numbers associated to the two 3-spheres are to be associated to each 2-hole, or to each principle number.

Chapter 23

MATERIAL-INTERACTIONS

1. Review of the descriptions of material interactions based on "spin-group" geometric structures, in relation to the geometry of interaction space-forms, on both a many-dimensional and discrete subgroup structures for the base space of a principle fiber bundle whose fiber groups are the spin-groups.

1. Review of the descriptions of material interactions based on "spin-group" geometric structures, in relation to the geometry of interaction space-forms, on both a many-dimensional and discrete subgroup structures for the base space of a principle fiber bundle whose fiber group is the spin-group.

** The force fields are defined in relation to either 1-differential-forms and/or 2-differential-forms, which in turn, are defined on the interaction space-form so that the dynamics which these force fields identify is (are) determined by the geometric relation that the 1-forms and 2-forms have to the spherical geometry which is defined in the fiber spin group. This "interaction space-form"—"fiber spin group" geometric relation identifies a geometric property between the spin-group's 2-dimensional and 3-dimensional spherical geometric properties (which can be identified in the fiber spin-group) and the 1-forms and 2-forms defined on the interaction space-form, and (but) the spin group is (of course) a group whose elements identify local coordinate transformations which act on the spatial face of the interaction space-form, which is a face which is one-dimension less than the dimension of the interaction space-form (or one-dimension less than the dimension of the [material containing] metric-space). Thus, if a particular direction is identified on a sphere (in the fiber spin-group) by a differential-form, then this direction identifies the direction in which the local coordinate transformations are defined by the elements in the spin group whose local coordinate transformations transform the coordinates in the base space in that well identified spatial (or Euclidean metric-space) direction.

That is, if a particular direction on the fiber group's geometry can be identified, then this, in turn, identifies a geometric direction of local coordinate transformations for the interacting materials in relation to the coordinates of the base space within which the material is contained. Namely, the coordinates which are transformed are the coordinates associated to the spatial-face of the interaction space-form, where the spatial-face determines the spatial relation that the material components of the interaction have to one another. Such a local coordinate transformation will change the geometric properties of the spatial part of the interaction space-form, thus changing the geometric relation (or changing the properties of separation) that the interacting materials have to one another in their containing metric-spaces (ie both Euclidean and hyperbolic, and in higher dimensions the higher oscillating-charge containing metric-spaces, which are characterized by the signature of their metric-function and the dimension [of their oscillating-charged-material containing] metric-space).

That is, a force-field is a Lie-algebra valued 2-form (though occasionally a 1-form) defined on the spatial face of the interaction space-form over a spin fiber-group.

This discrete process of change (of the spatial part of the interaction space-form) is related to the time period of a spin-rotation of metric-space states, in regard to the property of time (which is defined on hyperbolic metric-spaces, where the two states of time are forward moving time and backward moving time [the mixture of time states, which characterizes local coordinate transformations, are defined on complex coordinates; related to unitary fiber groups]).

The issue of higher dimensional time-subspaces, other than the one-dimensional time-subspaces which are related to hyperbolic space. These different time dimensions can either be related to the motion of a particular material-type, eg oscillating-charges of particular (but different) dimensions, or related to a velocity space similar to hyperbolic space, and thus in this hyperbolic-space, time can have a space-form structure, and this geometric structure can be related to directions of spatial displacements. This may, or may not, be consistent with identifying time-directions with the changes in motion of particular-types of oscillating-charges. Because time is related to energy, the relation that time might have to material displacement (different from constant motion, relative to some "near-by" center of mass) can be affected by the (relative) energy of an oscillating-charge.

** The dynamic "spatial displacement" of the spatial-face of the interaction space-form, is defined by local coordinate transformations of the spin-group which act locally on the metric-space coordinates. The directions of the local coordinate transformations is being determined in relation to how the geometry of the 1-forms and 2-forms . . . , which are defined on the interaction space-form . . . , are related to the 2-sphere and 3-sphere geometric structures of the spin group, respectively, so that the action of the local coordinate transformations will be on the spatial face of the interaction, ie down in the metric-space which contains the material which is interacting.

**

Note: The local (or tangent) geometry of a spin group is analogous to the local geometry of the differential-forms, which are defined on the base space over (upon) which the spin group acts. This is because the spin group is a subset of the Clifford algebra, where the Clifford algebra has the same dimension as the set of differential-forms defined on the same metric-space over which the Clifford algebra is also defined.

The above construction solidifies the idea of favoring metric-invariance, since this identifies dynamics . . . .

> not in regard to a general-metric (and the associated non-linearity of a general metric) and the subsequent relation that the metric-function has to determining geodesics on a manifold,

. . . . but rather with local coordinate transformations related to isometry (or unitary) groups which identify spatial displacements in an interacting system (in which the two time-states "of a metric-invariant metric-space" are mixed).

Newton identified and followed (or developed) the ideas of the derivative and its inverse structure the integral, where the derivative is a tangent structure on a function's graph which identifies a local-linear scale (or measurement) relation between different quantitative types, ie the fundamental linear (or multiplicative) property of measuring and re-scaling. He then set the inertial force (defined by Galileo) = (equivalent) geometric field force (defined by Kepler). This was very useful and widely applicable, in regard to defining a physical system's differential equation, and then solving that equation gives one enough information so as to control such classical systems, but it "is" not understood why such a differential equation is so useful, ie local measurability of a property is equal to an equivalent geometric property.

In particular, the differential equation is most useful (or only useful) in the context of linear, separable, metric-invariant (context for) differential equations.

Why is this context so important?

Why do non-linear differential equations have such limited uses?

The context of linear, separable, metric-invariant differential equations is still the main idea (or main context) which is guiding technical development, though now it is often about coupling classical systems to the observed properties of quantum systems, though still it is about controlling the classical system.

It should be noted that quantum physics cannot describe the properties of the general quantum systems, whereas linear, separable, differential equations defined on a metric-invariant space are always relevant as a descriptive structure which allows for a great deal of control over classical systems.

The main attribute of a quantum system's physical properties is the distinction between spin-½
and integer-spin components of statistical systems, ie Fermion or Boson system components, respectively. Integer-spin systems (components) are related to the regular measuring properties of a metric-space, while spin-½ properties seem to be outside the usual properties of geometric measurement, that is, geometric orientation of the spin components does not seem to alter their internal spin properties of up or down.

What is spin?

or

What properties of spin groups are (most) useful in regard to physical description?

Spin-(n), is a low-dimension representation of a simply-connected covering group of the Euclidean isometry groups of dimension-n,

or

Is it the square-root of the Laplacian operator defined on each metric-space, ie the Dirac operator, which is related to both the spin-½ properties of components of quantum systems and to the positive and negative energies of these components, ie matter and anti-matter?
R Penrose defines "twisters" as geometric structures which are related to fiber groups which are spin-groups, in order to define coordinate properties (or coordinate shapes) in general relativity, but this is suspicious, since general relativity is defined with diffeomorphism fiber groups. Penrose's idea seems to be a "change of variables" which is (usually) only useful if it identifies separable coordinates. That is, in the only problem associated to general relativity, "the 1-body problem in spherical symmetry" the spherical geometric decomposition of the spin groups might help in the context of spherical symmetry, ie the sphere is identified with a separable coordinate geometry.

So

(A) How do spin groups relate to existence?

Is it a math property which is correlated with spin-½ properties (and the fairly rare, matter and anti-matter states) of components in statistical systems, or of components in quantum systems which have spherical symmetry?

Or

Is it a math property which is related to spherical symmetry in the very limited descriptive range of general relativity?

Answer to (A): It seems that the strongest relation of spin groups to existence is in relation to both the descriptions of force fields in material interactions, and the spin-rotation of pairs of metric-space states.

And

(B) Why are linear, separable differential equations defined on metric-invariant coordinate spaces so useful?

And

(C) Why is the derivative, and their subsequent differential equations, so useful, ie what causes the differential equation to be so effective in identifying (or describing) physical properties in the context of classical physics?

In regard to today (2010), question (C) is given two contexts:

Today the derivative is placed into two contexts:

(1) that of classical physics where the derivative models a local, linear, change of scale,

and

(2) that of quantum physics where the derivative is an operator which represents the classically measurable property of motion, and it (along with other operators, such as the $1/r$ potential energy function, which represent classically measurable properties) act on a function-space which is Hermitian-invariant to unitary operators which represent classically measurable properties, such as the energy operator, ie Schrodinger's wave-equation, so that subsequently the "quantum interaction," related to the $1/r$ potential energy, is modeled, in an incoherent manner, by extending the idea of unitary invariance to both . . . particle-collisions modeled as unitary particle-state transformations

at a point (to model the interaction), ... and Poincare (spin-)representations of these unitary particle-state transformations on the function space of the quantum system, then renormalizing, and then extending the values of the (global) functions . . . , which had been collapsed to the collision-point . . . , back out to infinity so as to reform the function-space of the global wave-function of a quantum system in its Poincare representation, which assumes the spherical symmetry of both the quantum system and the particle-collision interactions, which in turn, assumes the collision point as being a singularity point of a 1/r energy expression of the particle-collision interaction.

Essentially, the structure of (2) is not been sufficient so as to be able to solve for the radial equation in quantum systems which are assumed to have spherical (or Lorentz) symmetry (usually associated to the observed bounded-ness of the quantum system), ie the principle-energy value of general quantum systems cannot be found. Furthermore, the quantum description is an incoherent descriptive language, as well as being without practical value.

The precision proclaimed (for two or three systems) as proof of the validity of both quantum physics and particle physics . . . , in reality cannot be considered to be anything other than an unbelievable epicycle structure used to fit a few data points. That is, this descriptive method fits a very limited amount of data.

It is really the great limitations of an incoherent descriptive language which leads one, who is really a scientist . . . , ie one who questions authority and who presents alternative descriptive languages . . . , to <u>not-take</u> the context of (2) as a valid, useful, widely applicable, descriptive language.

Measurable verification is really to be related to the usefulness of the quantitative description, ie how much control and creativity emerges from the description, and not in relation to its (very few) measurable verifications, since based on its measurable descriptions quantum physics has "not gotten the job done," ie most general quantum systems do not have a valid measurable description in relation to fitting the description (the prediction) with the observed measured properties of general quantum systems.

New descriptive language

The point of the new description is that it gives a model for interaction which (1) allows for quantum properties, (2) in 3-dimensional Euclidean space it is in agreement with the properties of spherical symmetry, and (3) it provides the geometric structure upon which the principle quantum number, the fundamental energy of the system, can be realized and (4) this is done in relation to the quantum system being in resonance with the spectral properties of its high dimensional containing space, which has many spectral properties with which any low dimensional quantum system in the containing space can be in resonance, and (5) the stable space-form geometry . . . , into which the correct energy of the quantum system can exist . . . , is given.

Thus, the act of fitting data in the new description, is really an act of determining the geometric properties which are fundamental to the "complete containment set" of existence (or of some existence), and this information is related to a more complete set of information about how different material (or more general) systems can be related to one another, and thus it is a new descriptive language which is related to more control, and more capacity to create, within a new "broader" descriptive context of existence.

(A), (B), and (C), above are three questions which have been given a vastly expanded new set of contexts by the: many dimensional, discrete isometry (or spin) or unitary subgroups, where the descriptions of material interactions are placed in the context of the spin-groups.

The second question (B) is answered by the assumption that the geometry of material and the geometry of interactions are determined by the geometry of space-forms, ie separable, and in the context of metric-invariance. Furthermore, the properties of space-forms can be defined by solutions to linear differential equations defined on differential forms.

Linearity is simply the idea that two measuring scales must be related to one another in a consistent (or in a coherent) way, and that way is defined by linear relations. That is, there is only defined a linearity in regard to scale change.

Thus, non-linearity means that a function's values and the values of its domain space cannot be consistently related to one another. Thus, in a non-linear context, new values discontinuously emerge between the two scales of measurement due to this incompatibility of the two number systems, so as to create chaos for systems defined with non-linear differential equations.

And

The first question (A) is answered above in (section) 1.

Furthermore, (C) is best answered in relation to the relation that (or association of) both the derivative and the differential equation have to interaction space-forms. That is, the derivative has been so useful because there exist discrete geometric structures which determine the properties of material existence, and the tangent structures of these space-forms (associated to differential-forms) determine the useful differential equations for both classical physics and this new more broadly expanded context within which to describe existence and its useful properties.

It is best to assume that particle-collisions are rare occurrence's rather than the basis for fundamental interactions in material systems.

That is, the properties of existence are best placed in a context of continuity for each dimensional level, but in living systems this continuity might be best described in relation to a higher-dimensional oscillating-charge space-forms.

3. A new definition for a manifold

The interaction space-form takes the descriptive structure of the classical system down to the level of the quantum system where it defines Brownian motion.

This Brownian motion is the basis for apparent quantum-randomness of physically observed properties.

However, the interaction space-form in a many-dimensional context, which is characterized by its discrete properties, also re-defines the manifold structure (which has traditionally been associated to geometry), so as to form a new local (topological) structure (ie conservation properties [based on either continuity, or invariance to displacements of the system (or its properties) in space or time] are maintained within any given dimensional level).

In the new local (topological) structure, which is placed within the spin-group context, there can be a description of both interactions (of force-field spatial displacements) and the formation of stable quantum systems (associated to resonances and if the interaction is in an "allowable" energy range, within the many-dimensional and discrete structure). The local linear structure (associated to local geometric measures, ie associated to differential forms) of a traditional manifold is placed in the new context of interacting space-forms which span dimensional levels.

Furthermore, the energy-structures of oscillating-charges of a given dimensional level (ie associated to lower-dimensional material systems which the oscillating-charge contains) can be temporary structures, due to a constant energy value (for an oscillating-charge) being re-distributed (or allowed to flow) between the oscillating space-form's faces, so that within, say, living systems, both the dynamics and the geometry of the molecular systems, which are affected by the structure and energy of higher-dimensional oscillating-charges, can change from time to time (including molecular shape).

During such an energy re-arrangement, energy is conserved on a higher-dimension oscillating space-form structure (which contains the [living] system which is being described), so that energy can be transferred from one face (or from one spectral-flow) to another, due to resonances and/or pushed energy flows in relation to a higher-dimensional oscillating-charge space-form (where a given energy is allowed to flow between different faces of the oscillating-charge of some given dimension, eg quasi-spatial faces of an oscillating-charge can change so as to change properties of energy resonance within the oscillating space-form, or the energy flow can be related to a time period defined by the spin-rotation of a time-state at a particular dimensional level, for a particular oscillating-charge space-form (which is associated to its own independent time-dimension), and of a particular metric-function signature).

Note: Professional science is about identifying an "in crowd" of "top scientists" who work on the projects which are of most interest to the owners of society, the owners want to control this knowledge so as to limit any competition, so what these "scientists" (or engineers) do is treated in the media as that which defines "legitimate" science, so that the outsiders, those not working on engineering projects, or those not in academia (academics are funded based on the projects upon which they work), do science which is amateurish and it does not measure-up to the "science" of the "master scientists" of real science's "in crowd," which is defined by salaries and access to equipment. That is, only people who are capable of understanding "science," as defined by the "in crowd," are let into the "inner circle" of "true science."

This is the same as a small clique of people defining an "in crowd," as occurs in high schools.

Indeed it is an organization of "science" which depends on an immature and arrogant behavior which opposes both reflection and careful analysis of the descriptive language which "scientists" use. The "scientists" of the "inner circle" have become "too important" for outside criticism, so that they maintain their descriptive structure in a fixed language, but the observed properties and the language which they use have become inconsistent. That is, the descriptive languages of: quantum physics, particle physics, general relativity, string theory, quantum gravity, are all languages which have never been widely applicable, and they have only been applicable to a very limited number of uses (if they have any relation to measurement at all, remember measurement is not about verification [Ptolemy's system was verified], rather it is about coupling to the system so as to use the properties of the system for some further creative uses). Defining science as an "in crowd" which can only use language in an orthodoxed manner results in a professional discussion about irrelevant illusions. It is all about making "knowledge" appear to be unobtainable, except for those few who can prove they belong within the "inner circle," and it is this definition of "science" as identifying social class which is now the true property which defines "science" today.

(see I M Benn and R W Tucker, in, An Introduction to Spinors and Geometry with Applications in Physics, 1987, Chapter 2)

Chapter 24

A FURTHER LOOK AT CUBICAL SIMPLEXES

The stable order of material is observed, and the professional science and math communities try to use the ideas (really, dogmas) of indefinable randomness and non-linearity to describe this order, but they cannot, eg the stable spectra of the general nuclei.

The control (and the stability needed to achieve this control) that a living system has over itself is observed, and the professional science and math community, again, try to describe these stable properties with the assumptions, that such descriptions must be based on indefinable randomness, and again they cannot describe these properties, eg the origins of life, and life's capability to intend and act etc.

Instead begin one's descriptions with a descriptive math language whose structure is based on "stable math properties," and see if the stable order of material systems can be described, and if the stable control that life has over itself can be described. [see first page, and see section, Getting serious about math and science, for the math and science part (model of molecular folding (which is highly controllable) is given).]

This is a paper mostly about how the properties of real and complex (fiber) isometry groups can be put in a relation to the stable (discrete) geometric shapes which fit naturally into metric-invariant (base) spaces, ie the stable shapes based on cubical simplexes.

The new context focuses on the math properties associated to stability, ie the shapes based on cubical simplexes, so that these cubical shapes are placed within a math structure of a dimensional layering of a set of containing spaces, so that these containing metric-spaces are, themselves, associated to the stable shapes of cubical-simplexes.

These metric-spaces (as well as Hermitian spaces) exists up to a space-time dimension of 12.

This (new) math structure is very simple, because the stable structures of mathematics are relatively simple, and it is consistent with the observed properties of material which appear (to us) to exist within a space-time dimension of 4 (but now there is a road-map into higher dimensions, so that these higher dimensions turn-out to be the (very real, and macroscopic) structure upon which life is built).

Currently a Hermitian-form invariant math structure fits into a 4-space-time structure, which nonetheless requires the structure of (non-linear) curvature.

The new context allows one to refine the structure of Hermitian-form invariance, where the refinements are geometric with the new dimensional structure much more rich in regard to (useable) geometric ideas, and it allows for more precise and accurate set of descriptions for the observed stable properties of physical systems, rather than an overly complicated description being centered on function spaces and subsequently focused on the useless ideas of indefinable randomness, non-linearity and involved in a process of carefully describing an inaccessible descriptive context, eg string-theory, ie materialism is assumed and the descriptions are outside of the bounds of materialism.

Thus, at best, these so called authorities, are trying to describe properties which only have an indefinable random relation to space-time, so that these properties exist down at the size-scale of mathematical point-particles. They have been trying to make this model useful for over 100 years, but, as yet, they have failed.

The best cultural position for society to consider, is that this attempt to describe the observed stable properties of material systems based on indefinable randomness and non-linearity has failed, and other math-science ideas should be considered. [But it is not the culture which is making such decisions within the US society, rather, it is the few owners of society who decide how language is used within society.]

This is the true failed state of the currently accepted (failed) descriptive structures of math-"science," if the currently accepted description of matter is science at all? It seems more like arbitrary authority, sort of like a religion of an indefinable "but highly valued" intellect (where social value is being determined by the owners of society), which, in turn, seems more like arbitrary authority upheld by violence. The current authorities have nothing about reality by which to recommend their ideas, they express (authoritative) ideas which are neither accurate nor of any practical value, but the probabilities of random particle collisions are related to rates of nuclear explosions, (thus the military business community makes them authorities), but this particle-collision probability model of reactions is also the 19th century model of chemical reactions, and despite the hype, the chemical engineers are also quite inept, (genetics is being used to put together chemicals whose reactions and structures are not understood) where the inept modern physicists contribute nothing except a useless intellectual hierarchy to the society.

New math structures, cubical simplexes in a containing space of layered dimensions (where the dimensional layers are associated to stable shapes which have stable spectral properties), in a new math context which emphasizes the math properties of stability, accounts for all the observed properties of material where its fundamental descriptions of material interactions are very similar to the math structure of classical physics yet it is a model which implies the atomic hypothesis so that it is a description which has many points in common with both statistical physics and quantum physics and its derivation of the property of quantum randomness are classical interactions (essentially, classical micro-collisions) defined in a statistical setting of atomic interactions resulting in Brownian motions and in the atomic context it is equivalent to quantum randomness (of small components interacting, as waves, between macro (or nano) geometries). However, many of these interactions can be resonant with the over-all high dimensional containing space which can lead to new stable systems (emerging from the interactions), thus it explains stable structures in a direct manner and the math is geometry, ie it is a useful description. Material interactions, in the new descriptive structure, are spherically symmetric in 3-Euclidean-space but not in 4-Euclidean-space, but spherical geometry is non-linear when perturbed and thus it is not useful, it cannot be used to understand the observed stable structures of material systems.

The simple axiomatic context from which this description emanates is from that of: numbers (counting), set containment (finding a context wherein the containing set is finite), operators (grouping, number-type [properties], order of operations), measuring (stable uniform unit), measuring a property as a function of a containment set (of independent) coordinates.

These ideas, though, can be claimed as the readers own, and used without fear of copyright infringement, except that they are original ideas, and thus as an originator of these ideas it must be stated . . .

(as to what the author's opinion of the public domain is, as opposed to a justice system wherein the supreme court negotiates the "letter of the law" with the owners of society, for whom the supreme court serves)

. . . that these ideas cannot be used for military purposes, or for purposes of domination, or destruction, or they cannot be used within a society whose law is "not" based on equality (theoretically US law is based on equality, ie as stated in the Declaration of Independence), where the power to govern is obtained from the people so that the structure of law serves the society (instead of society serving the owners of society), where, in the law, all people are equal (society does owe things to the people, not the people owing to the owners of society, nor should people be attacked by the selfish interests of others).

A dominant unequal group of people within a society cannot acquire selfish value from these ideas, though a creative group of people can use them but not to gain dominance within society.

These ideas (this representation of knowledge) are meant for practical creativity, where the creativity is motivated by selflessness.

People should be opposed to science and math in its current (violently) authoritative state, because of "how it is used within the US society."

It is used to justify the destruction of the earth in order to maintain the growth of unwise resource dependent big businesses.

It is used to divide the public into "the experts," who are required to be within absolute institutions which serve big businesses, and "those who are inferior."

However, because of propaganda, science and math have come to be neither accurate in its (their) descriptions nor is it useful for practical development. This remark is about the genius of the propaganda system, and the high-value identified (by the public, due to propaganda) with experts, ie the failure of the education system.

Where it should be noted that the education system should be based on equal and free-inquiry as well as the relation that equal and free-inquiry has to practical creativity and to new ways in which to build a (precise) language based on assumptions and new contexts.

The religious right uses science in the same manner as do the owners of society, they oppose its development because of their own agenda of domination within some (their own selfish) absolutist context, yet they use any part of it for their own petty needs, eg the fertilized human-egg is a human being.

Perhaps they should consider the equality of all life. But even science is in opposition to this idea, yet science has no valid description of life's origins.

In the west it is always the model of an overly domineering religious-owner-empire. (where religion is essentially absolutist dogmas, and this includes both material-based-science and religion)

The Declaration of Independence (of the US) was a split from the western culture of that time, a western culture dependent on domineering psychopathy and a (western) culture which is opposed to equal freedoms (western culture is always placed in a context of absolutists languages).

What is "the west?" Answer: The west is the Judeo-Christian-Islam religious based cultures whose theme, is a culture dominated by the few. In history, its defining point seems to be the Roman-Catholic-Empire begun by Constantine, and perhaps copied by Mohammed. Communism is also an oligarchic social structure, but it has the language of equality and perhaps it was vaguely an attempt to realize equality. Perhaps Lenin wanted to bring America to realize its historic intent of equality, but Lenin's actions were those of an Emperor. The US Declaration of Independence was a conscious effort to break-apart from the oligarchic social structure of Europe, and to realize a society base on equality and creativity, but the residuals of European economic exploitation seemed to be too prevalent in the US society around 1776. Furthermore, at the time of Lenin, the Robber Barons of the US society saw a threat to their selfish rule due to the rhetoric of equality associated to communism, and its natural resonance with the US society. Thus people of the type as J E Hoover got their jobs within the US justice system which has always been compliant with (or served the interests of) the ruling oligarchy of the US society, despite this being in opposition to US law, as sated by the Declaration of Independence.

If the US society should suddenly flip, perhaps a way to poetic justice would be for many of those in prison (eg the drug offenders) to be given the administrative positions of the 1-million, or so, key managers which are today (2012) within the US institutions, and allow those 1-million administrators to become homeless, and for them to then be treated as the justice system (which they depend on and manage) now treats the poor and "minorities," but do not allow the (hypothetical) new police-force to murder with impunity. [Perhaps one should seek this justice, and not the western oligarchic model of justice; "of slaughter," as expressed in the French revolution and by Lenin, if the society's social structure flips.]

Why is science and math not causing wide-ranging (new) technical development?

Why is the main form of technical development of our society based on 19th century classical physics? eg electronics, optics, thermal physics, and classically defined statistical physics.

Why cannot quantum physics describe, in a convincingly precise manner, the very stable definitive discrete spectra observed for general quantum systems? [Answer: The assumptions which are implicit in regard to containment sets, function spaces, domain spaces, and sets of operators are not sufficient to describe the observed patterns of the material world.]

It is also because indefinable randomness (and non-linearity) is not a valid descriptive basis for the descriptions of the ordered systems which are observed to exist, eg the stable definitive spectral properties of (1) atoms (2) nuclei (3) crystals (4) solar systems etc.

Is the set of overly authoritative assumptions which define the absolute institutions of western culture, such as western science, the basis for the endless failures (of the absolutist institutions of western culture) . . . ,

> (absolute institutions) such as the "peer reviewed (to ensure dogmatic purity)" and hierarchical authoritative sciences, whose research is associated to the businesses of the owners of society, eg particle-physics is about nuclear weapons engineering, [Is such hierarchical authority a search for truth? Answer: (no) hardly, it serves business interests, both production and the lack of other creative developments eliminates competition]], and fixed scientific authority is an expression of how the human behavioral trait of autism can be manipulated within the social structures to consolidate a fixed authoritarian basis for the endless failures of math and science, eg cheap clean fusion energy was supposed to be developed by 1955,

. . . , when it comes to both accurate descriptions . . . ,

> (of which very few exist for general quantum systems, neither the general nucleus nor the general atom have valid measurable descriptions, ie the spectra of these systems cannot be found from calculations (which are supposed to be derived

from physical law) thus the discrete, stable spectral properties of these systems are [or form] the indefinable set of random events, upon which both quantum and particle physics are, supposedly, based)

..., and when it comes to providing wide-ranging (new) technical development (creativity) on new scientific frontiers?

Why turn science and math into a jeopardy contest? A "jeopardy contest" is a set of dogmatic and fixed interpretations of (or contexts for) "facts." Narrowly focused attentions, which can be overly obsessive, win the contest, ie the computer as Jeopardy Champion.

Science and math are about developing new languages outside of dogmas and meaningless measurable verifications, remember the model of Ptolemy was measurably verified, furthermore, today the observed patterns are outside of the capacity of the dogmatic laws of physics and math to describe, which is also similar to the manner in which Ptolemy's model could not keep-up with the observed structures seen in regard to the properties of the planets (eg with the aid of a telescope, the phases of the planets, yet their might have been a way to also fit that data into the scheme of Ptolemy [everything fits into the indefinably random descriptive structure of quantum physics and particle-physics, eg function spaces and elementary particle are equivalent to epicycle structures within a descriptive context (which is based on indefinable randomness and non-linearity, and) which is practically useless]).

Education is a somewhat gentle process guided by inquiry and integrity, not absolutes and authoritative truths. Most of what is being said, in regard to our understanding of the world, is wrong, but it could also be interesting and educational.

Authoritative knowledge is a construct most useful to the owners of society who exist in an unequal and far too narrowly focused society, so that the narrowly defined experts will do what the owners of society want these obedient experts to do.

One uses languages in new ways to discuss "what seem to be interesting ideas." One does not speak so as to only express authoritative ideas which are absolute truths in a "perfectly structured" language. Rather, language changes at its simplest levels of:

> assumption and
> contexts and
> interpretations,
> so as to remain practically useful, so that one can create within a context of what
> one believes "existence to be."

The only aspects of the current western society which have great success are the social structures of wage-slavery and inequality, and its associated institutional violence, which lends itself to "overly authoritative" technical outlooks (as a [narrowly defined] culture) and a fixed (and development suppressing) means of expressing an authoritative belief structure, and the (old stand-by of western culture) extreme violence, which its dehumanized culture can generate, where this extreme violence is expressed by narrowly defined media and by mean narrowly fixed managers of communication institutions, where extreme violence (which) includes the capacity to uphold a narrow dogma for science through "peer review" so as to demand science authorities to publish within "peer reviewed" journals, a dogmatic version of science, through which scientific authority is defined within the society.

This scientific dogma is then used as an example of an authoritative truth which results in a demand that all publishing uphold a particular standard of accuracy (truth) so that the public will be protected from non-truths which the inferior people will try to instill in the un-witting public. It is, instead, the owners of society who mis-represent the world to the un-witting public.

This idea that inferior people want to interfere with science is usually construed as the expressions of the religious dogmatists.

As long as there is a split between two sets of descriptive language structures, which both claim to express truth . . . ,

> so that one language is based on materialism and the other language is based on an assumption of "another world," from which, so called, non-material experiences emerge, eg emotions, thoughts, life, etc (which the materialists deny, based on both their faith in materialism and their arrogance),

. . . , then one can exploit the idea of "protecting the truth" of one (all inclusive) dogma from the truth of the other (all inclusive) dogma.

Furthermore, this provides the context in which the idea of superior and inferior people can be expressed, where the superior people are the experts and authorities (the learned ones).

The correct conclusion concerning Godel's incompleteness theorem . . . ,

> (that precise languages have great limitations as to the types of patterns (which exist within the context of precise description) which they are capable of describing)

. . . , is not that the assumptions involved in the conditions of Godel's theorem, itself, should be questioned at length (rather one needs to consider the content of the incompleteness statement) but rather that it is the set of assumptions, interpretations, contexts, containments, well defined-ness, meanings, etc* which should be questioned for the whole of the "subject matter" which has been built upon a set of assumptions and interpretations. This can be: science, law, government, media, education etc.

That is, the language of a society should be at the very simple level of assumptions and interpretations concerning all social institutions, that is the public debate needs to exist at the level of an 8-year old, so as to both be meaningful and to facilitate education based on equal and free-inquiry and the relation of a described (measurable) truth to practical creativity. The property of being measurable is not so much about verification, though that is included, but rather measuring is about using a description used to build something, where building depends on measuring and planning and organizing practically useful information when putting together the components of a new system.

The context of assumptions and interpretations and word meaning are the fundamental aspects of forming a precise descriptive language upon which new observed patterns of existence, ie the data, can be described, and new contexts of relations (between properties) can be realized, and new creative developments explored.

One must change, reconfigure, re-organize, and both add new as well as re-word axioms, so as to change the descriptive language and to be able to change the patterns which one can describe and use in a practical measurable context.

Getting serious about math and science

If one criticizes the structure of math, its assumptions, contexts, and interpretations, then one aspect of a math-science description which is "passed over" far too quickly are the: number, geometric, and physical attributes of the property of stability.

Stable, reliable measures of (physical or mathematical) properties eg being able to define and count (random) events, and the property of being able to have within the description a stable uniform unit of measuring, or to have a description which remains quantitatively consistent as well as a descriptive language where the meanings of words are stable and fixed (this is clearly the point where the ideas of absolutes enters the descriptive structure so that limitations of a "language's descriptive range" enter the constructive process of a precise (measurable) descriptive language) thus a set cannot be "too big," otherwise "what are considered to be independent properties" can become mixed together within the set which is "too big," eg the plane filling curve blur's the distinction between the two different directions of the plane.

For example, Thurston-Perelman suggests that the stable shapes for a dynamic context are the spaces of non-positive constant curvature, where general (and/or non-linear) shapes either disintegrate or evolve towards the (just mentioned) stable shapes (though there may be a few exceptions).

While the useful aspects of the classical descriptions, ie the ideas upon which new creative developments depend, are related to the solvable math structures which are stable, linear, "separable geometries," which exist in a metric-invariant context for shapes which are often shapes of non-positive constant curvature, though in Euclidean 3-space the geometrically separable shape of spherical symmetry seems to be an important shape for relating spatial displacements of the objects to the material shapes (which determine the shapes of force-fields) which represent the equivalent forces (associated to the spatial displacements) etc.

Furthermore the stable shapes of the spaces of non-positive constant curvature (where spaces of positive constant curvature are the spaces of the non-linear spheres, which are non-linear since their metric-functions change as one changes one's position along a spherical shape, or in a spherical space), are shapes which are related to the "cubical" simplexes, ie (1) the cubes and (2) the cubes attached to one another at their vertices.

It should be noted that a 2-square can have its opposite 1-faces (or edges) associated to themselves so as to form first a cylinder and then a torus (or doughnut shape). This idea can be continued in a similar manner into higher dimensions (of higher dimensional cubes).

These shapes naturally lead up into higher dimensions, where both material and the metric-spaces, themselves, can both be given a stable shape (associated to themselves), so that these shapes are stable and they carry on themselves stable spectra. Thus these shapes form the model of stable material systems, but since the metric-space can also have these stable shapes (though at a different dimension from the dimension of the material shapes which they can contain), the metric-space can also have a stable spectra associated to their (or its) over-all high-dimensional containment set structure.

Furthermore, an odd-dimensional cubical simplex, which has an odd-genus number associated to itself, could well represent, a relatively stable, but unbalanced (charge), and thus it would be an oscillating, energy generating shape (since based on charge, and thus contained in hyperbolic space), ie a high-dimensional but simple model of life.

Note:

This is a mathematical context which the professional mathematicians do not consider, yet it is a context which gives more insight into the issue of "how to organize the math structure which is needed to be able to describe from whence the observed physical order comes?", ie the observed stable, discrete, definitive, spectral-orbital properties of: Nuclei, general atoms, molecules (and how they fold), crystals (now, often referred to as condensed matter), the solar system, where none of these general systems is capable of being described, to a valid level of precision, by using the currently accepted (and very dogmatic) laws of physics.

Yet modern physics claims these domains as if they possess valid descriptions, while modern physics acts as if the only questions open for careful consideration are about the structure of elementary particle (eg strings) and the relation that these elementary particles have to gravity (or to geometry) and about "what dark matter ands dark energy are?" ie what elementary particle are they to be related?

Everything that is being discussed in this article is concerned with the math patterns which are a part of the correct description of those math-material-containing-space structures which underlie the observed order of the world, and no discussion in Physical Review (or any other professional math-science journal) has much to say (which is of any value) in regard to this problem.

If one is going to describe a stable pattern, either mathematical or physical, then it must be a quantitative (measurable) language based on "cubical" simplexes in a linear descriptive context, in metric-invariant spaces which are also spaces of non-positive constant curvature.

This allows both quantitative consistency, and stability, for the descriptions of measurable patterns, upon which a creative structure can be planned, and parts measured so as to be put together.

Creative development depends on a stable context upon which descriptive patterns are constructed.

This assumes set containment of the pattern (or process) one is describing to be within a quantitative pattern (construction) of independent measuring directions in regard to the measurements of (for) some stable (substance) pattern, and that substance-pattern is the "cubical" simplex.

However, the "cubical" simplex can model both material and its stable containing metric-spaces, which can exist in many higher-dimensional contexts, so that the material simplexes in the different dimensional levels . . . , or the geometrically-independent geometric processes defined in a context of (on) a "cubical" simplex . . . , are constrained to their containing metric-spaces (which are also "cubical" simplexes) since higher dimensional shapes (or material) will interact with the stable shapes of the same dimension . . . ,

(higher-dimensional shapes (then the material shapes))

. . . , which are contained in higher-dimensional [material-containing (metric-invariant)] metric-spaces.

If this rule is not upheld, in a measurable context, which ensures stability of substance as well as the stability of the quantitative sets (which represent measurements of a system's property), which are used to describe the measurable patterns, then the description:

1. is not quantitatively consistent,
2. does not properly fit into a quantitative measurably distinguishable descriptive structure, so

3. the patterns described become meaningless and
4. are neither accurate
5. nor practically useful.

When considering the set of stable substances (stable math structures) which can be contained in a metric-space, then this set can be limited to a "finite set," related to the number of separate sub-spaces contained in the various dimensions, up-to Euclidean dimension-11, where 5-dimensional hyperbolic "cubical" simplexes are the last closed and bounded hyperbolic "cubical" simplexes [upon which are based discrete hyperbolic shapes], and furthermore, there is also a set of conformal (constant) multiplicative factors which exist between the different metric-spaces of either the same dimension (but different subspaces) or between adjacent dimensional levels.

This set (of subspaces and sets of conformal factors) would determine the allowed spectral set of the substances (closed and bounded discrete hyperbolic shapes) contained in the dimensionally-layered material containing metric-spaces, ie a mass-energy spectra where, mass = energy, is to be interpreted as the stable charged-system's energy, contained in hyperbolic space, (which) is related by resonances to the inertial properties contained in Euclidean space, which identify the local linear as well as geometric structure which are associated to material interactions through the action-at-a-distance properties

[of Euclidean "cubical" simplexes]

which are subsequently, also, associated to a geometric-local-coordinate-transformations of the spatial translations of inertial material (interactions) in Euclidean space, in regard to a set of discrete translations of material positions, where the transformations are defined by (both the just mentioned geometric properties and) the discrete time intervals, in turn, defined by the (time) period of the rotation of metric-space states.

of inertial material in Euclidean space in a set of discrete

The material interaction, if within energy bounds, resonates to and "enters into" the set of allowed stable structures, in regard to the spectra of the over-all containing space, due to the discrete subgroup structure of both the fiber group and the containing (base) space, where the brief transition, from one stable system to another stable system, inter-relates both the opposite metric-space states within the material containing space and the opposite vector structure of the fiber (semi-simple Lie) group's root systems, which is a part of the geometric structure (within the group of local coordinate transformations) which finds an extremum (usually to be thought of as a minimal energy [action]) of an expression for action (usually an integral expression, which involves local spatial displacement transformations of the object) in a closed bounded shape of an allowed shape for the "material end-point" of the material interaction. That is, the resonance and allowable energy values cause the opposite metric-space states (in the base space) and the opposite root system structures (in the fiber group) to become consistent with one another, so that this is related to the maximal torus of the fiber group.

One must consider the stable shapes which exist in a metric-invariant context. That is, the diffeomorphism group as a fiber group over a non-linear geometry is describing an unstable and quantitatively inconsistent context which cannot accurately describe stable structures and is unrelated to practical development.

Only the non-positive constant curvature spaces have metric-functions with constant coefficients and thus their local descriptions can be related to a linear, metric-invariant context wherein the stable geometries have the property of being parallelizable and orthogonal (or geometrically separable shapes, [as they have been called in these papers], ie the local coordinate directions are always perpendicular to one another). Being metric-invariant allows these shapes to be related to some of the classical Lie groups [compact Lie groups, and its desirable that they be connected lie groups, ie compact, connected Lie groups]

> eg $SO(s,t)$, where $s + t = n$, where s is the spatial subspace and t the temporal subspace

(Note: The maximal torus will be defined in regard to the, s, subspace), $SU(s,t)$, $Sp(2s)$, etc. where Euclidean space is $SO(n,0)$.

The simplest spaces to consider are $SO(n,0)$ and $SU(n,0)$ while $SO(n-1,1)$ and $SU(n-1,1)$, and $SO(s,t)$ and $SU(s,t)$ can also easily be considered.

If n is odd then $SO(n)$ is related to $SU((n-1)/2)$, if n is even then there is an extra (real) dimension in $SU(n/2)$ in regard to its dimensional relation to $SO(n)$.

The geometric shape which is being inter-related [between shapes in the base space and a maximal torus in the fiber group] is concerning (about) a geometric-coordinate relation between $R(n)$ and $C(n)$, where, in turn, complex circles, in $C(n)$, touch opposite faces of a cube, $R(n)/Z(n)$, where $Z(n)$ are the integer coordinates (or integer lattice), and the plane (which is defined by the complex circle) is to be normal to each n-direction in $R(n)$. Furthermore, the cube, $R(n)/Z(n)$, where $Z(n)$ are the integer coordinates, can also be thought of as a torus, $T(n)$, a toral shape which is contained in the base space (of a principle fiber bundle), but this torus (in the base space) can be related to an equal dimension torus in the maximal torus of the fiber group.

If one assumes that the rank (ie the dimension of a maximal torus, $T(n)$ (in)) of the isometry fiber group (over the base space) is n . . . , then n-cubes in $R(m)$ [one may assume m=n (to make the idea easier to grasp), though m>n] correspond to maximal tori, $T(n)$, in the fiber group.

The math of the fiber group must preserve metric-invariance, while the tori within the metric-space (base-space of the principle fiber bundle) can have different sizes. Otherwise the cubical tori in the two spaces (ie the fiber group and the base-space) are geometrically similar, and thus they also possess "conformally similar" spectra (or equivalent, but perhaps the different independent circles, which are components of a torus, are related to different spectral values [ie different size circles]).

[Note: Closed curves on the tori can be defined as lines with rational slope defined on R(n) which must, at some lattice point, intersect with an element of the lattice Z(n). These closed curves define the spectral lengths on a torus. However, one can think of the spectra as being the different edges of the cube R(n)/Z(n).]

(For a Lie group's associated Lie algebra) a maximal torus so that

How are a set of Lie algebra "vectors" (in the adjoint representation of the Lie algebra) . . . ,

which exist in Weyl chambers of the root space decomposition of the "vector space," and which are associated to an (n-dimensional) maximal torus subgroup (of the associated Lie group),

. . . ., related to the properties of arc length of the maximal torus's unit circle components, and in turn, these arc-lengths are related to (eg equal to) the arc lengths of a corresponding set of circle component of an n-dimensional torus in a (base-space) metric-space (of a principle fiber bundle)?

This could be related to either a metric-space toral shape or to a toral component of a "discrete hyperbolic metric-space geometric shape" which has toral components.

When one wants the arc-length of a unit circle (within a maximal torus) to equal to 10, then define the "angular argument," Y, of exp(iY) (ie on the unit circle), to be, [2(pi)/10]t, for t a real number, where t is the arc-length along a unit circle component of a (given) maximal torus, then when t=10 the argument will be 2(pi) and a full circuit (the first full circuit) around the given unit circle (component of the maximal torus) has been traversed.

Thus, the length around a unit circle (component of a maximal torus), when there is a factor of 1/10 in the argument of the circular component, is 10, where the factor, 1/10, is the value (or slope) of some Lie algebra vector, which is tangent to the maximal torus at the identity, and this "vector" is in some Weyl chamber. The slope of the Lie algebra vector can be determined as the tangent function of the angle which can be determined (using inner products) between a root vector of the set of Lie algebra "toral vectors" and another Lie algebra "toral vector." Whether, the value of this slope (or angle between the set of nearest root vectors in the Lie algebra of a maximal torus), as in this case 1/10, is in one Weyl chamber or another depends on whether the angle between the nearest root vector (in regard to some given original [or first] root vector) is less than 90-degrees. If it is, then more than one Weyl chamber needs to be considered.

Does this angular relation between root vectors (in the Lie algebra of a maximal torus) provide a valid pattern to consider in a plane, since the real irreducible "diagonal" subspaces..., which are related to a particular circular component of a torus..., are 2-dimensional?

This factor (or value of a tangent function at a particular angle measured between a vector and a given original root vector), eg in the above case, 1/10, corresponds to a length of 10 around a unit-circle component of a maximal torus. However, it can also correspond to the measure of a circle's circumference for a circle component of a torus contained in a metric-base-space, where in the metric-base-space the circumference of a circle component is 10 = 2(pi)r, and r = 10/2(pi), for the circle component of the torus contained in the metric-space (base-space).

Thus Lie algebra "toral vectors" can be related to tori contained in metric-spaces which have various sizes, where the "toral vectors" of the Lie algebra can be associated to particular conjugation classes of the Lie group.

This context seems to provide a description in regard to the range of arc-lengths..., for circles of a toral component (or a cubical component) defined in the base space..., which can stay within a given conjugation class of a maximal torus in the fiber group. Thus, if the spectral lengths (of circular components) stay within this range, then the corresponding stable geometric shape, in the base space, can have its shape related to a set of diagonal "local coordinate" transformations which "map out" the full geometric shape in the base space, ie the description of the stable shape in the base space stays within the simple context of diagonal local coordinate transformations. How can this be related to the math structures of opposites, where the structure of mathematical opposites (or involutions, also associated to a function [or operation] being its own inverse, eg multiplying by -1, or multiplying by the complex number, i, or multiplying by diagonal matrices which have 1's and -1's along the diagonal, defining reflections in a geometric context, etc) are so prevalent in the root space and in the cubical foundations of these stable shapes (in the base space), as well as to the structure of opposite metric-space states, through which the (physical) properties of metric-spaces, and spin rotations of metric-space states can both be described?

Is this the natural limit as to the types of spectra that a single stable shape can possess in the base space?

or

Can discrete conjugation (perhaps taking place at the distinguished point of the stable shape in the base space) between conjugation classes still be allowed for a stable shape in the base space to have a wider range of spectral values associated to itself?

If such a particular type of discrete conjugation take place so as to increase the spectral possibilities for stable cubical shapes in the base space then could these discrete conjugations (which change maximal tori in the fiber group) also be related to the folding structure of the stable geometric shape in the base space? This folding, based on discrete conjugation, at a specific geometric location within a stable shape, is a very good model of molecular folding, so that this folding could exist in a very controllable context.

Consider a Euclidean 3-cube, there are three sets of 4-edges, ie 3 x 4 = 12 edges on a 3-cube, where the length of the edges (in one of these three sets) is equal (to, say, 10), then one wants to associate the length 10 to a related length of a circle's circumference for a circle component on (of) a torus contained in a metric-base-space. In turn, one wants the torus contained in the base space to be related to the circle components of a maximal torus in the fiber group.
*The 3-cube is a 3-torus in a 4-dimensional Euclidean metric-space, where Euclidean 4-space is related to a fiber group SU(4) whose rank is three, ie the dimension of a (any) maximal torus in SU(4) is three. That is, the dimension of the 3-cube and the dimension of SU(3)'s maximal tori are consistent.

The spectra of Euclidean n-cube (or n-subcube) is exactly similar to the spectra of a maximal tori (of rank-n) in the fiber group of SO, while the geometry of the (very stable) discrete hyperbolic shapes in the base space of a SO(n-1,1) fiber group, is a shape which is (can be considered to be) equivalent to cubes (of the appropriate dimension [equal to the rank]) attached only at vertices, so as to form a diagonal of cubical shapes, where each cube in the diagonal would possess spectra similar to the spectra of a maximal tori in the fiber group, ie the edges of the cube. Note: The spectra of the discrete hyperbolic shapes are much more limited than are the spectra of the discrete Euclidean shapes, where the spectra of the discrete Euclidean shapes are all the rational sloped lines, whereas the spectra of the discrete hyperbolic shapes are (essentially) only the edges of the cubes.

This defines a context wherein the local transformations of a shape on the base space can be determined by a set of diagonal matrices in a particular maximal tori of a fiber group, where the circles (or higher-dimension faces) which compose the maximal tori correspond to the stable circular paths (or bounding shapes) around the tori on the base space.
The context of non-linearity cannot stay within a context of coordinates related to the maximal tori, ie non-linear shapes are transformed in a continuous path away from a diagonal context of local coordinate transformations associated to the non-linear shape.

The problem with toral-shapes in the Euclidean base-space is that either cubes or rectangles are equally "good" shapes upon which a torus can be defined (in a base space), and thus tori (in a base space) have the property of continuity but it is difficult to maintain the specific conditions for the stability of some specific rectangular shape (eg a rectangular prism) which exists in the base space, in relation to any other such rectangular shape. Whereas the discrete hyperbolic shapes, ie the shapes associated to a diagonal set of cubes attached at vertices, are very stable in hyperbolic space, ie associated to an SO(n-1,1) fiber group.

So how are the spectra of the diagonal cubes (or cubes in Euclidean space) defined in the base space, related to the spectra of maximal tori on the fiber group?

There is either a set of discrete local maps [f(g,t(i),x) to Lg(x,t)(dx) at t(i)] from the fiber group onto (changes of material positions, x, on) the base space. [in this set of conditions the "opposite roots"

> (dg and -dg) define local opposite directions (for the local maps) for a pair of separate (opposite moving) paths whose interaction structure both begins and ends.]

Or

Due to the system (in the base space) being at the "right" energy and there existing resonances, so that the "opposite roots" define a pair of opposite (and consistent) paths (orbits).

There is an orbit of a point, x, of a (material) system in the base-space of the principle fiber bundle due to a toral subgroup (of the fiber group) forming the orbit by the subgroup acting on x in a continual manner,

> (however, this orbit space could also be a result of locally discrete actions by the fiber group on the orbit of x in the base space).

Or

One can think of the (either discrete or continuous) action of the toral subgroup on x is expressed as a circular subgroup, S (of the subgroup T(n)), acting on x, as Sx, so as to form a circular orbit (or a stable closed orbit) in the base space. Thus, there could be a further (periodic) scalar function, h(x,t), (based on geometric scalar values, and physical constants) so that the orbit [of a material system] in the base space is given by h(x,t)Sx (along with its opposite orbit).

That is, the dynamics of material contained in a metric-space can be described by a mechanism which most often leads to non-linear patterns since the descriptive structure is so very similar to the (non-linear) connection-form related to the derivative (but in a non-linear context), where the derivative is a local linear and thus allowing a consistent measuring relation to exist between a function's values and the same function's domain values, yet the interaction structure between stable shapes can at certain times during the interaction enter into resonance with an equivalent (dimensional) stable discrete shape which is part of (or contained within) the over-all high-dimension containing metric-space (base-space), and thus the relation between that stable shape (if the interaction has the "correct" energy) and the containing (hyperbolic, Euclidean, Hermitian (unitary), symplectic) metric-space's fiber group is quite different (than is the interaction process, though they are consistent, but this consistency depends on the interaction suddenly being related to a new resonating structure, eg new dynamic pathways, which the containing space (itself) upholds or supports. That is, the material of the interaction gets related to a new material [or new metric-space] shape related to the fiber group's orbital structures, where the cubes (tori) within the group get related to the cubes (toral components) of [within] a containing-space shape [cubes related to the shape of a containing space]), ie the stable under-pinnings of a (the) containment set are built from the properties of the fiber groups, namely, their discrete subgroups, which are subsequently related to the fiber group's maximal tori.

Can it (a locally diagonal relation between a dynamical path and its representation as an action by a fiber group) stay within the diagonal context of a maximal torus, or does it need to continually depart from a diagonal relation?

Can it (a locally diagonal relation between a dynamical path and its representation as an action by a fiber group) stay within the diagonal context of a maximal torus, or can it discretely change to other maximal tori?

Sets of discretely distinguishable opposites (sets of opposite roots, or characters, or eigenvalues) defined on a maximal torus (or its tangent space, within the context of the adjoint representation), in turn, define a discrete partition of a Lie group's orbit structure into conjugation classes associated to these toral eigenvalues. This is observed to be due to the properties of reflections which can be defined for vectors [defined on the local tangent space of a maximal torus (at the identity)] (equivalent to eigenfunctions) through the hyper-planes which are defined by the pairs of opposite characters (or roots). Does this observed pattern mean that sets of opposite eigenvalues are central to the stable orbital properties of dynamic or material interaction systems in the base spaces of "isometry fiber bundles?" Is a system of opposite values, opposite directions, opposite states, and geometrically opposite orbital structures the basis for stability of an orbital system, or for (continuously) staying within a conjugation class?

Would such discrete changes in a conjugation class be related to an associated stable shape, in the base space, being folded (associated by means of orbital properties of the shape)?

For the diagonal array of cubes, composing a discrete hyperbolic shape, can each cube be related to the spectra of one of the maximal torus?

or

Does each cube need to be related, by conjugation, to the spectra of some of (all) the other different maximal tori?

Consider acting (by means of the fiber group) on the base space with the toral transformations of the fiber group, so that the rank (the dimension of a maximal torus), n, is the same dimension as corresponding cubes in the base space, then the diagonal (of a matrix in a torus of a particular conjugation class) is filled with the same value exp (i 2(pi) t(j)z(j)), where the j's are summed to n, and the t(j)'s are real numbers from the real number continuum of the diagonal Lie algebra elements corresponding to the "diagonal Lie group elements" (from the maximal torus) and the z(j)'s are from the integer lattice Z(n). For the value t(j)z(j)) . . . , where the j's are summed to n . . . , to represent a closed curve on a torus, then either the value 2(pi), in exp (i 2(pi) t(j)z(j)), needs to be multiplied by a constant 1/a(j), or simply letting the a(j) be arbitrary real numbers]. For if the a(j) are constant real numbers then the [z(j)/a(j)]x=y line on the (x,y)-plane will intersect an element of the Z(x,y), or integer lattice defined on the (x,y)-plane when x=a(j)/z(j), thus assuring that such a curve will be closed on the T(2), or 2-torus, (or 2-subtorus of T(n)). Note: x is equivalent to the values t(j).

If one considers the extra (constant) factor of 1/a(j) then the different cubes (in the base space) associated to different a(j) values, in exp (i 2(pi) [t(j)/a(j)]z(j)), can have different relative sizes in the (base) metric-space (so as to be easily related to the maximal tori), in turn, associated to the eigenvalues,

> exp (i 2(pi) [t(j)/a(j)]z(j)) (or conjugation class functions). This would allow for more variability in regard to the set of eigenvalues which can be defined on the base space.

On a maximal torus (or on a conjugation class within a fiber group) the function value,

> exp (i 2(pi) [t(j)/a(j)]z(j)), (which identify "values" on a circle) remains a constant (stays [or is well defined eigenfunction] on the circle).

Conjecture: This corresponds to the lines in the Lie algebra associated to the Weyl chambers. This means that the range of values associated to (or allowed by) a particular conjugation class is determined by the values of slope, z(j)/a(j), (of the vectors in the Lie algebra) which stay within the "bounding walls" of the Weyl chambers, which, in turn, are determined by the root system structures, ie vector relations characterized by sets of opposite roots (or opposite vectors), related to the adjoint representation.

Assume that there is a set of "values" of the form, exp (i 2(pi) [t(j)/a(j)]z(j)), which define the set of values which are allowed on a conjugation class. The different functions of the form exp (i 2(pi) [t(j)/a(j)]z(j)), identify the different conjugation classes [on the adjoint representation of a Lie group] of the group (as well as identifying the different set of conjugate tori which cover the Lie group). Are these different conjugation classes defined by ranges of a set of values (or set of functions from the torus into the circle)? [If so] What is the range of allowed values (of these constant values which is associated to one conjugation class)?

That is, is there a set of such values (functions onto a circle) which define the same conjugation class? Answer: Yes, this would be the set of values which range over (z(j)/a(j)) as this slope stays within the Weyl chambers.

This is important, since each such different value (on one conjugation class) can be associated to one of the different cubes in the (base space) structure of the diagonal of cubes (which touch at their vertices). What range of constant values can be associated to a single conjugation class of a classical Lie group?

If the (fixed) value associated to a cube (in the diagonal of cubes in the base space) is not the value (of the eigenvalue of the maximal torus, which one wants) of a particular conjugation class (if this is truly the case for conjugation classes), then one would have to (discretely) conjugate between the different maximal tori in the fiber group, by means of the Weyl group. Thus, the eigenvalues of the set of diagonal cubes in the base space can have a wide range of values associated to any of each of the separate cubes, if the eigenvalue of each cube is discretely related to the (various) eigenvalues of maximal tori in the fiber group.

Is it "one eigenvalue for each conjugation class?"

or

Is it "A combination of a both a continuous range of eigenvalues defined by a continuous group action within each Weyl chamber, ie within a conjugation class, and (also) discretely related eigenvalues, defined by a discrete group action between the Weyl chambers"?

If only one value (only one function [whose values are defined on a circle]) is allowed for one conjugation class to be associated to a cube, then the set of maximal tori (which the full set of conjugation classes identify) would then have to also be "the same set of values from whence the different cubes (in the base space, of the diagonal of cubes) are allowed to have the different values for their eigenvalues," (so that the eigenvalues of the other cubes in the diagonal of cubes in the base space are different). That is, for a different cube in the diagonal of cubes on the base space to have a different value for an eigenvalue (from the eigenvalues of the other cubes) one would have to (discretely) conjugate between the different maximal tori in the fiber group, by means of the Weyl group, W = [The normalizer of T(n)]/ The centralizer of T(n) = (N(T(n))/C(T(n))=(N(T(n))/T(n))). where the order of W (the number of elements in W) is finite.

For example, in SU(n) the Weyl group, W = [The permutations of n-things, the symmetry group of n-things, ie n! permutations, on the n-elements which are defined along the diagonal of T(n)].

Note: The normalizer of T is the biggest subgroup in G within which T is a normal subgroup. The centralizer of T are the elements of G which commute with all the elements (or with each element) of T (one might believe that for (a) T this is T itself, ie the centralizer of (a) T is T itself, (yes, this is correct)).

But (if it is true that each conjugation class can only have one eigenvalue) and if each different cube has a different eigenvalue, then this would mean that there can only be as many cubes along the diagonal of cubes (in the base space), each with different eigenvalues, as there are conjugation classes in the Lie group.

Major questions:

Does a description based on non-linear geometry have any meaning, other than it being a quantitatively inconsistent description?

Does indefinable randomness have any meaning?

Do descriptions which are based on sets which are too big have any meaning? Do distinguishable patterns blend into other, different, distinguishable patterns so that the original distinguished property has no meaning, because the set is "too big"? Are opposites truly distinguishable from opposites in sets which are "too big"?

Can these math properties of: non-linearity, indefinable randomness, and defining math patterns on sets which are "too big," be used to identify (describe) stable properties (stable patterns) in a accurate manner (or are the descriptions unstable), and can these properties be used to build something (based on inter-relating measurable properties of the new system's regions and/or components) which is practically useful?

The above description based on cubical-related stable geometric shapes is ultimately based on a finite set, namely, the finite set of stable spectra which can be fit into an over-all high-dimension containing space where the containing spaces have stable shapes. This finite set determines the set of material or metric-space types which can compose the containing space. The highest stable dimension is an 11-dimensional hyperbolic metric-space, but there are no 11-dimensional hyperbolic shapes.

There are two parts to the current authoritative viewpoint of physical existence, both parts are based on the idea that material determines existence. One part is classical physics which is a description which is widely applicable and useful, but it does not account for the spectral structure of microscopic material systems, and the other part of physics is quantum physics, which tries to describe the spectral (and apparently random) properties of micro-systems. Even though quantum physics is framed in the all inclusive language of probability (since the set of a system's events can always be adjusted to account for the events of any physical system within the language of probability) nonetheless quantum physics is only applicable in a narrow sense, and other than describing "in a general way" the stable spectral structure of micro-systems (which is really just an empirical fact), quantum physics has not shown itself to be all that useful. For example, the spectra of high atomic number atoms are not computable using the laws of quantum physics.

That is, it is classical physics which is still the basis for technical development, eg electronic information systems, eg computers, are more derivatives of the circuits of TV (and other aspects of classical physics) rather than it being a result of quantum physics. The quantum systems of semi-conductors can be used in circuits because the classically describable circuits can be controlled to such a high degree that this allows a circuit's properties to be coupled to the sensitive voltage properties of semi-conductor energy states.

There is an alternative basis for a mathematical description of (physical) existence, ie alternative basis for physics based on discrete isometry subgroups. Indeed it is a new description which encompasses classical and quantum physics, but it is also a new description which goes beyond these descriptions in a manner in which the descriptive language is coherent and it describes the properties of physical systems which the current language is not capable of describing. Furthermore, the ideas upon which the new language is based extend past the notions of materialism in a way which is applicable to describing the properties of life and mind, and it extends the notion of existence into higher dimensions, where the description could be very useful in extending human exploration into the unknown.

Discrete isometry groups imply the existence of space-forms. Whereas space-forms can also emerge from the mathematics of differential forms or Laplace operators, ie Newton's gravity, and electromagnetism. Space-forms are also related to the spectra of systems. The properties of both physical systems and space-forms depend on the implied spectral resonance between spectral sets which can be defined by the properties of space-forms, through the relation that "spectral-lengths" (or spectral areas, etc) can have to space-form geometries. Space-forms can be derived either from discrete isometry subgroups [and in the hyperbolic (metric-space) case, the spectral lengths of space-forms have very fixed and stable set of spectral properties] or from differential form solutions to metric invariant differential operators, ie Laplace or wave operators, or Dirac operators.

The crux of the new model is that both space and material can be modeled as, and contained within, space-forms. That is, both material and space can be modeled as a space-forms, thus space and material (space-forms) have resonance's with one another. Note: The values of physical constants can be related to this spectral-geometric structure.

Inertia is about displacement in both rotational and translational frames of Euclidean space. The spin states within a metric-space existence are related to metric-space properties of displacement (momentum) and the direction of time (energy), where physical displacement of material space-forms depends on rotating the metric-space states of time [(+), (-)-time states) in hyperbolic metric-spaces.

It should be noted that charge is contained in hyperbolic space and is related to electromagnetic fields, or equivalently, space-form structures found to be properties of differential forms, and the electromagnetic field has a relation to both the space-form structure of charge as well as being related to an associated inertial space-form structure. That is, different metric-spaces are closely associated with one another in the composition of physical systems.

Furthermore, interactions and motions of micro-material (or micro-system) space-forms and their interactions occur often in time and are random in nature (like Brownian motion), thus they are stochastic processes, therefore according to E Nelson the motions of micro-entities which interact often are wave-like in nature, ie quantum mechanical particle-waves. On the other hand space-forms have distinct points on themselves about which their interactions are centered, thus micro-entities modeled as space-forms appear to be both wave-like and particle-like.

However, in quantum physics, describing systems based on random particle events so as to identify spectral properties with operators acting on function spaces has not been conclusive, or all that useful, in relating the spectral relations of physical systems to probability. This is mostly because of the relation of a quantum system's function space to probability, has problems because of the relation of operators to function spaces is not understood, or has not been widely applicable, eg it has not been applicable to high atomic number atoms nor to nuclear orbit properties, nor has it been all that useful. That is, quantum physics follows data better than identifying uses, because the understanding of quantum physics has not penetrated the relations that operators have to function spaces, where most operators are found through analogy with classical physics properties (or descriptions), thus quantum physics has trouble extending itself past classical physics.

On the other hand the new description is a geometric-probability relationship, which is not fundamentally probabilistic, but rather is fundamentally geometric. Thus the new language is not logically inconsistent as the language of string theory is, ie where the language of string theory is both geometric and probabilistic, thus the new language is the type of descriptive language for which string theory is really looking.

Furthermore, thermal and wave-like (quantum) physics can now both be placed within the differential form structure of space-forms, where thermal system parameters, eg volume, number of atoms, and state and/or energy, can be placed in a space-form geometric structure. Thus space-forms fit from: classical to micro-system and they fit from: micro-system to thermal system, in a better (or more natural) way than regular quantum and classical thermal physics, since space-forms are differential forms and thermal systems are described by means of differential forms.

The particle-states in high energy particle collisions, that are seen in particle accelerators, are really properties of both material and material interaction systems' which have relationships (partly by means of unitary fiber groups) to higher dimensional space-form structures, so that material geometry and its interaction structure depend on particular dimensional (and geometric) properties. This causes the geometric and interaction properties of material space-form systems, within various dimensional levels of existence, to be independent of the higher dimensional structure of existence. Furthermore the geometric properties of higher dimensional interactions are not spherically symmetric, as such interactions are in 3-space (see below).

Review: Isometry groups, their discrete subgroups, and the base space upon which both the isometry fiber group representations depend, and upon which the metric invariant differential-form relationship between both material and space (through space-forms) are defined, so that these systems' properties "of dimensional and [their space-form] metric-space-state (material space-form) interaction structure" are geometric, and these dimensionally independent geometric properties of the different dimensional levels are the main structures of physical description.

These structures can be used to account for the macro-material, the micro-material, and the entropy -spectral-geometric properties of physical systems, and these structures allow such physical systems to be described in a multi-dimensional, spectrally inter-dependent structure of containment.

Thus the differential structure of manifolds, or the local transformation structure of atlases for manifolds, which can be used to model local system structure, so that on certain (or particular) manifolds there also exists a property which allows both the local (derivatives) and the global geometric structures to be contained within a flat (or metric invariant) metric space, eg within a space-form geometry, so that the flatness of these manifolds characterize classical systems, but the variation of this descriptive structure caused by the (containing) space's discrete isometry subgroup structure is central to all physical description, allowing quantum physics and classical physics to merge together.

That is, the isometry groups (of these new spaces) identify a new discrete group structure with which the spectral properties of physical systems can be described so that this description is applicable to higher dimensions so that each level (for any particular dimension level) is geometrically (and essentially interaction-ally) independent of the other dimensional levels. Yet the spectral properties of different dimensional levels can be spectrally quite relevant to any particular dimensional level. This descriptive structure provides a new structure, a local spectral structure, within which to describe both life and mind, as well as allowing the spectra of higher atomic-number atoms to be related to metric-spectral structures which are beyond the typical 3-dimensional metric-space within which we constrain our perceptions. However, the different dimensional levels are geometrically and interactively independent, or conservative.

This new descriptive context also provides a new context within which to describe or identify non-linear (differential equation) systems and their properties.

The spherical symmetry of 3-space results from the SO(3) part of the Euclidean isometry of fiber group, SO(3) x R^3, ie the differential structure of a Euclidean manifold, whereas the three dimensional Euclidean space's space-form structure comes mostly from the R^3 translation part. However, Euclidean 4-space has an SO(4)=SO(3) x SO(3) isometry group structure and thus cannot have a radial-spherical interaction geometry.

Chapter 25

CUBICAL SIMPLEX

"Cubical" simplexes revisited

The failed dogmas of modern (2012) science and math,
Namely, the failed attempts to try to describe the following observed properties of the world, that:

1. Ordered, stable "physical systems" emerge from a context of randomness,
2. Life developed into greater complexity by evolving within a random context (which, supposedly, leads to greater complexity).

These are failures which result from the idea (belief) that math and science need to have a descriptive basis which depends on indefinable randomness and non-linearity.

The new ideas expressed in this paper, can claimed by anyone as their own, and used without fear of copyright infringement, except that they are original ideas, and thus as an originator of these ideas it must be stated that they cannot be used for military purposes, or for purposes of domination, or destruction, or they cannot be used within a society whose law is based on property rights and not based on equality, and in a society where the law requires that the power to govern is obtained from the people, who are all equal. That is, a dominant, unequal, group of people within a society (the owners of society) cannot derive value for their selfish purposes from these new ideas. These ideas are meant for practical creativity, motivated by selflessness.

The failed dogmas of science and math,

Namely, that:

1. Ordered, stable "physical systems" emerge from a context of randomness,
2. Life developed into greater complexity by evolving within a random context (which, supposedly, leads to greater complexity) . . . ,

[Note: This is a statement about ideas in general, it is not a statement in favor of religion, since religion has virtually nothing to contribute to understanding the observed property of order which is observed to exist within the world, rather "how order emerges from the world" is to be considered a question about the math-assumption (or math basis) for a valid mathematical description. In an equal (and free) society, people should be allowed to question authority in a meaningful way (this is how learning develops).] (This disclaimer is needed because of the distorting affects on people's thinking (the belief that they belong to the learned class) brought on by the social condition of wage-slavery, in a society based on inequality.)

. . . , are ideas which result from a math and science which is based on indefinable randomness and non-linearity.

Such a mathematical basis (as indefinable randomness and non-linearity) can be used to describe many of the details which exist within an observed, measurable, descriptive context, but, these details are unstable and fleeting as are the math structures which are based on indefinable randomness and non-linearity. Apparently, indefinable randomness and non-linearity cannot be used to describe "from whence" the observed order of the world comes. Furthermore, indefinable randomness and non-linearity has not been able to do so during the 100 years within which indefinable randomness and non-linearity has been used to try to describe the observed order of the many fundamental material systems which are observed to possess the property of very stable order (nuclei, atoms, molecules, crystals, life, mind, planetary orbits, etc).

The failure of these assumptions (indefinable randomness and non-linearity) to be able to describe the observed order of matter (and life) is a result of the fact that indefinable randomness and non-linearity are math ideas which are quantitatively inconsistent, improperly associated with a quantitative system (by means of both a set structure, and the math structures which model measuring and/or counting), and dependent on very large sets (ie sets which are "too big") wherein the meaning of words do not remain definitive (ie the type of thing to which the word refers changes, [the meaning of the word changes] while it [the thing] still appears to be a quantitative value, as it is used in a mathematical description), eg there is a plane-filling curve (a 1-dimensional pattern is equivalent to a 2-dimensional pattern).

Though the properties of indefinable randomness and non-linearity are observed in the "transitional structures" of both physical and living systems, the very stable, definitive, discrete spectral-orbital structures upon which our existence depends, as well as the math context through which order and complexity in living systems develop, are not being adequately described. This suggests a need to consider "how it is that very stable (yet still complex) math structures, eg 'cubical' simplexes, can be related to indefinable randomness and non-linearity?" rather than considering ideas which develop exclusively along the opposite logical path (of language's logical structure), namely, that order emerges from indefinable randomness and non-linearity, ie the current assumption.

The assumption that "order emerges from indefinable randomness and non-linearity" is the basis of the current (2012) authoritative dogma, which is considered to be an "unchallengeable fundamental assumption," ie a fundamental dogma, but this assumption needs to be challenged in fundamental ways.

There are a large number of mathematical contexts which are not considered. In math certain contexts are explored, where if certain contexts (apparently) possess many describable patterns, then these contexts become traditional topics of study within mathematics. Yet these traditional contexts may not be the patterns which are best used to describe the observed patterns of existence in a precise and practically useful way. There are very many mathematical contexts most of which are not considered.

The new assumption that very stable math structures, such as "cubical" simplexes, are primary, and that randomness is a derived property, is about re-organizing the language of math so as to develop a measurable description of existence which can be used in both a descriptive context (which can adequately describe the observed stable properties) and in a practically creative context, ie used within the context of existence itself [not simply existing within un-useful descriptive dogmas] Note: The new idea of existence extends beyond the idea of materialism into higher dimensions. The new context is that, there are layers of (adjacent) dimensions, so that each dimensional level is a stable shape, and that each dimensional level can contain many lower-dimensional stable shapes (which can be considered to be either closed metric-spaces or material systems).

This is essentially this is the constant curvature metric-space spaces of: $SO(s,t)$ and $SU(s,t)$ as well as $Sp(2s,?)$, where $s + t = n$, and s is the dimension of the spatial subspace, while t is the dimension of the temporal subspace, apparently up to $SO(11,1) = H(11)$ (hyperbolic space).

This new context is not a traditional math context. Apparently, this is because the math structure with which this new context is composed are very simple, namely, closed, bounded, stable shapes associated to "cubical" simplexes. Spherical symmetry is non-linear, and thus unstable, and so it is not useful.

Whereas people will always enjoy creating in a context of art, ie a sense of balance, as well as developing the discipline for accurate replications (models) of the world's beauty, and also in relation to an abstract sense of beauty.

But now, in a culture which "focuses on markets" which serve big financial interests, art is an expression which requires a notion of being caught-up in a narrow context, where careful judgments can be easily made so as to distinguish the so called superior art and superior artists, from the inferior [Note: Van Gough was placed in the inferior category.]

Thus, the artist became a producer of a "creative" commodity, in a context of distinguishable value.

This idea of distinguishing superior art, within a narrow context of a market, has been used to create the idea that people are not equal.

In professional math the same focus on superior and inferior is also required (by the owners of society) thus math traditions which are filled with a lot of complication are the desired set of contexts, and this leads to indefinable randomness and non-linear math patterns.

The ideas expressed in this article are very much pro-science, or equivalently pro-practical-creativity, where practical-creativity depends on "practically useful" knowledge. Note: The natural context for the creativity of a human being (an equal human being) may not be the context of materialism.

By science "it is meant" being able to use a set of described "measurable properties" to build things.

As opposed to an authoritative science whose descriptions depend on math structures which are unstable, inconsistent, and whose meanings cannot be pinned-down, so that in such an authoritative scientific context in order to identify a 'precise' description of patterns the measurable properties of such patterns only exist in a fictional world, and thus the quantitative descriptions are not practically useful. These visions of an authoritative science provide creative descriptions which only relate to a body of literature which represents the authoritative dogmas, which describe a fictional world, (ie science is no longer science, rather science is now science-fiction).

Traditionally, the idea of "measurable science" has meant descriptions of the material world. Note: Even the esoteric string-theory requires a special math structure (local space-time manifolds) to accommodate and maintain the idea of materialism.

Nonetheless, these traditional (professional) descriptions of a fictional world are claimed to be measurably verified, but this verification represents a process of data-fitting similar to the data-fitting associated to Ptolemy's measurable verification of his fictional world, (where Ptolemy's model of his fictional data-fitting math structure was also capable of being measurably verified).

A true useful measurable description, based on the stable math of the geometric structures of "cubical" simplexes, seems to be pointing towards human creativity which transcends materialism, and enters into dimensions which exist beyond the material world, but it is a description which contains the material world as a subset.

That is, the higher dimensions are not exclusively composed of abstract mental constructs, ie they are ideas which are not "literary creativity," as the authoritative Physical Review is today, rather it is about a creativity which involves the intent of (knowledgeable) life interacting in a greater context of existence. It is a knowledge of the "true" structure of existence, which can be used in a context of measuring, in turn, used to form a new system, or so as to build something new.

But these new ideas about "how materialism (and a greater containing structure for existence) should be mathematically modeled" are outside of the current dogmas of science and math, ie outside their traditions of (absolute) authority, ie outside of what is considered to be the current belief, a belief which possesses an extremely high-cultural-value in regard to math and science (where science and math are represented as a difficult to understand descriptive structures which are used to define inequality amongst the people who compose a society). That is, science of authority is used in a similar manner as is commercial art, it is mainly used to identify superior people within society, so as to prove that people and not equal.

Simply because descriptions of non-linear curvature can be described as algebraic structures of local geometric measures (of curvature of a geometric shape), this does not mean that these algebraic patterns which are associated to curvature (which identify a local non-linear geometric measuring relation between a function and its domain space) are related to stable measurable system structures, or to stable descriptive math structures, eg the algebraic structures of systems of non-linear differential equations can be described, but they are inconsistent quantitative structures, which are also not practically useful.

On the other hand, "cubical" simplexes can be extended to higher dimensions in a stable manner, ie they remain stable geometries in higher dimensions. Furthermore, these high dimensional geometries are not only related to the structure of a material world, but they also are related to (or lead into) very interesting contexts, where both material and metric-spaces can have stable shapes, and many new math-physical properties can be defined, so as to be (or determine) a physical map which one can follow into higher dimensions. Both a map for the mind, and a map to practically, providing creatively useful patterns, including simple models of life, and its associated memory, where the intent of a living system can be modeled to interact with a further context of existence.

Perception can also be modeled.

Furthermore, as in the case of Copernicus, the new math-physical model has the capability to better predict (or fit) data. Remember, truth is about a measurable, precise description's usefulness, while a descriptive language (based on new assumptions) organizes how the observed patterns (or the facts) fit together.

Truth is not about absolute truths.

**Particle-physics is a bad joke, a tribute the manipulation of obsessive people and to the social control of all aspects of the language which is used in society. The idea of the existence of a "scientific consensus" is due to the relation that science and math have to wage-slavery.

What does the observed property of precise quantum discreteness imply? Answer: (obviously) System discreteness, and (some might be surprised) metric-invariance of the descriptive context, and the need for stable math patterns to be used to describe the stable discreteness of quantum systems.

Precise quantum discreteness implies metric-invariance. Otherwise, if quantum interactions were truly non-linear then the local properties of indefinable randomness, associated to local non-linear curvature, would interfere with a quantum system's stable discrete properties. [Note: The non-linearity of particle-physics is expressed in a math context where the fiber group is unitary, or Hermitian-form-invariant.] Thus, in order for the local properties of indefinable randomness to not interfere with a quantum system's observed stable discrete properties, the non-linear property would have to be both local (as all curvature can be represented as local) and global, and thus the need to search for a non-linear global function, which contains "all information," so that the local properties of indefinable randomness do not interfere with a quantum system's observed stable discrete properties. That is, the global non-linear function would be needed to compensate for the condition of local randomness which is associated to non-linearity. Otherwise the discrete properties of quantum systems would not be stable, and a quantum system, such as an atom, could not be identified by their definitive spectra (because in the case of the existence of local non-linearity, a quantum system would not have a stable spectra associated to itself, because the local condition of the quantum system would be a condition of randomness, and thus no two quantum systems would have the same properties).

That is, the discrete spectral properties of quantum systems would change from point to point within its containing space because of the local randomness, which is associated to any point in space within which the quantum system exists.

Good luck at finding such a function, which contains how the local discrete spectra change from point to point in space due to local non-linearity, since what is observed is the same definitive stable discrete spectra of a quantum system which is independent of its position in space.

It is observed that quantum systems have the property of non-locality, or action-at-a-distance. Thus, the property of inertia is a property of Euclidean space, where it should be noted that this is not a part of a (gravitational) manifold which is locally space-time.

However, material interactions in 3-space are spherically symmetric in the new descriptive structures.

> The physical sciences and
> mathematics, along with
> the foundations of biology, as well as (it also turns out)
> "the basis for understanding the spirit,"
> (all these subjects) need to be based on the math of:

1. Stability, the stable discrete shapes of spaces of non-positive constant curvature which are defined over (under):

   (a) $SO(s,t)$, so that $s+t = n$, and similar
   (b) $SU(s,t)$, and
   (c) $Sp(2s)$, and possibly
   (c2) $Sp(2s,2t)$ (?),

   principle fiber bundle spaces,

   The exotic Lie groups may also be important, but the above are the simplest of the stable structures, in regard to both measuring and geometry (especially in regard to lower dimensions).

2. Sets which are defined, so as to be consistent with an idea of descriptive confinement, so that ideas actually have meaning, ie sets cannot be "too big," and the "type of number" can remain well defined within the (simple) set structure, the limit process does not change the number-type, and the limit process (which is dependent on the order of numbers) is actually defined with a finite amount of information (so that it is not clear what the set structure "actually is"),

[Note: When sets are "too big" then questions about "number order," the "existence of a number," and "number-type," cannot be answered without an infinite amount of information (G Chaitin), ie the questions go unanswered. Yet answers are assumed to exist, but the meaning of the math patterns which are described, when one assumes that such answers exist, are quite questionable, ie such descriptive patterns are not to be trusted.

Either "finite sets" or some cut-off as to a limit to the capacity to measure within a quantitative structure, within which one's descriptions of math patterns exist, are needed in order for one to truly believe in the consistent structure of one's descriptions.]

3. Geometry, so that the description is both stable and use-able,

4. Each point of a coordinate space, upon which a stable system (both material as well as metric-spaces) is defined, is composed of a set of independent (and orthogonal) measurable sets (parallelizable and orthogonal, or geometric separability),
5. Local linear (models of local measures on systems), and the matrices, which are "local coordinate transformation matrices" of a stable system, are diagonal, though they are not so much (local) dynamic systems as they are stable geometric shapes,
6. The descriptive context transcends materialism (the description includes higher-dimensions) [this can lead to great complexity],
7. Physical constants are conformal factors which are defined between the different dimensional levels, which are (effectively) "geometrically independent" of one another, or constant factors defined between different subspaces of the same dimension,
8. Each metric-space can be identified with a (physical) property, a property which the stable shapes (of the metric-spaces) can preserve within a continuous context, where the continuous context is defined on the next dimensional level, ie defined on the metric-space within which the material property (or the discrete material shape) is contained, ie both metric-spaces and material are the same type of stable geometric construct (or shape) separated from one another by a dimension,
9. Along with a Real metric-space structure, $R(n)$, there is also a Complex coordinate structure, $C(n)$, so as to allow for the two independent "metric-space states" which can fit into a complex coordinate space for each dimensional level, ie the physical (material) property (associated to a stable geometric shape) has two states (see below), thus the stable geometric shapes, ie the stable material systems, are also a mixture (or built from "separated") opposite metric-space states, upon which spin-rotations of metric-space states are defined.
10. There is a spin-rotation between the pairs of metric-space states of each metric-space (or equivalently within each material system), and the time-period of the spin-rotation between the metric-space states can be used to identify a discrete process of change defined around the center-of-mass of (or between) two (or more) interacting material-shapes, which are relatively stable,
11. The (continuous) shapes involved in the force-fields of material interactions are tori, and for each new discrete time interval (defined by the period of the spin-rotation of metric-space states) these (interaction) tori re-fill the material-containing metric-space. For each such, re-invented, interaction associated to each such small time intervals, the relative positions of the interacting stable material shapes are changed in a relation to the local coordinate transformation fiber groups (of the principle fiber bundle). The tori of interaction can fill space because of an action-at-a-distance property which is a part of the material interaction process (or the tori form interaction shapes instantaneously) so as to define the context of a local measurement for each time interval defined by the spin-rotation of metric-space states (as well as providing values of geometric measurements and conformal factors [or physical constants]) which is (are) involved in describing a material interaction. These local and

global geometric measures on a material containing metric-space are also related to the local coordinate transformations of the interacting material positions in that metric-space (in Euclidean space the spatial displacements [or inertial changes] are defined).
12. The stable shapes of material systems are discrete hyperbolic shapes (which are defined over a range of dimensions, but which seem to no longer exist beyond hyperbolic-dimension-10 [due to Coxeter Theorem]).
13. Finitely defined elementary-event spaces, so that the elementary-events are both defined (identifiable) and stable, are the types of event-spaces which are needed in regard to the descriptions of random-event phenomenon (so that the calculated probabilities can be trusted in a meaningful context of randomness),
14. That is, indefinable randomness and non-linearity are ruled-out of the descriptive structures of stable systems, where it is assumed that "one only needs to describe the stable structures of systems, or of existing (math) structures," and therefore, one should deal with fleeting, unstable system structures in a qualitative manner, where only certain aspects of the description of an unstable system, (perhaps) defined as a set of differential equations (in a metric-invariant metric-space, in spaces of non-positive constant curvature), need to be considered in a descriptive structure, which is identified above.

Both the stability of a measurable description, and the stability of a measurable system, are both of great interest, in regard to the use of math patterns to provide an accurate and practically useful descriptions of a system.

The observed pattern is assumed to be stable (or has been seen to be stable) so that the described pattern needs to be stable and the descriptive words which identify properties and categories within the system's description need to have a fixed narrow meaning (for the description to be precise).

In math, this seems to require that the description fit into a set which can be defined "to be finite," but still within a continuous context of energy and material, for each dimensional level.

The above identified math structures which allow for stable math descriptions can be both:

(1) accurate for general systems within a sufficiently precise descriptive structure, and
(2) the description identifies a practically useful context, wherein technical development is widely related to the descriptive context (since it is a geometric descriptive context).

It should be noted that the idea of "peer reviewed" science and math is not valid, since it is exactly equivalent to requiring that the ideas of Copernicus be edited by the pope, but now it is the authoritative assumptions of "science."

That is, absolute authoritative descriptive structures are too narrowly defined so as to not be able to assure anyone, that these absolute ideas contain the truth, where "truth" means, sufficiently precise descriptions which can also be applied to the observed properties of the world in a practical creative manner.

"Peer review" is about narrowly defined categories which have been identified by business interests, and then these narrow, absolutely based descriptive (measurably verifiable) languages define an institution (defined by its fixed traditional structure, which enshrines authority) wherein the expert authoritative leaders, of such absolute institutions, possess the same personality characteristics as do the owners of society. Namely, these authorities are: competitive, narrow, obsessive, manipulative, and they seek to be domineering. However, these authorities (many of whom are psychopaths, eg E Teller), are represented, to society, as being the supreme examples of human intellectual capability, so as to form a notion (or a belief) within the public, that within our society, "our culture really does understand everything."

Scientific consensus is really the relation which scientific authority has to wage-slavery.

This is clear in the global warming debates, where the governing rule of the relation between carefully considered possibilities and public policy should be:
If problems are (truly) possible then society should correct itself, ie societies should be adaptable rather than being both rigid but (nonetheless) easily manipulated within its different levels of social authority.

(some of) The new mathematical descriptive structures

The two real, but opposite, metric-space states, naturally fit into the two subsets of $R(n)$ and $(iR(n))$ which are contained in $C(n)$ and are mixed by the spin-rotation of states in $C(n)$, so as to be a part of each time interval of the "dynamic changes" of the material interaction, which involve material displacements whose dynamic paths are contained in the two opposite spaces, in the two subsets $R(n)$ and $(iR(n))$.

At a point in $C(n)$, which is along a dynamic path, one needs two (independent) global position functions of a dynamic interaction, which are only locally related as (opposite) independent directions of spatial displacements.

For stable systems the entire descriptive structure in $C(n)$ needs to be "geometrically and temporally separable."

For general dynamic paths the local set of opposite directions do not have a strictly causal, retrace-able path direction at each point (ie the path can be non-linear), though they are caused by both the locally measurable structure of toral shapes (which have properties of action-at-a-distance) and their relation to local coordinate transformations which identify spatial displacements due to the interaction (this is similar to the structure of a connection, ie a generalized derivative).

**During a material interaction a "total time of interaction" can be determined so as to define (with the help of the speed-of-light) a spherical ball which confines the two (opposite) dynamic paths, so that in the two separate balls, in the two subsets of R(n) and (iR(n)) respectively, there are two independent and opposite dynamic paths, whose relative times can be identified from the mid-point of the total time (interval) of the interaction, [where the two dynamic paths meet (or originate) at the mid-point time of the total time interval of the interaction]. Thus, the opposite states (of the dynamic path) can be made to coincide in regard to time measurements along this time interval. That is, from the time mid-point the two opposite time directions can both be given relative directions of time so that the two time values coincide at the mid-point and then are related to one another by the "relative opposite time directions" measured from the mid-point in time, for the two opposite time directions.

Consider two ways in which organize a complex structure for C(n):

(1) related to a spherical space, eg the Riemann n-sphere associated to R(n) [which can also be thought of as being in (iR(n))], and
(2) a disc (or loop) space of a disc associated to each C in C(n), (or a C-plane normal to each direction of R(n)).

The sphere has positive curvature and it has different curvature properties at different points (the metric-function of a sphere does not have constant coefficients), while the "disc-space" can maintain constant curvature, which would be a non-positive curvature (which could be zero curvature or a constant negative curvature).

Does one extend dimensional independence in C(n) by a pair of independent subsets R(n) and (iR(n)) [each containing an independent dynamic curve] or does one take R(n) and attach a C-disc to each independent direction?

The disc-space (at high enough dimension) can represent either the, high-dimension spectral context of a material interaction (which is contained in an over-all high-dimension containing space), or the spectral context of the (lower dimensional) material-toral interaction structure.

Can both structures be simultaneously true? Apparently, Yes.

The most significant example of identifying a mid-point in the time interval of interaction would be in relation to when a new system forms.

This could be modeled as a classical collision of relatively stable shapes. (or in the opposite sense when a stable system breaks apart, due to extra added energy). Where the new system depends on forming into a set of independent discs which are in relation to a fixed stable spectral structure of the over-all high-dimension containing space.

A large spectrum can be defined on a maximal torus and a (fairly large) set of multiplicative factors (or physical constants) associated to the set of sub-tori for each different dimensional level.

The spectra of a maximal torus can be related to either the entire spectra of the high-dimension over-all containing space,

or

To the slightly changing spectra of a material interaction.

This toral relation would be defined on a cube or rectangular shape, the n-cube could be represented as an n-torus in R(n+1).

The coordinate structure of the two opposite metric-space states associated to a [C(n),t] coordinate system are locally exactly opposite dynamic pathways, but globally, in [C(n),t], the spatial displacement paths diverge from one another as time changes, but can be made globally (locally) independent to one another by using two representations of the two "opposite position paths." One a global spatial position function of the (spatial) displacements of material shapes contained in R(n), and the other time direction of the dynamic path identified as a "position path" in (iR(n)).

(the two directions of) time, t and -t, can be identified on the two subsets [R(n),t] and [(i(R(n)),-it] in space as containing oppositely directed paths.

In the complex coordinates, the (i(R(n))) subset is associated to (-t), and the (R(n)) subset is associated to (+t). Thus, there are the two sets which identify a pair of dynamic paths in C(n), during a material interaction, which can be identified as being [R(n),t] and [iR(n),-it], where for local measures of time, either dt or d(-t), (these local measures) identify oppositely directed paths, at the point where the displacements take place at a time interval so that the opposite path stays in both sets, [R(n),t] and [iR(n),-it].

The origin (or end-point) of these dynamic paths are either regions where the various material interactions are all about equal, or where, during a collision (or a break-up, or sudden change), the geometric-interaction-complex resonates with the containing (spectral) set, so as to form a stable system (contained in the next higher-dimension metric-space), where everything stabilizes and becomes "geometrically separable."

From these origins there is defined a well identified dynamic (or spatial displacement) path. The midpoint, in time, for these two opposite paths can be identified.

The pairs of opposite paths are present at each point in the interaction process. That is, the dynamic paths exist at each time interval of the two opposite dynamic processes, yet each separate state seems to be individually determined (in relation to the geometry which exists during that time interval) for each short time-interval wherein certain changes are (causally) realized on (along) the (continuous) dynamic path so that these causal relations can be determined for either of the pair of opposite paths in an independent manner for each path, associated to its time direction.

However, locally the two path directions are opposite, but if they were determined in a causal manner and the path was a result of a non-linear relation, then the two directions at "the given point in time" would not necessarily be in opposite directions, this is because non-linear relations are indefinably random so if the paths at the same point in time are determined in a casual non-linear context for each separate time direction then they would not necessarily identify two opposite path directions.

At a point in C(n) there need to be two global position functions, which are only locally related (as opposites), in regard to dt and d(-t), while globally they are separate.

These two global functions (dynamic paths) form a pair of local independent metric-space states, as well as a pair of local independent coordinate structures which are the same path but spatially directed along the opposite direction.

From the time mid-point a real time relation between the two opposite times can be inter-related.

But if there is a property of the dynamic path being a pair of local linear and independent orthogonal coordinates as well as metric-space states "geometrically separable as well as being time-separable, eg (i(t)) vs. (t)," is only assured in the linear separable conditions of a stable system which exists in two independent metric-space states. A (periodic) position and its opposite position within a stable system relate the spatial time-path in one metric-space state with the opposite (spatial) time-path.

Thus, there could be a relation between opposite roots (of the fiber group's Lie algebra root system) and the weights, or eigenvalues (or spectra), identified on the fiber group's maximal tori, but which are related to of the systems which are contained in the base spaces of the principle fiber bundle. Thus one would "conjecture" that the pairs of opposite roots, in a fiber group's root system, identify opposite dynamic paths and that these opposite paths naturally fit into orthogonal parallel geometric shapes so as to be geometrically consistent with the pairs of opposite roots in a fiber group's root system, so that the local spatial translations are closely associated to the diagonal matrices of the fiber group's maximal tori.

One-half of the roots in the root systems of fiber groups are opposites and this is consistent with a geometry where there also exists an equal balance of a set of opposite vectors which can identify the geometric shape, thus identifying a stable shape upon which can be defined dynamic material flows, so that there can exist a stable geometric spectral structure. That is, sets of opposite vectors identify stable geometric and stable mathematical (spectral) patterns.

Such a stable system depends on the system being resonant with the spectra of the over-all high-dimension containing space for the stable system, where that spectra can be represented on the maximal tori of the fiber groups.

A pair of globally opposite functions in [C(n),t], ie but time independent so as to be contained in R(n) and iR(n), requires a spectral memory in SU(n) related to a pair of "opposite" SO(n) fiber bundles.

symplectic group and the space of interactions

In symplectic space, ie x^p or Sp(2n), a material interaction is related to a geometric interaction structure of, T(n) x H(n-1)/#, which is contained in R(n+1) x H(n), or contained in an, n-cube x H(n), however, the dynamic path is contained in R(n) x H(n).

When n=2, then the interaction structure is a Euclidean discrete shape (or Euclidean space-form) is 2-dimensional, ie the dynamic interaction is contained in 3-space, that is the dynamic path of the interacting material system is contained in a symplectic space, and its spatial displacements and its changes in motion (or moving displacements) are contained in R(2) x H(2), ie the path of spatial displacement is in R(2), and thus the inertial spatial displacements are on a plane.

Does this also imply that the interaction will be based on a 2-body (material interaction) problem? No.

For a material interaction there is a Euclidean interaction discrete shape, T(n), and then there is the stable material discrete hyperbolic shapes, H(n-1)/#, where, #, is a discrete subgroup, which are the materials (which are interacting).

Thus, the interaction space, T(3) x H(2)/#, (where, #, is a discrete subgroup) is contained within, R(4) x H(3) so that the dynamic path is contained in R(3) x H(3).

Chapter 26

CUBICAL WORLD

"Cubical" simplexes instead of materialism

"Media Dissidents" (academics or best selling authors) who express a desire for a fairer society (they are not for "equality within society," rather they are for "a more watered down goal" of "a fairer society," ie these intellectual dissidents are essentially elitists), many of whom are willing to challenge the authority of religion, nonetheless, they do not consider challenging the dogmatic authority of math and science, despite the failure of current math and science "to describe the observed order, to a sufficient level of precision, of general physical systems, from the nuclei to the solar system." Not to mention the calculations of risk by these authorities should not be taken seriously by anyone (unless the elementary event space is: finite, well defined, and stable). [Yet, these authorities dwell on irrelevancies, such as black holes and Higg's particle, which seem to only have relevance to nuclear weapons research.]

These "media dissidents" serve the interests of the over-lords of society.

Revolution is about new ideas.

"Cubical" simplexes (which are associated to very stable discrete isometric shapes), placed in a many-dimensional context of (real) metric-spaces where these metric-spaces can only have non-positive constant curvature, and where each metric-space has a pair of opposite "metric-space states" associated to itself, eg positive and negative directed time-flow (ie in hyperbolic space), is to be the new fundamental set of assumptions concerning the math structure of physical descriptions. These "cubical" simplexes represent both the material (n-dimensional) as well as metric-spaces ((n+1)-dimensional), which are related to the various (real) SO(s,t) isometry fiber groups, [where s+t=n and s is the spatial subspace and t is the temporal subspace], as well as SU(s,t) unitary fiber groups (so as to account for the two opposite metric-space states in complex coordinates, and subsequently, a unitary fiber group, ie SU(s,t)).

Why should there be a new set of assumptions concerning the math structure of physical description? Because the assumptions of indefinable randomness and non-linearity (along with the idea of materialism) [which are the fundamental assumptions of current authoritative dogmas of math and science] do not form a valid basis for a (or [the] in the authoritative social context of science) fundamental set of assumptions concerning the math structure of physical description. This is because the current assumptions do not allow for the structure of measuring to be both stable and logically consistent, and this means that the properties of numbers, as well as the properties of algebra, and the properties of the local and regional operators of calculus, cannot be relied upon, in regard to providing valid descriptions of the patterns which are (physically or descriptively) measurable within a consistent structure of measuring. Being able to measure in a consistent manner is needed, if the patterns one is describing are going to be related to any practical creative efforts.

This is why the current (authoritatively dogmatic) descriptive languages have contributed so little to both technical development and accurate (stable) descriptions. Namely, because the descriptions, or the numerical values which compose the descriptions, are neither consistent nor valid. Furthermore, the descriptions are incoherent and have no meaning in regard to measuring, and subsequently they also have no meaning (or no predictive value) in the context of practical creativity, which is based on the information which is provided within a (physically) measurable description.

Math must remain consistent with its fundamental numerical basis (or its fundamental basis for measuring), so that measuring the "predicted" properties of the description is consistent with the [physical] measuring in the lab.
Namely, it must remain consistent with the uniform unit upon which a number system is based.
That is, the laws of algebra are only valid if this property of maintaining a uniform unit of measuring, which is stable for all the numerical values where these numerical values are used in a description, and where these values can be measured both "in the descriptive structure" and physically.

Another thing which needs to be preserved, is that "when measuring sets are related to one another (as valid measuring sets), the only relation which is valid is a local (linear) relation which depends only on multiplying one measuring set by a (local) constant so as to be consistently related to the values of the other measuring set. That is, only linear relations preserve a consistent numerical relation between the two numerical sets which are being related to one another in a ([physically] measurable) description.

Furthermore, in random patterns, the definition of an elementary random event needs to be:

> (or indefinable randomness can be identified by the negation of any one of the following properties):
>
> (a) random events can be defined (identified, or calculated by the laws of the descriptive structure),
> (b) the set structure is not too big,
> (c) the set (of elementary events) is not to be changed according to data (from the beginning the elementary particles need to be defined, or predicted, but new elementary-particles cannot be added, to an elementary event space, as new data is found [so one should wonder, "how the elementary events are to be counted in a valid manner if the set keeps changing?"]), and
> (d) the events cannot be unstable [ie the events must be stable] (elementary-particles are unstable, thus they are not events which can be properly counted ["who knows what the full set of unstable elementary-particle events actually is?"]),
>
> [the probabilities of random events can only be defined in such a context as defined by all of the properties of (a), (b), (c), and (d), ie there needs to be a basis by which to count a set of elementary events, and thus probabilities can be defined from a process of counting the elementary events].

Furthermore, simply being able to identify a distinguishable random property does not mean that there is a valid probabilistic structure associated to that property.

For general quantum systems, the spectra of the system cannot be represented, by means of calculation based on the laws of quantum physics, to a sufficient distinguishing level of precision, and therefore such a description cannot be considered a valid representation of a general quantum system, ie the random (ie spectral-particle) events cannot be defined for a general quantum system (using the laws of quantum physics).

Yet, the spectra of a quantum system is stable and precise, in regard to the structure of its observable spectral properties.

The conclusion should be, that within the quantum system its spectra is (now) to be related to a stable geometry [ie the stable properties of quantum systems depend on stable geometric properties, but this is an idea which is in opposition to the probabilistic assumption of quantum physics. Namely, that dual variables of a quantum description, eg a variable and its associated Fourier transformed variable, require that stable geometry cannot be a property of a quantum system.].

Stable geometries could also define a context of material interactions, so as to be able to understand (or to be able to describe) the property of the geometry of spherical symmetry in material interactions of physical systems, in both classical and quantum physics, which exists in 3-space).

Note: Though spheres may be useful, in math, for artificially bounding regions in a conceptual manner, based on the idea of bounding a region or a place, which is bounded in a uniform (or a very symmetrical) shape, however, as actual physical bounds, once perturbed, spheres are too unstable to be used to describe stable physical systems, which are contained in some measurable context. However, in 3-space, the natural structure of a material interaction (confirmed by the new geometric model of material interactions) is in the radial direction for sphere which bounds an interacting material system's center of mass (or for spheres which are centered at either of the two [or more] interacting material components).

That is, the random properties of quantum systems only need to be described in regard to the random quantum properties of particulate-material (being identified for small material components), whose randomness exists between the observed "relatively stable" classical (or macroscopic) geometries.

In other words, the stable macroscopic (nanometer and greater) geometries describe the context within which the (property of) randomness of small (quantum) components exists.

Consider that, in math, it is not clear that the property of order, in regard to real numbers, is maintained when the "axiom of choice" is added to the structure of numbers (since it now takes an infinite amount of information to identify a real number [G Chaitin], thus the numerical order of the real numbers is indeterminable [in real time]), where numerical order is a property of central importance in regard to the structure of convergence and the existence of limits.

This means that the statement "a bounded, increasing sequence of real numbers has a ** 'least upper bound'" has no meaning, since it [in general] requires an infinite amount of information to determine the numerical order of a candidate for a real-number being a "least upper bound," thus it is a property (of being the least) which is indeterminable.

Furthermore, the "least upper bound" (which may not be an element of the sequence itself) rather it may be in a set where the types of "independent directions," eg two independent direction, does not have the same meaning as in the set of numerical elements which compose the increasing sequence, where (say) each of the numerical elements which compose the increasing sequence is in a set which has only one independent direction.

This is an example of where, large sets can cause "what are considered to be independent sets" to no longer be independent, eg the plane filling curve (thus, a set with two directions associated to itself can subsequently be represented as a set with one direction associated to itself).

Thus one needs:

(1) a stable unit of measuring, which is
(2) consistent with the other measuring sets involved in the same description, and
(3) each of the number sets cannot be a set which is "too big" (of a set).

These three properties are all properties of numbers which are a part of a descriptive numerical structure which is measurable.

or, Alternatively:

The best set structure would be, if a description can be related to some fundamental property . . . of the description of a (physical) system's properties . . . whose structure (or whose values) within the descriptive context is a finite set.

If one wants to have a: stable, consistent, definable, coherent, and valid descriptive math structure, when measuring is involved in the math-science description, or when geometry is a part of the descriptive properties (that is, randomness is not a property of the description), then the description needs to be based on the geometries derived from "cubical" simplexes which are defined on metric-spaces which can only have non-positive constant curvatures, where these geometries define both component (or material) structures [contained in the metric-spaces] and the metric-spaces, themselves, and these structures exist in a many-dimensional context.

However, when randomness is assumed to be a part of the descriptions of systems, which also possesses stable properties, then the stable properties of these systems (ie the systems whose descriptions are assumed to be based on randomness) have not been accessible to valid descriptions. Thus, the descriptions of these systems, assumed to be based on randomness, need to be reconsidered, and they also need to be based on the geometries of "cubical" simplexes, rather than being based on randomness.

Furthermore, the various metric-spaces of non-positive constant curvatures, ie $R(s,t)$ [where $s+t=n$ and s is the spatial subspace and t is the temporal subspace], need to be associated with physical properties, which are relevant to the system being described, and are properties which are associated to the components contained in the metric-spaces.

"Cubical" simplexes exist and are geometrically very stable, where these "cubical" simplexes represent both material and metric-spaces, (this is analogous to both the assumption of materialism and to the natural assumption of higher dimensions which goes beyond the idea of materialism) so that these "cubical" simplexes are to be placed in a many-dimensional context of (real) metric-spaces, which can only have non-positive constant curvature, and where each metric-space has a pair of opposite "metric-space states" associated to itself, eg positive and negative directed time-flow . . . ,

> so that these pairs of metric-space states also require complex coordinates, and, subsequently, unitary fiber groups, which are related to the various (real) SO(s,t) isometry fiber groups, so that in SU(s,t), the description is a mixture of these two (real) metric-space states which are separated into the two real (metric-space) states, which exist in two separate SO(s,t) contexts,

4. . . , and as for material interactions, the n-dimensional, discrete Euclidean shapes probe the stable (n-1)-dimensional discrete hyperbolic shapes within the metric-space by means of action-at-a-distance properties in a (n+1)-dimensional Euclidean metric-space so as to define material interactions in hyperbolic n-space. (The 2-forms mentioned below are defined in Euclidean (n+1)-space), is to be the new fundamental set of assumptions concerning the math structure of physical description.

> [As opposed to: materialism, differential equations, indefinable randomness, non-linearity, and sets which are "too big," which are now the fundamental set of assumptions concerning the math structure of physical description.]

The new set of assumptions leads to a new way in which to describe the material interaction process.

First; To compare with the old (and/or current) models of material interactions:

I. The particle-physics interaction process (which is, essentially, the quantum interaction process) is too complicated, it is probabilistic for (what should be) a relatively few interacting components. Thus, (at best) it is a description which only provides probabilities which can (only) be used to "bet on," at gambling casinos. That is, it is a complicated description and it has shown itself to be incomprehensible (it is both probabilistic and geometric, ie the particle-collision) and thus it leads to no new technical developments. This is the same old problem, wherein one is trying to describe an ordered structure with a descriptive language which is based on indefinable randomness and non-linearity. This is not possible, even if one introduces a new identifiable random event which might point towards a state of order, eg a supposedly limited set of elementary particles for the particle-collision model of material interactions, or (in evolution) the survival of a better adapted organism, etc.

II. On the other hand, the classical picture of interaction has two main contexts in which it can be described.

Namely,

(1) optimizing a system's action by a variation process, or more intuitively minimizing a physical system's energy identifies the process (or state-path in phase-space) of the interaction, where one needs to identify both a system's energy and a method for finding its minimal "path" (in phase-space), and

(2) the original (Newtonian) classical descriptive structure, where one defines the inertial properties of a system after its force-field has been identified, where the locally measurable relation between an object's velocity and time, ie the slope of the velocity-time graph, is proportional to the (local) force-field defined at the object's position. The force field is a material-geometric property (which is defined globally) which "causes" the object's velocity to change (or the second derivative of the object's spatial displacements to change, as time changes, or the values of the object's spatial displacements in uniform time intervals to change as time changes).

Note: Inertia (or mass) is a measure of the resistance (by an object) to velocity changes.

In regard to the conceptually "easier to describe" second characterization of material interactions, the descriptive context is: material, position in space, time, and the object's motion being affected by the geometry of other material which exists "around" the given material object.

These are properties of material and its inertia in a context of space and time in either coordinate frames of absolute space or in coordinate frames of relative motion and time, where material and its force-fields cause changes in the velocity of a distant object. In the coordinate frames, the geometric measures on the coordinates are both well defined and uniform, within the coordinates (or within the frames). But the coordinate spaces are defined by the idea of materialism and (time can be defined in regard to) the motions of material.

Secondly, present the new set of assumptions (also see above):

In the new descriptive structure the material and the metric-spaces (or coordinate frames) are both shapes, where these discrete shapes are also discrete subgroups of the frame's isometry (and unitary) fiber groups (where these fiber groups are most easily distinguished by the R(s,t) real coordinate frames). The discrete (Euclidean and hyperbolic) shapes either fill the space (in time), or define the stable boundaries of the coordinate frame, where the coordinate frame is itself a shape (hence determines an absolute space within which there (can) exist relative frames of motion).

However, hyperbolic space is itself a coordinate space of motion, thus its properties of space are determined by the Euclidean spaces that are contained in space-time (where hyperbolic space is derived from space-time). The Euclidean shapes naturally fill space, up to its stable boundaries, which are composed of either the hyperbolic metric-space or the (lower-dimension) discrete hyperbolic shapes contained in the hyperbolic metric-space. The Euclidean shapes fill space instantaneously (by action-at-a-distance) with each new discrete time interval, where the discrete time intervals are determined by the time period of a spin-rotation of metric-space state (see below).

Differential-2-forms (based on geometric measures which are defined on these, many, space-filling discrete Euclidean shapes) are of the same dimension as the spin fiber-groups (associated to the metric-space within which the discrete shapes are contained) and thus these fiber groups can be related to an active process by which the local coordinate transformations of the spin-groups change the geometrically measurable relationships which exist between both the stable discrete hyperbolic shapes contained in a metric-space and the boundary of the metric-space.

All of the properties of:

1. contained shapes (within metric-spaces), and
2. metric-spaces,
3. the Euclidean 2-forms (of a discrete Euclidean shape of a material interaction), and
4. the system ([local]) changes caused by local coordinate transformations

are all math patterns which can (all) be thought of as fitting into a (the associated) fiber group; either as discrete subgroups, or as subsets, or as local geometric measures on discrete subgroups.

All aspects of the description fit into a set of isometry and/or unitary Lie groups, for the various sets of R(s,t) metric-spaces needed for the description.

The direction in which the local coordinate transformations act on the interacting system, are (in turn) determined by:

(1) the local directions and associated values (eg sectional curvature values) of the discrete Euclidean shapes, which in turn, are based on the stable boundaries which exist within the (hyperbolic) metric-space,
(2) the 2-forms which are determined by the discrete Euclidean shapes, as well as
(3) similar math structures for the different physical properties which get defined "based on the dimension of the time subspace, t," in R(s,t),
(4) all together when related to the local directions of a fiber (spin or unitary) group at a distinguished-point of a stable boundary shape (in the coordinate space) determine a direction and magnitude of the local coordinate transformation on the interacting system.

The local coordinate transformations at points (in the metric-space) on shapes which exist between the metric-space's stable bounding shapes can be quite varied, including unstable (non-linear) dynamics of material interactions but the underlying stable context of the systems and their interaction structures depends (only) on very simple geometric shapes which are a natural part (property, or pattern) of the fiber group.

This new descriptive structure, based on "cubical" simplexes, unifies a spatial displacement process (ie inertia) with a field structure and a definite simple geometric model of stable material systems (by means of the stable discrete geometric shapes which are involved in the interaction).

In classical physics, an inertial object's motion changes because of a distant material based geometric cause, but the cause (of inertial changes) appears to be by means of action-at-a-distance [even though for electric charges the force-fields, supposedly, propagate (in space-time) in a causal relation, but it is not always clear that electromagnetic waves are the mechanism by which the fields are propagated].

Thus one might still ask, "how does something which is far away from an object, cause certain effects on that object?"

The new description of material interaction allows one to generate a description of existence which emerges from the same source as does the local cause of motion (or spatial) changes of existence, ie the fiber group, wherein the local relation "of local coordinate transformations" of the fiber group to geometric shapes cause the geometric changes in a material interaction.

The "discrete geometric shapes" emerge from the discrete subgroups of the fiber group, and these shapes subsequently determine both metric-spaces and material systems as well as the interaction shapes of the descriptive structure.

Nonetheless, the new description is describing the same effects as the classical description (ie of changes in an object's motion caused by the geometric properties of distant material), where in either description both the motion and the geometric relations between material are (assumed to be) measurable, and thus explicitly describable (within a stable system of measurability which is associated to everything which is being described [in this context]).

When the new model of material interaction is applied to many small material components then one gets spatial displacements which are (can be) large for "a single time interval," and thus the material interaction for the next time interval [where time intervals are determined by the time period of the spin-rotation of state] exists in a new geometric configuration because the material geometry for these small components is very different. Thus, this set of material interactions for each time interval defines a form of Brownian motion.

This Brownian motion for the many small components is equivalent to quantum randomness (E Nelson, 1957, 1964, 1967).

That is, this new descriptive structure derives the property of quantum randomness for the many small components which exist between the stable macroscopic geometric bounds, defined by classical physical structures.

Pairs of opposite metric-space states for each metric-space

In regard to the two opposite metric-space states of time, consider a given point (in space) of a material-point's dynamic path, so that in a local neighborhood (region) of the given point there are two real spatial displacements (in Euclidean space) of the dynamic path which are in opposite directions to one another. One direction is in the positive-time "metric-space" state (our usual time direction), and the other (opposite) direction (of time) is in the negative-time "metric-space" state, where "the metric-space states" are both the metric-space and the two opposite directions of time, which together, form the opposite pair of metric-space states, which are the two time-states which are properties of the (hyperbolic) metric-spaces, but the spatial displacements are in Euclidean space.

Time has a discrete structure, which is related to the time-period of a (the) spin-rotation of metric-space states.

If one has the information about spatial displacements for each discrete time interval in one of the two time directions then one can determine the opposite (or inverse) spatial displacements for the opposite direction in time, for each discrete time interval in a sequence of time which is reversed.

For a given time direction, a spatial displacement for the discrete material shapes, which are involved in a material interaction in an n-dimensional space, is determined by the relation that the Euclidean 2-form . . . , of the discrete Euclidean interaction shape (which exists in (n+1)-dimensional Euclidean space) . . . , has to the local fiber spin-group (of the (n+1)-dimensional space) so as to determine the direction of the spatial displacement (caused by a local coordinate transformation of the fiber spin group), where the coordinate directions of these local coordinate transformations (determined by the isometry fiber group) are multiplied by the various values (of the sectional curvatures of the space-forms involved in the interaction) so as to determine both the direction and magnitude of the spatial displacement. This can be done for each time interval in a sequence of time intervals associated to the material interaction, while the spatial displacement associated to "time going in the opposite direction" is the inverse transformation for each time interval of the time-interval sequence.

In determining the magnitude of the spatial displacements for any given time interval, the sectional curvatures of the space-forms involved in the interaction are multiplied together as factors and they are also multiplied by a proportionality factor which relates the two different dimensional levels involved in the material interaction to determine the direction and magnitude of the spatial displacement.

Consider other points in space (different from the (original) given point), which exist along the (same) material point's dynamic path, these other points are separated from each other in both space and by a sequence of discrete time intervals, where a spatial displacement was associated to each time interval (of a particular time state), as well as its inverse (or opposite) spatial displacement in the opposite metric-space state, which is also associated to each time interval in the sequence of time intervals, which is associated to a material object's dynamic path. One can think of the over-all space as being associated to a mixture of opposite metric-space states so that this mixture can be represented in the set of complex coordinates, which has within itself a corresponding pair of sets of (opposite) real-space coordinates of opposite metric-space states (or opposite dynamic directions), namely, the sets of R(n) and (iR(n)).

If the geometric basis for a descriptive math structure is not based on "cubical" simplexes (which are very specific and very simple geometric structures) then the underlying number-system will "not" be stable, and the math descriptions (of a math structure) will not have a valid (basis for) meaning.

This can be seen (or can be believed), because in . . . :

>   general relativity,
>   quantum theory,
>   particle-physics (except related to randomly based particle-collision reactions),
>   string-theory,
>   grand unification,
>   supergravity,
>   Etc, which are all based on indefinable randomness and non-linearity,

. . . , there are no (or very few) practical applications which come from these descriptive structures, and the descriptions cannot be used to identify the stable (and most often very detailed, and precise structures of the) spectra of any general quantum system.

Nonetheless, the general quantum systems, which these theories are trying to describe, do have very stable and very specific measurable properties, and this means that there is a descriptive math structure which can account for these stable properties,
Thus, one can conclude that the logical foundations of indefinably random and non-linearity (or instability and inconsistency) are not correct.

Rather, the fundamental assumption of "materialism" needs to be changed, to the assumption that:

> "the logical foundations of physical description are to be based on "cubical" simplexes, where these "cubical" simplexes are to model both
>
> (1) the structures of stable (material) spectral-orbital-geometry, and
> (2) the stable containing metric-spaces,
>
> which are to be placed in a many-dimensional context."

Allowing equal and free intellectual expression, also means criticizing ideas and providing alternative ideas

The criticisms by the fundamentalist religious people . . . that "highly ordered systems cannot emerge from a descriptive language which is based on indefinable randomness" . . . are correct criticisms.

But nonetheless, the assumption of God, which such a fundamentalist might claim to be the correct basis for a description which the fundamentalist might claim to be "as accurate and as precise as one might want a description (about the properties of God) to be" . . .

> [in a sense this is very similar to quantum theory and particle-physics, which also claim a few "precise predictions" of observed properties, but only "predicted" after the property has already been observed, where these few calculations are supposed to satisfy the requirement of science, where science actually requires that a "true" theory be able to predict the properties of a general system, within a valid range of precision, over the entire class of systems related to the descriptive language (this is, this is the standard by which classical physics has been thought to be scientifically true)]

. . . , but for a religious fundamentalist, the "God assumption" is not related to a measurable description, and thus one cannot build anything of practical value [other than social institutions through social manipulation] from the descriptions based on this assumption.

But this is the same criticism, of today's overly authoritative science, by which one might be motivated to try to find an alternative math structure, which is accurate (for general systems), measurable, and useable for practical technical development.

That is, in a similar manner, since quantum theory and particle-physics as well as general relativity are based on indefinable randomness and non-linearity, they "too" are not measurable descriptions (since they are invalid and inconsistent math structures, which cannot be related to a stable uniform unit of measuring [or for an elementary-event set to have events which possess the properties necessary for having a valid process for the counting of these elementary events]), and thus one cannot build anything (of practical value) from the descriptions based on these theories (just as one cannot build anything of practical value from the information gained from the assumption that all observed properties are derived from God).

But since the descriptive structures of these theories cannot be used to accurately describe, "in a precise enough manner," a general, stable spectral-orbital physical system, one can be led to also classify these types of "math descriptions" (based on indefinable randomness and non-linearity) as also being authoritative dogmatic religions.

Note: Wherein God has become the math-physics authorities, themselves, ie a religion of personality cult, as well as a religion based on the authoritative dogmas derived from the assumptions of mathematics.

Yet the opposite is claimed, by science, that the descriptive languages of quantum theory, particle-physics, and general relativity, can describe any general system as precisely as one might want, "such as description of any general system, which is to be described" (contingent only on whether one only had enough time and enough money).

But in science, such a claim needs to be upheld by actual evidence of sufficiently precise descriptions of any general system's set of properties (the essential standard by which classical physics was judged to be true), and this evidence does not exist.

Religion is about arbitrary high-social-value (arbitrary authority), it is not about useful knowledge, rather it is about social manipulation, by manipulating language and social-value, and this is now (2012) also the main function of modern math and physics, ie math and science functions as a religion of both meaningless authoritative dogma and personality-cult.

This authoritative-science-religion (which is virtually without meaning) fits into a social structure whose primary focus is domination. It is about an elite set of people who can crush the knowledge and creative aspirations of everyone else in society who does not support the authoritative dogmas of science and math.

Unfortunately, the primary function of modern science and math is to define a domineering elite set of authorities who are encouraged to crush any knowledge which does not fit into their fixed authoritative dogmas.

Equality and free speech

When one discusses the different categories of descriptive knowledge the discussion needs to begin at the level of assumptions which are explicitly described:

Thus, discussion about the structure of US society and its governance, the main assumptions are about:
Equality vs. property rights.

Whereas in the subject of math and science, the main assumptions of measuring are (now, 2012) about:
Stability and consistency vs. indefinable randomness and non-linearity

or, equivalently,

Reliable measuring and subsequent building vs. an inability to relate a description to a valid model of measuring and an inability to build something new from a Given, dogmatically authoritative, descriptive language.

However . . . ,
this should also be the set of assumptions in regards to:

"Science and math" vs. "western religion."

. . . but now science and math are indistinguishable from (US) religion, because science and math are now (2012) defined by a fixed authoritative dogma.

APPENDIX I

At some given point on an object's dynamic path in space, the coordinates of the other (distant) spatial points (which are on the same dynamic path) are related to one another (or to each other) by means of a (spatial) path which, in turn, is related to discrete local coordinate transformations, which are path-connected (in time) by a discrete set of time intervals, so that a discrete spatial displacement is defined at the end-points of these discrete time intervals.

A direction of a spatial displacement (at the end-points of each time interval) is defined by a local geometrically related directional (or vector) structure which, in turn, is defined by:

>spatial directions, and
>directions defined in the space of 2-forms, (whose geometric measures of area are contained in the spatial domain space)
>so that these directions are related within the local fiber group (because the fiber group has the same dimensional structure as the 2-forms), in either a rotational or a unitary fiber group.

In turn, this spatial direction is multiplied by a scalar, which is associated to the multiplicative products of the various sectional curvatures of the discrete shapes which compose the material interaction structure, eg (separation distance, mass and electric charge, etc),

The set of opposite pairs of discrete coordinate transformations in $R(n)$ at each point in $R(n)$ identified by the discrete time-interval end-points (real coordinates of dimension-n) are best represented as mixtures of opposite states in C(n) (complex coordinates of dimension-n), where there are now two sets of R(n) in C(n), which these two sets are the sets R(n) and (iR(N)) in C(n).

Thus, the displacements which are locally considered to be in opposite spatial directions at the end-points of the discrete time intervals come to be represented as being in orthogonal (or independent) sets in C(n).

Thus, at some (any) given point (on an object's dynamic path) there are two opposite directions of spatial displacements associated to the two (metric-space) states of time (in hyperbolic space, or equivalently, space-time) but in the next discrete time interval the local opposite directions, at the (new) given point, the dynamic displacements on the (dynamic) path come (begin) to have independent directions.

It seems that only in a such a simple math (or algebraic) structure of a discrete set of locally opposite metric-space states can the separate sets of advanced and retarded wave-function solutions to the wave-equation (or potential functions) be defined, eg reflections of waves are discrete properties.

The "real" algebra is (in general) not diagonal but local coordinate transformations on C(n) do form an algebra of diagonal matrices, which mixes two spatial dynamic paths defined by real matrices which act within R(n) and (iR(n)), ie a 90-degree phase relation exists between the two opposite metric-space states in C(n).

This (also) means that dynamics is not a part of a continuous "fabric," ie it does not define a continuous shape (or continuous manifold), though it may have some continuous math structures associated to its (descriptive math) process, but in fact it is discretely defined by very short time intervals (though the time intervals can be put together into a continuous set.

APPENDIX II

Notes:
　Double spaces can mean a sudden new direction of the discussion without a new paragraph title.
　The *'s represent either favorites (of the author) or (just as likely) indecision and questions about (logical) consistency. Information and discussion about ideas is not a monolithic endeavor pointing toward any absolute truth, the wide ranging usefulness, in regard to practical creativity, might be the best measure of an idea's truth, it is full of inconsistencies and decisions about which path to follow (between one or the other competing ideas) are either eventually made or the entire viewpoint is dropped, but this can occur over time intervals of various lengths.

　The marks, ^, associated to letters, eg a^2, indicates an exponent.
　The marks, *, in math expressions can have various math meanings, such as a pull-back in regard to general maps which can, in turn, be related to differential-forms, defined on the map's domain and co-domain (or range),
　but in this book it usually denotes the "dual" differential-form in a metric-space of a particular dimension, eg in a 4-dimensional metric-space the 1-forms are dual to the 3-forms and the 2-forms are self-dual, etc.

　The main idea of thought (or of ideas) is that it is about either sufficiently general and sufficiently precise descriptions based on simple patterns, or it is about developing patterns (of description) which lead to particular practical creativity, or to new interpretations of observed patterns, or to directions for new perceptions.

Chapter 27

DIAGRAMS

This set of diagrams represents a symbolic-map which can be used to help identify a set of analogous higher-dimensional diagrams (or an analogous set of higher-dimensional constructs), where in lower-dimensions these diagrams are consistent with the observed material patterns, though now the ideas of either materialism . . . , or existence which is contained within a greater set of higher-dimension so that the higher-dimensional analogous constructs possess the properties of macroscopic geometries . . . , is given a new interpretive context.

> (in the Diagram section) The diagrams provide a succinct outline of the simple math, which is based on stable discrete (hyperbolic) shapes, and which is the basis upon which the stability of measurable description . . . , of the observed stable, definitive properties of physical systems . . . , depends.

The diagrams provide a clear picture (or clear analogy, or clear map in which to think about moving into the higher-dimensions) of the context within which these stable discrete shapes (of non-positive constant curvature) are organized, so as to form a many-dimensional context (whose higher-dimensional properties should be thought of as being macroscopic) of both component containment, and component interaction.

The many-dimensional containment set, possesses a macroscopic and stable geometric context, which is composed primarily of "discrete hyperbolic shapes," which, in turn, are contained in hyperbolic metric-spaces, (wherein it is true that each dimensional level, except the top dimensional level, has a discrete hyperbolic shape associated to itself).

The finite set of stable discrete hyperbolic shapes, which model both the different dimensional levels, as well as the different subspaces of the same dimension, is a geometric foundation upon which the construct of a finite spectral set depends (a finite spectral set for all existence, contained within a high-dimension containing metric-space, ie an 11-dimensional hyperbolic metric-space).

Each dimensional level (and each subspace of any dimensional level) is associated to a very stable "discrete hyperbolic shape," and each subspace (within the many-dimensional set) is characterized by a size-scale (determined in relation to the finite spectral set), where the size-scale of a dimensional level of a particular subspace is also determined by a set of constant multiplicative factors, which are defined both between dimensional levels and between different subspaces of the same dimension.

The fundamental properties of the high-dimension containing space are determined within an 11-dimensional hyperbolic metric-space.

Hyperbolic space is analogous (or isomorphic) to a general model of space-time defined for various dimensions, eg R(3,1)=[space-time], while generally, R(n,1), is a "general space-time."

However, there are also various other "metric-function signature" "types of metric-spaces" of the various dimensions and metric-function signatures, R(s,t), which are involved in the description {where s=space dimension, and t=time dimension, where s must be less than or equal to 11 (it seems (?)), and s+t=n}, most notably the "discrete Euclidean shapes," in R(s,0), which possess properties of continuity (of size), which is needed in the interaction process.

These diagrams identify the context in which "material" components exist within each dimensional level, and they identify the context of both "free" components (associated to both parabolic and hyperbolic second order partial differential equations in regard to inertial properties), as well as orbital components (associated to both elliptic and parabolic second order partial differential equations in regard to inertial properties).

These diagrams also (pictorially) show the basis for "material" interactions, and the relation that a new material system, which is emerging from a material interaction, has to being resonant with values of the "finite spectral set" which is defined for the total containment space.

These diagrams provide a context for the emergence of new stable systems from material interactions.

These diagrams of "'material' component interactions" can be identified at any dimensional level (dimension-2 and above).

Note: Interactions are constrained by:

1. the process itself,
2. dimension,
3. size,
4. subspace, and
5. a finite spectral set,

where the basic form for such "material" interactions has an analogous structure (or is "the same") for each dimensional level, though there are differences, in regard to the properties of material interaction, between the different dimensional levels, which can be due to dimension, subspace, and size.

The diagrams give low-dimension pictures of the very simple, quantitatively consistent, geometric shapes which are stable, ie most notably the discrete hyperbolic shapes, and it is these stable shapes upon which stable mathematical patterns can be described in a context where measuring is reliable, and because the description is geometric this means that the description can be very useful.

Note: Following "the diagrams themselves" there is a section in which the descriptions associated to the diagrams are re-written in reference to the number of the diagram, eg 1, 2, etc. given on each page of the diagrams.

If one cannot read the words on the diagrams, they are here (below), where they are associated to the numbers of the diagrams.

1. If the circle is rigid in its shape then the complex plane defines a commuting number field, as does the real-line.

    Linear measuring directions are perpendicular (or they are independent of both one another's measured values as well as measuring directions) a complex number, z, is represented as, $z=x+iy=r(\cos W, \sin W)=re^{\wedge}(iW)$, ($W$ is a measured angle) Line segments and circles are quantitatively consistent shapes.

2. The following (above) shapes are quantitatively consistent shapes, and locally their directions are independent, and form commutative algebraic constructs at each point over the entire shape, ie global commutative algebraic constructs, locally linear and invertible [one-to-one and onto] and this is true everywhere on the shape.

3. Cubical (or rectangular) simplexes are related to circle-spaces by means of "equivalence-relation topologies," or equivalently, by a "moding-out" process.

    On such shapes local geometric measures are based on either measuring rectangular shapes or (equivalently) by a measuring process based on tangents to the circle, which is used as a basis for measuring along a circle's curve, eg $rdW=dx+dy$ along a circle.

4. * Lattice in the "hyperbolic circle"
    This lattice is more restricted than would be rectangles attached at vertices

5. Discrete hyperbolic shapes are composed of toral components

The number of holes in a discrete hyperbolic shape is called the shape's genus

2-holes are surrounded and caught by 1-curves, and defined by 2-dimensional discrete hyperbolic shapes
3-holes are surrounded and caught by 2-surfaces, and defined by 3-dimensional discrete hyperbolic shapes etc

The faces on the fundamental domain (which result from the faces of a hyperbolic shape's rectangular (or "cubical") simplex) form very stable spectral measures on the very stable shapes of the hyperbolic space-forms.

Discrete hyperbolic shapes, or equivalently, hyperbolic space-forms, have open-closed metric-space topological properties, and they may be bounded or unbounded shapes, but all existing hyperbolic space-forms which are 6-dimensional or greater are unbounded shapes, and the dimension of the last known hyperbolic space-form is hyperbolic 10-dimensions (Coxeter).

Orbits on discrete Euclidean shapes and discrete hyperbolic shapes
Subsystems (or sub-metric-spaces) either occupy spectral orbits or they are "free"

6. The shapes obtained from this rectangular simplex for a 3-dimensional hyperbolic space-form . . . , which contains a 'free" 2-dimensional hyperbolic space-form . . . , are contained in a 4-dimensional hyperbolic metric-space.

7. The separation of two hyperbolic material components is, r, where, r, is defined between the two vertices. Take smallest toral component of each hyperbolic space-form, average the sizes of the two toral components, as is also done in center-of-mass coordinates. Represent the average value as a pair of equal oppositely positioned (rectangular) 2-faces, so as to define a right rectangular volume whose separation, r, is the distance between the vertices of the original interacting hyperbolic space-forms.

Then define an interaction differential 2-form on the geometry of this 3-dimensional Euclidean torus, which is contained in Euclidean 4-space. This determines the force-field, defined between the interacting material components.

The local vector geometry of the differential 2-form is relatable to the local geometry of the fiber SO(4) Lie group, since the 2-forms in Euclidean 4-space have the same dimension as SO(4) the geometry of the spatial displacement is determined in SO(4) by it geometric relation to Euclidean 4-space which is given by the 2-forms so that a local spatial displacement occurs due to a local coordinate transformation with SO(4) acting on the positions of the vertices [(in relation to center-of-mass coordinates) of the original pair of interacting hyperbolic space-forms] in Euclidean 4-space.

If the force is attractive and if the interacting (charged) material (the interaction structure) is contained within either 2-dimensional or 3-dimensional, or 4-dimensional Euclidean space then the force is radial and attractive, and r is made smaller. {Note: If the material is of a new type (oscillatory) and contained within Euclidean 4-space then the force-field has a new geometric structure contained in a higher-dimensional Euclidean space.}

Then the same type of process repeats, for time intervals determined by the spin-rotation period of the spin-rotations of opposite metric-space states (about 10^{-18} sec).

In this process the Euclidean torus which forms for each discrete time interval, forms in the context of action-at-a-distance.

Classical partial differential equations are defined within very confining and very rigid sets of both discrete hyperbolic shapes and action-at-a-distance material interaction Euclidean toral components which link the hyperbolic material together, so that the force-field differential 2-form is defined on the torus. (7)

The above interaction for material contained in 3-space results in a 4-dimensional descriptive context, but it can be symbolically represented in 3-space.

8. It should be noted that in this new descriptive context eigenfunctions would also be both discrete hyperbolic shapes and discrete Euclidean shapes (tori).

Forming new stable hyperbolic space-forms from a material interaction. Assume an attractive (or repulsive) interaction in 3-space, then the interaction of "free" material components would be similar to a collision of components,

if the material components get very close during the (collision) interaction then if the energy of the over-all interaction is within (certain) energy ranges and the closeness allows resonances (with the spectral set of the over-all high-dimension containing metric-space) to begin to form, thus forming a new state of resonance for the interaction simplex, so that the over-all energy, as well as the resonances, allow a "new" stable "discrete hyperbolic shape" (in the proper dimension of the interaction) to form, so a new hyperbolic space-form emerges.

9. Weyl-transformations between two maximal tori within a Lie group (rank-k compact Lie group) Two intersecting circles of the two maximal tori may be angularly related to one another by Weyl group transformations, where the Weyl group defines the conjugation classes of the maximal tori which "cover" the compact Lie group.

10. Forming angular changes between toral components of a discrete hyperbolic shape by using Weyl-transformations, which change the angular relations between circles which

The Mathematical Structure of Stable Physical Systems

compose a toral component of a hyperbolic space-form. These Weyl-transformations allow Envelopes of orbital stability for "free" subsystems (or sub-metric-spaces) to be defined.

11. * But rectangles attached at vertices and then moded-out, "without expanding the vertex," shows a model of a discrete hyperbolic shape's toral components, represented as separate tori attached at separate vertices. (This is to emphasize an apparent toral component structure of discrete hyperbolic shapes, which is an important aspect of these discrete shapes.)

12. Various types of unbounded 2-dimensional discrete hyperbolic shapes

13. The figure titled "The mathematical structures of stable physical systems," represents information similar to figures 9. and figure 10.

14. The figure titled "Partitioning a many-dimensional containment space" represents two different dimensional levels, where the 2-dimensional level is identified as an un-deformed rectangular lattice, where a deformed lattice shape (in 2-hyperbolic-dimensions) is given in figure 4. Whereas in this 2-dimensional figure there are contained representations of 1-dimensional shapes. The other (larger) representation of the 3-dimensional partitioning structure are the un-deformed "cubical," or right-rectangular, 3-dimensional shapes, which contain within itself the stable 2-dimensional discrete hyperbolic shapes, this process of partitioning space can continue up into higher-dimensions, where a deformed 3-dimensional lattice shape can be moded-out to form into a geometric-shape, which would exist in 4-dimensional hyperbolic metric-space.

15. The figure titled "Perturbing material-components on stable shapes:" shows an atomic orbital structure which is a stable geometric structure with electrons in the outer-orbits of concentric toral-components (folded into their stable shape) and the nucleus in the center small orbital shape, where the electron's orbit is mostly held stable by it (the electron) following the geodesic path, which is defined on its toral component, and the electron is also interacting as if in a 2-body interaction with the nucleus, so that this 2-body interaction (most noticeably) perturbs the orbit of the electron, eg perhaps causing the electron to possess an elliptical path, where these orbital deformation may result in the variations in the details of the atom's discrete energy structure, where the stable orbital shapes, perhaps related to various Weyl-angle shapes, are the basis for the atom's stable discrete energy structure.

16. The figure titled "Describing the dynamics of 'free' material components in higher-dimensions" is a diagram quite similar to figure 8.

Chart of the face structure for rectangular simplex geometry

2-rectangular-simplex
1-face
vertices

3-rectangular-simplex
2-faces
1-face
Vertices

DIAGRAMS

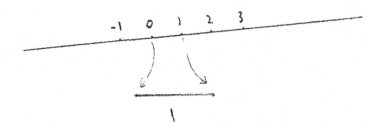

Uniform unit of measurement
(modeled on real-number line),
This unit must remain stable and consistent

If the circle is rigid in its shape
then the complex plane defines
a commuting number field, as
does the real-line.

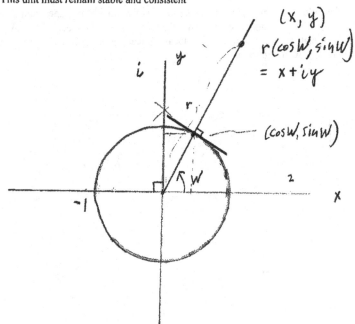

Representation of rectangular coordinates; (x,y)
as well as the complex-number plane,
Linear measuring directions are perpendicular
(or they are independent of both one another's measured values as well as measuring directions)
a complex number, z, is represented as, z=x+iy=r(cosW,sinW)=re^iW

Line segments and circles are quantitatively consistent shapes.

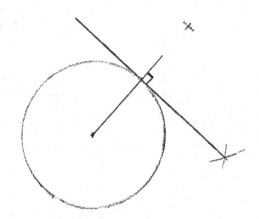

2-dimensions

The following (above) shapes are quantitatively consistent shapes, and locally their directions are independent, and form commutative algebraic constructs at each point over the entire shape, ie global commutative algebraic constructs, locally linear and invertible [one-to-one and onto] and this is true everywhere on the shape

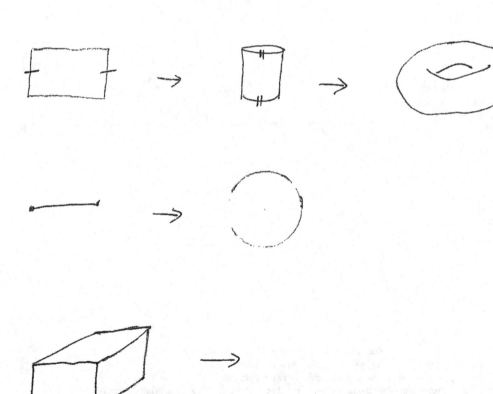

660

Cubical (or rectangular) simplexes are related to circle-spaces by means of "equivalence-relation topologies," or equivalently, by a "moding-out" process.

On such shapes local geometric measures are based on either measuring rectangular shapes or (equivalently) by a measuring process based on tangents to the circle, which is used as a basis for measuring along a circle's curve, eg $rdW=dx+dy$ along a circle.

Euclidean shapes

Euclidean lattice

Rectangles   moding-out   tori
Fundamnetal domains

2-dimensions   Contained in Euclidean 3-space

3-tori contained in Euclidean 4-space

3-dimensions

Hyperbolic shapes

Lattices (in hyperbolic 2-space)

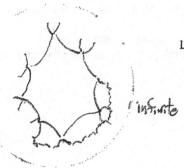
infinite

Fundamental domains   moding-out   discrete hyperbolic shapes
(contained in hyperbolic 3-space)

vertices

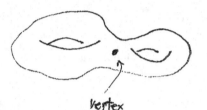
vertex

Rectangular simplexes   fundamental domains   discrete hyperbolic space-forms

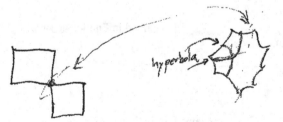
hyperbola

Edges of fundamental domain

The vertex is pulled apart   orthogonal pairs of hyperbolae

The Mathematical Structure of Stable Physical Systems

Discrete hyperbolic shapes are composed of toral components

The number of holes in a discrete hyperbolic shape is called the shape's genus
2-holes are surrounded and caught by 1-curves, and defined by 2-dimensional discrete hyperbolic shapes
3-holes are surrounded and caught by 2-surfaces, and defined by 3-dimensional discrete hyperbolic shapes
etc

The faces on the fundamental domain (which result from the faces of a hyperbolic shape's rectangular (or "cubical") simplex) form very stable spectral measures on the very stable shapes of the hyperbolic space-forms.

Discrete hyperbolic shapes, or equivalently, hyperbolic space-forms, have open-closed metric-space topological properties, and they may be bounded or unbounded shapes, but all existing hyperbolic space-forms which are 6-dimensional or greater are unbounded shapes, and the dimension of the last known hyperbolic space-form is hyperbolic 10-dimensions (Coxeter).

Orbits on discrete Euclidean shapes and discrete hyperbolic shapes
Subsystems (or sub-metric-spaces) either occupy spectral orbits or they are "free"

Euclidean case

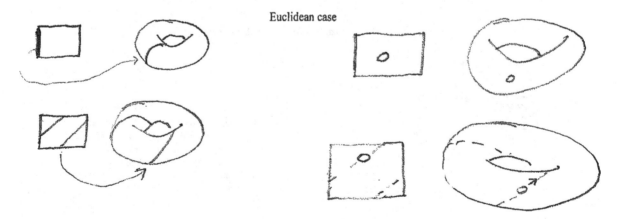

Stable orbits   "free" subsystems

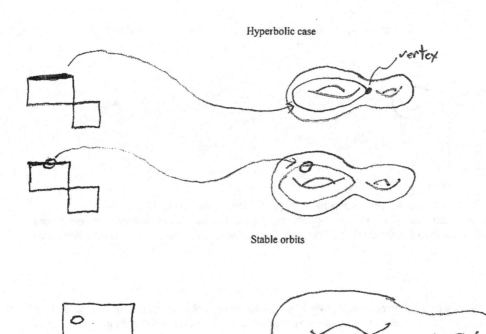

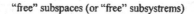

"free" subspaces (or "free" subsystrems)

The shapes obtained from this rectangular simplex for a 3-dimensional hyperbolic space-form..., which contains a 'free' 2-dimensional hyperbolic space-form...., are contained in a 4-dimensional hyperbolic metric-space.

2-dimensional "free" components (or subsystems) are contained in hyperbolic 3-space

The Mathematical Structure of Stable Physical Systems

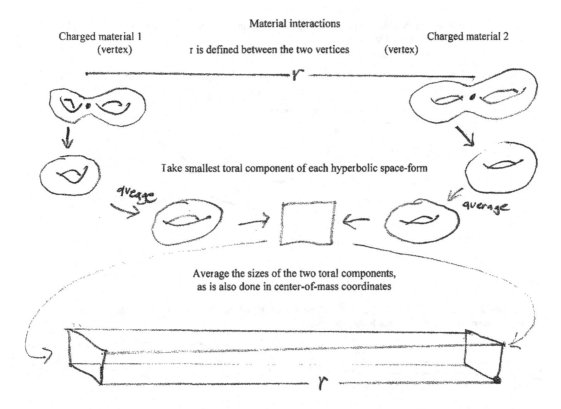

Represent the average value as a pair of equal oppositely positioned (rectangular) 2-faces
So as to define a right rectangular volume whose separation, r, is the distance between the
Vertices of the original interacting hyperbolic space-forms

Then define an interaction differential 2-form on the geometry of this 3-dimensional Euclidean torus, which is contained in Euclidean 4-space This determines the force-field, defined between the interacting material components.

The local vector geometry of the differential 2-form is relatable to the local geometry of the fiber SO(4) Lie group, since the 2-forms in Euclidean 4-space have the same dimension as SO(4) the geometry of the spatial displacement is determined in SO(4) by it geometric relation to Euclidean 4-space which is given by the 2-forms so that a local spatial displacement occurs due to a local coordinate transformation with SO(4) acting on the positions of the vertices [(in relation to center-of-mass coordinates) of the original pair of interacting hyperbolic space-forms] in Euclidean 4-space.

If the force is attractive and if the interacting (charged) material (the interaction structure) is contained within either 2-dimensional or 3-dimensional, or 4-dimensional Euclidean space then the force is radial and attractive, and r is made smaller. {Note: If the material is of a new type (oscillatory) and contained within Euclidean 4-space then the force-field has a new geometric structure contained in a higher-dimensional Euclidean space }

Then the same type of process repeats, for time intervals determined by the spin-rotation period of the spin-rotations of opposite metric-space states (about 10^{-18} sec).

In this process the Euclidean torus which forms for each discrete time interval, forms in the context of action-at-a-distance.

Classical partial differential equations are defined within very confining and very rigid sets of both discrete hyperbolic shapes and action-at-a-distance material interaction Euclidean toral components which link the hyperbolic material together, so that the force-field differential 2-form is defined on the torus.

665

The above interaction for material contained in 3-space results in a 4-dimensional descriptive context, but it can be symbolically represented in 3-space.

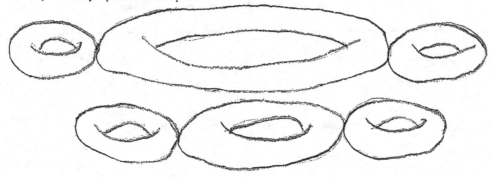

For an attractive interaction in 3-space

It should also be noted that in this new descriptive context eigenfunctions would also be both discrete hyperbolic shapes and discrete Euclidean shapes (tori)

Forming new stable hyperbolic space-forms from a material interaction

Assume an attractive (or repulsive) interaction in 3-space, then the interaction of "free" material components would be similar to a collision of components,

State of resonance   new hyperbolic space-form

if the material components get very close during the (collision) interaction then if the energy of the over-all interaction is within (certain) energy ranges and the closeness allows resonances (with the spectral set of the over-all high-dimension containing metric-space) to begin to form, so that the over-all energy, as well as the resonances, allow a "new" stable "discrete hyperbolic shape" (in the proper dimension of the interaction) to form.

The Mathematical Structure of Stable Physical Systems

Weyl-transformations
Representing two maximal tori within a Lie group
(rank-2 compact Lie group)

These two intersecting circles of the two maximal tori may be angularly related to one another by Weyl group transformations, where the Weyl group defines the conjugation classes of the maximal tori which "cover" the compact Lie group

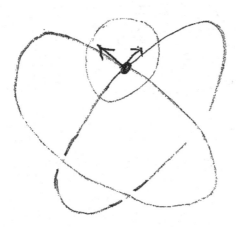

Forming angular changes between toral components of a discrete hyperbolic shape by using Weyl-transformations, which change the angular relations between circles which compose a toral component of a hyperbolic space-form.

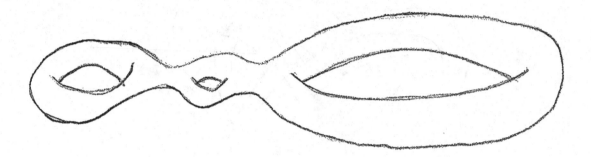

Envelopes of orbital stability for "free" subsystems (or sub-metric-spaces)

The Mathematical Structure of Stable Physical Systems

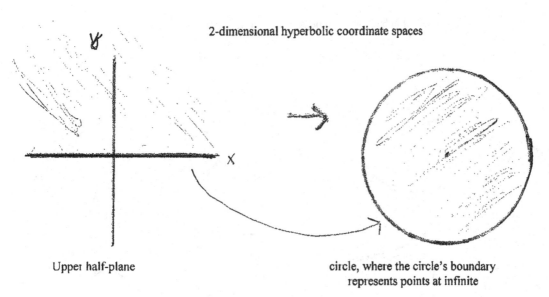

2-dimensional hyperbolic coordinate spaces

Upper half-plane

circle, where the circle's boundary represents points at infinite

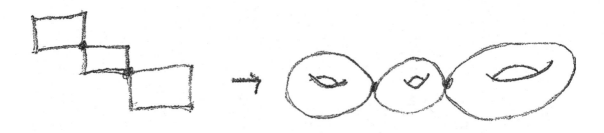

Rectangles attached at vertices and then moded-out, without expanding the vertex, shows a model of a discrete hyperbolic shape's toral components, represented as separate tori attached at separate vertices

Various types of unbounded 2-dimensional discrete hyperbolic shapes

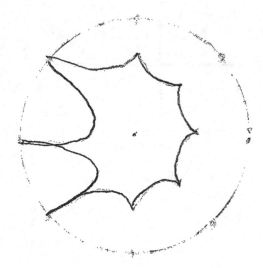
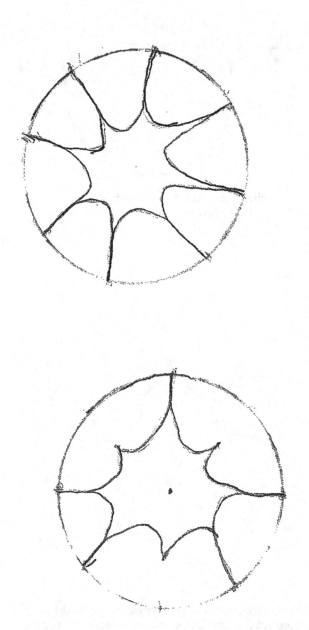

The Mathematical Structure of Stable Physical Systems

The stable geometric shapes

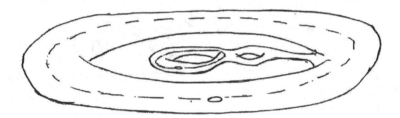

The same shape "folded" by Weyl-angles

Partitioning a Many-Dimensional Containment Space

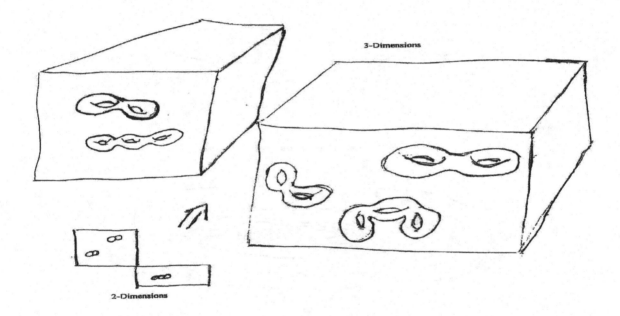

Perturbing Material-Components on Stable Shapes: How Partial Differential Equations Fit into the Descriptions of Stable Physical Systems

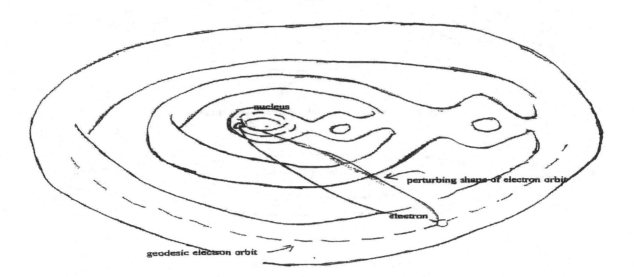

Describing the Dynamics of "Free" Material Components in Higher-Dimensions

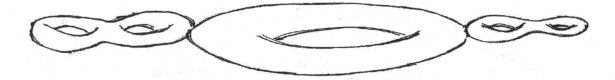

Interaction components and the interaction shape

The approximation of the interaction

REFERENCES

1. A New Copernican Revolution, Bill G P H Bash and George P Coatimundi, Trafford Publishing, 2004. www.trafford.com/03-1913,
2. The Authority of Material vs. The Spirit, Douglas D Hunter, Trafford Publishing, 2006. www.trafford.com/05-3038
3. Topology and Geometry for Physicists, C. Nash and S. Sen, Academic Press, 1983.
4. The Infamous Boundary, David Wick, Springer-Verlag, 1995.
5. Function Theory, C. L. Siegel,
6. Three-dimensional Geometry and Topology, W. Thurston, Princeton University, 1997.
7. Gauge Theory and Variational Principles, D Bleeker, Addison, 1981.
8. Geometry II, E B Vinberg, Springer, 1993.
9. Spaces of Constant Curvature, J Wolf, Publish or Perish, 1977.
10. Contemporary College Physics, Jones and Childers, Addison-Wesley, 1993 (High School text).
11. I M Benn and R W Tucker, in, An Introduction to Spinors and Geometry with Applications in Physics, 1987, (Chapter 2)
12. Representations of Compact Lie Groups, T Brocker, T tomDieck, Springer-Verlag, 1985.
13. Dynamical Theories of Brownian Motion, E. Nelson, Princeton University Press, 1967 (1957).
14. Quantum Fluctuations, E. Nelson, Princeton University Press, 1985.
15. Algebra, L Grove.
16. Electron magnetic moment from gonium spectra, H Dehmelt (Nobel prize winner) et al, Physical Review D, Vol 34, No. 3, Aug 1, 1986.
17. Newton's Clock, Chaos in the Solar System, I Peterson, W H Freeman and Company, 1993.
18. Quantum Mechanics, J L Powell and B Crasemann, Addison-Wesley Publishing, 1965.
19. The End of Science, J Horgan, Broadway Books, 1996.
20. Riemannian Geometry, L. P. Eisenhart, Princeton University Press, 1925.
21. Reflection Groups and Coxeter Groups, J Humphreys, Cambridge University Press, 1990.
22. Partial Differential Equations, J Rauch, Springer-Verlag, 1991.
23. The Foundation of the General Theory of Relativity, A Einstein, 1916, Annalen der Physik (49).

D Coxeter
Katok

Just as Copernicus, Kepler and Galileo provided a quantitative-geometric context for the properties of the solar system, which were then precisely identified by the solutions to (the) solvable differential equations of Newton; Martin Concoyle now provides the stable geometric structures which fit . . . , both macroscopically and microscopically . . . , into a many-dimension containment set (hyperbolic 11-dimensional), so that these shapes are the solutions, ie the geometries of the stable spectral-orbital properties, of all the fundamental stable systems which have stable spectra and orbits, and it is the basis for a quantitative system (the spectral set of a measurable existence) which is finite, and

These ideas are discussed in the following books: (available at math conference, 2013)

1. A New Copernican Revolution (p286), B Bash & P Coatimundi, Trafford, 2004.
2. The Authority of Material vs. The Spirit (p483), D D Hunter, Trafford, 2006.
3. Introduction to the Stability of Math Constructs; and a Subsequent General, and Accurate, and Practically Useful Description of Stable Material Systems, Concoyle, and G P Coatimundi, (p262), 2012, Scirbd.com.
4. A Book of Essays I: Material Interactions and Weyl-Transformations, Martin Concoyle Ph. D., (p234), 2012, Scribd.com.
5. A Book of Essays II: Science History, and the Shapes which Are Stable, and the Subspaces, and Finite Spectra, of a High-Dimension Containment Space, Martin Concoyle, (p240), 2012, Scribd.com.
6. A Book of Essays III: Elementary Topics, Martin Concoyle (p303), 2012, Scribd.com.
7. Physical description based on the properties of stability, geometry, and quantitative consistency: Short essays which are: simple, "clear," and direct Presented to the Joint math meeting San Diego (2013), (p208), Martin Concoyle, 2013, Scribd.com.
8. Describing physical stability: The differential equation vs. New containment constructs, Martin Concoyle, 2013, (p378), Scribd.com. (also equivalent to, VII 3, at Scribd.com) Old SD
9. Introduction to the stability of math constructs; and a subsequent: general, and accurate, and practically useful set of descriptions of the observed stable material systems, Martin Concoyle Ph. D., 2013, (p70), Scribd.com,

See scribd.com put m concoyle into web-site's search-bar

As well as in the following (new) books from Trafford:

1. The Mathematical Structure of Stable Physical Systems, Martin Concoyle and G. P. Coatimundi, 2013, (p449), Trafford Publishing (equivalent to 3. And 5. Above, Scribd)

2. Partitioning a Many-Dimensional Containment Space, Martin Concoyle, 2013, (p477), Trafford Publishing, (equivalent to 4. And 6. Above, Scribd)
3. Perturbing Material-Components on Stable Shapes: How Partial Differential Equations Fit into the Descriptions of Stable Physical Systems, Martin Concoyle Ph. D., 2013, (p234), Trafford publishing (Canada) (equivalent to 7. And 9. Above, Scribd, and new material)
4. Describing the Dynamics of "Free" Material Components in Higher-Dimensions, Martin Concoyle, 2013, (p478), Trafford Publishing (equivalent to 8. Above, Scribd, and new material)

Alternative title to any of 1-4 Trafford:
The Unbounded Shape, and the Self-Oscillating, Energy-Generating Construct

Copyrights

These new ideas put existence into a new context, a context for both manipulating and adjusting material properties in new ways, but also a context in which life and creativity (practical creativity, ie intentionally adjusting the properties of existence) are not confined to the traditional context of "material existence," and material manipulations, where materialism has traditionally defined the containment of material-existence in either 3-space or within space-time.

Thus, since copyrights are supposed to give the author of the ideas the rights over the relation of the new ideas to creativity [whereas copyrights have traditionally been about the relation that the owners of society have to the new ideas of others, and the culture itself, namely, the right of the owners to steal these ideas for themselves, often by payment to the "wage-slave authors," so as to gain selfish advantages from the new ideas, for they themselves, the owners, in a society where the economics (flow of money, and the definition of social value) serves the power which the owners of society, unjustly, possess within society].

Thus the relation of these new ideas to creativity is (are) as follows:

These ideas cannot be used to make things (material or otherwise) which destroy or harm the earth or other lives.

These new ideas cannot be used to make things for a person's selfish advantage, ie only a 1% or 2% profit in relation to costs and sales (revenues).

These new ideas can only be used to create helpful, non-destructive things, for both the earth and society, eg resources cannot be exploited to make material things whose creation depends on the use of these new ideas, and the things which are made, based on these new ideas, must be done in a social context of selflessness, wherein people are equal creators, and the condition of either wage-slavery, or oppressive intellectual authority, does not exist, but their creations cannot be used in destructive, or selfish, ways.

INDEX (KEY WORDS)

Alternating forms
base space
Bosons
"cubical" simplexes
conformal factor groups
Commutative
Conjugation
Conjugation classes
Constant curvature
Derivatives
diffeomorphism groups
differential equations
differential forms
dimensional hierarchy
Dirac operators
discrete isometry subgroups
Discrete (separated points, periodic translations, fundamental domains, reflective group)
Discrete Euclidean shapes
Discrete hyperbolic shapes
Dual differential-form
E&M, electromagnetism
Euclidean space
Euclidean metric-space
Equation
Fermions
fiber group
fundamental domains

Geometrically separable
Hermitian form (finite dimensions)
hyperbola
hyperbolic
hyperbolic metric-space
Independent
infinite extent space-forms
interaction
interaction potentials
Inverse
Invertible
Isometry
Isometry groups
Lattices
Linear
Lie algebra valued connection 1-forms
Lie group
Maximal torus
metric spaces
metric-space states
moding out
non-reducible
Orthogonal
Parallelizable
Physical properties and fundamental invariance's, eg translations and linear momentum,
principle fiber bundles
Pull-back (defined on maps between different spaces which possess differential-forms)
sectional curvature
signature of a metric,
Solvable
space-forms,
space-time,
unitary groups
Weyl group

This book is an introduction to the simple math patterns used to describe fundamental, stable spectral-orbital physical systems (represented as discrete hyperbolic shapes, ie hyperbolic space-forms), the containment set has many-dimensions, and these dimensions possess macroscopic geometric properties (which are also discrete hyperbolic shapes). Thus, it is a description which transcends the idea of materialism (ie it is higher-dimensional, so that the higher-dimensions are not small), and it can also be used to model a life-form as a unified, high-dimension, geometric construct, which generates its own energy, and which has a natural structure for memory, where this construct is made in relation to the main property of the description being, in fact, the spectral properties of both (1) material systems, and of (2) the metric-spaces, which contain the material systems, where material is simply a lower dimension metric-space, and where both material-components and metric-spaces are in resonance with (or define) the containing space. Partial differential equations are defined on both (1) the many metric-spaces of this description and (2) the lower-dimensional material-components which these metric-spaces contain, ie the laws of physics, but their main function is to act on either the, usually, unimportant free-material components (so as to most often cause non-linear dynamics) or to perturb the orbits of the, quite often condensed, material which has been trapped by (or within) the stable orbits of a very stable hyperbolic metric-space shape.

It could be said that these new ideas about math's new descriptive context are so simple, that some of the main ideas presented in this book may be presented by the handful of diagrams which show these simple shapes, where these diagrams indicate how these simple shapes are formed and folded, or bent, to form the stable shapes, which can carry the stable spectral properties of the many-(but-few)-body systems . . . , where these most fundamental-stable-systems have no valid quantitative descriptions within the, so called, currently-accepted "laws of physics," (ie the special set of partial differential equations associated to the, so called, physical laws) . . . . so that the diagrams of these stable geometric shapes are provided at the end of the book.

This new measurable descriptive context is many-dimensional, and thus, it transcends the idea of materialism, but within this new context the 3-dimensional (or 4-dimensional space-time) material-world is a proper subset (in a subspace which has 3-spatial-dimensions),

The, apparent, property of fundamental randomness (in a currently, assumed, absolutely-reducible model of material, and its reducible material-components) is a derived property, but now in a new context in which stable geometric patterns are fundamental,

The property of spherically-symmetric material-interactions is shown to be a special property of material-interactions, which exists (primarily, or only) in 3-spatial-dimensions, of Euclidean space, wherein inertial-properties are to, most naturally, be described,

It is both reductive, (to some sets of small material-components, but elementary-particles are most likely about components colliding with higher-dimensional lattice-structures, which are a part of the true geometric context of physical description) and unifying in its discrete descriptive contexts (relationships) which exist, between both a system's components, and the system's (various) dimensional-levels (where these dimensional-levels are particularly relevant, in regard to understanding both (1) the chemistry and (2) the functional organization of living systems),

But most importantly, this new descriptive language (new context) describes the widely observed properties of stable-physical-systems, which are composed of various dimensional-levels and of various types of components and interaction-constructs, so that this new context provides an explanation about both (1) "how these systems form" and (2) "how they remain stable," wherein, partial differential equations, which model material-interactions, are given a new: context, containment-structure, organization-context, interpretation, and with a new discrete character,

It provides a (relatively easy to follow, in that, the containment set-structure for these different-dimensional stable-geometries are simple dimensional relations) 'map' "up into a higher-dimensional context (or containment set) for existence," wherein some surprising new properties of existence can be modeled, in relation to our own living systems also being modeled as higher-dimensional constructs, and this map can shed-light onto our own higher-dimensional structure, and its relation to both existence, and to the types of experiences into which we may enter (or possess as memory) (or within which we might function), where because any idea about higher-dimensions is difficult to consider, and is relatively easy to hide and ignore these higher-dimensions, especially, if we insist on the idea of materialism.